3

678 DIC

LEARNING
services

01209 616259

Cornwall College Camborne
Learning Centre - FE

This resource is to be returned on or before the last date stamped below. To renew items please contact the Centre

Three Week Loan

DICTIONARY OF RUBBER

DICTIONARY OF
RUBBER

K. F. HEINISCH, Dr. techn., Dipl. Ing., F.I.R.I.

Malaysian Rubber Bureau
Vienna, Austria

APPLIED SCIENCE PUBLISHERS LTD
LONDON

APPLIED SCIENCE PUBLISHERS LTD
RIPPLE ROAD, BARKING, ESSEX, ENGLAND

*Originally published in German by
A. W. Gentner Verlag, Stuttgart, 1966*

ISBN: 0 85334 568 6

© APPLIED SCIENCE PUBLISHERS LTD 1974

Printed in Great Britain by Galliard (Printers) Ltd Great Yarmouth

APPLIED SCIENCE PUBLISHERS LTD
RIPPLE ROAD, BARKING, ESSEX, ENGLAND

Originally published in German by
A. W. Gentner Verlag, Stuttgart, 1966

ISBN: 0 85334 568 6

© APPLIED SCIENCE PUBLISHERS LTD 1974

Printed in Great Britain by Galliard (Printers) Ltd Great Yarmouth

DICTIONARY OF
RUBBER

K. F. HEINISCH, Dr. techn., Dipl. Ing., F.I.R.I.

Malaysian Rubber Bureau
Vienna, Austria

APPLIED SCIENCE PUBLISHERS LTD
LONDON

INTRODUCTION

This dictionary is a collection of terms, concepts and explanations originally compiled over the years for personal use from numerous technical journals, company publications and other sources of literature, and from personal experience, discussions and correspondence. The work was not put together systematically but developed rather arbitrarily during the course of many years of practical work. At the suggestion of the original publishers, A. W. Gentner, Stuttgart, it was condensed to its present size and published in the German language in 1966 under the title *Kautschuk-Lexikon*.

Everyone working in the field of rubber technology and reading the comprehensive technical literature will at times be confronted with questions concerning the composition and properties of various commercial products, equivalent materials, concepts, jargon, abbreviations, and so forth. Finding answers to such questions is usually very time-consuming. It is the purpose of this dictionary to provide a tool to help find these answers, and it therefore deals with individual entries in a rather more general way. For the same reason, this dictionary also includes obsolete products, outdated concepts and processes, as well as some historical and botanical data which earlier published literature may refer to.

The trade names in this book are, as a rule, registered, though not specifically mentioned as such in the text. In most cases entries containing commercial chemicals are followed by a number in parentheses referring to the list of suppliers on pages 535 to 545. Products marked with a plus sign (+) are, as far as can be established, no longer commercially available.

Chemical compounds are listed under their accepted chemical names, reference being made to trade names, abbreviations and synonyms. On the other hand, mixtures and inhomogeneous products are described under their trade names. Individual entries have been arranged in alphabetical order, but, with a few logical exceptions (such as Buna, Buna CB, Buna Hüls, which will precede Bunac), complex titles are treated as one word. Prefixes such as bis-, cyclo, di-, iso-, mono-, tetra-, tri-, etc. are considered an integral part of the chemical name and have thus been placed alphabetically, whereas prefixes which describe the organic structure, such as cis-, trans-, o-, m-, p-, n-, N-, do not affect alphabetical order.

This English edition represents in principle a straightforward translation of the German original. It does, however, contain new material which has been used to bring the information up to date.

Although the author has taken the opportunity to correct some errors, no attempt has been made to alter the method of presentation or nature of technical approach.

It is hoped that the comprehensive coverage of topics together with the simplicity of lexical approach, avoiding the cluttering of descriptive material with the paraphernalia of polymer science, will commend this dictionary in particular to people seeking ready answers to day-to-day problems.

The author would like to express his thanks to all commercial firms who have cooperated by providing informative material and to Mrs. J. Lang, B.A. for valuable cooperation.

A

A Group of older accelerators in current use no longer available under this name.

A-1 See Thiocarbanilide (6).

A-2-Z Sodium alkyl aryl sulphonate, wetting agent and emulsifier.

A-7 Group of aldehydeamine condensation products, accelerators (5).

A 5/10 Former trade name for anhydroformaldehyde-aniline.

A-7, A-11, A-40, A-50, A-77 Acetaldehyde aniline condensates.

A-10 Anhydroformaldehyde-aniline (5).

A-11 Acetaldehyde-aniline condensation product (5).

A-16 Acetaldehyde-butyraldehyde-aniline condensation product (5).

A-17 Anhydroformaldehyde-p-toluidine (5).

A-19 Acetaldehyde-aniline-formaldehyde condensation product (5).

A-20 Heptaldehyde-aniline (5).

A-22 Di-o-tolylthiourea (5).

A-32 Reaction product of butyraldehyde and butylidene-aniline, a

reddish-brown, viscous fluid, s.g. 0·98. Accelerator for NR, SBR, and NBR, activator for thiazoles. Easy processing, gives plateau vulcanisation curves, high tensile strength and high modulus, use of zinc oxide and fatty acid is usually desirable but they may be omitted for certain applications.
Uses: hard rubbers and mechanical goods (5).
Dosage: primary 0·5–2% with 2·5–4% sulphur, secondary 0·1–0·5%.

A-40 Acetaldehyde-aniline condensates (5).

A-50 Acetaldehyde-aniline condensates (5).

A-77 Acetaldehyde-aniline condensates (5).

A-100 Condensation product of a primary aromatic amine with acetaldehyde and butyraldehyde, reddish-brown, oily fluid, s.g. 1·04. Accelerator for hard vulcanisates of NR and synthetic rubbers.
Usage: 1–2% (5).

A 232-B Air-drying, unvulcanisable neoprene adhesive (14).

A-510 Condensation products of formaldehyde and aniline.

A-1010 Condensation products of formaldehyde and aniline.

A-4000 Adhesive based on silicone rubber (7).

AA Castor oil, plasticiser (9).

A.A Dolomitic limestone, filler (8).

1

AB 11-2 Thermoplastic resin, softener for NR and synthetic rubbers (357). Abbreviation for acetaldehyde-butyraldehyde-aniline condensate.

ABALYN *See* Methylabietate (10).

ABBA RUBBER Resin rich, low-grade wild rubber from *Ficus Vogellii* (Lagos, Nigeria).

ABC Plasticiser for vinyl plastics (11).

ABC-TRIEB Ammonium bicarbonate, blowing agent for sponge-rubber (32).

ABIARANA Gutta-percha-like product from *Lucuma lastiocarpa*.

ABIETIC ACID Dimethyl-isopropyl decahydrophenanthrenecarboxylic acid, $C_{19}H_{29}COOH$. Melting point 170–174. Main constituent of colophony (10–40%), obtained by distillation from pine resin. The structural formula can be split up into four isoprene units. Soluble in benzene, alcohol, chloroform, ether. Insoluble in water.

Use: Softener for natural rubber but reduces ageing properties.

ABITOL Colourless, sticky, viscous fluid. Mixture of di-, tetra- and dehydroabietyl alcohol (10).

ABR Abbreviation for acrylonitrile-butadiene copolymer, nitrile rubber.

ABRACOL Group of softeners (12).

203:	paratoluene sulphonamide
234:	dicresylglyceryl ether
777:	monocresylglyceryl ether-diacetate
789:	para-toluene sulphonamide
888:	dicresylglyceryl ether mono-acetate
1001:	tertiary butyl phenol
1011:	benzylglyceryl ether
S.L.G.:	glycerine monostearate.

ABRASION INDEX Index of abrasion resistance (loss of volume) of a vulcanised elastomer in comparison with a similar control sample tested under standard conditions. Often expressed as a percentage of the control value. Abrasion index = $S/T \times 100$.

ABRASION PATTERNS (Schallamach-patterns). Arrays of parallel ridges with sawtooth-shaped cross section which frequently appear on the surface of abraded rubber at right angles to the direction of abrasion. Their orientation on tyres shows the prevalent direction of abrasion during service.

ABRASION RESISTANCE The laboratory assessment of abrasion resistance is complicated by the poor reproducibility of current test methods and by their failure to correlate adequately with service conditions. Most methods employ the principle of abrading a test-piece in contact with a rough surface, the results being reported relative to a standard sample.

1. Test machines with fixed, stationary test-pieces:
Grasselli tester (Williams tester; Du Pont tester). Two block-shaped test-pieces about $2.5\ cm^2$ and $0.75\ cm$ thick are held against a vertical rotating abrasive wheel under a fixed

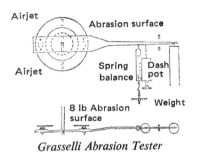

Grasselli Abrasion Tester

pressure. National Bureau of Standards tester. The test-piece is held vertically against an abrasive surfaced drum under variable pressure. Kelly–Springfield machine. A portion of the test-piece surface is abraded under a periodically changing pressure intended to represent the behaviour of a tyre.
2. Machines with rotating samples:
Akron machine (Goodyear tester); Lambourne (Dunlop) tester. The principle of these machines is that a rotating, disc-shaped test-piece is held with its periphery (or surface) against an abrasive disc or drum. Either the test-disc or the drum (or both) may be driven and test conditions may be varied, *e.g.* the angle

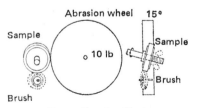

Akron Abrasion Tester

of rotation of the test-disc relative to that of the drum.
3. Machines using loose abrasive powder:
Test-pieces are rotated under controlled conditions in loose carborundum powder.
4. Other machines:
Pico abrasion tester (B. F. Goodrich). The original instrument, as described by E. Newton *et al.* made use of rotating knives which abraded the test-piece, the surface of which was also exposed to loose grit. Conditions are specified in ASTM D228-63T. *See also* AP-Abrasion machine.
Standards for abrasion testing:
ASTM D 394-59; D 1630-59T.
BS 903: 24: 1950.
DIN 53 516.

ABRIL Group of synthetic waxes. Various compounding uses and as release agents (13).

ABSON Group of acrylonitrile-butadiene-styrene terpolymers (14).

ABS-POLYMER Abbreviation for acrylonitrile-butadiene-styrene copolymers.

AC-165 Mixture of diphenyl-guanidine and mercaptobenzthiazole with zinc chloride.
Use: Accelerator (5).

A-C-A-K *See* Zinc isopropyl xanthate.

ACCEL Group of accelerators (16).
22:   2-mercaptoimidazoline
AA: acetaldehyde-ammoniate
BX: di-ortho-tolyl guanidine salt of dicatechol borate
CZ: N-cyclohexyl-2-benzthiazyl sulphenamide

DM: Dibenzthiazyl-disulphide
EZ: Zinc diethyl-dithiocarbamate
F: Mixture of accelerators
M 2-mercaptobenzthiazole
PX: Zinc ethyl-phenyl-dithiocarbamate
PZ: Zinc dimethyl-dithiocarbamate
S: Sodium dimethyl-dithiocarbamate
SL: Selenium diethyl-dithiocarbamate
TMT: Tetramethyl-thiuram-monosulphide
TS: Tetramethyl-thiuram-monosulphide

ACCELERANTE Group of accelerators.
FQ: activated thiazole
MBT: mercaptobenzthiazole (397).

ACCELERATED AGEING The relative durability of rubber may be tested by a process of artificial ageing under intensified conditions which should be comparable to normal service conditions. Since the various physical characteristics do not all change in the same way on ageing, characteristics required for specific uses should be tested in addition to tests for tensile strength, elongation at break, hardness and deformation under load. The results of accelerated ageing tests do not usually correspond to the results of natural ageing and are therefore used mainly for comparative judgements. In practical use, ageing is caused by a complexity of processes, a limited selection of which are operative during testing. Various ageing factors (*e.g.* the influence of ozone) exert a different influence under dynamic stress from that apparent under the static strain of the ageing tests.

Chemicals which protect the rubber against ozone under static strain have no effect when it is under dynamic strain. For testing against ageing, samples of the same type should be used, since dissimilar samples may influence each other through the presence of their component ingredients (softeners, antioxidants). The most frequently used methods of ageing are:
1. *Heating oven (Geer's).* The unconditioned samples are hung up and buffeted by air at a temperature of 70°C for approx. 7 days.
2. *Cell oven.* To prevent their influencing each other, the samples are placed in glass tubes in a suitable heating apparatus (metal block, oil bath, water bath) and are heated to between 120–150°C for 10, 20, 40, 70 and 168 hr.
3. *Air pressure ageing (Booth).* The samples are heated in a pressure bomb of 6 kg/cm^2 to 120°C ASTM and BS.
4. *Oxygen pressure ageing (Bierer–Davis).* The samples are heated in oxygen at a pressure of 21 kg/cm^2 at 70°C.
5. *Air/steam pressure ageing.* By filling a pressure bomb with 1/5 of its volume of water and heating to 180°C, a pressure of 10 kg/cm^2 is reached.
6. *Ozone ageing.* Determination of tear resistance under tensions, using a specific concentration of ozone.
7. *Light ageing.* Subjecting the samples to light rays of similar spectral distribution to that of the sun's spectrum, but with a higher ultraviolet content.
The relative ageing durability (*Z*) of the sample is expressed by:

$$Z = \frac{O - A}{O} \times 100\%$$

where O is the property of the unaged sample, and A the property of the aged sample.

Standards:
ASTM D 454, D 572, D 573, D 865, D 518, D 750, D 1149, D 1206.
BS 903: 13: 1950, A19: 1956.
DIN 53 508, draft ISO recommendation 171.

ACCELERATEUR Group of French accelerators:

3 RS:	Zinc dibutyldithiocarbamate (19)
80:	α-o-tolylbiguanidine (20)
B:	Triphenylguanidine (19)
D:	Diphenylguanidine (19)
DOTG:	Di-o-tolylguanidine
DPG:	Diphenylguanidine
DT:	Di-o-tolylguanidine (19)
G:	Mercaptobenzthiazole (19)
G.R.E.:	2-mercaptobenzthiazole (19)
HS:	Butyraldehyde-aniline-condensation product
L:	Thiocarbanilide (19)
LL:	Thiocarbanilide (19)
MBT:	Mercaptobenzthiazole (418)
MBTS:	Dibenzthiazyldisulphide (418)
MBTZn:	Zinc mercaptobenzthiazole (418)
PTX:	Phenyltolylxylylguanidine (19)
S:	Acetaldehyde-aniline-condensation product (19)
TB:	Tetramethylthiuram-disulphide (19)
TE:	Tetraethylthiuram-disulphide (19)
TM:	Tetramethylthiurammono-sulphide (19)
TS:	Reaction product of acetaldehyde, p-toluidine and aniline (19)
VS:	Butyraldehyde-aniline-reaction product (19).

ACCELERATEUR RAPIDE

3 R;	
3 R.N.:	Zinc ethylphenyl dithiocarbamate (19)
3 RS:	Zinc dibutyldithiocarbamate (19)
5 R:	Zinc isopropylxanthogenate (19)
5 R special:	Sodium isopropylxanthogenate (19)
200:	2-mercaptobenzthiazole (20)
201:	Di-2-benzthiazyldisulphide (20)
300 A:	Butyraldehyde-aniline-condensation product (20)
400:	Diethylbenzthiazylsulphenamide (20)
465:	Mixture of accelerators. Type (20)
500:	Tetramethylthiurammono-sulphide (20)
A:	undisclosed mixture (19)
ARA:	Dibenzthiazyldisulphide and dithiocarbamate (19)
ARF:	Dibenzthiazyldisulphide, diphenylguanidine and hexamethylenetetramine (19)
ARK:	Benzthiazyl-2-phenylethyldithiocarbamate (19)
ARP:	undisclosed mixture
ARS:	Mixture of dibenzothiazyldisulphide and organic salt of monoethanolamine (19)
ARU:	Dibenzthiazyldisulphide, diphenylguanidine and dithiocarbamate (19)
ARZ:	Dimethyl-2-benzthiazylsulphenamide (20)
ARV:	undisclosed mixture
DEDCZ:	Zinc diethyldithiocarbamate (419)
DMDCZ:	Zinc dimethyldithiocarbamate (419)

DTET: Tetraethylthiuram-
disulphide (419)
DTMT: Tetramethylthiuram-
disulphide (419)
G: 2-mercaptobenzthiazole
(19)
GR: Dibenzthiazyldisulphide
(19)
GS: Di-2-benzthiazyldi-
sulphide (19)
MTMT: Tetramethylthiurammono-
sulphide (419)
P: Piperidinepentamethylene-
dithiocarbamate (19)
R: Zinc methylphenyl-
dithiocarbamate (19)
T.B.: Tetramethylthiuram-
disulphide (19)
TC: Diethyldiphenyl-
thiurammonosulphide (19)
TE: N,N'-diethyl-N,N'-
diphenylthiuram-
disulphide (19)
TM: Tetramethylthiurammono-
sulphide (19)
X: Mixture of 60% diphenyl
guanidine and 40% zinc
oxide (19)
Z 200: Zinc 2-mercaptobenz-
thiazalate (20).

ACCELERATEUR SOLUBLE
Group of French produced soluble
accelerators for latex.
F: Monoethanolamine (19)
Lat. 2: Reaction product of
2-mercaptobenzthiazole
with monoethanolamine
(19)
Lat. 4 FL: Sodium dimethyl-
dithiocarbamate (19)
Lat. 5: Sodium isopropyl-
xanthogenate (19).

ACCELERATOR Material
added to a rubber compound to
increase the rate of vulcanisation and

to permit vulcanisation to proceed
at lower temperatures and with
greater efficiency.
Ref.: D. A. Smith, *Rubber Journal*,
1967, **149**, No. 2, 55; No. 3, 10.

ACCELERATOR Chemicals
which increase the speed of reaction
between sulphur and rubber, or
lower the vulcanisation temperature.
They can also decrease the quantity
of sulphur necessary for vulcanisation and considerably improve the
physical and technical properties of
the vulcanisate. Some accelerators
can also act as vulcanising agents.
Historical development: The use of
accelerators was discovered con-
currently with vulcanisation in 1839
by Goodyear. Goodyear was trying
to find a means of decreasing the
tackiness of rubber and investigating
combinations of sulphur with the
lead pigments used in the colour
industry, which had been used for
the 'drying' of rubberised textiles at
132°C. Soon afterwards the effectiveness of a series of metal oxides and
carbonates was discovered. In 1881
Rowley patented the use of ammonia
as a vulcanising auxiliary; this discovery was a result of his attempts
to discover how to produce sponge
rubber and how to vulcanise impregnated materials. Later, aliphatic
ammonia derivatives were investigated. Since these were too volatile,
the most simple aromatic derivative,
aniline, was tried. Raw aniline, also
known as 'red oil', appears to have
been used for the first time in 1902
at the 'Rubber Works at Wundt',
near Frankfurt (no longer in existence), and in 1906 by the Diamond
Rubber Co., USA. Oenslager, in
1906, discovered the accelerating
properties of aniline and of its
carbon disulphide compounds

(thiocarbanilides),and thus the organic accelerator. By 1904 Glidden had discovered the influence of oleic acids on litharge. In 1908 Wo. and Wa. Ostwald took out the first patent for aniline as an antioxidant for rubber; in 1911 Bayer patented the use of piperidine as an accelerator. In addition to the simple alicyclic and aromatic amines, guanidine derivatives belong to the group of effective, organic bases. In 1914–15 under British and American patents, Hoffmann and Gottlob show that methyl rubber vulcanises with sulphur only in the presence of organic nitrogen compounds, and refer particularly to aldehyde-ammonia and hexamethylene tetramine. In 1914 Peachey introduced nitrosodimethylaniline (BP 4263) in Gt. Britain under the trade name of Accelerene. The first thiurams and dithiocarbamates were used in the USA around 1914, and at about the same time zinc alkylxanthates were being studied in Russia by Ostromyslensky. In 1915 A. B. Molony introduced a reaction product of amines from beet molasses with carbon disulphide, under the name of Aksel, which was later identified as tetramethyl thiuram disulphide. Research on the effects of zinc oxide led to the discovery of two important products, diphenyl guanidine and mercaptobenzthiazole. Aminoguanidine, suggested in 1915, did not come into widespread use. Around 1920 organic accelerators came into common use. Thiocarbanilide and hexamethylene tetramine, used also around this time, were succeeded soon after by the more effective diphenyl guanidine and di-o-tolyl guanidine. It was discovered in 1921 that the reaction of thiocarbanilide and sulphur yielded an extremely active accelerator, the active constituent of which was identified as mercaptobenzthiazole. About 1919 a new type was created from the tetra-substituted thiuram disulphides. The rapid effect of dithiocarbamic acid zinc salts was discovered about 1920, and also that thiuram disulphides enable vulcanisation to proceed without the use of sulphur. In the USA in 1920 Vanderbilt began to produce tetramethyl thiuram disulphide (Tuads) and 'super sulphur' (a mixture of zinc dimethyl dithiocarbamate and alumina 1:15) and, soon after, lead dimethyl dithiocarbamate. Due to the supply of dimethylamine, methyl compounds were preferred in the USA, whereas diethyl compounds were preferred in Britain. Zinc dimethyl dithiocarbamate came into general use shortly afterwards. About 1930 scientists began to show interest in accelerators with a delaying effect, which reacted slowly and safely at processing temperatures but rapidly at vulcanisation temperatures. This led to the discovery of accelerator condensation products, from which the accelerator, and in some cases even the activator, could be released at high temperatures. The last step in the development of the accelerator was the discovery of the ultra-accelerator, used for vulcanisation at room temperature, and particularly in the processing of latex.

Around 1937 scientists began to explore the possibility of compounds having a combination of the vulcanisation speed of thiurams and the ability to delay vulcanisation. This led to the development of Santocure, a reaction product of cyclo-hexylamine with mercaptobenzthiazole (cyclohexylbenzthiazyl sulphenamide). Since 1945 many variants of this product have been used in place of mercaptobenzthiazole.

Classification. Accelerators have been classified arbitrarily as slow, medium-fast, fast, semi-ultra, and ultra, and as having direct or delayed action. More than 50 different accelerators are currently in use, and fall into the following chemical categories: 1, aldehyde-amines; 2, guanidines; 3, dithiocarbamates; 4, thiurams; 5, thiazoles; 6, xanthates; 7, thioureas. Of these, the thiazoles have the most widespread use. By combining accelerators from different groups, almost any desirable vulcanisation behaviour can be achieved, and they are characterised by the following: scorching period, vulcanisation time, plateau effect, vulcanisation temperature, physical properties of the vulcanisates.

Lit.: F. Jones, *India Rubber J.*, 1947, 112, 795, 839, **113**, 23, 56.
P. Schridowitz and T. R. Dawson, 'History of the Rubber Industry', Cambridge, 1952.

ACCELERATOR Group of English and American produced accelerators (some obsolete, others produced under different names).

Grades:

1:	p-nitroso-N,N'-dimethyl aniline (6)
2-MT:	mercaptothiazoline (6)
2P:	piperidine pentamethylene dithiocarbamate (22)
4:	original trade name for aniline, the first organic accelerator (6)
4P:	pentamethylene thiuram disulphide (22)
5:	anhydroformaldehyde-aniline (6)
6:	methylene dianilide (6)
8:	anhydroformaldehyde-p-toluidine (6)
11:	triphenyl guanidine (6)
12:	diphenyl guanidine (6)
15:	thiocarbanilide (6)
17:	di-o-tolyl thiourea (6)
18:	di-o-tolyl guanidine (6)
18X:	mixture of a formaldehyde-toluidine condensation product with a guanidine
19:	aldehyde-aniline condensation product (6)
21:	butyraldehyde-aniline condensation product (22)
22:	product of butyraldehyde-aniline and carbon disulphide. Accelerator for hot air vulcanisation. Quantity: 1% with 3% sulphur and 5% zinc oxide (6)
30:	aldehyde-amine reaction product (54)
40:	aldehyde-amine reaction product (54)
49:	$R_1NH.C(:NH).NH.R_2$, diaryl guanidine mixture. White odourless powder, m.p. 132–137°C, s.g. 1·20. Gives vulcanisates with a fairly high tensile strength and a high modulus; zinc oxide is necessary. Boosts and may be used in conjunction with thiazoles and other guanidines. Activity as primary accelerator lies between that of di-o-tolyl guanidine and diphenyl guanidine (21)
52:	mixture of ethyl and methyl tetra-alkyl thiuram disulphides. Contains 75% methyl and 25% ethyl groups.

8

52-0, 52-1 (with 2% mineral oil)
52-2, 52-3 (with 2% mineral oil)
52-6, tetramethyl thiuram disulphide (4)

57-0: zinc dimethyl dithiocarbamate (4)

60: aldehyde-amine reaction product (54)

62: tetraethyl thiuram disulphide (4)

66: selenium diethyl dithiocarbamate (4)

67: zinc diethyl dithiocarbamate (4)

77: zinc dibutyl dithiocarbamate (6)

85: potassium-2-mercaptobenzthiazole (6)

87: potassium pentamethylene dithiocarbamate (6)

89: potassium pentamethylene dithiocarbamate (6)

108: mixture of 66·7% tetramethyl thiuram disulphide and 33·3% mercaptobenzthiazole (33)

108 PDR: mixture of 2 parts Tuex and 1 part MBT (23)

113: 1 part dibenzoyl quinone dioxime with 2 parts clay (23)

117: 1 part p-quinone dioxime with 2 parts clay (23)

122: aqueous solution of 40% potassium salt of mercaptobenzthiazole and 10% potassium pentamethylene dithiocarbamate. Extremely fast curing accelerator generally used in latex compounds, active above room temperature. Gives good ageing properties. Pale brown liquid, s.g. 1·22.

0·5–2% solution diluted with an equal amount of water. Has no effect on the stability of latex. Damp latex films discolour slightly when handled (6)

531: zinc-2-mercaptobenzthiazole (mixture with anhydroformaldehyde-p-toluidine and di-o-tolyl guanidine) (6)

552: piperidine pentamethylene dithiocarbamate (6)

576: acrolein-amine reaction product (408)

737: aldehyde-amine accelerator (obsolete) (6)

737–50: aldehyde-amine accelerator (obsolete) (6)

774: cyclohexyl ethylammonium cyclohexyl dithiocarbamate (408)

808: butyraldehyde-aniline condensation product (6)

833: butyraldehyde-butyramine condensation product (6)

1000: α-o-tolylbiguanidine(408)

A: N-cyclohexyl-2-benzthiazole sulphenamide

AZ: N-diethyl-2-benzthiazole sulphenamide (Vulkacit AZ) (408)

B: mixture of 1 part 2-mercaptobenzthiazole with 2 parts tetramethyl thiuram disulphide. Pale brown powder, s.g. 1·45. *Uses:* accelerator for butyl rubber (5)

BB: thiocarbanilide (14)

CT-N: tricrotonylidene tetramine (Vulkacit CT-N) (408)

CX: 2-mercaptobenzthiazole (6)

9

CZ: N-cyclohexyl-2-benz-thiazyl sulphenamide (408)

D: diphenyl guanidine (408)

DB-1: double salt of zinc ethyl phenyl dithiocarbamate and cyclohexylethylamine (Vulkacit DB-1) (408)

DM: 2,2'-dibenzothiazyl disulphide (408)

DOTG: di-o-tolyl guanidine (408)

DZ: sulphenamide accelerator (Vulkacit DZ) (408)

F: mixture of dibenzthiazyl disulphide with basic accelerators (408)

FP: anhydroformaldehyde-p-toluidine (408)

GZ/GR: mixture of sulphenamides (408)

H: hexamethylene tetramine (408)

H30: hexamethylene tetramine with 3 % anticaking agent (408)

HX: cyclohexylethylamine (Vulkacit HX) (408)

J: dimethyl diphenyl thiuram disulphide (Vulkacit J) (408)

L: zinc dimethyl dithiocarbamate (408)

LDA: zinc diethyl dithiocarbamate (408)

LDB: zinc dibutyl dithiocarbamate (408)

Mercapto: 2-mercaptobenzthiazole (408)

MDA/C: mixture of 2-mercapto-benzthiazole and zinc diethyl dithiocarbamate (408)

MOZ: N-oxydiethylene-2-benzthiazyl sulphenamide (408)

MT/C: mixture of 2 parts tetramethyl thiuram disul-phide and 1 part mercaptobenzthiazole (408)

NP: triazine derivative (408)

P: zinc ethyl phenyl dithiocarbamate (408)

P Extra N: zinc ethyl phenyl dithiocarbamate (408)

R: water soluble catalyst, used to harden formalde-hyde-nitrogen resins (24)

Thiuram: tetramethyl thiuram disulphide (408)

Thiuram
MS: tetramethyl thiuram monosulphide (408)

TR: polyethylene polyamine (408)

U: mixture of 2-benzthiazyl-2:4-dinitrophenyl thio-ether and diphenyl guani-dine (Vulkacit U) (408)

WL: sodium cyclohexyl ethyl dithiocarbamate (Vulkacit WL) (408)

X 28: diphenyl guanidine

Z-51: zinc dithiocarbamate (45)

ZM: zinc-2-mercaptobenz-thiazole (408)

ZP: zinc pentamethylene dithiocarbamate (408).

ACCELERENE VI Mixture of p-nitrosodimethyl aniline and β-naphthol in equimolecular quantities (60).

ACCINELSON p-nitroso-dimethyl aniline.

ACCO POLYMERS Group of acrylic polymers and copolymers (17).

ACCOSPERSE Group of aque-ous pigment dispersions for latex colours (21).

ACCRA PASTE (LA GLU) Trade name for a natural rubber

which comes from the Ivory Coast in the form of bales weighing 2–3 kg. Has a high resin content and is a poor quality wild rubber from *Carpodinus hirsuta Hua.*

ACCURAY Systems used for continuous measurement of thickness and production control on calenders, extruders and spreading machines (18).

ACELE American produced cellulose acetate fibre, s.g. 1·32 (6).

ACETALDEHYDE CH_3CHO. Colourless liquid, b.p. 21°C, s.g. 0·8. Miscible with water, alcohol, ether, benzene. Easily polymerisable to paraldehyde and metaldehyde; intermediary product for plastics. Occurs in small quantities in *Hevea* latex (approx. 5–6 mg/l).

ACETALDEHYDE-AMMONIA (1-aminoethanol, α-aminoethyl alcohol) $CH_3CH(OH)NH_2$. Colourless crystals; m.p. 97°C; b.p. 111°C. Soluble in alcohol and ether. Vulcanisation accelerator; needs zinc oxide and a fairly high sulphur content. Gives slow cures with a long plateau, a low modulus and good ageing properties. Non-staining. Added in the form of a masterbatch at a fairly low mixing temperature. *Uses:* rubber thread, rubber bands, golf ball centres.
TN: Aldamine
Aldehyde ammonia
Ammonia
Grasselerator 101
Velosan (365)
Vulkacit A (43).

ACETALDEHYDE-ANILINES CONDENSATION PRODUCTS $(C_6H_5N=CH—CH_3)_x$. *Accelerators:*

Viscous, brown alkaline liquids or dark brown powders; s.g. 1·05. Soluble in benzene, toluene, chloroform, acetone, partially soluble in alcohol, insoluble in water, dilute acids and alkalis. Booster for thiazoles. In the absence of zinc oxide give vulcanisates with a low modulus and high elongation at break. Should not be used in the presence of magnesium and calcium oxides. Vulcanisates have a high surface glaze. Suitable for hard rubber and rubber thread.

Quantity: 1–2%.
TN: A-7 (5)
A-11 (5)
A-19 (5)
A-40 (5)
A-50 (5)
+ A-77 (5)
Accélérateur S (19)
+ Crylene Base (23)
+ E.A. (6)
+ Ethylidene-Aniline (6)
F-C-B
Nocceler K (274)
RH 40
RH 50D
Tensilac 40, 41 (202)
+ Vulcafor PT (60)
+ Vulcafor RN (60)
+ Vulcone (6).

Antioxidants: Brown, resinous powder, m.p. 60–80°C, s.g. approx. 1·15, or brown liquids. Soluble in acetone, benzene and ethylene dichloride, insoluble in water. Have a boosting effect on vulcanisation. Do not bloom and do not stain in contact with vulcanisates. Give good protection against normal oxidation and fairly good protection against heat. Offer no protection against tensile fatigue. Stiffen unvulcanised compounds and thus improve shape retention. Useful in enabling a specific hardness to be attained and retained in roller coverings.

Uses: gum stocks, mechanical goods, solid rubber tyres, roller coverings. *Quantity:* 1–2%, may be increased with advantage up to 5%.

TN: Anti-Age 55 (3)
VGB (23)
EA (6)
Crylene (23).

ACETALDEHYDE-ANILINE-p-TOLUIDINE CONDENSATION PRODUCT

$(CH_3.C_6H_4N.CHCH_3)_x$, (ethylidene-p-toluidine). Pale brown, resinous powder, m.p. approx. 85°C, s.g. 1·15. Fairly slow accelerator, used for reclaim compounds with good ageing properties. Booster for acid accelerators, active above 140°C. Retarded by clays and inorganic fillers. Causes dark staining.

Quantity: primary 0·5–2%, mostly boosted by thiurams and thiazoles; secondary 0·3–0·5% with thiazoles and thiurams.

TN: Vulcanex (6)
Accélérateur TS (19)
Tensilac 50 (202).

ACETALDEHYDE-BUTYR-ALDEHYDE-ANILINE CONDENSATION PRODUCT

Dark brown, viscous liquid, s.g. 1·06. Miscible with benzene, chloroform and acetone, immiscible with water, dilute acids and alkalis. Accelerator. Suitable for use with reclaim; in the absence of zinc oxide, compounds have a low modulus and a high elongation at break.

TN: A 16 (5)

ACETANILIDE

$C_6H_5NH.COCH_3$, N-phenylacetamide, acetoaminobenzene, acetaniline. White crystalline powder, m.p.

113–116°C, b.p. 305°C, s.g. 1·211. Soluble in alcohol, acetone, benzene, ether, chloroform, hot water. Accelerator.

ACETEX Polyvinyl acetate latex (23).

ACETIC ACID CH_3COOH. Colourless liquid with unpleasant odour, m.p. 16·7°C, b.p. 118°C, s.g. 1·055. Coagulant for latex; as 2% solution.

Quantity: 8–12 g concentrated acid/kg rubber. Slow to react, promotes fermentation. Is being replaced increasingly by formic and oxalic acids.

ACETINE Glycerine monoacetate.

ACETO Group of accelerators, antioxidants and blowing agents (25):

AD:	azodicarbonamide
AN:	aldol α-naphthylamine
AZIB:	azobis (isobutyronitrile)
DIPP:	N,N'-di-β-naphthyl-p-phenylene diamine
DNPT:	N,N'-dinitrosopentamethylene tetramine (with 40, 80 and 100% active constituent)
HMT:	hexamethylene tetramine
PAN:	phenyl-α-naphthylamine
PBN:	phenyl-β-naphthylamine
POD	polytrimethyl dihydroquinoline
TETD:	tetraethyl thiuram disulphide
TMTD:	tetramethyl thiuram disulphide
TMTM:	tetramethyl thiuram monosulphide
ZDBD:	zinc dibutyl dithiocarbamate
ZDED:	zinc diethyl dithiocarbamate

ZDMD: zinc dimethyl dithiocarbamate.

ACETONE CH_3COCH_3 Dimethyl ketone, ketopropane, 2-pro-
panone. Colourless liquid with
characteristic odour, m.p. $-94.2°C$,
b.p. $56.1°C$, s.g. 0.7972 ($15°C$).
Miscible with water, alcohol, ether,
chloroform, oils.
Uses: solvent for cellulose acetate,
extraction agent for non-rubber
constituent (resins) and free sulphur
in vulcanisates.

ACETONE EXTRACT Acetone
totally extracts certain non-rubber
constituents and fatty acids, among
them quebrachitol, sterols and resins
from natural rubber; emulsifiers,
antioxidants etc. from synthetic
rubber; resins, sulphur, mineral oils,
paraffin wax, degradation products,
from vulcanisates. Also partially
extracts substances such as fatty
oils, bitumens, oxidised rubber. The
acetone extract of raw rubber consists of: sterol 8.2%, sterol ester 2.8%,
sterol glucoside 17.5%, 1-methylin
sitol 0.2%, stearic acid 5.5%, oleic
and linoleic acids 46.1%, unidentified
constituents 19.7%.
ASTM D 297–59T,
BS 1673: F2: 1954.
DIN 53 557.

ACETONITRILE $CH_3C{\equiv}N$
Methyl cyanide, cyanomethane,
ethane nitrile. Poisonous liquid with
ether-like smell, m.p. $-45°C$, b.p.
$81.6°C$, s.g. 0.787 ($15°C$). Miscible
with water, methyl alcohol, ethyl
acetate, acetone, ether, chloroform,
carbon tetrachloride, ethylene chloride and many unsaturated hydrocarbons. Forms an azeotrope with
16% water, b.p. $76°C$.
Use: solvent, *e.g.* in the extraction
distillation of butadiene, removal of

tar and phenols from petroleum
hydrocarbons.

ACETONYL ACETONE
$CH_3COCH_2CH_2COCH_3$ 2:5-di-
ketohexane,1:2-diacetylethane.
Colourless liquid, m.p. $-5.4°C$, b.p.
$192.2°C$, s.g. $0.971–0.975$.
Use: solvent for cellulose acetate.

ACETOPHENONE
$C_6H_5CO.CH_3$ Phenyl methyl
ketone, acetyl benzene. Colourless
liquid or laminated crystals, m.p.
$20.5°C$, b.p. $202°C$, s.g. 1.033.
Soluble in alcohol, chloroform, ether,
glycerine.
Use: deodorant for NR, SR and
latices.

ACETYLENE BLACK Semi-
reinforcing, electrically conducting
black used for antistatic products
(427).

ACETYLENE DICHLORIDE
$CHCl{=}CHCl$ Dichloroethylene.
Liquid which is difficult to ignite,
b.p. $55°C$, s.g. 1.28.
Use: solvent for rubber resins, oils
and fats.

ACETYL SALICYLIC ACID

Salicyl acetic acid, acetosal, aspirin,
etc. White, crystalline powder, m.p.
$135°C$, s.g. 1.36. Soluble in alcohol
and ether, sparingly soluble in water
(1:300). Retarder, prevents thiazole
containing compounds from scorching during processing.
Quantity: $0.25–0.5\%$.
TN: Retarder ASA (with 10%
stearic acid), Retarder A, and many
others.

ACETYL TRIALLYL CITRATE
$CH_3COOC_3H_4$
　　　　$(COOCH_2CH=CH_2)_3$.
Colourless liquid, b.p. 142–143°C,
s.g. 1·146. Miscible with acetone,
benzine, chloroform, dioxane, dimethylformamide, immiscible with
water. Cross linking agent for polyesters. Polymerises to a clear hard
resin, which solidifies on heating,
b.p. 120–150°C. Also forms copolymers, *e.g.* with acrylonitrile,
acrylic and methacrylic acids, vinyl
esters.
Use: for coatings.

ACETYL TRIBUTYL CITRATE
$CH_3COOC_3H_4(COOC_4H_9)_3$. Colourless, odourless liquid, m.p.
$-80°C$, b.p. 173°C, s.g. 1·046.
Immiscible with water. Softener for
vinyl polymers, NR and SR; improves flexibility at low temperatures.
TN: Citroflex A-4 (144).

ACETYL TRIETHYL CITRATE
$CH_3COOC_3H_4(COOC_2H_5)_3$. Colourless, odourless liquid, b.p. 131–
132°C, s.g. 1·135. Softener for NR,
SR, cellulose plastics and polyvinyl
plastics. Has no effect on vulcanisation.
TN: Citroflex A 2 (155).

**ACETYL TRI(2-ETHYL
HEXYL)CITRATE**
$(C_8H_{17}OCOCH_2)_2$.
　　　　$C(OCOCH_3)COOC_8H_{17}$.
B.p. 225°C (1 mm), s.g. 0·983 (25°C).
Softener for natural and synthetic
rubbers and vinyl resins.
TN: Citroflex A 8 (155).

ACETYL TRIHEXYL CITRATE
$(C_6H_{13}OCOCH_2)_2C(OCOCH_3)$.
　　　　$COOC_6H_{13}$.
Colourless liquid, m.p. $-80°C$, b.p.
215°C (1 mm), s.g. 1·005 (25°C).

Immiscible with water. Softener for
natural and synthetic rubbers, improves flexibility at low temperatures.
TN: Citroflex A-6 (155).

ACHARD, FRANZ KARL
German chemist, 1753–1821. Carried
out the first dry distillation of
rubber, obtaining a 'rubber oil' which
could be used as a rubber solvent.
Was the first to attempt the determination of the mechanical properties
of rubber by loading rubber strips
of equal length with increasing
weights; this experiment resulted in
the first approximation to a stress/
strain curve. In 1777, published an
article on 'Experiments with elastic
resin'. Known for work on production of beet sugar.

ACHRAS ZAPOTA W. Indian
Mispel, Sapodilla, Sawo Manilla.
Member of Apocynaceae (tropical
S. America, S.E. Asia), the unripe
fruits of which contain rubber. Used
in Guatemala to produce chicle gum,
by evaporation of the sap.

ACID ACCEPTOR Substance
which reacts chemically to bind
traces of acids in elastomers and
plastics, and thus acts as a stabiliser.
The acids may be formed during
production or may occur because of
internal decomposition.

ACID GAS PROCESS Process
patented by Goodyear in 1837 for
the treatment of rubber with nitric
acid and nitric acid vapour.

ACID NUMBER The number of
mg of potassium hydroxide required
to neutralise the free fatty acids in
1 g of test substance.

ACINTOL Group of tall oils and derivatives (26).

ACOFOR Group of distilled tall oil fatty acids (28).

ACOLIN Refined tall oil fatty acids. Plasticisers (28).

A-C POLYETHYLENE Polyethylene of low mol. w. (approx. 2000), s.g. 0·92. May be dispersed easily in synthetic elastomers and NR.
Uses: reduction of the tackiness of compounds, lubricant, reduction of shrinkage, improvement of the dispersion of fillers. Also used in emulsified form for latex. Prevents sticking on mill or calender rolls (29).

ACRAWAX Group of synthetic waxes. Diluents, prevent sticking (30).

ACRI-FLO Styrene acrylate latices with 47% total solids, s.g. of latex 1·04, pH 8·5–10. Type 10, viscosity 50–150 cp, particle size 1500–1800 Å. Type 151, viscosity 150 cp, particle size 900 Å.
Uses: colours, paper and carton coatings (31).

ACRILAN Formerly Chemstrand acrylic fibre, staple fibre of a copolymer of at least 85% acrylonitrile and max. 15% vinyl acetate, vinyl pyridine or other vinyl compounds. In experimental production since 1950, produced commercially since 1952 (36).

ACRIN Condensation product of hexamethylene tetramine, benzyl chloride and mercaptobenzthiazole. Accelerator (6).

ACRO Group of aqueous dispersions of polymethyl methacrylate and copolymers (59).

ACROGOMMA Trade name for polymethacrylate (59).

ACRONAL Aqueous dispersions of polyacrylates and copolymers with vinyl compounds. Total solids 40–50%. Form soft, tacky to hard films. *Use:* impregnation (32).

ACRONAL 700 Copolymer of acrylic acid butyl esters with vinyl isobutyl ester. Bonding agent for compounds containing chlorinated rubber and nitrocellulose. Resistant to light and normal ageing (32).

AC RUBBER Abbreviation for anticrystalline rubber. Natural rubber vulcanisates stiffen at low temperatures (between −20 and −40°C) and are rendered unusable. By adding small quantities of a thiolic acid (R.COSH), *e.g.* thiobenzolic acid, crystallisation at low temperatures may be retarded almost completely. The thiolic acids may be added either to the solid rubber or to the latex prior to coagulation. Treatment of rubber at approx. 140°C with sulphur dioxide (from butadiene sulphone) and cyclohexylazocarbonitrile as a catalyst, gives similar results. The retardation of crystallisation is caused by a cis-trans isomerisation of the rubber molecules and even a small degree of isomerisation is sufficient to achieve this effect.
Lit.: *NRPRA Techn. Bull.,* 1962, 4.

ACRYLAN Acrylonitrile synthetic fibre (5).

ACRYLIC ACID

$CH_2=CH.COOH$ Propenic acid, vinyl carbonic acid, ethylene carbonic acid. Colourless liquid with pungent smell, m.p. 13°C, b.p. 142°C, s.g. 1·051. Miscible with water. Polymerised very easily by heat, light and peroxides and must be stabilised with inhibitors (*e.g.* hydroquinone, phenylene diamines).

Uses: water soluble polymers and copolymers, thickening, binding and glueing agents, raw material for plastics.

ACRYLIC ACID AMIDE

$CH_2=CH—CONH_2$. In flake form, m.p. 84°C.

Uses: polymers and copolymers which may be crosslinked with aldehydes. The low polymer products are soluble in water, the high polymers swellable; they become insoluble through crosslinkage.

ACRYLIC ACID ESTERS

Properties of the lower esters are as follows: methyl b.p. 77·4°C, s.g. 0·956 (20°C); ethyl b.p. 99·4°C, s.g. 0·941; propyl b.p. 123°C; butyl b.p. 138°C, s.g. 0·899 (20°C); the esters of C_{12}–C_{18} alcohols are wax-like products. The esters may be polymerised by light or by heating in the presence of organic peroxides, or copolymerised with vinyl compounds or dienes.

ACRYLIC ACID ETHYL ESTER

$CH_2=CH.COOC_2H_5$. Colourless liquid, b.p. 98·5°C, s.g. 0·914. Polymerises on distillation. Solvent for rubber and nitrocellulose, raw material for synthetic resins.

ACRYLOID

Group of polyacrylates and polymethacrylates (38).

ACRYLON

Elastomers based on acrylic acid esters and acrylonitrile. Have good heat resistance and resistance to hot oils, boiling water, high pressure lubricants, oxygen, ozone and light.

Uses: sealing materials, conveyor belts, hose lines, protective paints, adhesives.

Grades:

EA-5, copolymer of 95:5 ethyl acrylate and acrylonitrile

BA-12, copolymer of 88:12 n-butyl acrylate and acrylonitrile (34, 37).

ACRYLONITRILE

$CH_2=CHCN$. Colourless, pungent smelling liquid, m.p. −82°C, b.p. 78°C, s.g. 0·797. Unstable when pure but can be stabilised with copper oleate or dioxidiphenyl.

Uses: raw material for production of polyacrylonitrile (fibres) and butadiene-acrylonitrile copolymers (rubbers). The highly polar nitrile group is responsible for the relatively high inter-chain cohesion and for the insolubility of the rubbers in hydrocarbon solvents. Vulcanisates swell to a limited extent in petroleum solvents but this swelling is usually sufficiently small to be acceptable in service. Raw material also for styrene containing co- and terpolymer plastics materials.

ACRYSOL

Group of aqueous solvents and emulsions of sodium and ammonium polyacrylates.

Uses: dispersing and thickening agents for latex compounds, latex emulsion colours, adhesives, paints and spraying compounds (33).

Grades:

GS, sodium salt

GA, ammonium salt.

16

ACRYVIN Methyl methacrylate resins (35).

ACS Abbreviation for American Chemical Society, 1155 16th Street, Washington, D.C. Founded 1876.

ACS COMPOUNDS Compounds suggested by the ACS, Division of Rubber Chemistry, for use in research into the physical properties of rubber. Natural rubber is often deficient in fatty acids and thus gives unsatisfactory results. In such cases ACS II, which has a higher fatty acid content, may be used.

	ACS I	ACS II
Rubber	100	100
Zinc oxide	6	6
Stearic acid	0·5	4
Mercapto-benzthiazole	0·5	0·5
Sulphur	3·5	3·5

Cure: 5–225 min at 141°C.
Lit.: *Rubber Chem. & Tech.*, 1939, **12**, 633; 1944, **17**, 529.

ACT 3 Rubber obtained from skim latex by a chemical and enzymic deproteinisation process. The crumb form is compressed hydraulically into blocks after drying; it may be used as natural rubber (38).

ACTIFAT Activated fatty acids. *Use:* activator (39).

ACTIVATION ENERGY A measure of the energy required to effect a chemical reaction. Derived from the Arrhenius equation

$$\ln k = \ln A + \frac{E}{RT}$$

where k is the appropriate rate coefficient, A the pre-exponential factor, R the gas constant and T the absolute temperature. The graph of $\ln k$ *v.* T^{-1} should be linear with slope E/R so that E, the activation energy, is the temperature coefficient of reaction rate. Over a small temperature range, the reaction rate will increase approximately linearly with the temperature difference (not its reciprocal) and this has given rise to the concept of a vulcanisation coefficient, the factor by which the cure rate must be multiplied, or the cure time divided, to compensate for a 10°C increase in curing temperature. For accelerated sulphur systems in the temperature range 135–160°C, the overall activation energies for first order crosslinking lie in the range 20–35 kcal/mole, and the vulcanisation coefficients in the range 1·6 to 2·3.

ACTIVATOR Material or mixture added to an accelerated curing system to realise its full potential. In accelerated sulphur vulcanisation, metal oxides (usually zinc oxide) and saturated fatty acids (usually stearic or lauric acids) are used, commonly in the range 3–10 phr oxide and 1–2 phr fatty acid. Increase of fatty acid level usually slows the rate of cure but results in a higher concentration of more stable crosslinks, *i.e.* a more tightly cured and thermally stable rubber network. Unsaturated fatty acids, such as oleic, are to be avoided in curing elastomers, such as butyl rubber, with limited main chain unsaturation.

ACTIVATOR 1102 *See* Di-butylammoniumoleate (22).

ACTIVATOR DN Surface coated urea. Promoter for use with

nitrogen type blowing agents. Reduces decomposition temperature (6).

ACTIVEX Activator of undisclosed composition (40).

ACTIVIT Thiocarbanilide.

ACTO Group of petroleum sulphonates. *Uses:* latex emulsifying agents, dispersing agents (41).

ACTOR Group of vulcanising agents (16).
Grades:
CL: tetrachlorobenzoquinone
DQ: p,p'-dibenzoyl quinone dioxime
Q: p-quinone dioxime
R: morpholine disulphide.

ADAM Aqueous solution of sodium dimethyl dithiocarbamate.

A D C Abbreviation for ammonium diethyl dithiocarbamate.

ADDITION POLYMERISATION Polymerisation reaction by the simple addition of monomers through the dissociation of a double bond, whereby two free bonds are released, which may then unite with the equivalent bonds of the neighbouring molecules. The presence of a further double bond in dienes is important in facilitating further reaction during vulcanisation. For dienes, the following addition structures may arise:
1. symmetrical dienes, *e.g.* butadiene

$$-CHCH_2- \quad\quad -CH_2CH=CHCH_2-$$
$$\overset{|}{\underset{\overset{\|}{CH_2}}{CH}}$$

1:2-Addition   1:4-Addition

2. non-symmetrically substituted dienes, *e.g.* isoprene or chloroprene

$$-CH_2-\overset{CH_3}{\underset{\overset{\|}{CH_2}}{\overset{|}{C}}}- \quad\quad CH_3-\overset{-CHCH_2-}{\underset{\|}{\overset{|}{C}}}$$

1:2-Addition   3:4-Addition

$$-CH_2-\overset{CH_3}{\overset{|}{C}}=CHCH_2-$$

1:4-Addition

ADDITIVE PASTE Additives to blowing agents. Produce a finer cell structure and reinforcement of the blowing action, eliminate the smell of cellular rubber and facilitate the introduction of fillers (42). Type 1100, finely dispersed urea with addition of activating dispersing agents. Type 1600, combination of activating substances with dispersing agents.

ADDUCT RUBBER Diene polymers, produced by emulsion polymerisation at the double bond in the presence of aliphatic mercaptans. The products have a much lower unsaturation level but can be vulcanised in the normal way.

ADELE RUBBER Okala, rouge de Kassai, Congo. Good, 'nervy', natural rubber from *Landolphia owariensis* (Toga).

ADIMOLL Softener based on adipic acid esters. A.BB, benzyl butyl adipate, A.BO, benzyl octyl adipate, A.DB, dibutyl adipate (43).

ADIPOL Softener based on adipic acid esters.

Grades:
2EH: di-2-ethyl hexyl adipate
10 A: diiso-octyl adipate
BCA: dibutoxyethyl adipate
DIBA: diisobutyl adipate
ODY: octyl decyl adipate
XX: didecyl adipate.

ADIPRENE Group of poly-urethanes, reaction products of poly-ethers (*e.g.* polytetrahydrofuran) with diisocyanates.

Grades:
B: millable type with terminal active hydrogen atoms. Can be crosslinked with iso-cyanates.
C: millable type; s.g. 1·07. Partially soluble in tetra-hydrofuran, methyl ethyl ketone and dimethylforma-mide, swells in chlorinated solvents and in petroleum hydrocarbons. An unsatu-rated character is obtained through the insertion of double bonds into the pre-polymer. This makes vul-canisation of the product possible, as for natural rub-ber, by means of sulphur and accelerators. Processing by conventional machinery. Reinforcement is with carbon black and other fillers. Vul-canisates are abrasion resist-ant, resistant to ozone, oxy-gen, and to a number of solvents.
L: liquid stabilised prepolymer; s.g. 1·06, viscosity 14 000–19 000 cp (30°C). Soluble in toluene, acetone, carbon tetra-chloride, methyl ethyl ketone and some aromatics. Chain extension and network for-mation is achieved by using a diamine (methylene

bis-o-chloraniline) 11–12%, or glycol (ethylene diamine tetraisopropanol).
L-100 (previously, L): s.g. 1·06, viscosity 14 000–19 000 cp (30°C).
L-167 (previously, LD-167): s.g. 1·06, viscosity 5000–7000 cp (30°C), of higher isocyanate content and more reactive than L-100 (6).

ADMEX Group of softeners of the epoxy type, s.g. 0·91–0·99, mol. wt. 350–950. Insoluble in water (44).

ADM VEGETABLE ACID Group of distilled vegetable fatty acids. Activators and softeners for natural and synthetic rubber (44).

A.D.S. Abbreviation for air dried sheet.

ADVAGUM Plasticised high polymer. Extender and tackifier (45).

ADVAPLAST Group of epoxy softeners (45).

ADVARESIN Thermoplastic resins, compatible with NR, SR and vinyls.
Uses: cements, improvement of the adhesion of rubber to rayon (45).

ADVASTAB Group of antioxi-dants, light protecting agents and stabilisers (45).

ADVASTAT Group of antistatic agents (45).

ADVAWAX Group of lubri-cating agents (45).

ADVAWET Non-ionic esters. Wetting agent for latex.

19

AERO Group of rubber auxiliary products.
Grades:
Aero: sulphur and zinc stearate
Aero AC 50: reaction product of 2 mol. di-o-tolyl guanidine and 1 mol. zinc chloride.
Uses: accelerator and activator.
Aero AC 165: guanidine derivative.
Uses: accelerator and activator.
Aero DOP
and
Aero DOPI: diocytyl phthalate.
Use: plasticiser (21).
Aero-X: aniline diisopropyl dithiophosphate.

AEROLASTIC Joint sealant made from an oil resistant bitumen/rubber mixture.

AEROPREEN Polyurethane foam materials (46).

AEROSIL Highly active white filler for tear and rupture resistant NR and SR vulcanisates (156).

AEROSOL Group of esters and dicarboxylic acids.
Uses: mould release agents, wetting and dispersing agents for latex (21).

AERUM NO. 5 Mixture of 40% urea with N-cetyl betaine; s.g. 1·42. Activator for blowing agents (136).

AETERNAMID Group of plastics (104).
Grades:
A: adipic acid-hexamethylene diamine condensation product
B: polyamide based on caprolactam
S: sebacic acid-hexamethylene diamine condensation product
U: polyurethane elastomers.

AETNA Levigated alumina. Filler (47).

AF- Group of silicone oils.
Uses: antifoam and mould release agents (198).

AFA Abbreviation for anhydroformaldehyde-aniline condensate.

AFB Phthalic anhydride. Vulcanisation retarder (6).

AFCOTHENE Low pressure polyethylene (48).

AFCOVYL Group of vinyl polymers and copolymers (48).

AFFENHAARE Fossil rubber of plant origin which has been vulcanised by the natural infiltration of sulphur. Found in the tertiary lignite layers of central Germany.

AFLUX Compound in powder form with a plasticising and dispersing effect on NR and SR; combination of surface active substances with polar compounds bound to active silica. S.g. 1·25. Improves plateau effect and flow properties, protects against dynamic fatigue. Non-discolouring (42).
Quantity: 3–15%.

AFPT Abbreviation for anhydroformaldehyde-p-toluidine.

AGCHEM V Ketone-aldehyde condensation product. Antioxidant and accelerator (49).

AGEBEST 1293-22A Antioxidant of undisclosed composition (50).

AGE REGISTER Group of antioxidants (440):

Grades:

ADD: substituted aromatic amine. Wax-like lumps. Soluble in benzene, chloroform and acetone, insoluble in water. Does not migrate below 1%. Has no influence on vulcanisation

AL: N-Aldol-α-naphthylamine

MB: mercaptobenzimidazole.

AGERITE Group of antioxidants (51).

Grades:

AK: polytrimethyl dihydroquinoline

Alba: hydroquinone monobenzyl ether

DPPD: diphenyl-p-phenylene diamine

Excel: combination of diphenyl-p-phenylene diamine and isopropoxydiphenylamine

Gel: combination of 75% octylated diphenylamines (Agerite Stalite) with 25% petroleum waxes. Light brown, wax-like mass, m.p. 40–50°C, s.g. 0·91–0·97. *Use:* antioxidant for mechanical goods and neoprene. *Quantity:* 0·5%, excess causes wax-like bloom.

Geltrol: alkylated phenol. Brown, viscous liquid.

Hipar: combination of 50% phenyl-β-naphthylamine, 25% diphenyl-p-phenylene diamine and 25% p-isopropoxydiphenylamine. Brown powder, m.p. 65–75°C, s.g. 1·13–1·19. Used with natural and synthetic rubber, protects against oxidation and heat, improves tear resistance. Causes serious discoloration in light-coloured compounds. *Quantity:* 1%, excess produces bloom.

HP: combination of 65% phenyl-β-naphthylamine and 35% diphenyl-p-phenylene diamine. Brown powder, m.p. 96–98°C, s.g. 1·18–1·24. For NR, SR and latex, protective agent against oxidation and tear initiation. Causes serious staining. Suitable for heavy duty latex articles.

HPX: N,N'-diphenyl-p-phenylene diamine

ISO: p-isopropoxydiphenylamine

OD: alkylated diphenylamine

Powder: phenyl-β-naphthylamine

Resin: Aldol-α-naphthylamine resin

Resin D: polytrimethyl dihydroquinoline

Spar: combination of mono-, di- and tristyryl phenols

Stalite: combination of octylated diphenylamines. Reddish brown viscous liquid, s.g. 0·97–1·01. Soluble in alcohol, benzene, carbon disulphide and benzine. Normally used with natural rubber, synthetic elastomers and latices. Particularly effective in neoprene, also as vulcanisation retarder during processing. Relatively non-staining and non-discolouring. *Quantity:* 1%, in heatproof SBR combinations 1% with 1% polytrimethyl dihydroquinoline

Stalite S: combination of octylated diphenylamines. As Stalite but in solid form. Light

21

brown powder, m.p. 75–90°C, s.g. 0·94–1·00. Soluble in alcohol, benzene, carbon disulphide and benzine. *Uses:* as Stalite

Superflex: reaction product of diphenylamine and acetone. Dark brown liquid, s.g. 1·08–1·12. Soluble in acetone, benzene, chloroform, carbon disulphide. Discolours. Good protective agent against tear formation and weather. *Uses:* tyres and heavy duty mechanical goods

Superflex Solid: diphenylamineacetone reaction product. Dark brown powder, s.g. 1·33.

Superlite: polyalkyl polyphenol. Brown liquid, s.g. 0·945–0·965. Soluble in benzene, chloroform and benzine. Non-staining and non-discolouring antioxidant for natural and synthetic rubber and latices. Normally used for white and light-coloured articles and foam rubber, also for articles in contact with varnishes and paints. *Quantity:* 1–3%.

Superlite Solid: solid form of Superlite; s.g. 1·26.

Syrup: condensation product of a ketone and amine (obsolete)

White: di-β-naphthyl-p-phenylene diamine

XPX: diphenyl-p-phenylene diamine.

AGILENE Polyethylene (52).

AGROPUR 2:4-dichlorphenoxyacetic acid.

AGRUNOL WO tetramethyl thiuram disulphide.

AGUM Mixture of thiazole and guanidine. Accelerator (23).

AIBN Abbreviation for azoisobutyronitrile.

AIDON 30 Specially treated urea. Sponge blowing agent (407).

AIR DRIED SHEET ADS air dried sheet, PAUS pale amber unsmoked sheet. Light coloured, yellow-brown unsmoked sheet rubber, dried at approx. 60°C. To avoid enzymatic discoloration the wet sheets can be treated for a short time with steam. Small additions of enzyme inhibitors and fungicides may be used, *e.g.* p-nitrophenol.

AIRDRIRUB Trade name for a light, air dried, unsmoked sheet rubber.

A.I.R.I. Abbreviation for Associateship of the Institution of the Rubber Industry. Awarded to qualified specialists in the field of rubber, usually on the basis of an examination or thesis.

AKROFLEX Group of antioxidants for NR, SBR, butyl rubber, and neoprene (6).
Grades:

A, B and D: combination of naphthylamine derivatives and secondary aromatic amines.

AZ: mixture of p-phenylene diamines; m.p. 82°C, s.g. 1·18. Non-blooming antiozonant for neoprene, with good protection against heat ageing. Heightens the injectability of compounds without crosslinking and checks tearing during shaping. Reduces the increase of

22

compound viscosity due to premature gelation. *Quantity:* 2–3%.

C: combination of 65% phenyl-α-naphthylamine and 35% diphenyl-p-phenylene diamine. Black wax-like mass, s.g. 1·18. Does not influence vulcanisation, prevents stiffening of SBR rubber during mastication. Effective against tear formation, high temperatures and normal ageing. *Quantity:* 1–2% for normal use, 3% and 2% phenyl-α-naphthylamine for heat resistant neoprene combinations. Bloom likely in concentrations above 1%.

CD: combination of 65% phenyl-β-naphthylamine and 35% diphenyl-p-phenylene diamine. Dark grey granulate, s.g. 1·12. *Quantity, properties and uses:* as Grade C

F: combination of phenyl-β-naphthylamine, diphenyl-p-phenylene diamine and a booster. Grey black powder, s.g. 1·16. Does not influence plasticity or vulcanisation. *Quantity:* 1–2%. Discolours in sunshine. No bloom until 1·75%.

AKRO GEL Sodium salt of a fatty acid ester; s.g. 1·01–1·10. Non-poisonous, non-staining wetting agent. Does not influence vulcanisation. Mould release agent, dipping and spraying agent for unvulcanised compounds to prevent self-adhesion.
Quantity: 5–15% solution (54).

AKRON Group of organic and inorganic pigments; masterbatches,

in paste form, water solutions and dispersions (54).

AKSEL One of the first organic accelerators of vulcanisation. Developed by A. B. Molony in the USA, 1915; the active component is now known to be tetramethyl thiuram disulphide.

AKTICIT Group of filler activators.
Grades:
A: mixture of fatty acid derivatives and higher alcohols. Brown liquid, s.g. 1·07. *Dosage:* 1–5% on the filler content, improves dispersion of fillers.
B: secondary amines. Yellow-brown liquid, s.g. 0·91, specially for highly active silica fillers. *Dosage:* 1–5% on the filler (43).

AKTIPLAST Peptising agent for NR and polyisoprene, processing aid for SBR; zinc salt of high mol. unsaturated fatty acids, m.p. 80–85°C, s.g. 1·08. Active as a peptiser from about 60°C. At the usual loadings of 1·5–3% may influence curing behaviour.

AKTIVATOR DN *See* Urea (6).

AKTIVATOR GL Accelerator/activator based on a polyester/isocyanate system.

AKTIVATOR 1987 RHEINAU Cure activator for NR and SBR, particularly useful in transparent rubbers.

AKTIVATOR 2009 RHEINAU Combination of surface active

materials with activated silica. Particularly useful because of the storage stability of compounded mixes.

AKTONE Urea complex. White flakes, m.p. 105–115°C, s.g. 1·39. Activator of SBR and NR, especially for thiazole and thiuram accelerators. Deodorant for sponge rubber. *Dosage:* 2–5% (40).

AKZELERIN Group of accelerators (396).

AL Russian produced stabiliser; reaction product of acetaldehyde and aniline hydrochloride.

ALAMASK Group of deodorants which also finds applications in the rubber industry. Added to raw compounds and latex in dosages of 0·01–0·5% (20).

ALAMINE Group of primary fatty acid amines. Activators and release agents for natural and synthetic rubbers (55).

ALATHON Group of polyethylenes (6).

ALBALITH 73 Lithopone made hydrophobic by fatty acid treatment. Filler (56).

ALBAN Probably $C_{20}H_{32}O_2$. Crystalline constituent of guttapercha resin, m.p. approx. 160°C. Probably occurs in other types of resinous rubber. A resinous component of Balata known as balalban is identical.

ALBASAN β-naphthol-o-toluidine condensation product. Light brown, waxy substance, m.p.

80–100°C. Formerly used as a stabiliser (23).

ALBERTONI TESTER Horizontal stress/strain tester with facilities for autographic recording.

ALCOFLEX CONCENTRATE Water soluble urea-formaldehyde resin. Hardening agent for latex compounds (57).

ALCOGARD 354 Condensation product of alkylated phenols. Stabiliser (57).

ALCOGUM Group of polyacrylates. Latex thickeners and stabilisers (57).

ALDAIR PROCESS Process developed by the Goodyear Company, for the production of semipermeable rubbers.

ALDEHYDE-AMINE Substance formed by condensing an aldehyde with a primary amine; examples are formaldehyde, acetaldehyde, butyraldehyde, with aniline or β-naphthylamines. Used as an accelerator. When organic accelerators were first used, the aldehyde-amines were valued for their concomitant antioxidant behaviour.

ALDOL $CH_3CH(OH)CH_2CHO$. Acetaldol, β-hydroxybutyraldehyde. Colourless, viscous liquid, b.p. 83°C, s.g. 1·11. Miscible with water, alcohol and ether. Intermediate in the synthesis of butadiene, aldehyde resins and vulcanisation accelerators.

ALDOL ANILINE Yellowish powder. Slightly discolouring antioxidant (obsolete). *TN:* Resistox (5).

ALDOL-α-NAPHTHYLAMINE RESIN Reddish-brown resin with faint, sweet aroma, softening p. 70–100°C, s.g. 1·15. Soluble in acetone, ethyl acetate, benzene, methyl chloride, carbon tetrachloride, practically insoluble in alcohol, insoluble in water. Antioxidant. Has the same effect in rubber as Aldol-α-naphthylamine, but dissolves more easily. The processing and durability of unvulcanised compounds, and particularly synthetic rubber, is improved by the softness and tackiness of the resin. However, at high temperatures, it acts as a strong activator of mercapto and thiuram accelerators. Its use is limited because of its unpleasant smell. Causes discoloration on exposure to light; colour migrates on contact with other vulcanisates.
TN: Agerite resin (51)
Antioxidant AH (43)
Antage A (16)
Antioxigène RES (19).

ALDOL PROCESS Production of butadiene through hydrogenation of acetylene to acetaldehyde over a mercury catalyst, the formation of acetaldol with dilute potassium hydroxide, hydrogenation to 1:3-butanediol and conversion to butadiene.

ALFIN RUBBER Butadiene polymers and copolymers, *e.g.* butadiene and styrene with a mol. w. of several millions. Produced by using an alfin (alcohol plus olefin) catalyst in solvent polymerisation. The catalyst is a three-component system consisting of a combination of sodium salts of secondary alcohols containing at least one methyl group, together with sodium olefin, and finely dispersed sodium chloride,

e.g. sodium isopropylate and allyl sodium. n-Heptane acts as solvent, to which the catalyst is added as a dispersion in paraffin oil. Polymerisation occurs at 30°C within a few minutes. It is complicated, however, by a secondary reaction of the polymer with the sodium in the catalyst; this can lead to undesirable gel formation. Changes in the composition of the catalyst influence the mol. w., gel formation and the speed of polymerisation, and allow a certain degree of control over the reaction. Tough polymers, precipitated by methyl alcohol, are difficult to process alone, but are suitable for the production of highly oil-extended rubber.

ALGINATE Sodium, calcium, magnesium, ammonium salts of algin acids, polymannuron acids $(C_6H_8O_6)_n$. Obtained from brown algae (*Macrocystis pyrifera, Laminaria digitata, Laminaria Cloustonii, Ascophyllum nodosum*) by extraction with 1 % soda solution and precipitation with hydrochloric acid. Mol. w. approx. 150 000.
Uses: thickening agent, bonding agent and stabiliser for colloidal solution, creaming and thickening agent for latex.
TN: Keltex (236)
Amalg (72).

ALGOFLON Trade name for polytetrafluoroethylene (59).

ALI COHEN'S GUTTA Gutta Alco. Gutta-percha and balata substitute. Mixture of latex with calcium and aluminium stearate, produced according to BP 313 373 and USP 1 739 566. The product fractures quickly and dissolves in water; no practical use has been found for it.

ALIPHAT Trade name for a group of rubber auxiliary products.
Grades:
6:     palmitic acid
6A: combination of palmitic and stearic acids
7:     stearic acid
7A: combination of stearic and palmitic acids
16B: combination of oleic and linoleic acids
      Uses: emulsion polymerisation of synthetic rubber
26B: animal fatty acids. *Uses:* dispersing and emulsifying agents
45B: combination of 65–70% resin and 30–35% fatty acids. *Uses:* emulsion polymerisation of synthetic rubber, tackifier (55).

ALKALOIDS Various alkaloids, *e.g.* quinquonine, quinquonidine, quinine, quinoidine and emetine, act as vulcanisation accelerators.

ALKATERGE-C Substituted oxazoline. Liquid, s.g. 0·93. Antifoam agent for latex (62).

ALKATHENE High pressure polyethylene. Produced by polymerisation of ethylene at 1000 to 3000 atm. Miscible with rubber above 60°C; it does not stick to rubber, and is used as inner lining in paper sacks for packing high grade raw rubber, also as intermediate layer in packing sole crêpe (60).

ALKAZENE Group of solvents.
Grades:
3:   triethylbenzene
13: triisopropylbenzene
21: 1-chloro-4-ethylbenzene
24: 1:2-dichloroethylbenzene
25: dichlorethylbenzene
31: pentachlorethylbenzene

40: bromethylbenzene
42: dibromethylbenzene
47: bromocumene (61).

ALKOFEN MBF Russian produced, 1-di-(α-methyl benzyl)-4-methyl phenol. Heat and light stabiliser for vulcanisates of NR, SBR and PB.

ALKYD Group of alkyd resins. *Uses:* extending agents in natural rubber, synthetic rubber and latices (50).

ALKYDAL ST Yellow viscous fluid polyester based on phthalic acid, glycerine and ricinic fatty acids, s.g. 1·05. Softener for synthetic rubber; improves extrusion and calendering. Increases elasticity while decreasing hardness and tear resistance.
Loading: 5–20%.

ALKYL DIPHENYLAMINE Brown to reddish, viscous liquid with faint odour, b.p. above 300°C, s.g. 1·08. Highly soluble in organic solvents, insoluble in water. Gives neutral reaction. Effective antioxidant for natural and synthetic rubbers, improves flex-cracking resistance, heat resistance and is effective against copper and manganese. Causes a yellowish discoloration.
Quantity: 0·5–2%.
Uses: for coloured articles in which the slight discoloration is of no importance, inner tubes, bicycle tyres, sponge rubbers. Has no effect on odour or taste, suitable for use in articles which come into contact with food.
TN: Antioxidant DDA and DDA-EM (emulsion for latex) (43).

ALLYL CYANIDE
$CH_2=CHCH_2CN$, vinyl acetonitrile, 3-butene nitrile. Liquid, m.p. $-86°C$, b.p. $119°C$, s.g. $0·8341$.
Use: crosslinking agent in polymerisation reactions.

ALMEIDINA RUBBER Potato rubber, Cassoneira, potato gum. Resin-like, easily fractured product from *Euphorbia rhipsaloides*, or *Fockea multiflora* (W. Africa). Contains 10–25% rubber and 60–80% resin, and has occasionally been used as a rubber additive and in the bonding of rubber to metal. Named after the Portuguese explorer, d'Almeida, who, in about 1880, introduced the product from Angola, *via* Lisbon and London. The alternative name is derived from its potato-like appearance.

ALOXITE Aluminium oxide Filler (63).

ALPCO POLYRUBBER Polyurethanes in the form of a three-component system (66).

ALPEROX C Technical lauroyl peroxide (65).

ALPEX Cyclo-rubber (58, 67).

ALPHA Chemically isolated soya albumen.
Uses: protective and stabilising agent for latex (64).

ALPHA S Terpene resin.
Use: tackifier (28).

ALPOREX Polyurethane foam material (68).

ALRESEN 214 R Non-solidifying terpene-phenolic resin, m.p. 63–70°C.

Improves building tack of NR and SR compounds.
Loading: 2% (58).

ALRO AMINES High mol. w. amines.
Uses: antiscorch agents and stabilisers for latex (69).

ALROSOL Group of fatty acid amides.
Uses: wetting agents for latex, emulsifying agents and softeners (69).

ALS Abbreviation for activated low structure (black).

ALUBRAGUM Sodium and potassium polyacrylates; water soluble resins.
Uses: thickening agents for latex (70).

ALUM
$K_2SO_4.Al_2(SO_4)_3.24H_2O$, potassium aluminium sulphates. White powder or translucent octahedra, s.g. $1·75$.
Uses: coagulating agent for natural and synthetic latices. Formerly used instead of acids in the plantations, since the slabs of coagulum have a drier and less slimy surface. About 12–15 g/kg dry rubber (or per 3 l latex) is used. Now used as an alternative to the brine/sulphuric acid coagulant for latex in SBR preparation.

ALUMINA GEL
$Al_6(OH)_{16}(SO_4).H_2O$, TEG, aluminium hydroxyl gel. White powder containing 55–60% Al_2O_3. White reinforcing filler. Produced by reacting aluminium sulphate with a soda solution. The water is bound water of crystallisation and thus makes

incorporation into the rubber difficult. The alumina gel loses its high activity on drying. It has roughly the same effects as semi-reinforcing blacks.

ALUMINIUM STEARATE
$Al(C_{17}H_{35}COO)_3$. White powder, m.p. 115°C, s.g. 1·07. Soluble in alkalis, insoluble in water, alcohol and ether.
Use: in chewing gum.

AMALG Ammonium alginate.
Uses: protective and thickening agent for latex.

AMARILLO RUBBER Resinlike product from *Euphorbia fulva Stapf* (Palo amarillo, Mexico). Of no commercial use.

AMAX NO. 1 Compound of N-oxydiethylbenzthiazyl-2-sulphenamide with dibenzthiazyl disulphide. Light flakes, m.p. min. 70°C, s.g. 1·40. Extremely soluble in benzene and chloroform, slightly soluble in methyl alcohol and acetone. Delayed action accelerator for natural rubber and SBR. Used as secondary accelerator to decrease tendency to scorch. Non-discolouring, non-staining on contact.
Uses: tyres, mechanical goods (51).

AMBER Mould release agent and surface lubricant for rubber and unvulcanised compounds (73).

AMBEREX Brown factice (74).

AMBEREX SOLUTION Solution of vulcanised vegetable oils in organic solvents.
Uses: Extender and plasticiser in fluid rubber adhesives (74).

AMBERLITE Aqueous solution of a synthetic resin.
Uses: thickening agent for latex (33).

AMBEROL Group of modified alkyd resins.
Grades:
M-93: modified phenol-formaldehyde resin. Use: tackifier for butyl rubber
ST-137: modified phenol-formaldehyde resin. Use: tackifier for SBR and butyl rubber (33).

AMBERSIL Silicone oil release agent (75).

AMERICAN SOCIETY FOR TESTING MATERIALS (ASTM) 1916 Race Street, Philadelphia 3, Pa, USA. American standards organisation, founded in 1898 as the American section of the International Association for Testing Materials. The society has existed under the present name since 1902. Group D 11 for rubber and rubber-like elastomers was founded in 1912. Standards for rubber and elastomers are summarised in 'ASTM Standards on Rubber Products. Methods of Testing, Specifications, Definitions'.

AMERIPOL Group of butadiene-styrene copolymers. Includes types from the ASTM series 1000, 1500, 1600 (R/F black master batch), 1700, and 1800, and also various special types (76). High styrene content types:
1902, 48/52 styrene/butadiene
1903, 54/46 styrene/butadiene
1904, 60/40 styrene/butadiene.

AMERIPOL CB Cis-polybutadiene, 98% cis content (76).

Grades:
220: light-coloured, non-staining and non-discolouring. Gives good resistance to cold, good abrasion resistance and elasticity
441: diluted with 37·5 parts highly aromatic oils. Discolours. *Uses:* tyres, conveyor belts, mechanical goods
442: diluted with 50 parts oil. Non-staining and non-discolouring. *Uses:* for light-coloured articles
880: CB 220 with 14 parts processing aid.

AMERIPOL SN (SN: synthetic natural). cis-1:4-polyisoprene with the same molecular structure as natural rubber (76).

AMINOGUANIDINE BICARBONATE
$[H_2N.NH.C(:NH).NH_2]H_2CO_3$.
Sponge blowing agent; yields a relatively coarse cell structure.

AMINOX Diphenylamine-acetone condensation products, m.p. 85–90°C, s.g. 1·15. Extremely soluble in acetone and ethylene dichloride, less soluble in benzene, insoluble in water. General purpose: antioxidant for natural rubber, neoprene, NBR and SBR. Has practically no influence on vulcanisation. Causes a slight yellow discolouring of light-coloured compounds on exposure to light. Does not cause bloom. Gives good protection against oxygen, heat and tear initiation.
Loading: 0·25–2% (23).

AMIZEN Group of accelerators (395).
Grades:
DM: dibenzthiazyl disulphide
M: mercaptobenzthiazole

TMTD: tetramethyl thiuram disulphide.

AMMONIA NH_3. Pungent-smelling, colourless gas, m.p. −77·7°C, b.p. −33·4°C, s.g. 0·597. In aqueous solution, there is a partial combination with water to form the hydroxide (NH_4OH); dissociation to ammonium and hydroxyl ions produces an alkaline reaction. Solubility in water at:

0°	15°	20°	25°	30°	50°C
47%	38%	34%	31%	28%	18% NH_3

Uses: preserving agent for latex concentrates (0·7–1·2% ammonia); infiltrated into the latex as gas. Anti-coagulant in fresh latex in concentrations of 0·35–0·75 g ammonia per kg of latex, alone or in combination with formalin. Addition to fresh latex as ammonia water in the storage tanks before transport from the plantation.

AMMONIUM ALGINATE Alginic acid is a colloidal, cellulose-like mass with a fibrous structure, mol. wt. approx. 250 000. Occurs in large quantities in brown algae, and is extracted mainly from *Laminaria digitata*. The ammonium compound is used as a stabiliser, thickening agent and creaming accelerator for latex; yellow powder, viscosity of a 1% solution approx. 100–130 cS. Effective at concentrations of 0·1–0·2% of the water phase.

AMMONIUM BENZOATE $C_6H_5COONH_4$. White, crystalline powder, m.p. 198°C, s.g. 1·25. Soluble in water and alcohol. Preservative for latex and adhesives.

AMMONIUM BICARBONATE NH_4HCO_3. White crystals, s.g. 2·4. Decomposes at approx. 65°C into ammonia, carbon dioxide and water.

Uses: blowing agent for hollow articles.

AMMONIUM CARBONATE

$(NH_4)_2CO_3$, Hartshorn salt. Decomposes at approx. 60°C into ammonia, carbon dioxide and water.
Uses: blowing agent for hollow articles.

AMMONIUM CASEIN

Eucasin water soluble powder.
Uses: as for casein.

AMMONIUM PERSULPHATE

$(NH_4)_2S_2O_8$. White colourless, crystalline powder, s.g. 1·98. Soluble in water. Strong oxidising agent, decomposes at 100°C.
Uses: catalyst in the emulsion polymerisation of diolefins with styrene or acrylonitrile.

AMOCO Group of antioxidants (78).
531: N-butyl-p-aminophenol
532: N, N'-disec-butyl-p-phenylene diamine
533: 2:6-ditert-butyl-4-methyl phenol.

AMOPOL 11C Hydrocarbon polymer.
Uses: reclaiming oil (77).

AMPAR Trade name for Guayule rubber marketed by the Intercontinental Rubber Company.

AMSCO Group of solvents; hydrocarbon petroleum fractions (71).

AMYL OLEATE

$C_{17}H_{33}COOC_5H_{11}$, m.p. $-40°C$, b.p. (20 mm) 200–240°C, s.g. 0·876. Softener.

AMYL STEARATE

$C_{17}H_{35}COOC_5H_{11}$, m.p. 16°C, b.p. (30 mm) 230–270°C, flash p. 187°C, s.g. 0·858. Softener.

N-TERT-AMYL UREA

$H_2N.CO.NH.C(CH_3)_2.C_2H_5$. Accelerator.
TN: Cardamide 783 (147).

AN- Group of antioxidants.
Grade:
2: 4:4'-methylene bis-(2:6-di-tert-butyl phenol)
3: 2:6-ditert-butyl-4-dimethyl-amine methyl phenol
4: condensation product of heptanol and p-anisidine
6: 2:2'-thio bis-(6-cumyl-4-methyl phenol)
9: 2:2'-thio bis-(6-tert-butyl-p-chlorophenol)
25: lauroyl-p-aminophenol.

ANATASE Titaniumdioxide. (Types E, LF.) White pigment (22, 80, 81).

ANCABLO A Zinc-amine complex. White powder, s.g. 2·8. Blowing agent. Non-staining. Replaces a part of the necessary zinc oxide in a compound.
Quantity: 1–7% according to the degree of inflation, produces a fine, regular, cellular structure up to 900%, below 150% the cells are not interconnected (22).

ANCALL Group of factices; sulphur vulcanised, glyceride oils combined with bituminous substances and waxes. Hard, flexible solid, s.g. 1·05–1·09, acetone extract 30–50%, free sulphur 1·2–2·5%. Thermoplastic, organic filler and softener for natural rubber and

synthetic elastomers. Types, 790A, 799B, 825, 839.
Quantity: up to 100%.
Uses: for articles used out of doors, *e.g.* in window sealing materials, footwear, good protective against weathering and ozone (22).

ANCAZATE Group of dithiocarbamate accelerators.
Grades:
AE:　30% solution of ammonium diethyl dithiocarbamate
BU:　zinc dibutyl dithiocarbamate
BZ:　zinc dibenzyl dithiocarbamate
EHP: zinc ethyl phenyl dithiocarbamate
ET:　zinc diethyl dithiocarbamate
ME:　zinc dimethyl dithiocarbamate
Q:　activated zinc dithiocarbamate. Slightly cloudy, viscous, brown liquid, s.g. 1·10. Soluble in hydrocarbons, acetone, alcohol, carbon tetrachloride, chlorinated solvents, insoluble in water. Ultraaccelerator for vulcanisation at 20–25°C for natural rubber latex, solutions and coating mix of natural rubber and SBR. *Uses:* In extrusion and calendering compounds, needs great care to prevent prevulcanisation. *Loading:* rubber 1·75–2·25% with 1·5% sulphur, latex 1% with 1% sulphur
WSB: 48% solution of sodium dibutyl dithiocarbamate
WSE: 30% solution of diethylammonium diethyl dithiocarbamate
XX:　activated butyl dithiocarbamate. Brown liquid, s.g. 1·0. Soluble in hydrocarbons, ketones, alcohols, carbon tetrachloride, and chlorinated hydrocarbons. Ultra-accelerator with temperature range 20–100°C. For solutions, coating mixtures, calendering and extrusion compounds based on natural rubber, butyl rubber, SBR and acrylonitrile copolymers. Should be added to the warm mix before calendering or extruding. Vulcanisation at room temperature, 2–4 days, 50–60°C, approx. 12 hr. *Loading:* solution and coating mixtures 3–4% with 1·5% sulphur, dry rubber compounds 2–3% with 0·25–1·0% dibenzthiazyl disulphide and 1·5–0·75% sulphur (22).

ANCAZIDE Group of thiuram accelerators and guanidines.
Grades:
1 S:　tetramethyl thiuram monosulphide
DOTG: di-o-tolyl guanidine sulphide
DPG:　diphenyl guanidine sulphide
ET:　tetraethyl thiuram disulphide
IS:　tetramethyl thiuram monosulphide
ME:　tetramethyl thiuram disulphide (22).

ANCHOID Sodium salt of a polyalkyl aryl sulphonic acid; s.g. 1·56. Anionic dispersing agent for latex compounds. Has no stabilising effect on latex.
Loading: approx. 4% on filler (22).

ANCHORACEL Piperidine pentamethylene dithiocarbamate (22).

ANCOPLAS Mixtures of sulphonated petroleum products. Practically odourless, dark liquids. Type ER (with paraffin oils), s.g. 0·85,

type OB (with a water-repellent alcohol), s.g. 0·92.
Uses: softeners and plasticising agents. *Loading:* 1–2% for natural rubber, 2–3% for reclaim, 3–5% for neoprene (22).

ANGHIERA, PIETRO MARTIRE D' Petrus Martyr

Anglerius. Spanish missionary, in *De orbo nuovo* (1530) mentioned trees yielding a milky liquid which solidified to a resin-like mass when left to stand. This was the first printed reference to rubber.

ANGOLA RUBBER African

root rubber of average quality from *Euphorbia rhipsaloides* and *Raphionacme utilis*; appeared in the form of small balls for trade. Contains about 9% resin and has a loss by washing of 18–26%. Almeidina rubber was often called Angola rubber.

ANHYDROFORMALDEHYDE ANILINE

(Formaldehyde aniline), yellowish-white amorphous powder, density 1·14, sinters at 70°C, melting point above 131°C, soluble in benzene and methylene chloride, slightly soluble in alcohol, acetone, ethyl acetate and benzine, insoluble in water. Slow accelerator with plasticising action from the aldehyde amine group, active above 143°C, easily dispersed. Has mild boosting effect on accelerators of the mercapto and guanidine groups and a softening effect in the unvulcanised mix, which, however, disappears after vulcanisation. Whiting and other mineral

fillers delay acceleration. Good ageing properties, causes discoloration during vulcanisation, articles darken in the light.

Quantity: Primary 1–2·5% with 4–5% sulphur; Secondary 0·3–1% with guanidine, thiazoles or thiurams. For SBR 2–3%. At higher concentrations, appearance of bloom, as well as contact discoloration. (BP 7.370/1914 Peachy.)
TN: A 5/10
+ A-10 (5)
F.A. -dur (202)
+ Formaniline (6)
+ L.S.H. (43)
Plasticiser A (19)
+ Vulcafor MA (60)
+ Vulcacit LS H (43)
Methylene dianilide

ANHYDROFORMALDEHYDE-p-TOLUIDINE

(Formaldehyde-p-toluidine, Methylene-p-toluidine, AFPT.) Aldehyde-amine accelerator, density 1·11, no sharp melting point, sinters at approximately 130°C and is molten at 170°C. White to yellowish powder, soluble in benzene, ethanol, ethyl acetate, acetone, carbon tetrachloride, insoluble in water. Safe accelerator with wide vulcanisation range, active above 143°C, a good booster for acid accelerators for articles with a high modulus, high elasticity and heat stability; requires fairly high addition of sulphur and fairly long curing time, shows tendency to decompose at over 70°C, has good dispersing properties and safety in processing. Has a softening effect in the mix, which disappears after vulcanisation. Mixes darken on vulcanisation, vulcanisates are discoloured in the sunlight. Suitable for natural rubber, SBR, reclaim and natural latex. Has a slight tendency to precuring in latex.

Quantity: Primary 1·5–2·5% with 4–5% sulphur; Secondary 0·5–1% as activator for thiazoles and thiazoline; in latex 0·25–1% with 1–3% sulphur.

TN: + A-17 (5)
Accelerator 8 (6)
+ Akbar
MPT
Plasticiser B (19)
Vulcacit FP (43)
Vulcafor MT (60).

ANHYDROUS ABIETIC ACID
Stable, resinous acid. Non-oxidising, tackifier and softener for natural rubber, neoprene, styrene-butadiene copolymers and reclaim; compatible with many elastomers, resins and solvents. Activator for blowing agents in sponge rubber. Reinforces compound containing calcium silicate fillers. Has a greater oxidation resistance than that of abietic acid (colophony) because of the higher level of saturation.
TN: Galex (208).

ANIC Abbreviation for Azienda Nazionale Idrocarburi, Ravenna. Subsidiary of the State petroleum concern, ENI, which produces Europrene, synthetic rubber.

ANILINE $C_6H_5NH_2$, aminobenzene, phenylamine. Colourless liquid, m.p. −6°C, b.p. 184°C, s.g. 1·022. Combines with ethanol and most organic solvents, 1 g soluble in 29 ml water. The first organic accelerator, probably used as early as 1902 in the rubber works in Wundt (near Frankfurt), as 'red oil'. In 1906 its use as an accelerator was discovered in the USA by Oenslager and used by the Diamond Rubber Co. The first patent for using it as an antioxidant in rubber was taken out by Wo. and Wa. Ostwald. Vulcanisation: 2–3% with 6% sulphur at 120–140°C, does not need zinc oxide.
Uses: raw material for plastics, vulcanisation chemicals.

3-ANILINE METHYL-2-BENZOTHIAZOLTHION Reaction product of mercaptobenzthiazole and aldehyde-amine. Accelerator—obsolete (23).
TN: BJF.

ANILINE RESIN Previously known as Iganil resin. Group of thermoplastic, artificial resins. Produced by condensation of aniline and formaldehyde in acid media with subsequent polymerisation.

ANKOLOR Group of organic pigments for rubber (22).

ANNEAL Temper. Reheating of moulded and extruded articles below the critical temperature to improve mechanical properties and decrease the internal moulding strains.

ANNEX Unpurified hexamethylene tetramine.

ANNULEX Mixture of tri-substitute phenols. Light yellow liquid with phenolic aroma, b.p. 240–255°C, s.g. 0·966. Soluble in most organic solvents. Antioxidant, non-staining, suitable for white and light-coloured compounds (82).

ANODEX Unvulcanised and prevulcanised latex compounds (83, 14).

α-ANOMALY Original term for the glass transition temperature.

ANOXIN Condensation product of an aromatic amine and an aliphatic ketone. Viscous, s.g. 1·1. Soluble in ethanol, acetone, benzene and dichloroethane, insoluble in water. Staining antioxidant for NR and SR. No effect on vulcanisation. Does not migrate.
Uses: tyres and mechanical goods (440).

ANSICO Trade name for rubber colours from the Ansbach–Sieger Corporation, USA.

ANSUL ETHER 181 Neoprene solvent (84).

ANTACOL Tetraethyl thiuram disulphide.

ANTAGE Group of antioxidants (16).
Grades:
A: aldol α-naphthylamine resin
C: aldol α-naphthylamine, powder
CP: phenol derivative
D: phenyl-β-naphthylamine
DAH: ditert-amyl hydroquinone
F: di-β-naphthyl-p-phenylene diamine
NBC: nickel dibutyl dithiocarbamate
SP: phenol modified with styrene.

ANTAGONISTS Chemicals with opposite effects.

ANTAROX Group of wetting agents based on polyglycol ethers. Stabilisers for latex (85).

ANTHRANYL-9-MERCAPTAN Peptiser, chemically acting softener which accelerates the oxidative breakdown of natural and synthetic rubber, and vulcanisates. The following process is recommended for reclaiming:
100 parts finely ground vulcanisates
2 parts anthranyl-9-mercaptan
10 parts tar fat oil and
336 parts 10% caustic soda solution are heated for 4 hours at 9 atm steam pressure. The reclaimed product has a firm, plastic consistency and can easily be processed with fresh rubber. DRP 908 298 (1954).

ANTI-AGE Group of antioxidants.
Grades:
33: amine reaction product
34: naphtholamine reaction product
44: naphtholamine reaction product
55: acetaldehyde-aniline condensation product
66: amine reaction product
77: combination of Grades 33 and 66
99: wax-like product (3).

ANTICOAGULANTS Substances used to hinder spontaneous precoagulation and coagulation of latex. Because of the action of bacteria on the proteins, acting as protective colloids, and the formation of free fatty acids and insoluble magnesium soaps, the latex remains stable only for a short time after emerging from the tree. It begins to flocculate and coagulate after a few hours. Latices of different clones reveal variations in stability, because of the difference in magnesium and phosphate content. Anti-coagulants are used to retard bacterial and

enzyme action. They are placed either in the collecting cups on the plantations or in the containers used for transportation, to keep the latex in liquid form until it can be processed. In 1853 ammonia was first patented as a stabiliser for latex (BP 467, William Johnson). The following anticoagulants are in current use on plantations:
ammonia 0·01–0·1 %
sodium sulphite 0·05–0·15 %
sodium carbonate 0·08–0·1 %
formaldehyde 0·02–0·15 %
formaldehyde + sodium carbonate 0·02–0·1 % and 0·02–0·1 %.

ANTIFIX D N-nitroso-diphenylamine.

ANTIGEN Group of unidentified antioxidants from various producers and originating from Japan.
Grades:
4010: N-phenyl-N-cyclohexyl-p-phenylene diamine
A: phenyl-α-naphthylamine (86)
BC: aldol α-naphthylamine (86)
C: compound of phenyl-α-naphthylamine and m-tolylene diamine (86)
D: phenyl-β-naphthylamine
F: di-β-naphthylamine-p-phenylene diamine
HP: mixture of 62·5 parts N-phenyl-β-naphthylamine with N,N'-diphenyl-p-phenylene diamine
MB: 2-mercaptobenzimidazole
MBZ: zinc mercaptobenzimidazole
P: N,N'-diphenyl-p-phenylene diamine
PA: N-phenyl-α-naphthylamine
PC: mixture of N-phenyl-α-naphthylamine with m-tolylene diamine
RD: 2:2:4-trimethyl-1:2-dihydroquinoline

W: 4:4'-dihydroxydiphenyl-cyclohexane.

ANTILUX Group of non-discolouring agents protecting against the influence of light, ageing and weathering for NR and SR. Moderate quantities have no effect on physical properties.
Grades:
Antilux: protective against light, ozone, effects of weathering
Antilux L: for food qualities
Antilux AO: combined protective agent against light, tear initiation and ozone, also contains antioxidant components. Protective against dynamic stress, suitable for food qualities. S.g. 0·91.
Loading: 1·5–5 % (42).

ANTIMONY OXIDE Used in non-flammable rubber articles, together with a chlorine source, *e.g.* a chloroparaffin, which reacts with the oxide to form the flame suppressing antimony oxychloride. Chlorine-containing rubber (neoprenes) contain sufficient chlorine for this reaction. Non-flammable NR compounds:
NR 100
antimony oxide 18
chloroparaffin 13
calcium stearate 1·2.
TN: Stibiox-Gi (417)
Timonox (416).

ANTIMONY PENTASUL-PHIDE Sb_2S_5, antimony-V-sulphide, golden sulphur. Orange-coloured powder, s.g. 4·12. Insoluble in water or alcohol. Can be dissociated into antimony trisulphide and sulphur by heating to 220°C or in sunlight. Formerly used as a vulcanising agent and inorganic pigment. Articles vulcanised by this means acquire a characteristic red colour.

ANTIMONY TETRASULPHIDE

Golden yellow powder, s.g. 2·6–3·0. Inorganic pigment.

ANTIMONY TRISULPHIDE

Sb_2S_3. Orange coloured powder (red modification), s.g. 3·0–4·5. Free sulphur content 0·2–approx. 17%. Inorganic pigment, suitable for press and open steam vulcanisation; for colouring ebonite.

ANTIMYKOTIKUM A Fungicide and bactericide used against micro-organisms in rubber articles, such as bath mats, shoes (43, 408).

ANTIOXIDANT A substance, when added to rubber in a small quantity, retards ageing and protects from internal and external influences. Classified according to the type of protection; they (1) Retard or prevent atmospheric oxidation and its effects (antioxidants). (2) Retard or prevent the formation of tears caused by ozone attack under static or dynamic stress (antiozonants). (3) Protect against certain types of ageing (specialised additives), *e.g.* against light (light ageing preservatives), against harmful metals (copper and manganese inhibitors), against flexing fatigue, and also heat and weather.

In 1865, the British chemists Miller and Spiller discovered the detrimental effect of oxygen on rubber and the first patent (USP 99 935) was granted in 1870 for the preservation of rubber with phenol or cresol.

Paraffin, inorganic and organic reducing agents, and organic nitrogenous substances were later suggested by Ostwald (EP 10 361). The development of the modern antioxidant is based on the experiments of Moureau and Dufraisse, who in 1918 discovered that the oxidation of many organic compounds can be retarded by the addition of substances which they called 'antioxygènes'. However, their explanation of the reaction mechanism as a negative catalysis is no longer valid.

The first effective antioxidants were aldehyde-amine reaction products from aliphatic aldehydes and primary aromatic amines, which were quickly superseded by secondary amines and ketone-amines. These products, though, caused staining and were unsuitable for light-coloured articles. Substances which did not cause discoloration were eventually discovered among the phenols. There are about fifty substances in current use as antioxidants. Classification is generally as follows:

aldehyde-amines,
ketone-amine reaction products,
primary aromatic amines,
secondary aromatic amines,
alkyl-aryl amines, and
phenols.

Most antioxidants give little or no protection against ozone, but sufficient protection under static conditions has been achieved with microcrystalline waxes. The waxes migrate to the surface area and form a protective film, thus preventing atmospheric contact. Because of their limited flexibility, they are hardly suitable for dynamic stress. While the waxes only impart physical protection, chemical antiozonants have recently been developed, *e.g.* 6-ethoxy-1:2-dihydro-2:2:4-trimethyl quinoline; nickel dibutyl dithiocarbamate (for dark-coloured SBR compounds only) and also dialkyl derivatives of p-phenylene diamine, which are suitable for SBR, natural rubber, and nitrile rubber, in dark-coloured compounds both under static and dynamic stress.

Oxygen and ozone are the principal destroyers of rubber. The mechanisms of their action are different, which explains the ineffectiveness of antioxidants against ozone.

Oxidation is regarded as an autocatalytic chain reaction caused by free radicals:

$$
\begin{aligned}
RH + v &\rightarrow R' + H' \\
R' + O_2 &\rightarrow ROO' \\
ROO' + RH &\rightarrow ROOH + R' \\
ROOH &\rightarrow RO' + OH' \\
RO' + RH &\rightarrow ROH + R' \\
'OH + RH &\rightarrow H_2O + R'
\end{aligned}
$$

The original free rubber radical has produced three new free radicals and has set up a chain reaction, which is continued either in the breaking of a chain (softening) or in network formation (hardening).

Chain breaking:

$$
\begin{aligned}
R' &\rightarrow 2R'H \\
ROO' &\rightarrow R'C{=}O + R'OH \\
& \qquad\;\; | \\
& \qquad\;\; H
\end{aligned}
$$

Crosslinking:

$$
\begin{aligned}
2R' &\rightarrow RR \\
ROO' + R' &\rightarrow ROOR
\end{aligned}
$$

Both reactions take place simultaneously, whereby under the influence of unknown factors, one reaction is, to a larger degree, predominant. The antioxidants have the task of stopping the chain reaction and of removing the existing peroxides:

$$
\begin{aligned}
R' + AH &\rightarrow RH + A' \\
ROO' + AH &\rightarrow ROOH + A' \\
RO' + AH &\rightarrow ROH + A' \\
ROOH + AH &\rightarrow \text{harmless products}
\end{aligned}
$$

Because of low activation energy, the free radical A' is incapable of a chain reaction. The antioxidant is therefore slowly exhausted. The mechanism explains the effectiveness of small amounts of antioxidants. Amines can destroy both free radicals and peroxides but phenols destroy free radicals only.

The mechanism of the ozone reaction is not yet clearly known, but it is thought that initially formed peroxides, and iso-ozonides are unstable and cause chain breakage through their degradation. As protective measures, ozonide formation can be prevented, alternatively the decomposition of ozonides retarded.

Lit.: *Rubber Age*, July 1957, 623–27. D. A. Smith, *Rubb. J.*, 1966, 148 (6) 28.

1.
$$
\begin{array}{c}
R'\;\; H \\
RC{-}CR + O_3 \\
\downarrow
\end{array}
$$

$$
\begin{array}{cc}
R'\;\; H & R'\qquad H \\
R{-}C{-}CR \;\;\text{or}\;\; & RC{-}\!\!-\!\!-\!\!-CR \\
(-)\;|\;\;| & |\qquad\quad| \\
O{-}O{-}O & O\diagdown\qquad\diagup O \\
(+) & O \\
\text{Ozonide}
\end{array}
$$

2.
$$
\begin{array}{c}
R'\;\; H \\
(-)\;\; RC{-}CR \\
O{-}O{-}O \\
(+)
\end{array}
$$

spontaneous | degradation ↓

$$
\begin{array}{c}
\qquad\qquad O \\
R'\quad (-)\quad \| \\
RC{-}OO + RC{-}H \\
(+)
\end{array}
$$

3.

$$R'\quad(-)$$
$$RC{-}OO$$
$$(+)$$
$$\downarrow$$

$$\frac{1}{2}\left[\begin{array}{c} R' \\ \diagdown \\ R \end{array} \underset{O-O}{\overset{O-O}{\diagup C \diagdown}} \begin{array}{c} R' \\ \diagup \\ R \end{array}\right]$$

Peroxide

4.

$$R'\quad(-)\qquad\qquad O$$
$$RC{-}OO + RC{-}H$$
$$(+)$$
$$\downarrow$$

$$R' \diagdown C \underset{-O-}{\overset{O-O}{\diagup}} C \diagup H$$
$$R \diagup \qquad\qquad \diagdown R$$

Iso-ozonide

Grades: 4010, N-phenyl-N'-cyclohexyl-p-phenylene diamine
4010 NA, p-phenylene diamine derivative
4020, N-(1:3-dimethyl butyl)-N'-phenyl-p-phenylene diamine
AFC, benzofuran derivative. s.g. 1·25, m.p. approx. 160°C, soluble in methylene chloride. Antiozonant for CR and blends with NR and SBR. Non-staining. Suitable for white and pastel products, (43)
AH, aldol α-naphthylamine. Reddish brown resin, s.g. 1·15
AP, aldol α-naphthylamine. Brownish powder, s.g. 0·98
BKF, 2:2'-methylene bis(4-methyl-6-tertbutyl phenol)
BKL, compound of tertbutylated phenols. Orange-coloured liquid, s.g. 1·04. Non-staining, non-discolouring antioxidant. Gives good heat stabilisation, suitable for cold vulcanisates and for latex, unsuitable for food qualities. *Loading:* solid rubber 0·3–1·5%, latex 1–3%, cold vulcanisates 0·3–0·8%
DDA, DDA-EM, alkylated diphenylamine. Reddish-brown viscous liquid,

b.p. 300°C and over, s.g. 0·97. Soluble in benzine, benzene, carbon tetrachloride, methyl chloride, ethanol, acetone, and in most organic solvents, insoluble in water. Effective antioxidant for natural and synthetic rubber. Can cause yellow discolouring of vulcanisates on exposure to light. Improves tear resistance, protects against copper and manganese. *Uses:* coloured articles with light, reinforcing fillers, bicycle tyres, tubes, articles for the food trade, cellular rubber
DDA-EM, 30% aqueous emulsion for latex compounds, heat sensitive latex and foam rubber
DNP, di-β-naphthyl-p-phenylene diamine
DOD, 4:4'-dihydroxydiphenyl
KB, 2:6-di-tert-butyl-4-methylphenol
KSM, mixture of alkylated phenols. Red liquid, b.p. (4 mm) 130°C, s.g. 1·06. Soluble in acetone, ethyl acetate, alcohol, methyl chloride, carbon tetrachloride, benzine, benzene, non-discolouring, medium-active antioxidant. *Uses:* light-coloured articles, food qualities, mechanical goods. Effective in light-coloured, heat-stable, neoprene goods. Has no influence on physical properties and processing. Loading: solid rubber 0·6–2%, latex articles 1·8–5%
MB, 2-mercaptobenzimidazole
PAN, phenyl-α-naphthylamine
PBN, phenyl-β-naphthylamine
RR 5, resorcinol derivative. Reddish-brown liquid. Antioxidant practically non-discolouring. (Obsolete)
RR 10, viscous, brown resorcinol resin. Discolours slightly (Obsolete).
RR 10 N, alkylated phenols. Brown liquid, b.p. 220°C and above, s.g. 1·09. Soluble in benzene, acetone, ethyl acetate, carbon tetrachloride, benzine. *Uses:* light and coloured

articles, hot and cold vulcanisation, transparent sulphur chloride vulcanisates.

TD-EM 50%, 50% water emulsion from TSP. Used for latex articles.

Quantity: 2–4%

TSP, compound of alkylated and aryl alkylated phenols. Red, viscous liquid, b.p. (10 mm) 170–180°C, s.g. 0·99. Non-staining, non-discolouring antioxidant. Effective against normal ageing, heat and crazing. Soluble in rubber and suitable for transparent articles. Has no effect on physical properties or processing. *Uses:* coloured, transparent articles, rubberised articles, toys, mechanical and dipped goods.

Loading: 0·6–2%

ZKF, dioxydiphenylmethane derivative. White crystalline powder, m.p. 180°C. Non-discolouring antioxidant for white and coloured articles. Gives good protective action against crazing and rubber poisons, suitable for solid rubber, solutions and latex.

Uses: food qualities, medical articles, toys, foam rubber, cold vulcanisates. Loading: 0·4–1·8%, cold vulcanisates 0·3–0·8%

ZMB, zinc-2-mercaptobenzimidazole (43).

ANTIOXIDANT Trade name of several British and American produced protective agents.

Grades:

1: 2:2'-methylene bis(4-methyl-6-tertbutyl phenol) (428)

3: styrenated phenol (428)

4: 2:6-ditertbutyl-4-methyl phenol (428)

5: polyalkylated bisphenyl-alkane (428)

6: substituted derivative of phenyl phosphite (428)

15: diphenylamine-acetone reaction product (21)

16: anisidine derivative (428)

30: 2:6-ditert-butyl-p-cresol

108: phenyl-β-naphthylamine (22)

116: phenyl-β-naphthylamine (22)

123: di-β-naphthyl-p-phenylene diamine (22)

167: antioxidant of unknown composition (81)

184: polytrimethyl dihydroquinoline

235: methylenebismethyl butyl phenol (54)

256: diphenyl sulfide derivative

415: alkylated diphenylamine (22)

423: bis(3-methyl-4-hydroxy-5-tert-butyl benzyl) sulphide (246)

425: methylene bis-ethyl butyl phenol (22, 21)

431: alkylated phenol (246)

439: thiobisphenol. Brown viscous fluid, s.g. 1·04 (246)

555: alkylated phenol (289)

654: antioxidant of unknown composition (87)

701: 2:6-ditert-butyl phenol (88)

702: 4:4'-methylene bis(2:6-ditert-butyl phenol) (88)

703: 2:6-ditert-butyl-α-dimethylamino-p-cresol (88)

712: 4:4'-bis(2:6-ditert-butyl phenol) (88)

720: 4:4'-methylene bis(6-tert-butyl-o-cresol) (88)

733: mixture of 75% 2:6-ditert-butyl phenol with other tert-butyl phenols (88)

736:     4:4'-thiobis(6-tert-butyl-o-cresol) (88)

762:     2:6-ditert-butyl-α-methoxy-p-cresol (88)

2246:   methylene bismethyl butyl phenol (21)

4010:   N-cyclohexyl-N'-phenyl-p-phenylene diamine (408)

4010NA: N-isopropyl-N'-phenyl-p-phenylene diamine (408)

AH: aldol α-naphthylamine (43)

AL: aldol α-naphthylamine

AP: aldol α-naphthylamine (43)

BKF: 2:2'-methylene-bis-(4-methyl-6-tert-butyl phenol) (43)

BKL: alkylated phenol (43)

CD: N-phenyl-N-isopropyl-p-phenylene diamine (428)

DDA: octylated diphenylamine (43)

DNP: N,N'-di-β-naphthyl-p-phenylene diamine (43)

DOD: 4:4'-bisphenol (43)

KSM: alkylated phenol (43)

MB: 2-mercaptobenzimidazole (43, 428)

M-24:   2:4-dimethyl-6-tert-butyl phenol (289)

OP-substituted aromatic amine: White crystalline powder, m.p. 104–106°C. Soluble in benzene, chloroform and acetone. *Staining antioxidant for NR and SBR. Uses:* tyre stocks, mechanical goods, shoes (440)

PAN: phenyl-α-naphthylamine

PBN: phenyl-β-naphthylamine (43, 428)

RR-10-N: alkylated phenol (43)

SP: styrenated phenol (22)

TSP: mixture of tert-octyl and tert-butyl cresols, s.g. 0·99 (43)

ZKF: alkylated bisphenols (43)

ZMB: zinc-2-mercaptobenzimidazole (43).

ANTIOXYDANT Group of antioxidants (428).
Grades:
MB: 2-mercaptobenzimidazole
PBN: phenyl-α-naphthylamine.

ANTIOXYGEN Term used by Moureau and Dufraisse to denote substances which only retard the action of atmospheric oxygen on rubber.

ANTIOXYGENE Trade name for some French produced antioxidants.
Grades:
A: phenyl-β-naphthylamine

AFL: condensation products of a ketone and an amine

AN: aldol α-naphthylamine

BN: β-naphthol

CAS: phenyl-α-naphthylamine

DIP: diphenyl-p-phenylene diamine

INC: aldol-α-naphthylamine

M2B: 1:1'-methylene bis-2-naphthol

MC: phenyl-β-naphthylamine

MTB: mercaptobenzimidazole

MTBZ: zinc-2-mercaptobenzimidazole

RA: aldol α-naphthylamine

RES: aldol α-naphthylamine

RM: aldol α-naphthylamine

STN: phenyl-α-naphthylamine

WBC: 1:1'-methylene bis-2-naphthol mixed with resin (19).

ANTIOZONANTS Substances which protect against the destructive effect of ozone. Static antiozonants: amorphous and microcrystalline waxes which migrate to the surface and form a physical, protective surface layer. Secondary aromatic amines belong to the group of chemically active antiozonants, *e.g.* N,N'-diphenyl-p-phenylene diamine, quinolines and furan derivatives.

ANTISCORCH T Calcium hydroxide dispersed in alumina. Vulcanisation retarder (6).

ANTISOL Protective agent against light, wax (89).

ANTISTATICUM RC 100 Nonstaining antistatic agent which prevents electrostatic charging of NR, SR, and p.v.c.; improves conductivity, especially of black-free vulcanisates. Also used for articles which come in contact with food (conveyor belts, tubes, driving belts, roller coatings, floor coatings). Viscous liquid, gives no change in physical properties, slight acceleration of vulcanisation.
Loading: 3–8% (42).

ANTOX Butyraldehyde-aniline condensation product. Accelerator and antioxidant for rubber and latex (6).

ANTOZITE Group of antiozonants based on p-phenylene diamine derivatives (51):

1:   N,N'-di(2-octyl)-p-phenylene diamine
2:   N,N'-di-3(5-methyl heptyl)-p-phenylene diamine
67: N-(1:3-dimethyl butyl)-N'-phenyl-p-phenylene diamine
67 S: Type 67 with an inert base (in granule form)
MPD: N,N'-bis(1:4-dimethyl pentyl)-p-phenylene diamine

ANTRAX Group of furnace blacks: 52SRF, 53HMF, 54MAF.

AO-3161 p-isopropoxydiphenylamine.

AP ABRASION MACHINE Machine developed by Continental Rubber Works AG, Hanover. In principle, consists of a roller, eccentrically placed, clamped with abrasive paper on which a specially vulcanised cylindrical test piece of 18×6 mm is turned. Several test pieces may be turned simultaneously. An improved design, the DIN abrasion machine (DIN 53 516), allows the test piece to be circulated over the roller.

APCOTHANE 1400 Polyurethane prepolymer (90).

APEX Group of softeners.
Grades:
1: dibutoxyethyl phthalate
2: diethylene glycol propionate
3: dibutylethoxy succinate
4: butyl stearate
5: lauryl butoxyethoxy acetate
6: dimethylethoxy phthalate (91).

APF Abbreviation for All Purpose Furnace Black.

41

d-APIOSE

CHO
|
HC—OH
|
C—OH
/\
HOH$_2$C　CH$_2$OH

Tetrahydroxyvaleraldehyde. Soluble in water. Identified chromatographically in extracts from the bark of *Hevea brasiliensis*, dried at 90°C, and characterised as phenyl osazone. As apiose and isoprene have the same carbon skeleton, it is suspected that in isoprene synthesis during plant metabolism apiose is converted to C$_5$-intermediates by condensation and reduction reactions. Other plants related to the *Hevea* species also contain apiose.

APT RUBBER Ethylene-propylene terpolymer. Produced from AP rubber by including small quantities of unsaturated structures. Properties resemble those of natural rubber but have optimum ageing and weathering.

APTA Abbreviation for acetaldehyde-p-toluidine-aniline condensate.

AQUAPLANING At high speeds and on roads which are awash, tyres may be lifted from the road surface on a thin wedge of water, the wheels may then come to a standstill and the vehicles can no longer be steered. The critical velocity for aquaplaning depends on the tread design, the loading, the thickness of the water film on the road, and other related factors. Recently, attention has been given to the design of passenger car treads which exert a 'squeegee' action intended to disperse the water film immediately in front of the tyre 'footprint' (*i.e.* the contact area with the road).

AQUAREX Group of surface active substances and wetting agents for latex and rubber.
Grades:

D:　CH$_3$(CH$_2$)$_x$CH$_2$OSO$_3$Na. Sodium salt of sulphate monoester of a mixture of higher fatty alcohols, white powder, s.g. 1·28, anionic. Active content 70–75%. Has no influence on ageing or on vulcanisation. Non-staining. *Loading:* release agent 0·25–0·5% solution, prevention of adhesion in unvulcanised articles 2–5% solution. *Uses:* stabiliser, emulsifying agent, wetting agent for latex, in low concentrations modifies the foam and gel properties of latex, release agent. Gives a smooth surface finish, miscible with silicone emulsions. Dilution for latex 1:9.

G:　32% aqueous solution of sodium alkyl sulphonate. Clear, brown liquid, s.g. 1·09, anionic, stable in either acid or alkaline medium. *Uses:* emulsifying agent for the emulsion polymerisation of synthetic elastomers. *Loading:* 2–6% on the monomers.

L:　ammonium salt of an alkyl phosphate, pH stable. *Loading:* 0·5–1% solution, effect improved by the addition of 1–5% talc. *Uses:* release agent, acts simultaneously as corrosion inhibitor, prevents adhesion of unvulcanised compounds

MDL: pale yellow paste, composition, characteristics and uses the same as for Grade **D**. Active content 38–42%, low electrolyte content. *Loading:* as stabiliser and wetting agent 0·5–2%, solution as mould release agent 0·5–2%, solution as antitackifier for extruded, unvulcanised articles 4–10%. Used in latex as a 25% solution

ME: white powder, s.g. 1·28. Composition, characteristics and uses as for Grade **D**. Active content 90–95%, low electrolyte content. Solubility in water at 24°C, approx. 1:10. Increases penetration of latex in textiles. Basic formula for 50% emulsion: oil, wax, resin etc.—100, Aquarex ME—1, 10% casein solution—30, water—69. *Loading:* stabilisers 0·2–2% on the elastomer, emulsifier 0·1–1%, mould release agent 0·25–0·75% solution.

NS: C-cetylbetaine. Slightly cloudy liquid, s.g. 1·05. Active content 25–30%. Amphoteric wetting agent, stabiliser for most latex systems at low pH and low temperature, unsuitable as an emulsifier. Does not influence ageing or discolouring, has slight retarding effect on vulcanisation. Gives significant improvement of the foam stability of neoprene foams. *Loading:* 1·25–4% stabiliser, 0·02–0·1% neoprene foam

SMO: sodium salt of sulphurised methyl oleate. Reddish brown liquid, s.g. 1·08. Active content 30–35%. Does not influence vulcanisation, ageing or discolouring. *Loading:* 2–3%. *Uses:* prevents formation of streaks in heavily filled latex compounds, especially in articles for casting and dipping. Chiefly used for articles of neoprene latex, is less effective in SBR, NBR, and natural latices.

WA: pale yellow paste, s.g. 1·06, composition, characteristics and uses as for Grade **D**. Active content 27–30%, low electrolyte content. *Quantity:* 0·5–3%

WAQ: sodium alkyl sulphate. Brown paste, s.g. 1·04, anionic. Stabiliser for latex, wetting agent, emulsifier, release agent. *Loading:* stabiliser 0·25–3% aqueous solution, dispersing agent 0·5–1% on solids, release agent 0·5–1% aqueous solution, for prevention of compound adhesion 4–10% solution (6).

AQUASOL Sulphonated castor oil.
Uses: extending and plasticising agent (21).

AQUASPERSE 30 Aqueous dispersion of casein.
Use: with latex colours (50).

ARBONITE Group of poly vinyl chloride plastisols (95).

ARCC Aqueous dispersions of natural, synthetic and reclaimed rubber and synthetic resins (50).

ARCCO Group of synthetic resin emulsions.

Uses: tackifier, reinforcing agents for NR, SR and latex compounds (50).

ARCO AC 50 di-o-tolyl guanidine zinc chloride. Secondary accelerator for thiazoles (21).

ARDUX 120 Synthetic resin tackifier for bonding rubber to metal (94).

ARESCAP 50 Sodium butyl-o-phenyl phenol sulphonate. Wetting agent for latex (5).

ARESCET 240 Sodium butyl diphenyl sulphonate. Wetting agent for latex (5).

ARESKLENE Disodium dibutyl-o-phenyl phenol disulphonate. Light brown paste, pH 7·5–8·5, stable in acid, alkaline and neutral solutions. Wetting agent and stabiliser for latex.
Uses: tyre cord and textiles.
Loading: 0·1–0·5% (5).

ARISTOWAX Refined, high melting wax. Softener for rubber (96).

ARKOPAL N Group of alkyl aryl polyglycol ethers. Non-ionic wetting and dispersing agents, in particular for blacks (217).

ARLACEL Group of non-ionic, surface active esters of sorbitan (monoanhydrosorbitol, $C_6H_8O(OH)_4$) and fatty acids.
Grades:
20: sorbitan monolaurate
40: sorbitan monopalmitate
60: sorbitan monostearate
80: sorbitan mono-oleate
83: sorbitan sesquioleate
85: sorbitan trioleate (96).

ARMAC Group of acetates of primary fatty acid amides. Soluble in water and in organic solvents.
Grades:
Armac C: coconut fatty acids, m.p. 50°C, iodine No. 11
Armac T: tall oil fatty acids, m.p. approx. 55°C, iodine No. 27–37.
Uses: activators for thiazole accelerators.
Loading: 0·2% as emulsifier for waxes (98).

ARMEEN Group of aliphatic, primary, secondary and tertiary amines of vegetable oils and fats.
Uses: retarders, softeners and processing aids (98).

ARMEEN HTD Mixture of primary aliphatic amines, s.g. 0·794. Retarder and antiscorch agent for NBR and butyl rubber.
Loading: 0·5–1·0% (98).

ARMI Abbreviation for the Association of Rubber Manufacturers of India.

ARMID Group of wax-like fatty acid amides. Soluble in oils, esters, ketones, alcohol, insoluble in water.
Uses: prevention of tack, release agents (98).

ARNEEL Group of nitriles of higher fatty acids with C_{18}–C_{24} structures. Softeners (98).

ARNIET Trade name for hard rubber used for lining coagulation tanks for natural latex. Primarily replaced now by aluminium.

AROCHEM Synthetic resin. Plasticiser (99).

AROCLOR Group of polychlorinated polyphenols, s.g. 1·03–1·38.
Uses: plasticisers, softeners and tackifiers for natural and synthetic rubber; increases flame resistance of synthetic rubber, improves the resistance of neoprene to water; adhesives and coating (5).

AROFENE Phenolic resin. Plasticiser and reinforcing agent (99).

AROGEN GPF GPF furnace black (oil), s.g. (in rubber) 1·77, pH 9·3, particle diameter 60 mμ, surface 45 m^2 g^{-1} (40).

ARO LENE Aromatic polymer. Plasticiser and softener for natural and synthetic rubbers (100).

AROMEX Group of furnace blacks. Aromex: HAF, Aromex 115: CF, Aromex 125: ISAF (40).

ARON-GOMU Japanese produced acrylonitrile copolymers. Stable against heat, light, ozone and oil (99).

AROPOL Group of polyester resins (99).

AROTHANE Group of polyurethanes (44).

AROVEL FEF black, s.g. 1·77, pH 8·4, particle diameter 55 mμ, nigrometer value 92 (40).

ARQUAD Group of alkyl trimethylammonium chlorides; cationic, surface active substances. Soluble in water and alcohol (98).

ARQUAD 12–50% 50% solution of dodecyl trimethylammonium

chloride in isopropyl alcohol. Foam stabiliser (98).

ARRCONOX SP Antioxidant for NR, SR, and especially for latices; styrenated phenol, yellowish liquid; s.g. 1·08. Soluble in acetone, ether, ethanol, chlorinated and aromatic hydrocarbons and alkalis. Nonhydrolysable.
Quantity: 1–2% (432).

ARROW TX MPC black, s.g. 1·77, pH 4·1–4·5, particle size 27 mμ, nigrometer value 85 (40).

ARUBREN II Plastic material. Improves the flame resistance of rubber compounds in combination with suitable fillers, *e.g.* magnesium carbonate, kaoline.
Loading: 40–100% with rubber (43).

ARUBREN CP Chlorinated aliphatic hydrocarbon with approx. 72% chlorine. Transparent, yellowish, viscous solid, s.g. 1·52. Increases flame resistance; likewise inhibits inflammability of vulcanisates. Higher loadings decrease hardness, rebound resilience and elongation, tear resistance is hardly affected.
Loading: 20–50%, addition of antimony trioxide improves effect.
Uses: conveyor belts, cable coverings, and non-flammable rubber goods (43).

ARUWIMI RUBBER Natural rubber of average quality (Congo). Handled commercially in the form of large bales.

ARW Abbreviation for Arbeitsgemeinschaft Regenerierwerke, Mannheim–Rheinau, P.O. 84/104. German association of independent producers of reclaimed products.

ARWAX Mixture of paraffin with butyl rubbers or polyethylene. *Use:* impregnation of paper (50).

ASBESTINE Fine-fibred asbestos powder, mainly magnesium silicate. Filler.

ASBESTOS Fibrous mineral silicates of different composition. Principal groups: chrysotile (serpentine) and amphibole (which includes the subdivision amosite and crocidolite). Chrysotile is the common material and is used in 90% of all products requiring asbestos fibre. It may be represented by the empirical formula $3MgO.2SiO_2.2H_2O$. Long fibre grades ($\frac{3}{4}$ in staple and longer) are of particular value for the reinforcement of rubber and flexible resins because the asbestos fibres themselves are relatively soft and flexible; care must be taken in mixing to avoid excessive breakdown of the long fibres.
Uses: reinforcing filler in asphalt and vinyl floor tiles (about 200 000 tons per year are used for this purpose in the USA), filler for rubbers and plastics used in missile application, in powder form as a bulking filler (extender), in fibre form blended with other fibres (*e.g.* glass, nylon, viscose) to produce textiles in sealants and caulking compounds, in brake linings.
Lit: M. S. Badollet, *Encyclopaedia of Chemical Technology* (Kirk-Othmer, Ed.), 2nd edn, Wiley, 1963, p. 734.

ASEPTOFORM p-hydroxy-benzoate.
Use: latex preservative (81).

ASOLVAN Factice with good stability towards mineral oils and benzine. Suitable for synthetic rubbers and articles resistant to swelling (42).

ASP Trade name for aluminium silicate pigments. Group of white pigments with excellent extrusion qualities: 45·4% SiO_2, 38·8% Al_2O_3, 1·5% TiO_2 with small quantities of Fe_2O_3, Na_2O, and K_2O (102).

ASPHALT Black mass, softening p. 70–160°C, s.g. 1–1·2. Soluble in benzene, chloroform, carbon disulphide. Natural product (*e.g.* Gilsonite, Syrian A, Cuban A), or can be produced artificially by air oxidation of petroleum residues. *Uses:* tackifier, extender.

ASRC Abbreviation for American Synthetic Rubber Corporation, 500 Fifth Avenue, New York, N.Y., USA.

ASRC POLYMERS Group of butadiene-styrene copolymers, numbered according to the ASTM code (103).

ASSINEE RUBBER Natural, African rubber of average quality, which came into trade *via* London.

ASSOGOMMA Abbreviation for Associazione Nazionale fra le Industrie della Gomma, Cavi Elettrici ad Afini. Association of Italian Rubber Manufacturers, Milan, Via S. Vittore 36/1.

ASTM Abbreviation for the American Society for Testing Materials, 1916 Race Street, Philadelphia 3, Pa, USA.

ASTM HARDNESS Impression of a hemispherical indentor with a diameter of 0·0938 in under specified conditions, expressed in thousandths of an inch.

ASTM OILS Oils with varying aniline points, standardised according to D471-59T, and used to determine the swelling of elastomers.

	Aniline p. (C°)	Swelling
No. 1	123·9 ± 1	low
No. 2	93 ± 3	medium
No. 3	69·5 ± 1	high

ASTM STANDARD MOTOR FUELS Motor fuels standardised according to ASTM D 472–59T, used to determine the degree of swelling of elastomers.
Grades:
A: 100% iso-octane
B: 70 vol. % iso-octane and
    30 vol. % toluene
C: 50 vol. % iso-octane and
    50 vol. % toluene.

ASTYR cis-1:4-polybutadiene (59).

ATHIE RUBBER Resin rich natural rubber from the Ivory Coast. Came into trade in the form of bales formed from wound-up strips.

ATLANTIC Group of reinforcing blacks.

75:	MPC
95:	MPC
98:	HPC
109:	MPC
115:	SRF
115 HM:	SRF–HM
118:	GPF
120:	HMF
125:	FEF(MAF)
130:	HAF
135:	ISAF
150:	SAF
156:	CC
170:	CC
CF:	CF
E 42:	EPC
TF:	FT
TM:	MT (105).

ATRACTYLIS RUBBER Product similar to rubber, comes from the root sap of *Actractylis gummifera* (Mediterranean coast, North Africa). Used locally as a substitute for glue.

ATV-TYRES Abbreviation for all-terrain vehicle tyres.

AUBLET, J. French botanist, 1762–64 studied the flora of Guiana and named the rubber tree described by Fresnea, *Hevea guayanensis.* His book *Histoire des Plantes de la Guiané Francaise* (Paris; London, 1775), contains the first exact botanical description of the species.

AUTOFORM Modern vulcanising press for car tyres; similar to the Bag-O-Matic.

AVARA SERINGA Local term for *Micranda siphonoides Benth* (Amazon region). The sap is orange yellow, and tastes bitter. The rubber is of excellent quality, containing about 91% pure rubber. According to Ule, the sap cannot be mixed with that of *Hevea brasiliensis.* The trees are not tapped often.

AVROS *Algemene Vereniging van Rubberplanters ter Oostkust van Sumatra,* Medan. Association founded in 1910 by the rubber planters of the east coast of Sumatra. The rubber growing concerns in E. Sumatra were established in 1906 with plantations in Asahan, Langkat and Serdang. In 1913, AVROS opened testing laboratories to deal with problems of cultivation and with technical problems. In 1958 the name was changed to Sumatra Planters Association (SPA). The Association amalgamated with the Deli Planters Vereniging in 1952 and controls about 250 plantations (rubber, palm oil, tea, sisal), covering 800 000 ha, owned by over 100 companies, and employing 200 000 plantation workers. The activities include organisation and consolidation of the plantations, treatment of social and agro-political problems, and also agricultural, chemical and technological development and experiments. Lit: *Mededelingen van het Algemeen Proefstation der AVROS*, since 1958. Communications of the Research Institute of the Sumatra Planters Association.

AW Abbreviation for atomic weight.

AXF Polyphenylethylene; reaction product of an ethylene dihalide with an aromatic hydrocarbon $R_1C_6H_4R_2$, in the presence of aluminium chloride. Saturated, dark brown, elastic, resin-like mass, s.g. 1·04. Soluble in chlorinated solvents. Has a retarding effect on vulcanisation. Introduced during World War II in America (small quantities).
Uses: substitute for factice in semi-rigid compounds (10–20%) or neoprene and thiokol, decreases swelling power.

AZOCEL Azodicarbonamide.

AZOCYCLOHEXYL NITRILE Colourless crystals. Blowing agent. Activation temperature 114°C.
TN: Genitron CHDN (209).

AZODICARBONAMIDE

$$H_2N-\overset{\overset{\displaystyle O}{\|}}{C}-N=N-\overset{\overset{\displaystyle O}{\|}}{C}-NH_2$$

Pale yellow, odourless powder, s.g. 1·56. Decomposes by splitting off nitrogen, dissociation temperature 195–200°C. Insoluble in water.
Uses: blowing agent for natural and synthetic rubber, and thermoplastics. Non-staining, forms a fine, evenly closed cell structure, extremely safe to process.
TN: Aceto AD (25)
Azocel
Celogen-AZ (246)
Genitron AC (209)
Kempore R-125 (237)
Kempore 150 (237)
Lucel ADA (65)
Porofor ADC (43)
Porofor K-1074 (43).

AZODICARBOXYLIC ACID DIAMIDE
$H_2.N.CO.N:N.CO.NH_2$. Blowing agent, dissociation temperature 140°C.
TN: Genitron AC (209)
+ Porofor 505A (43).

AZODICARBOXYLIC ACID DIETHYL ESTER
$C_2H_5O.CO.N:N.CO.OC_2H_5$. Blowing agent, dissociation temperature 105–110°C. 1 g yields 129 cm³ nitrogen.
TN:+ Porofor 254 (43).

AZODOX Deaerated zinc oxide (106).

AZOHEXAHYDROBENZONITRILE
$(CH_2)_5C(CN).N:N.C(CN)(CH_2)_5$. Blowing agent, dissociation temperature 103°C. 1 g yields 84 cm³ nitrogen.
TN: + Porofor 254 (43).

AZOISOBUTYRONITRILE

$$CH_3—\underset{\underset{CN}{|}}{\overset{\overset{CH_3}{|}}{C}}—N=N—\underset{\underset{CN}{|}}{\overset{\overset{CH_3}{|}}{C}}—CH_3$$

White powder. Blowing agent. Dissociation temperature in rubber 100–110°C, somewhat lower in plastics. 1 g yields 136 cm³ nitrogen. After dissociation poisonous tetramethyl-succindinitrile remains. Gives a fine, even pore structure, unsuitable for articles in the food or pharmaceutical trades. Mills and presses must be well ventilated during use.

Uses: hard rubber compounds (rubber/phenolic resin compounds), hard pvc.

TN: Aceto AZIB (25), AIBN, Genitron AZDN (209), Porofor N (43).

AZO ZZZ Group of lead-free zinc-oxides, equivalent to ASTM D 79–44 (106).

B

B-25 Anti-foam agent (6).

B 150 Polyisobutylene of medium mol. wt., approx. 150 000. Extender and tackifier (376).

B 200 Polyisobutylene of medium mol. wt., approx. 200 000. Extender and tackifier (376).

BA Abbreviation for butyralde-hyde aniline condensate.

BAC Latex from a modified butadiene copolymer. Carrier for resorcinol-formaldehyde resins.
Use: improves bonding between rubber and textiles (6).

BACKRINDING Rind-back, flash-back. Torn or gouged condition, usually at the mould parting line, due to localised contraction of the vulcanisate.

BAG-O-MATIC Modern vulcanising apparatus used for tyres with an integral rubber cylinder in place of the usual curing bar, which simultaneously shapes the unvulcanised rubber. Some versions are constructed so that the rubber cylinder can be retracted at the end of the vulcanisation period and mechanical devices then eject the finished tyres.

BAKA RUBBER Product resembling bird-lime which comes from *Ficus obligua* (Fiji).

BAKERS Group of softeners:
Grades:
1: methyl ricinoleate
2: ethyl ricinoleate
3: butyl ricinoleate
4: methyl acetyl ricinoleate
4C: methyl glycol acetyl ricinoleate
5: ethyl acetyl ricinoleate
6: butyl acetyl ricinoleate
7: methyl undecylenate
8: acetylated castor oil
9: polymerised, acetylated castor oil
11: polymerised methyl ricinoleate
12: polymerised ethyl ricinoleate
13: polymerised butyl ricinoleate
14: acetylated No. 11
15: acetylated No. 12
16: acetylated No. 13 (9).

BALANCING Centring of a tyre so that both axis of rotation and centre of gravity are the same.

BALATA Thermoplastic product similar to gutta-percha; the dried sap of *Mimusops globosa* (*Mimusops balata*) of the *Sapotacae* family. The tree (bully tree, bullet tree) grows to a height of up to 30 m, has a broad crown, and is a common species of the Amazon and Panama areas, in Venezuela and in Guiana. The wood is named 'horse meat' because of its red colour. Various other Mimusops types, such as *Sapota Mülleri, Achras balata, Manikara balata*, give an inferior balata product, or a product similar to balata. Balata became known in Europe in 1857 through Bleekrode (Leiden), and was known as Surinam gutta-percha after Wildeboer had discovered its possible uses in 1856. The sap is thicker than that of the *Hevea* latex, and contains about 50% solids. The balata is gained by natural coagulation through boiling or vaporisation. Balata can also be extracted from the leaves and wood by the Frank–Marckwald process, by digestion with alkali hydroxides or carbonates, under a pressure of 5 atm. This product is pure and free from resin.

The balata used to be taken from indiscriminately felled trees, and within a few years about 36 million trees were destroyed in this way. The tapping of trees was later introduced; around half of the trunk, V-shaped incisions are made, up to a height of 10 m. These are linked by a channel. To ensure good recovery the cambium must not be harmed. The sap lies primarily in the outer layers of the bark, the inner layer gives only a watery latex rich in tannin. After each tapping, producing on the average 3–4 l of latex containing 1·5–2 kg of balata, the tree must be left to recover for up to 5 years before it can be tapped again. As the growing period of the trees is long and tapping can only begin after 20–25 years, cultivation is uneconomical and has never been attempted.

Balata contains about 40% hydro-carbon, 40% resin, 7% moisture and impurities. The hydrocarbon is dispersed in latex in the form of small, globular particles, which probably have a surface layer of adsorbed resin. Formerly Harries cyclic empirical formula for the hydrocarbon was used, whereas today the linear formula $(C_5H_8)_x$ is universally accepted and balata is recognised as a trans-isomer of natural rubber, from which it is clearly differentiated by its physical properties. At normal temperatures, balata is tough, hard and fairly inelastic, but softens in hot water (at approx. 60°C). It has good crystallisation properties and low water absorption.

Uses: driving belts, covers for golf balls (deresinated balata), insulation of submarine cables (together with gutta-percha).

TN: Amber- and Surinam Sheet Colombia and Ciudad Bolivar Blocks Manaos Block (white).

Lit.: E. A. Hauser, *Latex-Chem. Catalog Comp.*, New York, 1930, pp. 51, 85.

E. J. Fisher, *Guttapercha and Balata, Allgemeine Industrie Verl.*, Berlin, 1933.

C. C. Davis and J. T. Blake, *The Chemistry and Technology of Rubber*, Reinhold, New York, 1937, p. 705. J. v. Wiesner, 'Die Rohstoffe des Pflanzenreiches', 4 vols, Leipzig, 1928.

K. Memmler, *The Science of Rubber*,

English translation, Reinhold, New York, 1934, bibliography, p. 721.

BALATA SYNTHETIC Synthetic trans-1:4-polyisoprene.

BALDWIN-TATE-EMERY MACHINE Apparatus for determining the stress in rubber under extension.

BALSAM Storax, Peru balsam, Canada balsam, Copaiva balsam, Dragon's blood. Natural, viscous solution of resins in volatile oil. Hardens slowly as the solvent evaporates in air.

BANBURY INTERNAL MIXER Machine invented in 1916 by Fernely H. Banbury for the production of rubber compounds. The closed mixing chamber has an outer cooling or heating jacket, and is in the form of a divided, double cylinder in which two hollow, spiral rotors of drop-shaped cross-section rotate in opposite directions. The rubber is kneaded and mixed between the chamber walls and the rotors, and

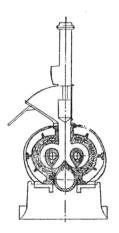

between the rotors themselves. The compound is kept under pressure by means of a pneumatic ram. The machines have a chamber capacity varying from 1–500 l and built for loads of up to 1875 kw and fitted with a universal gear.

BANBURY–LANCASTER PROCESS Method for reclaiming rubber, particularly manufacturing scrap, by using the Banbury internal mixer at increased pressure, temperature of up to 260°C and at a higher speed; peptising agents are sometimes added.

BANTEX DN Compound of zinc-2-mercaptobenzthiazole with diphenyl guanidine, s.g. 1·57. Soluble to a limited extent in alcohol and diluted caustic soda. Accelerator, may be used without zinc oxide.

BANTEX M Mixture of zinc-2-mercaptobenzthiazole and tetramethyl thiuram monosulphide. Accelerator (5).

BAREBACK BALES Rubber bales packed in sheets of the same type of rubber and protected with a talc layer.

BARIUM CHROMATE $BaCrO_4$, barium yellow, Baryta yellow. Yellow, rhombic crystals, s.g. 4·5.
Use: yellow pigment.

BARIUM STEARATE $(C_{17}H_{35}COO)_2Ba$. White, crystalline product. Insoluble in alcohol and water.
Uses: dusting agent, lubricant, anti-tack agent for uncured compounds, p.v.c. stabiliser.

BARIUM SULPHATE $BaSO_4$, Barytes, Barytes white, blanc fixe, Schwerspat, s.g. 4·5. Inert filler used either as the ground natural mineral, or as precipitated barium sulphate (blanc fixe) from barium chloride and sulphuric acid. A constituent of lithopone.

BARK SPECKS Impurities in rubber caused by particles of bark.

BARRETT Group of softeners.
Grades:
10 P: diiso-octyl phthalate
28 P: di(2-ethyl hexyl) phthalate
50 B: butyl cyclohexyl phthalate
DCHP: dicyclohexyl phthalate
         (107).

BARRIERS, PIERRE Botanist, 1690–1755. During 1721–24 was in French Guiana and in his work on its natural resources mentioned De La Neuville's reference to rubber rings and balls, and that rubber was obtained from a liane of the *Apocynanceae* family.
Lit.: *Nouvelle Relation de la France Equinoxialé*, Paris, 1743, Vol. I, p. 139.

BASRM Abbreviation for the British Association of Synthetic Rubber Manufacturers, 93 Albert Embankment, London, S.E.1, England.

BASSIA Pseudo gutta percha, gutta shee. Gutta percha-like product of the *Butyrospernum parkii* (Nile and Niger regions). Unsuitable for use as a gutta-percha substitute.

BATATA RANA Local name for *Impomoea bona nox L.*, the sap of which is used in the region for coagulating Castilla rubber.

BATEX Czechoslovakian produced adhesive for shoes; based on 30% polychloroprene latex with a solution of colophony in benzine (1:1) and an aqueous lime solution.

BAUMANN, PROF. PAUL Born 1897, died 1967. Chairman of the Chemische Werke Hüls AG Works, 1953–64. Developed the arc process for producing acetylene from hydrocarbons and is noted for his research on acetaldehyde and acetic acid. Founder of Bunawerke Hüls, 1955–58.

BAUXITE Naturally occurring aluminium oxide: 30–75% Al_2O_3, 8–30% H_2O, 3–25% Fe_2O_3, 2–8% SiO_2, 1–4% TiO_2, s.g. 2·0–2·6.
Uses: plastic and rubber filler.

BAVICK II Acrylic polymer (108).

BAYER-DEFO-PLASTOMETER Defo apparatus improved by Bayer, Leverkusen, in which the time consuming weight measurement is eliminated. The test piece, 10 mm thick, is compressed at an even speed, and the load is continuously measured by means of a mercury capillary manometer.

BAYER-SCHNEIDER PLASTO-METER Modified Williams plastometer by which recovery can be measured under a reduced applied stress. Load adjustable between 100–1000 g.

BAYERTITAN Group of white pigments based on titanium dioxide (43).

BAYOL White mineral oil, types D and F. Plasticiser (41).

BAYPREN Former Perbunan C. Group of polychloroprene rubbers and latices equivalent to Neoprenes in compounding, properties and uses (43).

Types 321 and 331 are further treated with thiuram.

Grading code: the first figure gives an indication of the crystallisation rate and processing.

Mercaptan types:

1. low crystallisation rate
2. medium crystallisation rate
3. high crystallisation rate

Thiuram types:

6. low crystallisation rate
7. medium crystallisation rate
8. high crystallisation rate

The second figure denotes the plasticity: ML-4/100°C:

1. approx. 30–60
2. approx. 60–90
3. approx. 90–120

The third figure shows special properties (after treatments, discolouring, unusual viscosity).

Grade	Crystallisation Rate	Mooney–Viscosity (approx. range)	Equivalent to
Mercaptan types			
110	low	48	Neoprene WRT Danka S 40 Butaclor MC 10
112	low to medium	48	
210	medium	48	Neoprene W Denka M 40 Butaclor MC 30
211	medium	38	
214	low to medium precrosslinked	55	
230	medium	100	Neoprene WHV Denka M 120 Butaclor MH 30
320	high	87	Neoprene AD Denka A 90 Butaclor MA 40
321	high	82	Neoprene AC
330	high	102	Neoprene AD Denka A 90 Butaclor MA 40
331	high	97	
Thiuram types			
610	low	48	
710	low to medium	48	

Baypren latices (43).

Grade	Crystal-lisation rate	Properties
4R:	medium	room temperature curing, virtually non-discolouring
GK:	low	
GKM:	low	low tack, non-discolouring
MK:	medium	
MKB:	medium	soft, special type for mining and road construction
SK:	high	high adhesiveness

Latices are stabilised with alkali and suitable emulsifiers at pH 11–13. Mean particle size range 110–160 mμ. *Uses:* as for neoprene latex.

BAYPREN LATEX Polychloroprene latices (Bayer).
B: 58% TS low crystallisation, staining antioxidant
MKB: 58% TS low crystallisation
SK: 55% TS, strong crystallisation
T: 58% TS, very low crystallisation
4R: 50% TS crosslinking with zinc oxide after drying, no curing.

BAYTOWN MASTER BATCHES Compounds of 100 pbw styrene-butadiene copolymers with 50 pbw HAF black (109).

BBP Abbreviation for butyl benzyl phthalate.

BC-105 Liquid barium/cadmium combination. Stabiliser for vinyl resins (45).

BD Abbreviation for 'Bodjong datar' clones (Java). BD 2, BD 5, BD 10, BD 17, cultivated 1918.

BD-8 Trade name for butanediol dicaprylate glycol ester. Softener for synthetic rubber, particularly at low temperatures (110).

BECKAMINE Urea-formaldehyde resin. Soluble in alcohol (111).

BECKMAN PROCESS Process for producing microporous rubber sheets or lamina from latex.

BEDACRYL Group of polymethacrylic acid esters (60).

BEESWAX Cera alba, Cera flava. Pale yellow to brown solid, m.p. 60–66°C, s.g. 0·95. Mixture of esters of high mol. wt., acids with high mol. wt., alcohols (*e.g.* palmitic acid ester of myricyl alcohol), free acids (melissic and cerotic acids), free alcohols and hydrocarbons.
Use: in expensive hard rubber.

BEIRA RUBBER Trade name for a poor quality natural rubber traded through Beira (East Africa).

BELGER Group of accelerators.
Grades:
D: diphenyl guanidine
DM: dibenzthiazyl disulphide
M: mercaptobenzthiazole (394).

BENGUELLA RUBBER An impure rubber (Angola and Benguella).

BENTONITE Naturally occurring aluminium silicate (main constituent montmorillonite) with high powers of adsorption and good swelling capacity, pH of an aqueous dispersion 9–10.
Uses: stabiliser for latex foam, especially for use with sodium silicofluoride as gelling agent, filler.

BENZENE-1:3-DISULPHONYL HYDRAZIDE

$$NH-SO_2 \diagdown \bigcirc \diagup SO_2-NH$$
$$\quad |_{NH_2} \qquad\qquad |_{NH_2}$$

Blowing agent, also participates in cure at activation temperature, 120°C.
TN: Porofor B-13 (43).

BENZENE SULPHON-HYDRAZIDE

$C_6H_5.SO_2.NH.NH_2$, BSH. Grey white powder, m.p. above 100°C with decomposition, s.g. 1·41. Active sponge blowing agent, liberates nitrogen. 1 g yields approx. 120 cm^3 nitrogen at 100–120°C. The cell size can be varied from very fine to very coarse. Suitable for press vulcanisation, expansion process, and open cure. Should be added last to a compound, not processed by mixing with other ingredients. Basic ingredients can lower the decomposition temperature considerably and decomposition can take place even at room temperature in certain circumstances. Dibenzthiazyl disulphide is recommended as accelerator. Articles produced with benzene sulphonhydrazide are odourless and non-toxic.
Loading: 2–10% according to the type of vulcanisate.
Uses: include cellular and microcellular articles, sponge rubber, matting, cushions, rollers, extruded profiles with a smooth closed surface,

cellular ebonite. Suitable for natural and synthetic rubber.
TN: Celogen BSH (246)
Genitron BSH (209)
Porofor BSH (43)
Treibmittel BSH (5).

BENZOFLEX Group of softeners:
Grades:

2-45:	diethylene glycol dibenzoate
9-88:	dipropylene glycol dibenzoate
E-60	ethylene glycol dibenzoate
P-(No.):	polyethylene glycol dibenzoates of varying mol. wt.
T-150:	triethylene glycol dibenzoate.

BENZOIC ACID C_6H_5COOH. White crystalline powder, s.g. 1·26, m.p. 120°C, sublimates above 100°C. Soluble in alcohol, benzene, carbon tetrachloride, ether and hot water, does not dissolve easily in acetone or benzine, 1:270 soluble in cold water. Softener for unvulcanised compounds, improves their tackiness and flow properties, also the dispersion of filler. Retards the vulcanisation of compounds with accelerators of the mercapto type. Has an activating effect on thiazoles, increases the hardness of vulcanised, filled compounds.
Loading: as a hardening agent in filled compounds: 2–8%, as softener: 0·5–2%, as retarder: 0·25–1%, as activator: 0·5–1·5%.
TN: Retarder BAX (54)
Retardex (100).

BENZONITRILE C_6H_5CN, phenyl cyanide. Colourless, oily liquid, m.p. −13·1°C, b.p. 190·6°C,

s.g. 1·005. Soluble in alcohol, ether and hot water.
Uses: solvent for vinyl polymers.

BENZOPHENONE
$(C_6H_5)_2CO$, diphenyl ketone, benzoylbenzene, diphenylmethanone. Colourless, rhombic prisms, m.p. 48·5°C, b.p. 305°C, s.g. 1·11. Soluble in alcohol, ether and chloroform. Substituted benzophenones have a high absorption capacity for ultraviolet rays and can be used in plastics and rubber articles as physical agents for protection against light, *e.g.* 2:4 dihydroxybenzophenone, 2-hydroxy-4-methoxybenzophenone, tetrahydroxybenzophenone, dihydroxydimethoxybenzophenone.
TN: Uvinul (85).

BENZOYL PEROXIDE
$(C_6H_8CO)_2O_2$, dibenzoyl peroxide, benzoyl superoxide. White crystals, m.p. 103–106°C. Soluble in benzene, chloroform and ether, practically insoluble in water and alcohol. Active oxygen 6·5%.
Uses: oxidising agent in redox polymerisation systems for cold rubber, crosslinking agent for plastics.
TN: Lucidol (171).

2-BENZTHIAZYLTHIO-1:4-DIHYDROBENZENE Reaction product of 2-mercaptobenzthiazole and p-benzo-quinone.
Use: rubber antioxidant.
Lit.: BP 668 952 (1952).

BENZYL BENZOATE
$C_6H_5.CH_2OOC.C_6H_5$. Colourless liquid, m.p. 18·8°C, b.p. 325°C, s.g. 1·119. Soluble in alcohol, chloroform and ether. Softener.

BENZYLBIS(DIMETHYL-DITHIOCARBAMATE)
$C_6H_5CH[S.CS.N(CH_3)_2]_2$. White powder, m.p. 174–184°C, s.g. 1·365. Ultra-accelerator, zinc oxide necessary.
Loading: 0·5–0·8% with 1·5–2% sulphur.
TN: + Novex (23).

BENZYL BUTYL PHTHALATE

$$C_6H_4 \Big\langle \begin{matrix} COOC_4H_9 \\ COO.CH_2.C_6H_5 \end{matrix}$$

Colourless liquid, b.p. (20 mm) 280–290°C, s.g. 1·09. Insoluble in water. Low volatility softener and elasticator for synthetic rubber, esp. SBR and NBR compounds. Increases elasticity and low temperature resistance; tensile strength, modulus and hardness decrease as concentration increases.
Loading: softener 5–10%, 20–60% for increasing elasticity and low temperature stability.
TN: Palatinol BB (43).

BENZYL BUTYRATE
$C_3H_7COOCH_2C_6H_5$. Colourless liquid, b.p. 240°C, s.g. 1·015. Soluble in alcohol. Softener.

BENZYL CELLOSOLVE Trade name for benzyl glycol (112).

BENZYL CHLORIDE
$C_6H_5CH_2Cl$, α-chlorotoluene. Colourless to yellowish liquid with an unpleasant smell, m.p. −43 to −48°C, b.p. 179°C, s.g. 1·105. High refractive index n_D15 1·5415. Miscible with alcohol, ether and chloroform.
Uses: vulcanising agent for synthetic rubber.

BENZYLIDENE RUBBER
White to yellow, amorphous substance, sinters at 180°C, s.g. 1·10.
Insoluble and hardly swells in organic
solvents. Produced by treating a
solution of rubber in carbon tetrachloride with benzyl chloride in the
presence of aluminium chloride.

BENZYL OCTYL ADIPATE
Yellowish liquid, b.p. (10 mm) 230–
260°C, s.g. 1·0. Softener and elasticator for nitrile rubber, increases the
elasticity and low temperature resistance, decreases tensile strength and
hardness.
Uses: highly elastic articles, in aeroplanes, refrigerators.
Loading: 5–10% for elasticity, 20–
40% for cold-stability.
TN: Adimoll BO (43).

BETANOL Group of polyethyl
glycol esters.
Uses: emulsifiers and thickening
agents for latex (113).

BET SURFACE Surface area of
fillers may be determined by nitrogen
absorption isotherms. Method according to Brunauer, Emmett and
Teller.
Lit.: *J. Am. Chem. Soc.,* 1938, 60,
309.

BEXIN Antioxidant of undisclosed composition (28).

BEXOL Solutions of resins in
organic solvents (243).

BFE POLYMERS Abbreviation
for bromotrifluoroethylene polymers.

BHA Abbreviation for butyl
hydroxyanisole.

BHC Abbreviation for benzene
hexachloride.

BHT Abbreviation for butylated
hydroxytoluene.

BICERA APPARATUS British
Internal Combustion Engine Research
Association apparatus for research
on dynamic fatigue in rubber cylinders and rubber mountings.

BIDDIBLACK Mineral black
produced at Bideford, Devon
(England) and used as a cheap filler.
The size of the particles is unsuitable
for reinforcement purposes. The
product contains approx. 25%
mineral substances.

BIERER–DAVIS AGEING TEST
Accelerated determination of the
ageing properties of rubber in a
bomb at 70–90°C under 20 atm
oxygen pressure for 1, 2 and 7 days;
procedure according to J. M. Bierer
and C. Davis. The advantage of this
test is the speed with which it can be
carried out, but the results do not
particularly correlate with those from
natural ageing.
ASTM D 573-53,
BS 903: 13: 1950,
DIN 53 508.
Lit.: *Ind. Eng. Chem.,* 1925, **16,** 711.

B-I-K Urea, surface-treated, for
use in the rubber industry. The surface layer is insoluble in solvents, but
soluble in rubber. B-I-K OT: dust
free (23).

BILTAC Group of softeners,
plasticisers and tackifiers for natural
and synthetic rubber (114).

BILTAX Accelerators of undisclosed composition (obsolete) (115).

BIN CURE Slow vulcanisation of compounds during storage.

BIOCIDE Use of biocidic agents for the prevention of bacteria and fungi on plantations is well known. However, biocide has also been successfully added to SR and NR vulcanisates.
Lit.: *Gummi, Asbest, Kunststoffe,* 1963, p. 718.

BIOS REPORTS Comprehensive reports of Allied teams on the progress of German industry during 1939–45. The report *Rubber Industry in Germany during the Period* 1939–45, BIOS Overall Report No. 7, London, 1948, was compiled by T. R. Dawson of the Intelligence Division of RABRM.

BIS(DIISOPROPYL THIO-PHOSPHORYL)DISULPHIDE b.p. 85–90°C. Insoluble in water, soluble in alcohol, benzene, chloroform and petroleum. Vulcanisation accelerator for latex, similar in structure to tetramethyl thiuram disulphide.

$$(RO_2)\underset{\underset{S}{\|}}{P}-S-S-\underset{\underset{S}{\|}}{P}(RO)_2 \quad R=(CH_3)_2CH-$$

Acts as accelerator and sulphur donor at low temperatures. With thiourea or diphenyl thiourea, rapid vulcanisation occurs at 100°C; addition of sulphur is unnecessary. Vulcanisates do not stain with traces of copper, *e.g.* in contact with textiles. *Loading:* 1·5–3% with 0·5–1% thiourea, 0–0·5% MBTS and 0–0·5% sulphur.
TN: DIPDIS (431).

N,N'-BIS-(1:4-DIMETHYL PENTYL)-p-PHENYLENE DIAMINE

Reddish-brown fluid, s.g. 0·90. General purpose antiozonant.
TN: Antozite MPD (51)
Cyzone DH (21)
Eastozone 33 (193)
Tenamene 4 (193).

4:4'-BIS(2:6-DITERT-BUTYL PHENOL)

Yellowish lumps, m.p. 186°C, s.g. 1·03. General purpose antioxidant.
TN: Ethyl Antioxidant 712 (88).

N,N'-BIS(1-ETHYL-3-METHYL PENTYL)-p-PHENYLENE DIAMINE

General use as antiozonant. Dark brown liquid, s.g. 0·87–0·93, protects against flex cracking.
TN: Antozite 2 (51)
Eastozone 31 (193)
Santoflex 17 (5)
UOP 88 (352).

BISETHYL XANTHOGEN

$(C_2H_5OCSS)_2$. Yellow crystals, m.p. 28–32°C. Soluble in benzene, ether, petroleum ether.
Uses: vulcanising agent.

N,N'-BIS-(1-METHYL HEPTYL-p-PHENYLENE DIAMINE)

Dark brown fluid, s.g. 0·90. General

purpose antiozonant, protects against flex cracking.
TN: Antozite 1 (51)
Eastozone 30 (193)
Santoflex 217 (5)
UOP 288 (352).

BIS(3-METHYL-4-HYDROXY-5-TERT-BUTYLBENZYL) SULPHIDE

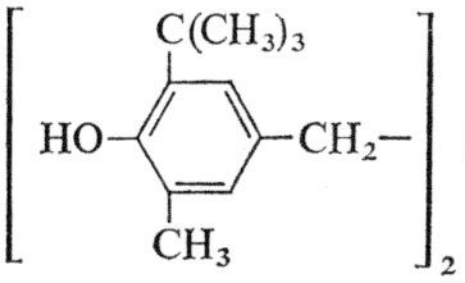

$$\left[HO-\underset{CH_3}{\overset{C(CH_3)_3}{\bigcirc}}-CH_2- \right]_2 S$$

General purpose antioxidant.
TN: Antioxidant 423 (23).

BISMUTH DIMETHYL DITHIOCARBAMATE

$[O(CH_3)_2N.C(S).S]_3Bi$. Yellow powder, m.p. above 230°C, while decomposing, s.g. 2·06, mol. wt. 570. Soluble in chloroform, sparingly soluble in benzene and carbon disulphide. Ultra-fast accelerator.
Uses: in natural rubber and butadiene-styrene rubber for fast cures and for cures above 160°C.
TN: Bismate
Rodform Bismate (51)
PBD-75 (75% with 25% polyisobutylene) (282)
Van Hasselt MTBi (355).

BISMUTH TRIOXIDE Bi_2O_3.

Yellow powder, m.p. 817°C, s.g. 8·9. Recommended together with litharge, as absorbent of X-rays in rubber [FP 986 868 (1951)]. A mixture of 70% bismuth and 30% rubber should be equivalent to lead protecting against X-rays.

BISOFLEX Group of softeners, mostly alkyl or alkaryl adipates, phthalates, sebacates:

9A: dinonyl adipate
47: butylbenzylphthalate
79A: adipate of C_7–C_{10} alcohols
79S: diester of sebacic acid and C_7–C_{10} alcohol compound
81: phthalic acid
91: dinonyl phthalate
100: diisodecyl phthalate
100A: diisodecyl adipate
102: triethylene glycol dicaprylate
102A: triethylene glycol dicaprylate
103: undisclosed composition
104: phthalate of an alcohol compound
108: phthalate of an alcohol compound
791: phthalate of a C_7–C_{10} alcohol compound
DBS: dibutyl sebacate
DNS: di(3:5:5-trimethyl hexyl)-sebacate
DOA: di-2-ethyl hexyl adipate
DOS: di-2-ethyl hexyl sebacate (116).

BISONIDE Group of polymers produced from reclaimed natural rubber and SBR, and modified with unsaturated organic acids and derivatives. Soft to hard, tough, thermoplastic substances containing 45–50% rubber hydrocarbon. Polar, having the characteristics of other polar, synthetic rubber types. Oil resistant and have good ageing qualities.
Uses: extenders for nitrile rubber and neoprene (117).

4:4-BISPHENOL

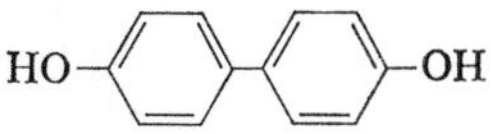

$$HO-\bigcirc\!\!-\!\!\bigcirc-OH$$

Greyish powder, m.p. 260°C, s.g. 1·22. General purpose antioxidant, used for latex.
TN: Antioxidant DOD (43).

2:5-BIS(TERT-BUTYLPER-OXY)-2:5-DIMETHYLHEXANE

White powder, s.g. 1·50. Cross linking agent for polyethylene.
Use: as 50:50 mixture with a mineral carrier.
TN: Varox (51).

BIT Abbreviation for black incorporation time: time required to mix blacks or black masterbatches. Varies with mixing behaviour.

BITUMEN Semi-solid or solid hydrocarbon compound occurring naturally or remaining as residue after distillation of petroleum. Softener, tackifier, processing aid in dark-coloured compounds; improves extrusion speed and the electrical properties of cable compounds. Softener for heavily filled compounds, in soles or in ebonite; has little influence on the hardness of the vulcanisates.

BL
Grades:
353: N,N′.dinitroso-N,N′-dimethylterephthalamide. 70% with 30% mineral oil. Blowing agent
425: additive for BL-353 (6).

BLACK BALATA Brittle balata. Mediocre quality product from an undetermined *Mimusops* type in British Guiana. Contains approx. 30% rubber-like substance, 68% resin and 1% protein.

BLACK GUM Rape oil factice. Vulcanised with sulphur chloride and extended with petroleum tar.

BLACK-OUT Trade name for a group of solutions of modified elastomers.

Use: surface finish for rubber ware (51).

BLANC FIXE Precipitated barium sulphate.

BLANDILE Factice produced from oxidised linseed oil with sulphur chloride; extended with a silicate filler.

BLANDITE Rubber substance, invented and promoted around 1890 by Blandy, in London. Type of factice produced by heating a mixture of oxidised linseed oil, asphalt, carbon disulphide and sulphur chloride.

BLANKET Thick crepe sheets.

B-LATEX A natural latex, where the enzymes have been inactivated by adding an equal volume of boiling water to the latex. Such latex remains stable for days, but coagulates immediately with the addition of small quantities of fresh latex. This method was previously considered as proof of enzyme action on coagulation.
Lit.: *J. Soc. Chem. Ing.*, 1918, 37, 48T, 262T.

BLE-25 Reaction product of diphenylamine and acetone at high temperatures. Dark-brown, viscous liquid, s.g. 1·09. Antioxidant for NR, SBR, NBR, and CR and for latices and has a negligible influence on vulcanisation. Causes dark brown staining, migrates during vulcanisation. Discolours light articles and stains on contact with light-coloured vulcanisates. Does not bloom. Very good protective action against heat, oxygen, flex fatigue.
Uses: used generally where colour is unimportant, *e.g.* mechanical

goods, tyres, heavily filled compounds, air tubing (23).

BLEEDING OUT Migration of dye pigments from one area to another.

BLEMISHES Non-classified defects, *e.g.* patches, in smoked sheet.

BLE POWDER Compound of 65% complex diaryl amine-ketone-aldehyde reaction product and 35% N,N'-diphenyl-p-phenylene diamine. Brown powder, m.p. 70–85°C, s.g. 1·095. Soluble in acetone, benzene and ethylenedichloride, insoluble in water and benzine. Antioxidant for NR and SBR, has a slight activating effect. Causes dark brown discoloration and stains light-coloured articles on contact. Migrates during vulcanisation. Blooms at 1% and above. Excellent protective agent against heat, oxygen, flex cracking, copper, manganese and the weather. *Loading:* 0·2–1%, in goods where blooming is unimportant up to 2%. *Uses:* natural rubber treads, mechanical goods, soles with a high resistance to flex-cracking, insulation in direct contact with copper conductors, reclaim compounds, SBR—to reduce heat build-up (23).

BLIC Abbreviation for Bureau de Liaison des Industries du Caoutchouc de la CEE. Founded 1959, Milan.

BLISTERS Small air and gas bubbles in sheet rubber, caused by air collecting during coagulation, if the latex coagulates too quickly or the latex concentration is too high. Also caused by slight fermentation, by free carbonic acid from the water used for dilution and from the soda used as anticoagulant.

BLOCK BALATA Trade name for balata blocks approx. 80 cm in length and 60 cm in width.

BLOOMING Efflorescence. Migration of the soluble components of a rubber compound to the surface, with the formation of visible crystalline.

BLOWING AGENT
Grades:
15:	mixture of biuret and urea. White, odourless powder, s.g. 1·466 (119)
BSH:	benzene sulphonhydrazide (5)
CP-975:	blowing agent for sponge rubber (119)
OB:	p,p'-oxybisbenzene sulphonylhydrazide (5).

BLOWING AGENTS Products which decompose into gaseous substances at high temperatures or which evolve gases, and thus cause rubber compounds to expand before vulcanisation. Used for the production of cellular rubber articles and hollow articles. They must be well dispersed to obtain an even cell structure, should not produce any toxic products on decomposition and should have a minimal effect on vulcanisation. The first blowing agents were inorganic carbonates, bicarbonates and nitrites, which are currently used only in the production of hollow articles, and because of their undesirable properties (poor dispersion, activation of vulcanisation) have been replaced by organic substances which evolve nitrogen. The substances now used include dinitrosopentamethylene tetramine, benzene sulphonhydrazide, azodicarbonamide, p-p′-oxybisbenzene sulphonhydrazide, azodicarboxylic acid diethyl ester, azohexahydrobenzonitrile, diazoaminobenzene, and azonitriles.

B-MERCAPTAN Trade name for tert-dodecyl mercaptan (120).

BOKALAHY Local term for *Marsdenia verrucosa* (Madagascar). Fruits yield a good rubber.

BONDOGEN Mixture of a high mol. wt. sulphonic acid, soluble in oil, hydrophobic alcohol with a high boiling point, and paraffin oil. Dark, practically odourless liquid, s.g. 0·02, acid value 42. Peptising agent and plasticiser for all rubber types.
Quantity: 0·3% in natural rubber, 0·5% in SBR, in concentrations above 2% as peptising agent and processing aid for SBR, neoprene and nitrile rubber (51).

BONE OIL Dippel's oil. Yellow, staining oils with extremely unpleasant odour, s.g. 0·75–0·85. Miscible with water, soluble in ethanol, ether and fatty oils. Obtained by dry distillation of bones and animal refuse, with distillation of the resulting bone tar. Contains basic nitrogen compounds, *e.g.* methyl amine, pyridine, aniline and quinoline. Bone oil was first discovered by Dippel in 1700 and was used as one of the first solvents for rubber. (L. A. P. Herissant and P. J. Macquer, 1763.) Rectified turpentine oil has been suggested as a much cheaper substitute.

BORAX $Na_2B_4O_7.10H_2O$, sodium tetraborate, s.g. 1·7. Soluble in water, gives alkaline reaction.
Uses: anticoagulant and preservative for latex, preservative for concentrated latex with a low ammonia content.

BORNEO RUBBER Getah susu (susu: milk), getah gerip (gerip: liane), getah Kalimantan. Trade name for wild rubber from Borneo, mostly from *Willoghbeia frima Bl.*

BORNESITE
$C_6H_6(OH)_5.OCH_3$, d-methylinosite, mesoinosite-1-methyl ether. Crystallises in prisms from water and methyl alcohol; m.p. 203°C, D + 31° (in water). Discovered in Borneo rubber (primarily *Ficus elastica*). Isomeric with quebrachitol, but dextro-rotary.

BORRACHA FINA Borracha defumado. Original trade name for first quality natural rubber from *Manihot Glaziovii*. Extracted from the sap by the Para-Smoking process.

BOUASSE, HENRY French physicist, 1866–1953. Conducted research on the elasticity of rubber (1903–04), forming the basis of modern rubber physics.

BOUCHARDAT, GUSTAVE French chemist, 1842–1918. Researched the dry distillation of rubber, discovered that the original hydrocarbon and the fractions derived from it, had the same empirical composition. In 1895 suggested that a substance which appeared in small quantities, liquid isoprene, was the basic constituent of rubber.

BOULINIKON Mixture of rubber and cork flour. Precursor of linoleum.

BOUNCING PUTTY Kneadable material with the consistency of window putty, which can be bounced on a hard surface. On sharp impact the material becomes brittle and splinters into tiny pieces. Produced by heating a compound of dimethyl silicone with a boron compound.
Uses: golf ball centres, by adding zinc stearate as a softener, the tendency to fracture on hard impact is decreased.
Lit.: USP 2 431 878, 2 451 851, 2 609 201.
Silicones and their Uses, McGraw-Hill, 1954, pp. 186–7.

BOUND RUBBER Carbon gel. Part of the rubber in unvulcanised rubber/filler (black) compounds, which is insoluble in benzene. The quantity of bound rubber is characteristic of the filler type and proportional to the reinforcing effect. It increases with the quantity of the filler, the filler surface area and the mixing temperature. Bound rubber is similar to a polymer gel, with three-dimensional network structure, and is dependent on the reaction between filler and rubber, caused by the production of free radicals in the rubber during the mixing process.

BPI Abbreviation for Banbury processability index.

BPPK Abbreviation for Balai Penjelidikan dan Pemakaian Karet, Bogor, Djalan Salak 1, Java (Indonesian Institute for Rubber Research and Development). Successor of INIRO (Indonesisch Instituut voor Rubber Onderzoek). For a short time it existed under the name Jajasan an Stelle von Balai (JPPK).

BR Butadiene rubber, as listed in the ASTM catalogue D-1418-59T.

BR Abbreviation for butyl ricinoleate.

BR Abbreviation for 'Bogoredjo' clones, Java.

BR-66 N-cyclohexyl-N′-phenyl-p-phenylene diamine. Antioxidant.

BRAAT MACHINE Semi-automatic machine with six pairs of rollers; used for milling sheet rubber.

BRABENDER PLASTOGRAPH Apparatus with a continuously regulated motor; used to determine the flow properties, mixing and extrusion, temperature/viscosity relationship and other properties of rubber and compounds.

BRAZE Compound of chlorinated rubber and hypochlorinated rubber. Bonding agent for rubber/metal (51).

BRC Group of solid hydrocarbons from coal-tar.
Uses: processing aids for elastomers intended for articles which must have high hardness; reclaiming of rubber (29).

BREON NITRILE LATICES Group of butadiene-acrylonitrile latices (BP Chemical International).

	Bound acrylonitrile %	TS %	pH
1512	31·5–33·5	38–40	10·0
1561	39–42	39–42	10·5
1562	31·5–33·5	39–42	9·5 general purpose
1571	37·5–40·5	39–42	8·5 carboxylate
1574	27–29	39–41	8·5 carboxylate
1577	27–29	37–40	10·0 modified

BRH 2 Asphalt product. Viscous liquid, s.g. 1·0. Softener and tackifier for electrical insulating tapes, friction compounds, adhesives, reclaiming oil, particularly for the Pan process (29).

BRIJ Polyoxyethlene lauryl ether. Wetting and dispersing agent for latex (97).

BRITENKA Viscose rayon (121).

BRITISH STANDARDS INSTITUTION BSI. Founded 1901 as the British Engineering Standards Association, together with the Institute for Civil Engineering, Electrical Engineering, Machine and Shipbuilding, and the Iron and Steel Institute. These organisations merged in 1918 and acquired the present name in 1932, when Chemical Standards were accepted in the industrial field. In 1938 a branch was proposed for the rubber industry (Rubber Industry Committee). The first Standard regarding the rubber industry was BS 7: 1904, Dielectrics for Cables.

BRITISH THERMAL UNIT (BTU). The quantity of heat necessary to increase the temperature of 1 lb water by 1°F (at the point of greatest density, or as near as possible to it). 1 btu = 252 cal.

BROMOBUTYL RUBBER Brominated butyl rubber with 1–3·5% bromine in the polymer chain. Has the characteristic properties of butyl rubber, but shows a greater vulcanisation speed and requires only a quarter to three-quarters of the accelerator quantity normally used for butyl rubber. As vulcanisation is not influenced by the combination with unsaturated polymers, bromobutyl rubber can be cured with natural rubber and SBR. Shows good adhesion to metals and to

	Bromobutyl Rubber	Butyl Rubber
Tensile strength, kg/cm^2	161	208
Modulus, 300%	73	36
Modulus, 400%	108	57
Elongation at break %	580	770
Hardness (Duro A)	77	61
Vulcanisation	40 minutes, 138°C	

other elastomers, has excellent resistance to ozone, good heat and ageing stability, good electrical properties and low permeability to gas.
TN: Hycar 2202 (14).
Lit.: Goodrich Service Bull., H-18.
Ind. Eng. Chem., 1955, 47, 1562.
USP 2 631 984, 2 681 899, 2 698 041.

N-BROMO-N-CHLORO-DIMETHYL HYDANTOIN White powder with a slight halogen-like aroma, m.p. 163°C, 33% active bromine and 14% active chlorine. Halogen carrier; reacts with unsaturated compounds by introducing hydantoin rings, bromine and chlorine to the chain. Modifying agent for elastomers and extender oils; increases polarity and ease of vulcanisation. With this additive butyl rubber can be vulcanised by the metal oxide system.
TN: Dihalo Modifier (182).

BROMOPRENE 2-bromo-1:3-butadiene. Can be polymerised.

BROOKFIELD VISCOMETER Rotating cylinder viscometer used for determination of latex viscosity.

BROWN CREPE Thin brown crepe, thick brown crepe, sometimes also known as Compo. Light to dark brown, thin or thick crepe rubber, produced either from waste rubber on the plantations or in so-called remilling factories from wet rubber slabs, which are produced locally. The waste from the plantations (10–20% of the total production), consists of strips dried on the tapping areas (panel scrap, tree scrap, curly scrap, sernamby), and residue latex which coagulates in the tapping cups. The dark discoloration is the result of oxidation and the formation of melanine. The scrap is extremely impure owing to bark, sand, etc., is washed in the so-called scrap-washer (a kind of internal mixer with serrated rollers), softened in detergent for 1–2 days, washed again on heavy, ribbed rollers (10–15 passes) and rolled out into thin sheets. The slab rubber is likewise washed in the remilling factories on heavy rollers, and rolled out to thick or thin sheets. It is dried at a max. temperature of 37°C, since at higher temperatures tackiness occurs. The thick sheets usually require 20–30 days to dry. The amines resulting from protein decomposition give this type of rubber a high vulcanisation rate.

BRPRA Abbreviation for British Rubber Producers' Research Association. In 1960 renamed the Natural Rubber Producers' Association.

BRS 700 Refined coal-tar. Dark viscous liquid, s.g. 1·17–1·22. Softener and extender for NR and SR. Gives good tensile strength and elasticity, very slight heat build-up.
Uses: tyres and mechanical goods (29).

BRT Coal-tar distillate with a high b.p., s.g. 1·14–1·18. Softener, plasticiser, dispersing agent for NR and SR, reclaiming oil for highly filled rubber (29).

BRUPA CREPE Bijzondere Rubber uit Pancreatine. Rubber produced from skim latex, where part of the normal protein content is separated by treatment with pancreatin. Contains approx. 0·6–1% nitrogen. Superficially, it is similar to Thin Brown 1x and 2x, modulus at 600%, approx. 50 kg/cm^2 (122).

BS Abbreviation for British Standard.

BS 3 TYRES BS: abbreviation for Battistrada Separato, tyres developed by Pirelli, the running surface forms a separate, removable element, comprised of three rings. The tyres consist of a carcase, on which three treads are set, each strengthened with steel and held in position by longitudinal rubber beads. The tread bands have a slightly smaller diameter than the carcase and remain firmly fixed once inflated. Can easily be removed after the air has been released.

BS-45 AK Russian produced butadiene-styrene rubber with 45% bound styrene, 1·2–2·2% Neozon D as antioxidant. Does not need to be plasticised.
Uses: shoes, cellular rubber.

BSH Abbreviation for benzene-sulphonhydrazide.

BS HARDNESS Abbreviation for British Standard Hardness.

BSI Abbreviation for British Standards Institution.

BUCAR Trade name for butyl rubber (159).

BUCAR Group of butyl rubbers.
1000 S: s.g. 0·92, unsaturation 0·6–1·0 mole %. ML-8/100°C: 41–49. Staining stabiliser. Rec.: maximum ozone resistance
1000 NS: as 1000 S with non-staining stabiliser

5000 S: s.g. 0·92, unsaturation 1·5–2·0 mole %, ML-3/127°C: 50–60, staining stabiliser. Rec.: intermediate ozone and heat resistance, inner tubes, airbags, extrusions, bladders
5000 NS: as 5000 S with non-staining stabiliser
6000 S: s.g. 0·92, unsaturation 2·0–2·5 mole %. ML-8/100°C: 41–49, staining stabiliser. Rec.: maximum heat resistance, steam hoses, curing bags
6000 NS: as 6000 S with non-staining stabiliser.

BUDENE Trade name for polybutadiene.

BUIST-KENNEDY Hardness meter. Instrument for determining the hardness of rubber; measures the load required to produce a pre-determined indentation BP 617 465 (1949).

BULK DENSITY The density of a fine, granular or powdered material in a loose form; expressed as the ratio, weight:volume, *e.g.* g/cm^3 or kg/m^3.

BULK FACTOR The ratio of the volume of loose, powdered materials to the volume of the same mixture in compacted form (the ratio of the apparent to the true density). With plastics, for example, it is the ratio of the density of the moulding powder to that of the moulding.

BULLET TREE Local term for *Mimusops globosa*, used by the major suppliers of balata. So-called because of its round to oval-shaped fruits.

BUNA German produced synthetic rubber developed in 1926–30. Originally produced by block polymerisation using metallic sodium (later potassium) as a catalyst. The name is formed from the first two letters of each of the compounds: *bu*tadiene and *na*trium (sodium). The various polybutadiene types were given number indices as a measure of the polymerisation degree (Buna numbers): 32, 85, and 115 (multiplied by 1000 they represent an average mol. wt. The original batch production process used stirred autoclaves with 0·15–0·2% sodium in the form of a 10% dispersion in Buna 32. Vinyl chloride, dioxane or vinyl ether, 0·5–1% acted as modifier, 2% phenyl-β-naphthylamine as stabiliser. Buna 32 and 85 were later continuously produced in spiral reactors with potassium as catalyst. Emulsion polymerisation, developed later, led to copolymerisation, with acrylonitrile (Buna N: Perbunan) and with styrene (Buna S), the products being distinguished by letters. As original Buna S 1 had at that time not been modified and was tightly cross linked, processing was difficult. In 1937 a thermal degradation treatment was developed, and by the end of 1937 the regulation of the chain length of the polymers, using mercaptan, was introduced.

Grades:

32: low viscosity. *Uses:* non-migrating softener for Buna S and Perbunan N (Plastikator 32) solutions with a high sulphur content for non-corrosive, hard rubber linings

85: for hard rubber linings with good resistance to chemicals

115: tyre rubber with good low temperature characteristics. Has poor resistance to abrasion and is no longer produced

D: copolymer of butadiene and ethylidene dimethyl vinyl carbinol. Produced by the IG Farben Industry during World War II

N: later known as Perbunan. Solvent resistant butadiene-acrylonitrile copolymer. 26% acrylonitrile, conversion 75%, continuous polymerisation 25–30 hr at 30°C, 0·2% potassium persulphate, 0·3% diisopropyl xanthogen disulphide (Diproxide), 3·6% Nekal BX as emulsifier, 0·3% tetrasodium pyrophosphate, 3% phenyl-β-naphthylamine, Defo plasticity 2400–2800

NN: later known as Perbunan Extra. Like Buna N, 40% acrylonitrile, batch polymerisation 60 hr at 24°C, conversion 75%. Increased resistance to solvents as compared with Buna N

OP: slightly modified Buna S. Extended before coagulation with 25%, 37·5%, or 50% Naphtholen ZD

R: modification of Buna S. Conversion 50% at 50°C and finally up to 96% at 90°C. Highly cross linked product with undesirable technological characteristics. *Uses:* additive for Buna S

S-1: monomer ratios 70:30 to 68:32, conversion 60%, Defo plasticity 4300, 4·4% fatty acids, Nekal BX as emulsifier, potassium persulphate as activator, linseed oil as modifier, phenyl-β-naphthylamine as short stop agent and stabiliser. Practically impossible to masticate, processing requires thermal degradation. Produced at Schkopau

S-2: like S-1, but with a higher styrene content and 55% conversion. Produced at Hüls

S-3: monomer ratio 68/32, conversion 58–60%. Developed 1942 as a result of linseed oil deficiency. Synthetic saturated fatty acids (Fischer-Tropsch process) and with alkali as additional emulsifier, Diproxide as modifier (0·06–0·1%), polymerisation 30 hr at 45–50°C, phenyl-β-naphthylamine 2–3%. Has high thermoplasticity, but lower elasticity and lower temperature stability than Buna S1 and S2. Produced by VEB Schkopau

S-4: like Buna S-3, but softer because of the larger addition of Diproxide (0·24%), Defo plasticity 700–800, thermal degradation unnecessary

S-4L: pre-plasticised rubber for light-coloured articles

S acid: developed as Buna S-3, under acidic conditions, with Esteramine as emulsifier. Monomer ratio 70:30, conversion 60%, polymerisation 15 hr at 35°C, 2% phenyl-β-naphthylamine, Defo plasticity 2500–3000. Low hysteresis and high elasticity

SR: blend of Buna S and Buna R, Defo plasticity approx. 3500. Improved processing qualities

SW: (S 10, W: soft) monomer ratio 90:10, 60% conversion, polymerisation 60–70 hr, Defo plasticity approx. 600–800. Good low temperature characteristics, tensile strength 190 kg/cm^2, phenyl-β-naphthylamine as antioxidant

SS: increased styrene content as compared with S-3, Defo plasticity 4500. Improved processing and electrical characteristics, lower resistance to abrasion, less elastic. *Uses*: pneumatic tubes

SSE: (E: free of iron; for pharmaceutical purposes) similar to SS, with oxycresyl camphane, non-discolouring stabiliser and short-stop agent, conversion 60–62%, polymerisation 30 hr at 40–45°C. Relatively unstable, cyclisation takes place in a benzene solution with precipitation of a product insoluble in benzene, high tackiness in solutions

SSGF: (GF: odourless) monomer ratio 50:50, production as with S 3, 3% phenyl-β-naphthylamine as antioxidant, conversion 62% in 30 hr at 40–45°C, Defo plasticity 3000. Can be separated from unreacted monomers by washing with calcium chloride. *Uses*: in the soft drink and food industry.

BUNA CB German produced cis-polybutadiene rubber (189). 98% cis, 1% trans, 1% 1:2 units, Ziegler catalyser, Mooney 45–55.

BUNAC K-17 Activator of undisclosed composition (125).

BUNA HÜLS Butadiene-styrene copolymer (cold rubbers) produced by Bunawerke Hüls (124):

Types:
150: (1500). Staining, Mooney 50
152: (1551). Non-staining, Mooney 50
153: (1502). Non-staining, Mooney 50
170: (1507, formerly BT 17). Non-staining, Mooney 30
180: (1500 series). Staining, Mooney 115
181: (1500 series). Non-staining, Mooney 130
190: (1500 series, formerly BT 24). Non-staining, Mooney 40
302: (1711). Staining, Mooney 49, 37·5 highly aromatic oil
321: (1712, formerly BT 21). Staining, Mooney 50, 37·5 highly aromatic oil
351: (1778, formerly 352). Non-staining, Mooney 37, 37·5 naphthenic oil
352: (1778, now 351). Mooney 49
362: (1778). Non-staining, Mooney 49, 37·5 naphthenic oil
363: (1713, formerly BT 13). Non-staining, Mooney 50, 50 naphthenic oil
372: (1707). Non-staining, Mooney 55, 37·5 naphthenic oil
373: (1713). Non-staining, Mooney 52, 50 naphthenic oil
610: (1600 series, formerly BT 104). Non-staining, Mooney 88 (compound), 15 naphthenic oil, 100 SRF

630: (1600 series). Non-staining, Mooney 68 (compound) 5 naphthenic oil, 50 FEF
651: (1601, formerly BT 101). Non-staining, Mooney 62 (compound), 50 HAF
670: (1610). Staining, Mooney 73 (compound), 10 highly aromatic oil, 52 ISAF
751: (1808). Staining, Mooney 60 (compound), 47·5 aromatic oil, 76 HAF
760: (1805, formerly BT 103). Non-staining, Mooney 67 (compound), 37·5 naphthenic oil, 75 HAF
770: (1818). Staining, Mooney 75 (compound), 37·5 highly aromatic oil, 70 ISAF
BT 4: (obsolete). Staining, Mooney 120
BT 12: (1500 series). Non-staining, Mooney 80 (compound)
BT 13: now 363
BT 16: (1507). Non-staining, Mooney 40
BT 17: now 170
BT 21: now 321
BT 24: now 190
BT 25: (1714). Staining, Mooney 60, 50 highly aromatic oil
BT 27: (1500, obsolete). Mooney 50
BT 101: now 651
BT 103: now 760
BT 104: now 610.

Latices:
200: (2111). Staining, stabiliser, concentration 23%, Mooney 50
220: (2111). Non-stabilised, concentration 23%, Mooney 50
K60: (2105). Concentration 62%, Mooney 140
K70: (2114). Concentration 67%, Mooney 150.

BUNALITE Chlorinated Buna-S, produced in Germany during World War II. Similar types to chlorinated rubber. Types N and H (different viscosity).
Uses: protective coatings, rubber/metal bonding agents.

BUNAREX Group of dark-coloured coumarone-indene resins, m.p. 100–115°C, s.g. 1·15. Softeners, plasticisers, tackifiers and reinforcing agents for NR and SR (127, 126).

BUNATAK Group of plasticising agents, softeners and tackifiers of undisclosed composition. For NR, SBR and chloroprene rubber.
Types:
21: light brown, pleasant smelling liquid, s.g. 0·92. Loading 15–30%. Addition on the mill or in the mixer. Improves processing ability. Excessive loading results in a spongy structure
90: brown, slightly aromatic liquid, s.g. 0·94. Improves dispersion, does not stick to the roll. *Quantity:* 10% for natural rubber, up to 50% for SBR.
210: light brown liquid with a faint, pleasant smell. Has no influence on vulcanisation. Suitable for non-black compounds. *Quantity:* 10–35%
AH: light-coloured liquid, s.g. 0·91. Gives a smooth, tacky surface, free of bubbles. *Quantity:* 10–20% for light-coloured articles
N: light brown liquid, aromatic, s.g. 1·032. Softener and tackifier for nitrile rubber and Hypalon. *Quantity:* 10–30%
U: light brown liquid with a faint, pleasant smell, s.g. 0·987. Improves dispersion of fillers and surface finish. Non-staining.

Quantity: 10–35% with 2–2·5% sulphur (128).

BUNAWELD POLYMER 780 Light brown, viscous liquid, s.g. 1·04. Plasticiser and tackifier for natural rubber and SBR. Improves calendering. Does not decrease the physical properties. Compound does not stick to the ròlls. *Quantity:* 5–20% (128).

BUNDESMANN TEST Determination of water repellent capacity of waterproof textiles, using a rain simulator and simultaneous rubbing of the underside of the texiles.

BUNNATOL Softener of undisclosed composition. For natural, synthetic and reclaimed rubber. Has no effect on vulcanisation. Type G, viscous, light brown liquid, soluble in water (113).

BUPS Russian produced antioxidant. 4:4'-dibutyl-2:2'-dihydroxydiphenyl sulphide.

BURCHARD REACTION Bordeaux red to violet red colour reaction which appears in a solution of rubber in chloroform with the addition of a few drops of acetic anhydride and concentrated sulphuric acid (confirmation of sterols).

BURMA RUBBER Natural rubber (Burma) obtained in small quantities from *Ecdysantera micranta D.C.* (Nwedo), *Parmeria pendunculosa Benth* (Kamano), to a lesser degree from *Urceola esculenta* and *Rhynchodia Wallichii*, and from *Chonemorpha macrophylla*. The sap is normally coagulated by adding hot water.

BUTAC Modified resin acid. Plasticiser and tackifier for SBR (40).

BUTACLOR French produced polychloroprene rubber (434). Similar in compounding, properties and uses to neoprene.

Grades:

MA 40: equivalent to neoprene AD
　　　　Perbunan C 320, C 330
　　　　Denka Chloroprene A 90
MA 41: equivalent to neoprene AC
　　　　Perbunan C 321, C 331
MC 10: equivalent to neoprene
　　　　WRT
　　　　Perbunan C 110
　　　　Denka Chloroprene S 40
MC 30: equivalent to neoprene W
　　　　Perbunan C 210
　　　　Denka Chloroprene M 40
MC 31: equivalent to neoprene
　　　　WM 1
　　　　Perbunan C 211
　　　　Denka Chloroprene M 30
MH 30: equivalent to neoprene
　　　　WHV
　　　　Denka Chloroprene M 120
MH 31: like MH 30, but with a lower crystallisation rate
SC 10: equivalent to neoprene GRT
SC 21: equivalent to neoprene GNA
　　　　Denka chloroprene PM 40
SC 20: equivalent to neoprene GN.

1:3-BUTADIENE

CH_2=CH—CH=CH_2, divinyl, vinyl ethylene, biethylene, pyrrolylene, erythrene. Colourless gas, m.p. $-113°C$, b.p. $-4.5°C$, b.p. (2 atm) $15.3°C$, b.p. (5 atm) $47°C$, b.p. (10 atm) $76°C$, b.p. (20 atm) $114°C$, b.p. (40 atm) $160°C$, s.g. 0.65, critical temperature $161.8°C$, critical pressure 42.6 atm. Soluble in organic solvents. Stabilised with o-dihydroxybenzene and with aliphatic mercaptans. Can be produced by approximately 100 processes. Technical production: from acetylene by a four-stage process or by the Reppe synthesis, from alcohol according to the methods of Lebedev and Ostromislensky, by thermal or catalytic cracking of petroleum, dehydrogenation of butene or butane, dehydrochlorination of chlorinated butenes. Polymerises easily with sodium, polymerisation capacity was discovered by Lebedev. Commercial polymerisation by sodium has been in use since 1926, polymerisation with styrene and acrylonitrile since 1929. *Uses:* synthetic rubber, copolymers with styrene and acrylonitrile.

BUTADIENE-ACRYLONITRILE COPOLYMERS

Nitrile rubber, NBR, GR-A. General term for rubber-like copolymers of unsaturated nitriles and dienes. Most products are copolymers of butadiene with 15–40% acrylonitrile. Produced in a similar way to SBR, as warm or cold polymers in an emulsion process. Polymerisation at 5°C gives a regular structure. Such polymers were first produced in Germany between the wars, as Buna N (later Perbunan, then Perbunan N). Because of the strong polarity of the acrylonitrile, NBR is insoluble in most non-polar solvents and resistant to swelling. NBR is insoluble in aliphatic hydrocarbons, mineral oils, animal and vegetable fats, but is soluble, and may also swell, in aromatic and chlorinated hydrocarbons, ketones and esters. The oil resistance is proportional to the acrylonitrile content, but the elastic properties and low temperature resistance decreases as the acrylonitrile content rises. Vulcanisation properties of NBR lie between those of natural rubber and those of SBR. NBR is usually vulcanised with sulphur, but as it has lower unsaturation than NR, less sulphur is necessary (1–2%). Other vulcanising agents can be used instead of sulphur, *e.g.* tetramethyl thiuram disulphide (approx. 3%). Vulcanisation with peroxides (1·25%) yields a vulcanisate with improved quality. To shorten the vulcanisation time carboxyl groups can be introduced into the molecule, thus increasing the number of crosslinks. Such polymers may be crosslinked with zinc oxide at low temperatures and show a decrease in set and an improvement in resistance to solvents. NBR is tougher and less plastic than natural rubber, and harder to process because it is less thermoplastic and tacky. Cold polymers and types containing vinyl benzene show improved processing characteristics. The use of plasticising agents, such as dioctyl phthalate, dibutyl phthalate, sebacates, tricresyl phosphate, tributoxyethyl phosphate, ensures good processing and improves flexibility at low temperatures. Processing and compounding is carried out as with NR. Although less sulphur is necessary than with NR compounds, a higher percentage of accelerators is required. Vulcanisates have good ageing properties and resistance to heat. Ozone resistance is similar to that of NR. Gum vulcanisates have poor physical properties. To gain maximum physical properties, reinforcing blacks or white fillers must be used. Filled vulcanisates have a better resistance to abrasion than NR. NBR has a lower electrical resistance and is suitable for compounding with conductive blacks for the production of antistatic articles. NBR may be blended with p.v.c., polysulphide rubber, polyacrylic acid esters, reactive phenol-formaldehyde resins. Apart from the hardening effect, good resistance to abrasion and swelling may be achieved with phenoplasts, without a large quantity of filler. Normally, NBR cannot be mixed with natural rubber, but a homogeneous two-phase compound may be achieved by combination of black master batches.

Uses: include mechanical goods which are resistant to swelling, sealings, roller coverings, conveyor belts, driving belts, fuel hoses, joint sealing compounds, brake linings, adhesives, hard rubbers, additive to plastics to increase impact resistance. NBR is marketed with varying nitrile contents, and with varying Mooney values, in solid or liquid form (Hycar 1312), or as latex.

TN: Breon
Butakon
Butaprene
Chemigum
Hycar
Krynac
Paracril
Perbunan N (Buna N, Igetex NN)
Polysar Krynac (Polysar N).

BUTADIENE-STYRENE COPOLYMERS SBR, GR-S.

Group of emulsion copolymers of butadiene and styrene, with varying styrene content up to a maximum of 85%. The properties of the polymers may be varied over a wide range according to the styrene content, the type of reaction, the activator, the emulsifier, the modifying agent, the short-stop agent, and the conversion rate. As the styrene content rises, the polymers increasingly resemble thermoplastics. The polymers were originally developed in Germany under the name of Buna S, and were

72

further developed in the USA during World War II. Polymerisation is achieved continuously or batch-wise in tanks connected in series. According to the polymerisation recipe and the temperature of reaction, the polymers are classified as hot, produced at 50°C (hot rubber), or cold, at 5°C (cold rubber). Production is based on the emulsion of the purified monomers in water, adding an initiator to start the reaction, and a modifier to limit mol. wt. and prevent crosslinking. At a pre-determined conversion (between 60 and 72%), the reaction is interrupted by adding a short-stop agent. Unreacted monomers are separated from the latex and recovered, the latex is coagulated continuously by means of salt solutions, sulphuric acid, acetic acid or glue, the crumbled coagulum is washed, dried and pressed into bales.

Initiators (activators): Polymerisation is started by means of a radial reaction. Potassium and ammonium persulphate are particularly suitable, but are effective only at 30°C and above. Introduction of the redox system reduces the polymerisation temperature (to −20°C) and thus improves physical properties. Examples of suitable systems are organic hydroperoxide/iron sulphate/alkali pyrophosphate with or without reducing sugar, hydroperoxide and polyamine with only slight traces of iron (peroxamine formula), sodium-formaldehyde/sulphoxylate-ferrous sulphate complexes based on ethylene diamine tetra-acetic acid (sulphoxylate formula). The last initiator formula is independent of critical temperatures and times, and achieves greater uniformity. Complex iron compounds increase the speed of reaction. Suitable peroxides include: cumyl hydroperoxide, p-menthane hydroperoxide, diisopropyl benzoyl hydroperoxide, and cyclohexyl benzoyl peroxide.

Modifiers (regulators): When conjugated dienes of a conversion of 25–30% gel point and over are polymerised, three-dimensional cross-linking takes place, resulting in the formation of undesirable insoluble polymers. Substances are therefore added which regulate the average mol. wt. and make the formation of shorter, more linear chain structures possible. Such polymers are much easier to process. In the USA, aliphatic mercaptans with a long chain structure are used, *e.g.* n-dodecyl mercaptan, while in Germany diisopropyl xanthogen disulphide (diproxide) is preferred. Sodium linoleate, p-nitraniline, thioethers, chlorinated hydrocarbons, among others, are also used as regulators.

Short-stop: As products with too high a conversion give rise to insoluble materials, which are difficult to process, the reaction is interrupted at a predetermined point, usually at a conversion of 60–72%. Sodium hydrosulphide is used as a stopper in hot polymers, hydroquinone in batch polymerisation, and phenyl-β-naphthylamine (PBN) in continuous polymerisation. PBN also acts as a stabiliser and antioxidant. For light-coloured types trialkyl phenol phosphites, 2:6-ditert-butyl-p-cresol, and others are used. These short-stops are inadequate for redox systems, and in order to interrupt such reactions, water soluble sodium dimethyl dithiocarbamate is used.

Stabilisers: To protect the latex from oxidation, antioxidants, *e.g.* PBN, must be added before coagulation, if they have not already been used as short-stops.

Crosslinking: To improve the dimensional stability and processability of compounds, some types are polymerised with small quantities of divinyl benzene, which causes intensified crosslinking.

Other monomers: In place of styrene, other substituted styrenes have been tried as a second monomer, *e.g.* vinyl naphthalene, halogenated styrene (*e.g.* dichlorostyrene), p-cyanostyrene, nitrostyrene, hydroxy, acetoxy and carboxy-styrene, as well as ether, chloro-substituted isopropyl benzene, halogenated and cyanoolefins, esters, and vinyl pyridines. No practical products have evolved from these experiments.

Cold rubber: The physical properties of the polymers can be improved enormously by lowering the temperature of polymerisation to approx. 5°C. The use of an active initiator is necessary, as a decrease in temperature also decreases the formation of free radicals, and the speed of reaction. Redox systems are used to lower the temperature to as low as −20°C, depending on the system (*see above*, Initiators). Cold rubber is polymerised continuously in tanks connected in series at 5°C, up to a conversion of 60%, and requires twelve hours. Sodium dimethyl dithiocarbamate is used as a shortstop. Latex is coagulated in the same manner as hot rubber. The production of highly concentrated cold latices makes the use of electrolytes necessary, as otherwise a paste is produced at low temperatures. The particle size is increased to 200–300 mμ by the use of electrolytes. In black compounds cold rubber has the same mechanical properties as natural rubber, although it has better resistance to abrasion and a lower elasticity. It is more easily processed than hot rubber, has a greater tackiness, has a lower heat build-up, greater tensile strength and flex-cracking resistance, and greater structural stiffness. Styrene may be introduced more uniformly, is less cross-linked, which improves flow behaviour and increases trans-1:4 content. Cold rubber which has been modified only slightly may be blended with large quantities of oil without causing deterioration of the mechanical properties.

Polymers with a high styrene content: By increasing the styrene content to over 50%, substances are produced which are thermoplastic, resin-like, easy to vulcanise, and have a hardening effect when used in rubber compounds with loading over 70%. *Uses:* include leather substitutes, coverings for floors, latex paints.

SBR types: In the case of styrene-butadiene, polybutadiene, and polyisoprene rubbers, the International Institute of Synthetic Rubber Producers, Inc., has established a program for the assignment of number for new polymers. The styrene-butadiene and solution rubbers in the various tables include those which carry regular Institute numbers as well as those which do not carry these numbers. For styrene-butadiene rubbers, the Institute numbering system is the same basic numbering system established by the Office of Rubber Reserve, R.F.C. and continued by the American Society for Testing and Materials.

The system is arranged as follows:

1000 series
  Hot non-pigmented rubbers
1500 series
  Cold non-pigmented rubbers

1600 series
　Cold black masterbatch with 14 or less parts of oil per 100 parts SBR
1700 series
　Cold oil masterbatch
1800 series
　Cold oil black masterbatch with more than 14 parts of oil per 100 parts SBR

1900 series
　Emulsion resin rubber masterbatches
2000 series
　Latex hot polymerisation
2100 series
　Latex cold polymerisation

SBR producers have been assigned ranges of code numbers for designating new rubbers or latices.

No.	Producer	Trade Name
3 000–3 499	American Synthetic Rubber Corporation	ASRC
3 500–3 999	Copolymer Rubber & Chemical Corporation	COPO, CARBOMIX
4 000–4 499	Firestone Synthetic Rubber & Latex Co.	FR-S
4 500–4 999	Ameripol, Inc.	AMERIPOL
5 000–5 499	Goodyear Tire and Rubber Company	PLIOFLEX, PLIOLITE
5 500–5 599	ANIC—Italy	EUROPRENE
5 600–5 699	International Synthetic Rubber Co., Ltd	INTOL, INTEX
5 700–5 799	Compagnie Francaise des Produits Chimiques Shell	CARIFLEX S
5 800–5 899	Shell International Chemical Co., Ltd	CARIFLEX S
5 900–5 999	Chem. Werke Hüls A.G.	DURANIT, BUNATEX BUNAHÜLS
6 000–6 499	UNIROYAL, Inc.	
6 500–6 999	Phillips Petroleum Co.	PHILPRENE
7 000–7 499	Polymer Corp. Ltd	POLYSAR, KRYMIX, KRYLENE, KRYNOL, KRYFLEX
7 500–7 999	Shell Chemical Co.	S
8 000–8 499	Texas-US Chemical Co.	SYNPOL
8 500–8 999	Ashland Chemical Co.	
9 000–9 499	General Tire & Rubber Company	GENTRO, GENTRO-JET, JETRON
9 500–9 599	The Japanese Geon Co., Ltd	NIPOL
9 600–9 699	Japan Synthetic Rubber Co., Ltd	JSR

continues

No.	Producer	Trade Name
9 700–9 799	Australian Synthetic Rubber Co., Ltd	AUSTRAPOL
9 800–9 899	Petrobas Quimica S.A.	PETROQUISA, PETROFLEX
9 900–9 999	PASA Petroquimica Argentina S.A.	
10 000–10 499	Hules Mexicanos, S.A.	

For producers with blocks of 500 numbers, each block should be further divided by type according to the following:

Producer's Code Number	Product
0–49	Hot non-pigmented polymers
100–149	Cold non-pigmented polymers
150–199	Cold black masterbatch with 14 or less parts of oil per 100 parts SBR
200–249	Cold oil masterbatch
50–99 and 250–299	Cold oil black masterbatch with more than 14 parts of oil per 100 parts SBR
300–349	Hot latex
350–399	Cold latex
400–499	Unassigned

The most important types are listed on pages 77–81. However, there are still numbers used by individual firms which do not coincide.

SBR Rubber—ASTM Types

SBR Type	Former Name	Temperature, °C	Bound styrene	Conversion, %	Mooney, ML 4/100	Emulsifier	Antioxidant	Coagulant	Carbon Black type	Carbon Black amount	Oil type	Oil amount	Remarks
1000	GR-S	50	23·5	68	44-52	FA	ST	SA					General purpose, staining
1001	GR-S-50	50	23·5	68	44-52	FA	SLST	SA					Less staining than 1 000
1002	GR-S-10	50	23·5	68	50-58	RA	ST	SA					
1004	GR-S-AC	50	23·5	72	46-54	FA	ST	AS					
1006	GR-S-25	50	23·5	72	46-54	FA	NST	SA					Non-staining
1007	GR-S-65	50	23·5	77	45-55	FA	ST	GA					
1009	GR-S-60	52·8	23·5	77	115-135*	FA	NST	SA					Crosslinked with divinylbenzene
1010	GR-S-30AC	50	23·5	72	25-35	FA	NST	AS					
1011	X-645	50	23·5	72	50-58	RA	NST	SA					1 006 slower vulcanisation, more tacky
1012	GR-S-86	50	43	77	95-105	FA	NST	SA					1 006 type with high Mooney
1013	GR-S-40AC	50	43	80	40-50	RA	SLST	SA					Less water absorption
1014	X-491	65·6	3·5	72	55-85	RA	ST	SA AS					Special impregnation
1015	X-489	51·7	23·5	72	55	FA	ST	GA					Crosslinked with divinylbenzene, special impregnation
1016	GR-S-65 SP	50	23·5	77	46-54	FA	NST	GA					Special impregnation
1018	GR-S-61 SP	52·8	23·5	77	115-135*	FA	NST	GA					Crosslinked with divinylbenzene, special impregnation
1019	GR-S-66 SP	50	23·5	77	46-54	FA	NST	GA					Special impregnation
1020	X-549-SP	52·8	23·5	77	80*	FA	NST	GA					Crosslinked with divinylbenzene, special impregnation
1021	X-238-SP	51·7	45	86	85	FA	SLST	SA AS					Special impregnation
1022	X-634-SP	51·7	13	72	70-90	RA	NST	GA					Special impregnation
1023	X-627-SP	50	23·5	72	46-54	FA	ST	GA					Special impregnation
1061		50	23·5	68	44-52	FA	NST	SA					Special impregnation
1100	GR-S Black I / GR-S Black IV			68	50*	FA	ST	SA	EPC	33·3			1 006 with non-staining stabiliser (Polyg.)
1103	X-579	50	23·5	72	48*	FA	ST	SA	MAF	33·3			
1104	X-419	50	23·5	72	48*	FA	SLST	SA	MAF	33·3			
1500	X-647 / X-670	6·1	23·5	60	46-58	RA	ST	SA					General use
1501	X-637	6·1	23·5	60	46-58	RA	SLST	SA					Wing stay 100 as stabiliser, less staining than 1 500
1502	X-625	6·1	23·5	60	46-58	FA RA	NST	SA					Non-staining, non-discolouring
1503	X-565-SP / X-620-SP	6·1	23·5	60	46-58	FA	NST	GA					Special impregnation

Butadiene-Styrene Copolymers

SBR Rubber—ASTM Types

SBR Type	Former Name	Temperature, °C	Bound styrene	Conversion, %	Mooney, ML 4/100	Emulsifier	Antioxidant	Coagulant	Carbon Black type	Carbon Black amount	Oil type	Oil amount	Remarks
1 504	X-601-SP	6·1	12	60	45-59	FA	NST	GA					Special impregnation
1 505	X-697-SP	6·1	9	60	34-46	RA	ST	SA					1 509 with lower Mooney
1 506	X-600	6·1	23·5	60	25	FARA	NST	AS					
1 507		6·1	23·5	60	30-40	FARA	NST	SA					
1 508		6·1	23·5	60	46-58	FA	NST	SA					
1 509		6·1	23·5	60	30-38	FARA	NST	SA					
1 510		6·1	23·5	60	30-35	FA	NST	SA					
1 551		6·1	23·5	60	46-58	RA	NST	SA					
1 570		6·1	23·5	60	125	FARA	ST	SA					Non-discolouring type 1 500
1 600	X-580, X-581	6·1	23·5	60	65-80*	RA	ST	SA	HAF	50			High molecular weight; 1 500 + Carbon Black
1 601	X-582; X-598, X-607; X-733	6·1	23·5	60	62-74*	FARA	ST	SA	HAF	50			1 502 + Carbon Black
1 602		6·1	23·5	60	77-92*	FARA	ST	SA	HAF	50	HIAR	10	1 502 + Carbon Black
1 603		6·1	23·5	60	50-65*	FA	NST	GA	EPC	60			1 503 + Carbon Black
1 604		6·1	23·5	60	65	FA	ST	SA	ISAF	60			1 500 + Carbon Black
1 605		6·1	23·5	60	52-66*	RA	NST	GA	FEF	52			1 503 + Carbon Black
1 606		6·1	23·5	60	56	FA	ST	AR	HAF	60	HIAR	10	1 500 + Carbon Black
1 607		6·1	23·5	60	52-72*	RA	ST	SA	HAF	52	HIAR	10	1 500 + Carbon Black
1 608		6·1	23·5	60	56*	RA	ST	Special	ISAF	40	HIAR	12·5	1 500 + Carbon Black
1 609		6·1	23·5	60	61				SAF	52	HIAR	5	1 500 + Carbon Black
1 610		6·1	23·5	60					ISAF			10	1 500 + Carbon Black
1 611		6·1	23·5	60					HAF	62·5		12	1 500 + Carbon Black
1 703	X-745	6·1	23·5	60	50-70	FARA	NST	SA			NAPH[1]	25	[1] Heavy processing oil; Non-staining, general purpose
1 704	X-743; X-744	6·1	23·5	60	60	RA	NST	SA			AR	25	Staining, general purpose
1 705	X-739; X-740	6·1	23·5	60	50-65	FARA	ST	SA			AR	25	
1 706	X-741; X-742	6·1	23·5	60	60	FARA	ST	SA			HIAR	25	
1 707	X-735	6·1	23·5	60	45-65	RA	NST	SA			NAPH	37·5	
1 708	X-718	6·1	23·5	60	50-70	FA	NST	GA			NAPH	37·5	

Butadiene-Styrene Copolymers

No.	Polymer												Remarks
1 709	X-724	6·1	23·5	60	45-65	RA	ST	SA			AR	37·5	
1 710	X-725	6·1	23·5	60	48-62	FARA	ST	SA			AR	37·5	
1 711	X-736	6·1	23·5	60	55	RA	ST	SA			HIAR	37·5	
1 712	X-737	6·1	23·5	60	45-65	FARA	ST	SA	HIAR		HIAR	37·5	
1 713	X-727, X-729, X-731	6·1	23·5	60	44-60	FARA	NST	SA			NAPH	50	1 712 + 12·5 oil
1 714	X-726, X-728, X-730	6·1	23·5	60	52	FARA	ST	SA			HIAR	50	
1 715	X-709	6·1	23·5	60	53-67	FA	NST	SA			NAPH	50	
1 773	X-721	6·1	23·5	60	48-72	FARA	NST	SA			NAPH	23	
1 800		6·1	23·5	60	55	FARA 1:1	NST	SA	ISAF	82·5	NAPH	37·5	1 703 + light oil
1 801		6·1	23·5	60	65*	FARA 1:1	ST	SA	HAF	50	HIAR	51·3	1 707 (F/H) + light oil
1 802	X-629, X-746, X-686	6·1	23·5	60	49	FARA 1:1	ST	SA	HAF	82·5	HIAR	25	1 712
1 803		6·1	23·5	60	55-70	FARA 1:1	ST	SA	HAF	50	HIAR	56·8	1 703 + Carbon Black
1 804			23·5	60			ST	SA	HAF		HIAR	25	1 712
1 805	Baytown B 151	6·1	23·5	60	45-60	FARA 1:1	NST	SA	HAF	75	NAPH	10	1 706 + Carbon Black; Changed to 1 607
1 806	3 757, 4 753, 8 778, 9 254	6·1	23·5	60	50	RA	NST	SA	FEF	60	NAPH	37·5	1 708 + Carbon Black
1 807	3 751, 4 750, 6 682	6·1	23·5	60	68	RA	ST	SA	ISAF	75	AR	37·5	1 707
1 808	4 751	6·1	23·5	60	48	FARA 1:1		A	HAF	75	HIAR	50	1 709
1 809	3 756, 8 759	6·1	23·5	60	62*	FARA 1:1	ST	Special	HAF	75	HIAR	37·5	1 712
1 810		6·1	23·5	60	57	FA	NST	Special	FEF	100	NAPH	50	1 712 + Carbon Black
1 811	3 753, 9 251	6·1	23·5	60	43	RA	ST	Special	SRF	75	HIAR	17·5	1 708
1 812	3 758	6·1	23·5	60		FARA	ST	SA	ISAF	60	HIAR	37·5	1 500 + Carbon Black + oil
1 813	3 760, 8 779	6·1	23·5	60	63*	FARA 1:1	ST	Special	ISAF	75	HIAR	37·5	1 712 + Carbon Black
1 814	9 250	6·1	23·5	60	60	FARA 1:1	ST	Special	ISAF	75	NAPH	50	1 712 + Carbon Black + 12·5 oil
1 815	9 252		23·5	60	45*	FARA 1:1	NST	Special	HAF	70	HIAR	50	1 778 12·5 naphthene oil
1 816		6·1	23·5	60	55	FARA 1:1	ST		ISAF	55	HIAR	45	1 712 + Carbon Black + 7·5 VO
1 817		6·1	23·5	60	50		ST		SAF	75	AR	45	1 712 + Carbon Black + 7·5 VO
1 819		6·1	23·5	60					HAF	75	AR	37·5	
1 822		6·1	23·5	60					ISAF	82·5	HIAR	37·5	
1 823		6·1	23·5	60					HAF	82·5	HIAR	62·5	
1 824		6·1	23·5	60					ISAF	75	HIAR	62·5	
1 828			23·5	60					SRF		NAPH	17·5	
3 014		w	43	60	60-80	RA	NST	SA					1 014 non-staining, antioxidant

SBR Latex—ASTM Types

SBR Latex	Former No.	Temperature, °C	% Solids	Bound styrene, %	Conversion, %	Mooney, ML 4/100	Emulsifier	Remarks
2 000	III, IV	65·6	40	46	90	75	HIAR	
2 001	X-381	65·6	38	46	90	30	HIAR	
2 002	X-446	60	48	46	95	65	HIAR	
2 003	V	65·6	59	29	98	140	HIAR	
2 004	X-653	65·6	58	0	90		HIAR	
2 005	II, VI	65·6	61	43	95	60	HIAR	
2 006	X-523	50	27	23·5	72	50	FA	
2 076	X-547	10	48	25	60	100	FA	
2 100	X-617	6·1	24	23·5	60	High	FAHIAR	
2 101	X-667	10	60	14	70	High	FA	
2 102	X-710	10	58	25	60	150	FA	
2 103	X-711	10	61	0	60	140	FA	
2 104	X-758	10	48	25	60	140	FARA	
2 105	X-760	10	61	25	60	140	FARA	
2 106	X-767	6·1	40	44	60	140	FARA	
2 107	X-765	10	39·5	23·5	60	High	FA	
2 108	X-800	10	60	40	60	High	FA	
2 109		10	20	25	60	52	FA	
2 110			40	23·5	60	52	RA	
2 111			48	23·5	60		RA	2 101, 40% 1 g
2 112		6·1		44	80	140	FARA	

SBR Rubber—ASTM Types

SBR Latex	Former No.	Temperature, °C	Solids, %	Bound styrene, %	Conversion, %	Mooney, ML 4/100	Emulsifier	Remarks
3 852			53	24		High	SA	
5 300-X		65·5	52	46		73	RA	Type 2 000 concentration
5 301-X		65·5	56	46		73	RA	
5 350			63	25			FARA	
5 352			69	25			FA	
5 353			54	25			FA	
8 174			69	25		135	FARA	

Abbreviations:

AR	Aromatic	P	Process Oil
AS	Aluminium sulphate	RA	Resin Acid
FA	Fatty Acid	SA	Hydrochloric Acid
GA	Glue Acid	SLST	Slightly staining
HIAR	Highly Aromatic	ST	Staining
NAPH	Naphthenic	*	Mooney of compound
NST	Non-staining		

81

BUTADOR Antistatic polychloroprene rubber with good mechanical properties and good flexibility (191).

BUTAKON Group of butadiene copolymers.
Types:
A: butadiene-acrylonitrile copolymers with an acrylonitrile content of 20–40%, s.g. 0·95–1·0. Oil resistant. *Uses:* conveyor belts, oilproof articles, tubes, fuel containers, softeners in p.v.c.
M: butadiene-methyl methacrylate copolymers. *Use:* in dyeing industry (60)
S: butadiene-styrene copolymers with a high styrene content, s.g. 1·04–1·05. Reinforcing agent for NR and SR, may be easily processed. *Uses:* soles, floor coverings, material for suitcases, ebonite.

BUTALENE
Grades:
B: benzoquinone dioxime. B-33, 33% active substances with kaolin
BD: p,p'-dibenzoyl quinone dioxime. BD-33, 33% active substance with kaolin
DN: poly-p-dinitrosobenzene. DN-25. 25% active substance with inert wax
TEL: tellurium diethyl dithiocarbamate (19).

1:3-BUTANEDIOL
$CH_2OH . CH_2 . CHOH . CH_3$. Colourless liquid, b.p. 20°C, s.g. 1·006. Soluble in alcohol and water. Intermediate in butadiene synthesis by the Aldol process.
TN: Butol.

BUTANEDIOL DICAPRYLATE
M.p. 10·5°C, s.g. 0·929, viscosity 19 cp. Softener at low temperatures. Resistant to extraction, compatible with natural and synthetic rubber and with resins, improves extrusion and ageing properties.
TN: BD-8 (110).

1-BUTANOL
$CH_3CH_2CH_2CH_2OH$, n-butyl alcohol, propylcarbinol. Liquid, b.p. 117–118°C, s.g. 0·81. May be mixed with alcohol, ether and organic solvents. Solubility in water 1:15.
Uses: solvent for controlling viscosity in cements.

2-BUTANOL
$CH_3CH(OH)CH_2CH_3$, secondary butyl alcohol, methylethyl-carbinol, 2-hydroxybutane. Colourless liquid, b.p. 100°C, s.g. 0·808. May be mixed with ether and alcohol. Solubility in water 1:12.
Uses: solvent.

BUTANOX Group of methyl ethyl ketone peroxides (171).

BUTAPRENE N Group of butadiene-acrylonitrile copolymers with varying acrylonitrile contents.
Types:
NF: Mooney 30–50. Medium grade oil resistance, low acrylonitrile content, high resistance to cold, good elasticity
NH: Mooney 40–60. High acrylonitrile content, little 'nerve', maximal resistance to solvents, little resistance to cold
NI: latex
NL: Mooney 50–70. Medium level acrylonitrile content, good oil and cold resistance

NNA: Mooney 50–70. Good oil resistance and processability
NXM: Mooney 80–100 (or 50–70). High acrylonitrile content, good oil resistance, poor resistance to cold (129).

BUTAPRENE PL Group of styrene-butadiene latices (129).

BUTAPRENE S Group of butadiene-styrene copolymers with varying styrene contents (129).

BUTAREZ Liquid polybutadiene. Softener which can be vulcanised (130).

BUTON Group of butadiene-styrene resins.
Uses: for surface coatings and as an additive for plastics which are hardened thermally (131).

BUTOXYETHYL LAURATE
Oily liquid with aroma similar to coconut, b.p. (4 mm) 166–220°C, s.g. 0·88, iodine No. 8. Softener for natural and synthetic rubber, particularly effective at low processing temperatures.

BUTOXYETHYL OLEATE
Oily liquid, b.p. (4 mm) 200–230°C, s.g. (25°C) 0·886. Softener.

BUTOXYETHYL PHTHALATE
Colourless, oily liquid, b.p. (4 mm) 212–221°C, solidification p. −55°C, s.g. 1·06, viscosity 30 cp. Softener for natural and synthetic rubber and vinyl resins.

BUTRAX Accelerator of undisclosed composition (obsolete) (115).

BUTVAR Group of polyvinylbutyral resins. Mol. wt. 32 000–225 000, viscosity 75–157 cp (10% solution in alcohol, 25°C). Type BR, 50% anionic dispersion with 40% softener. Types B-72A, B-73, B-76, B-90, B-98, as powder.
Uses: moulded articles, coatings, films, insulation, safety glass (133).

BUTYL ACCELERATOR 21
Compound of 67% tetramethyl thiuram disulphide and 33% 2-mercaptobenzthiazole. Yellow brown powder, s.g. 1·3. Compound accelerator for butyl rubber. Processes safely and vulcanises rapidly at normal vulcanisation temperatures. Does not scorch, needs 3% zinc oxide, 0·5–2% sulphur and 1–2% stearic acid. A low sulphur concentration gives a low modulus. Activation with 0·2% poly-p-dinitrosobenzene, this acts simultaneously as a stiffening agent.
Quantity: 2% in compounds with large addition of certain fillers or softeners, for good heat resistance 2·5–3% (6).

n-BUTYL ACETATE
$CH_3COO.CH_2CH_2CH_2CH_3$. Colourless liquid, m.p. −77°C, b.p. 125°C, s.g. (20°C) 0·8826. Can be mixed with alcohol.
Uses: solvent for polar rubbers and plastics, nitrocellulose, and for controlling the viscosity of cements.

BUTYL ACETOXY STEARATE
$C_{17}H_{34}(OCOCH_3)COOC_4H_9$, m.p. −7°C, flash p. 207°C, s.g. 0·992. Softener for plastics and elastomers.
TN: Paricin 6 (9).

n-BUTYL ACETYL RICINOLEATE
$C_{17}H_{34}(OCOCH_3)COOC_4H_9$. Clear yellow liquid with mild aroma, m.p. −32 to −55°C, b.p. (1 mm)

195°C. Combines with most organic solvents. Softener for natural and synthetic rubber and latices, improves cold resistance and stretching ability. Emulsifier.

TN: Flexricin P-6 (9)
PG 16
Staflex BRA (330).

N-BUTYL-p-AMINOPHENOL

$$HO-\!\!\!\langle\ \rangle\!\!\!-NH-C_4H_9$$

Brownish liquid, s.g. 0·91. General purpose antioxidant.
TN: Amoco 531 (307)
Tenamene 1 (193).

BUTYL BENZTHIAZYL SULPHENAMIDE

$$C-S-NH-C(CH_3)_3$$

N-tert-butyl-2-benzthiazyl sulphenamide. Pale yellow powder, b.p. above 105°C, s.g. 1·29. Soluble in alcohol, acetone, benzene, ether, carbon tetrachloride. Very effective accelerator with a delayed action, used particularly for compounds with SAF blacks having a high pH, activated by substances with an acid reaction.
TN: Delac NS (246)
Santocure NS (5)
TBBS (21).

BUTYL BENZYL PHTHALATE
$C_6H_4(COOC_4H_9)$
$$(COOCH_2C_6H_5),$$
BBP. Clear, oily liquid, b.p. 37°C, s.g. 1·11–1·12. Softener for plastics and elastomers.
TN: Morflex 145 (264)
Santiciser 160 (5).

BUTYL BENZYL SEBACATE
$(CH_2)_8(COOC_4H_9)$
$$(COOCH_2C_6H_5).$$
Pale yellow liquid, b.p. (10 mm) 245–285°C, s.g. 1·023. Softener for plastics and elastomers.
TN: Morflex 245 (264).

BUTYL 'CARBITOL' PELARGONATE
Clear, pale yellow liquid, s.g. 0·93. Softener for use at low temperatures, suitable for SBR, NBR, CR and IIR.
TN: Plasticiser 3497-A.

p-TERT-BUTYL CATECHOL
$(CH_3)_3C.C_6H_3(OH)_2$, 4-tert-butyl-1:2-dihydroxybenzene. Colourless crystals, m.p. 56–57°C, b.p. 285°C, s.g. 1·049. Soluble in alcohol, acetone, and ether. Strong reducing agent.
Uses: polymerisation inhibitor for butadiene, styrene, chloroprene (quantity: 0·005–0·15%); removed from butadiene by washing in alkali before polymerisation. Antioxidant in NR and SR, with dinitrobenzene in the vulcanisation of neoprene foam.

BUTYL CELLOSOLVE
Trade name for butyl glycol (123).

BUTYL 'CELLOSOLVE' PELARGONATE
Clear, light brown liquid, s.g. 0·910. Softener for use at low temperatures, suitable for SBR, NBR, CR and IIR.
TN: Plasticiser 3425-A (132).

BUTYL 'CELLOSOLVE' STEARATE
Colourless, oily liquid, m.p. 12–15°C, b.p. (1 mm) 155–200°V, s.g. 0·88, Iodine No. 2. Softener for natural, synthetic, chlorinated rubbers, vinyl and cellulose plastics, improves processability and has no effect on vulcanisation (132).

TERT-BUTYL-m-CRESOL

Liquid, s.g. 0·96. Protects against flex-cracking.
TN: MRMC (176).

BUTYL CYCLOHEXYL PHTHALATE

$C_6H_4(COOC_6H_{11})(COOC_4H_9)$. Clear liquid, b.p. 190–220°C, s.g. 1·078. Softener for NR, SR and vinyl polymers.
TN: Elastex 50-B (107).

BUTYL DECYL PHTHALATE

$C_4H_9OCOC_6H_4COOC_{10}H_{21}$, m.p. −45°C, b.p. (5 mm) 220°C, flash p. 193°C. Softener for plastics and elastomers.
TN: Elastex: 40-P (107)
Morflex 135 (264)
PX-114 (305)
Santiciser 603 (5)
Staflex BDP (330).

BUTYL EIGHT Activated dithiocarbamate accelerator used for vulcanisation at low temperatures or at room temperature. Reddish brown liquid, s.g. 1·01. Soluble in acetone, alcohol, benzene, carbon disulphide, chloroform, partially soluble in water and benzine.
Uses: self-vulcanising adhesives, impregnation and coating of textiles (51).
Quantity (in %):

	Butyl Eight	Thiazole	Sulphur
NR	3–4	0–1·5	1·5–0·75
SBR	4–6	0–2	3–2
Butyl	1–2	0·5–2	2–1

BUTYLENE OXIDES S Mixture of 1:2-butylene oxide, cis-2:3-butylene oxide and trans-2:3-butylene oxide. Suitable for polymerisation and polycondensation (61).

BUTYL EPOXY STEARATE

Colourless liquid, s.g. (20°C) 0·910. Softener for vinyl polymers, improves low temperature flexibility.

TERT-BUTYL HYDROPEROXIDE

$(CH_3)_3.COOH$. Liquid, m.p. −8°C, decomposes at 75°C, s.g. 0·986. Soluble in alcohols, esters, ketones, aliphatic, aromatic and chlorinated hydrocarbons.
Uses: polymerisation catalyst, activator in coagulating baths used for spinning rubber solutions.

3-TERT-BUTYL-4-HYDROXY-ANISOLE

2-tert-butyl-4-methoxyphenol.
White pellets, m.p. 48°C. General purpose antioxidant.
TN: Sustane BHA (352a)
Tenox BHA (193)
BHA.

4:4′-BUTYLIDENE-BIS(3-METHYL-6-TERT-BUTYL PHENOL)

$C_3H_7.CH[—(4)—C_6H_2(OH)$
$.(CH_3).\{C(CH_3)_3\}—1,3,6]_2$

White powder, m.p. 209°C, s.g. 1·04. Non-staining antioxidant for natural and synthetic rubber and latices, good protective against oxidation, on exposure to sunlight causes minor yellow discoloration in pure white articles.
TN: Santowhite powder (5).

BUTYL ISODECYL PHTHAL-ATE Colourless liquid, b.p. (5 mm) 190–255°C, s.g. 0·997. Secondary softener for NR, SR and vinyl polymers.
TN: Elastex 40-P (107).

BUTYL KAMATE 50% solution potassium dibutyl dithiocarbamate. Straw-coloured liquid, s.g. 1·08–1·12 (51).

BUTYL LAURATE
$C_{11}H_{23}COOC_4H_9$. Liquid, b.p. (5 mm) 130–180°C, s.g. (25°C) 0·885. Insoluble in water. Softener.

n-BUTYL LITHIUM C_4H_9Li. Liquid at room temperatures, s.g. 0·68–0·70. Soluble in most organic solvents. The trade product is usually dissolved in C_5–C_7 hydrocarbons.
Uses: catalyst for the stereospecific polymerisation of isoprene and butadiene.

n-BUTYL MYRISTATE
$C_{13}H_{27}COOC_4H_9$. Oily liquid, b.p. 180°, m.p. 3°C, s.g. 0·861. Soluble in methanol, acetone, chloroform, toluene, mineral oils, insoluble in water. Softener.

BUTYL NAMATE 47% solution of sodium dibutyl dithiocarbamate in water, s.g. 1·09. Ultra-accelerator for natural rubber latex and synthetic latices (51).

BUTYL OCTYL PHTHALATE Colourless liquid, s.g. 0·991–0·997. Softener for natural and synthetic rubber, vinyl polymers and plastisols.
TN: Elastex 48-P (107).

BUTYL OLEATE
$C_{17}H_{33}COOC_4H_9$. Oily liquid with characteristic odour, m.p. −26°C,

b.p. (6 mm) 190–230°C, s.g. 0·87. Soluble in alcohol, ether and oils, insoluble in water.
Uses: solvent, softener for neoprene, Types WRT and GRT, retards vulcanisation unless a fast vulcanising system is present—addition of 0·5% 2-mercaptoimidazoline overcomes retardation. Low temperature plasticiser and processing aid for NR, SR and, particularly, CR.

TERT-BUTYL PERBENZOATE
$C_6H_5CO.OOC(CH_3)_3$. Liquid, s.g. (25°C) 1·04. Soluble in alcohol, esters, ethers and ketones.
Uses: polymerisation catalyst for acrylates, styrene, vinyl-acrylate copolymers, crosslinking agent for polyesters, silicone polymers, and polyethylene.

TERT-BUTYL PERPHTHALIC ACID
$(CH_3)_3COO.CO.C_6H_4COOH$. White crystalline powder, m.p. 95–99°C. Soluble in alcohol, partially soluble in chlorinated hydrocarbons.
Use: polymerisation catalyst.

TERT-BUTYL PHENOL
$(CH_3)_3C.C_6H_4OH$.
Grades:
ortho, yellowish liquid, b.p. 224°C, m.p. −65°C, s.g. (20°C) 0·982. Soluble in alcohol, toluol,
para: white crystalline powder, b.p. 230°C, m.p. 98°C, s.g. 1·03.
Use: softener for plastics.

BUTYL PHTHALATE BUTYL GLYCOLATE
$C_4H_9OCOC_6H_4COOCH_2$
　　　　　　　　　　　$COOC_4H_9$.
Colourless liquid with characteristic odour, decomposes above 290°C, b.p.

(5 mm) 219°C, s.g. 1·097. Softener for NBR and SBR, improves processing without affecting vulcanisation.
TN: Santiciser B-16 (5).

BUTYL RICINOLEATE

$C_{17}H_{32}(OH)COOC_4H_9$. Colourless to yellowish oily liquid, m.p. $-30°C$, b.p. (13 mm) 275°C, flash p. 207°C, s.g. 0·917. Soluble in alcohol and ether. Softener.
TN: Flexricin P-3 (9)
Staflex BR (330).

BUTYL RUBBER Modified polyisobutylene, a copolymer of isobutylene with 1·5–4·5% isoprene. Polyisobutylene contains no unsaturated structure and cannot therefore be vulcanised. In 1937, however, W. J. Sparks and R. M. Thomas of the Standard Oil Co. discovered that isoprene and other dienes may be built into the polymer chain to give a copolymer which may be vulcanised with sulphur. Since 1943, butyl rubber has been produced in large quantities and now forms 6% of world rubber production. Butyl rubber is formed by cationic solution polymerisation which is propagated through ion chains. The isoprene content is 1–3%. By increasing the diene constituents, the product is more elastic and easier to vulcanise, but acquires a tendency for the chain structure to break, reducing mol. wt., and rendering the polymer more vulnerable to oxidation. Mol. wt. increases as the polymerisation temperature is decreased. Polymerisation takes place in methylene chloride as solvent at temperatures as low as -50 to $-100°C$. Friedel-Crafts catalysts, such as aluminium chloride or boron trifluoride, are added in methylene chloride as 5% solution in small quantities and act as initiators. An exothermic reaction begins immediately after adding the catalyst and is quickly completed. Liquid ethylene is used as a cooling agent and reduces the reaction heat by evaporation. The reaction product is washed in hot water and mixed with 0·15% stabiliser and 2% zinc stearate to prevent tackiness. The polymer is precipitated in flakes, dried, homogenised, rolled into long strips on a mill and cut into sheets.

Butyl rubber has a molecular structure which is distinct from other synthetic rubbers because of the high degree of linearity in the molecules and the high degree of extension crystallisation which occurs. It also has a much lower level of unsaturation than other elastomers (3% of natural rubber) and therefore vulcanises very slowly, making fast accelerator systems necessary. The best vulcanisation conditions for sulphur compounds may be achieved by a combination of tetramethyl thiuram disulphide with mercaptobenzthiazole (2:1) or with selenium, tellurium or bismuth dialkyl dithiocarbamate. In order to achieve optimum physical properties, addition of up to 5% zinc oxide or some other metallic oxide is necessary. Higher amounts (up to 25%) give considerable improvement in heat resistance, but have no effect on the modulus. For types with a lower degree of unsaturation, quinone dioxime with the addition of red lead, lead dioxide or dibenzthiazyl disulphide is preferred as the vulcanising agent; this also yields products with improved heat resistance. Compounds with 2% p-quinone dioxime and 4% dibenzthiazyl disulphide vulcanise rapidly and are further accelerated by zinc oxide (for continuous vulcanisation), but these have a tendency to scorch. A combination of 6% dibenzoyl quinone dioxime with 10% red lead yields compounds which vulcanise quickly, have delayed vulcanisation start and sufficient processing safety. Active fillers do not have considerable reinforcing effect on butyl rubber but increase modulus, tensile strength, abrasion resistance and tear strength. 30–60% SRF black, or a mixture of SRF and HMF blacks, ensure a good balance or physical properties. Channel blacks and alumina have a retarding effect on vulcanisation.

Butyl rubber is fairly soft and requires only little mastication. Pentachlorothiophenol and its zinc salt are suitable peptising agents. The addition of softeners (dioctyl phthalate, dibutyl sebacate, diisobutyl adipate) improves the rubbery characteristics at low temperatures. Unsaturated softeners retard vulcanisation to a considerable extent. Butyl rubber does not blend well with natural rubber, SBR and nitrile rubber, as the sulphur is used by the other polymers before the butyl rubber begins to vulcanise. Good vulcanisation, however, can be achieved, by using 5% lead oxide, 3% sulphur, and 1% tetramethyl thiuram disulphide. Butyl rubber blends well with neoprene, Hypalon, polyethylene and styrene copolymers with a high styrene content. Chemical resistance and low gas permeability (10% of that of natural rubber) are the most desirable attributes of butyl rubber. Butyl rubber has good stability in regard to ageing, tensile fatigue and heat, is resistant to oxygen, ozone, the effects of weather, and to salts of copper, manganese, and cobalt; is unaffected by acetone,

alcohols, ethyl glycols, animal and vegetable oils, concentrated sulphuric, hydrochloric, nitric acids, and alkalis, but is not resistant to paraffinic hydrocarbons. It has a low water absorption and good electrical properties. Vulcanisation speed, adhesion to metals and other properties may be improved by bromination. Brominated butyl rubber contains 1–3·5% bromine. Chlorinated butyl rubber also has improved vulcanisation properties and may be more easily blended with other elastomers. Butyl latex has 55% solid constituents, s.g. 0·96, pH 5–6.

Butyl rubber currently has four grades of unsaturation and 3 different mol. wt., both staining and nonstaining types are produced.

Uses: inner-tubes, inner-linings of tubeless tyres, hose for steam and chemicals, conveyor belts for use with hot articles, protective coverings against acids, cable insulations and sheaths, seals, sealing materials (mastics), rubberised textiles.

TN: Bucar

Chlorbutyl HT (1066)

Enjay Butyl

Enjay Butyl Latex 8021

GR-I (former US Government term)

Hycar 2202 (Hycar HH, brominated butyl rubber)

MD 551 (butyl chlorinated rubber)

Polysar Butyl

Socabutyl

Lit.: *The Vanderbilt Rubber Handbook*, New York, 1958, p. 88.

S. Bostrom, *Kautschuk-Handbook*, Berlin, 1959, BD. 1, p. 395.

G. S. Whitby, *Synthetic Rubber*, New York, 1954, p. 838.

BUTYL SETSIT Dithiocarbamate of undisclosed constitution (51).

n-BUTYL STEARATE

$C_{17}H_{35}COOC_4H_9$. Odourless, colourless liquid, b.p. (25 mm) 220–225°C, m.p. 16°C, flash p. 188°C, s.g. 0·85–0·86. Softener for natural and synthetic rubber, and latices, with the exception of butyl rubber and rubber hydrochloride.

Uses: mould release agent.

TN: RC B-17 (110)
Witciser 200–201 (215).

BUTYL TEN Mixture of zinc dibutyl dithiocarbamate with an organic amine and an inert mineral filler (12).

N-TERT-BUTYL THIAZOLINE-2-SULPHENAMIDE

A recently available ultra-accelerator with a retarding filler.
USP 2 779 809.

N-TERT-BUTYL UREA

$H_2N.CO.NH.C(CH_3)_3$. White powder, b.p. 96°C, decomposes on heating. Soluble in water, alcohol, ether. Accelerator.

TN: Cardamine 784 (147).

BUTYRALDEHYDE

$CH_3.CH_2CH_2.CHO$, butyl aldehyde. Clear liquid, m.p. −99°C, 74·7°C, s.g. 0·805. Base for vulcanisation accelerators and plastics.

BUTYRALDEHYDE-AMMONIA

$CH_3(CH_2)_2—CH=NH$. Colourless liquid with ammoniacal odour. Accelerator, mostly used in combination with diphenyl guanidine, needs zinc oxide.

Quantity: 1·5% with 5–6% sulphur.
Uses: sponge rubber.
TN: Vulcamel (363).

BUTYRALDEHYDE-ANILINE CONDENSATION PRODUCTS

$(C_6H_5N=CH.C_3H_7)_x$. Red-brown liquids, s.g. 0·95–1·04. Soluble in acetone, benzene, ethyl chloride, partially soluble in benzine, insoluble in water. Active aldehyde-amine accelerators and activators for acidic accelerators, antioxidants against heat and normal ageing. Good storage properties in closed containers, often protected against oxidation by means of a stabiliser.

Accelerator. Medium-fast with a medium to long vulcanisation range, active at 120°C and above, activity increases with temperature rise. Activator for thiazoles, thiurams, good accelerator with guanidines; zinc oxide and fatty acid are preferable, but not always essential. Gives a high modulus and tensile strength in the presence of zinc oxide, otherwise has a low modulus and high elongation at break. Compatible with channel black, reclaim and acidic softeners, alumina has a retarding effect. Safe processing but causes a dark discoloration during vulcanisation and irradiation, and is unsuitable for white or light-coloured articles. Has a slight softening effect in natural rubber, good ageing properties.

Quantity: natural rubber and SBR, primary 0·5–2% with 2·5–4% sulphur, secondary 0·25–0·5% with tetramethyl thiuram disulphide or dipentamethylene thiuram tetrasulphide.
Neoprene, Type W. 0·25–0·75%, free from scorch.
Type IG, 1–4% with 5–20% litharge for fast vulcanising cements.

Uses: tyre treads, carcases, driving belts, hard rubber, inner tubes, footwear, mechanical goods.

Antioxidants. A condensation product with s.g. 1·04 is recommended against heat and normal ageing in light-coloured articles. Has no effect on plasticity, acts as a strong activator on thiurams, thiazoles, and thiazolines. Causes only slight discoloration in sunlight.
Quantity: 0·5–2%.
TN: Accelerator:
A-32
Accelerator 21 (22)
Accelerator 808 (6)
Accélérateur VS (19)
Accélérateur Rapide 300A (20)
Beutene (= Butene) (23)
Eveite 101 (59)
Nocceler 8 (274)
Soxinol 808 (328)
Vulcaid III (3)
Vulcafor BA (60)
+ Vulkacit CT (43)
Antioxidants:
Antox (6).
+ Vulkacit CT (43).
Lit.: USP 1 417 970, 1 780 326, 1 780 334, 1 908 093, 1 939 192.

light, unsuitable for light-coloured articles. Used for solid rubber and latices and self-vulcanising neoprene cements.
Quantity: natural rubber 0·25–0·75% with 1–3% sulphur and activator, SBR 1–3% with 1–3% sulphur, reclaim 0·5–1·25% with 2–4% sulphur, butyl rubber 1–2% with 0·5–2% sulphur and 0·25–0·5% tetramethyl thiuram disulphide or dipentamethylene thiuram tetrasulphide, neoprene Type W 0·5–2%, Type CG 2–8% without sulphur, in Type KNR the product acts as an ultra-accelerator together with litharge for self-vulcanising cements.
TN: Accelerator 833 (6).
Lit.: USP 1 417 970, 1 780 326, 1 780 334.

BUTYRALDEHYDE-BUTYL-IDENE ANILINE Properties as for butyraldehyde-aniline. Accelerator (obsolete).
TN: + A32.

BUTYRALDEHYDE-BUTYL-AMINE CONDENSATION PRODUCT Red-brown, translucent liquid with aromatic smell, s.g. 0·86. Fairly strong basic accelerator, used either alone or with acidic accelerators. Active at 115°C and above. Compounds show practically no tendency to scorch, vulcanisates show good physical and ageing properties, butyl rubber compounds without activators have a relatively low tensile strength. In NBR, butyraldehyde-butylamine activates mercaptobenthiazole and dibenzthiazyl disulphide. Alumina and mineral fillers have a retarding effect. Causes dark discoloration during vulcanisation and products discolour in sun-

BUTYRILIDENE- p-AMINO DIMETHYL ANILINE
$C_3H_7CH:N.C_6H_4N(CH_3)_2$. Brown liquid, s.g. 0·95–1·02. Accelerator and antioxidant (obsolete).
Uses: ebonite.
Quantity: 0·75% with 5% zinc oxide and 2·25–3% sulphur, secondary with hexamethylene tetramine.
TN: + BB (23)
+ XLM.

BUTYROACETALDEHYDE-ANILINE Accelerator (obsolete).
Quantity: 1% with 2·5–3·5% sulphur and 5% zinc oxide.
TN: + A16
+ Waxens.

N-BUTYROYL-p-AMINO-PHENOL

$$HO-\langle\!\!\!\!\!\!\!\bigcirc\!\!\!\!\!\!\!\rangle-NH-\overset{\overset{O}{\|}}{C}-C_3H_7$$

White powder, m.p. 135–139°C. General use antioxidant.
TN: Suconox 4 (333).

BVE Abbreviation for butyl vinyl ether.

BVS Barium ricinoleate. Stabiliser for vinyl products (9).

BWH Abbreviation for Buna-werke Hüls.

BWH-1 Dark brown oil, s.g. 1·028. Soluble in acetone, benzene, ethylene dichloride, relatively soluble in benzine, insoluble in water. Plasticiser for natural rubber and SBR, improves processing behaviour and extrusion, particularly with reclaim. Reclaiming oil used for plasticising scrap, and for reclaiming vulcanised scrap of natural and synthetic rubber, including neoprene. Has a retarding effect on vulcanisation and causes slight staining (23).

BX-25 Antigel agent (113).

BXA Diaryl amine-ketone-aldehyde reaction product. Brown powder, m.p. 85–95°C, s.g. 1·10. Soluble in acetone, benzene, ethylene dichloride. Antioxidant, can be used for latex (23).

BXDC Butoxyethyl diglycol carbonate. Plasticiser (134).

BYERLITE Petroleum hydrocarbon, mineral rubber. Solid, black product, m.p. 325°C, s.g. 1·02. Filler

and softener for NR, SR and latices, improves dispersion of pigments and fillers, increases hardness and stiffness of compounds. Has no effect on vulcanisation (135).

BZP-98 Benzoyl peroxide (354).

C

C-2.3 RUBBER Copolymers of ethylene and propylene.

C 5 Modified cumarone-indene resin. Softener and dispersing agent (136).

C-6 (-28, -33, -42) Group of reclaiming oils (126).

C-10 75% solution of a styrene copolymer in mineral spirit.
Uses: compounds for coating (137).

C 27 Furnace black with properties similar to FEF (138).

C-255 Softening oil for butadiene-styrene copolymers (100).

CA Abbreviation for crotonaldehyde-aniline condensate.

CAA PLANT Extraction plant used to separate butadiene from the crude butadiene of the Houdry plant. Process depends on a copper ammonium acetate solution in which butadiene dissolves much more easily than butane and 1-butylene.

CAB Abbreviation for cellulose acetate butyrate.

CABFLEX Group of softeners for plastics and elastomers.
Grades:

DDA:	dodecyl adipate
DDP:	dodecyl phthalate
DOA:	di-2-ethyl hexyl adipate
DOP:	di-2-ethyl hexyl phthalate
Di-BA:	diiso-butyl adipate
Di-OA:	diiso-octyl adipate
Di-OP:	diiso-octyl phthalate
Di-OZ:	diiso-octyl azelate
ODA:	iso-octyl decyl adipate
ODP:	iso-octyl decyl phthalate
TTP:	tritolyl phosphate (139).

CABLE INSULATION Insulation of electric wires with gutta-percha was first carried out in 1847 by W. V. Siemens. In 1860 the insulation of copper wires was attempted with vulcanised rubber.

CABOL 100 Aromatic hydrocarbon condensation product, m.p. 5°C, b.p. 300–370°C, s.g. 1·06–1·09, mol. wt. 230–2500. Softener (139).
Identical to: Sovaloid C (327)
Nevillac 10 (269).

CAB-O-SIL Colloidal silica. Reinforcing white filler (139).

CACALOXOCHITL Natural rubber produced from the crushed branches of *Plumiera rubra L* (Mexico). Resin rich product of poor quality.

CADMATE Trade name for cadmium diethyl dithiocarbamate (51).

CADMIUM DIETHYL DITHIO-CARBAMATE Ultra-accelerator for butyl rubber.
Quantity: 1% for press cured articles,

1·5–3% for open steam curing. Safe to process, good ageing properties.
TN: Cadmate (51).

CADMIUM PENTAMETHY-LENE DITHIOCARBAMATE $[(CH_2)_5N.CS.S.]_2Cd$ Yellowish-white, odourless powder, m.p. 240–245°C, s.g. 1·82. Insoluble in organic solvents and water. Accelerator with slight delayed action, non-staining. Does not impart taste or odour to vulcanisates.
Uses: fast curing articles for use in contact with food, coatings, impregnating, light-coloured articles, suitable for press, hot air and steam vulcanisation.
TN: Kuracap Cadmium PD (241) Robac CPD (311).

CADMIUM RED Cadmium selenide lithopone. Group of light red to dark red pigments. Produced by precipitating a solution of cadmium sulphate with barium sulphide in the presence of selenium, when a compound precipitate of cadmium sulphide, cadmium selenide and barium sulphate is obtained, this is then heated. The pigments are resistant to light, heat, alkalis and acids.

CADMIUM RICINOLEATE White powder, m.p. 105°C, s.g. 1·11.
Uses: light and heat stabiliser for p.v.c. and copolymers.

CADMIUM SELENIDE CdSe. Red powder, m.p. 1350°C, s.g. 5·8. Pigment, resistant to light, heat, alkalis and acids, increases the abrasion resistance to rubber.

CADMIUM SULPHIDE CdS, cadmium yellow, jaune brilliant,

capsebon. Pale yellow to orange-coloured powder. Inorganic pigment. Occurs naturally as greenockite, a yellow pigment, s.g. 4·8.

CADMOFIXE Group of cadmium pigments, compound precipitates of cadmium sulphides, selenides and sulphoselenides with barium sulphate. Yellow to Bordeaux red, colour fast, non-blooming, resistant to light, softeners, alkalis and during vulcanisation (43).

CADMOPUR Group of cadmium pigments (sulphide, selenide, sulphoselenide). Yellow to Bordeaux red pigments, colour fast, non-blooming, resistant to light, softeners, alkalis and vulcanisation (43).

CADOX Group of organic peroxide catalysts:
B 160: 55% benzoyl peroxide with butyl benzyl phthalate. Paste
BC: 50% benzoyl peroxide with camphor
BCP: 35% benzoyl peroxide with calcium phosphate. Powder
BDP: 50% benzoyl peroxide with dibutyl phthalate. Paste
BPO: benzoyl peroxide (140)
BSA: 95% benzoyl peroxide with stearic acid
BSD: 50% benzoyl peroxide with silicone oil. Thick paste
BSG: 50% benzoyl peroxide with silicone oil. Thick paste
BTP: 50% benzoyl peroxide with tricresyl phosphate. Paste
MDP: 60% methyl ethyl ketone peroxide with dimethyl phthalate
PS: p,p′-dichlorobenzoyl peroxide
TBH: 70% tert-butyl hydroperoxide with ditert-butyl peroxide. Clear liquid

TDP: 50% 2:4-dichlorobenzoyl peroxide with dibutyl phthalate. Paste
TS-40: 40% 2:4-dichlorobenzoyl peroxide with silicone oil
TS-50: 50% 2:4-dichlorobenzoyl peroxide with silicone oil (140).

CALATON CA Nylon type. Soluble in a mixture of alcohol and water.
Uses: Textile finisher (60).

CALCENE T Technical calcium carbonate with 2·1% tall oil, s.g. 2·70. Filler and pigment for natural and synthetic rubbers and latices. Causes a slight activation of some accelerators (134).

CALCITONE Mixed solvent of acetone and methyl acetate.

CALCIUM CARBONATE $CaCO_3$, carbonic acid lime. White powder. Insoluble in water. Filler.

CALCIUM HYDROXIDE $Ca(OH)_2$, caustic lime, slaked lime. Filler and activator.

CALCIUM STEARATE $Ca(C_{18}H_{35}O_2)_2$. Fat-like white powder, m.p. 150–155°C, s.g. 1·04. Insoluble in water. Release agent, mould release agent, stabiliser for vinyl polymers.

CALCO RETARDER PD Modified phthalic acid anhydride. Vulcanisation retarder (obsolete) (21).

CALENDER Machine used to produce smooth, uniform sheeting and to coat and friction textiles. Consists of 2–5 chilled cast-iron or steel rolls, 500–2000 mm in length.

The nip between the rolls may be adjusted; the rolls are hollow and may be heated or cooled. To overcome deflections caused by pressure the rolls are cambered, the amount of camber depending on the type of process. The order in which the rolls are placed distinguishes them as I, F, L, and Z calenders.

Calender types
(Farrel, Birmingham)

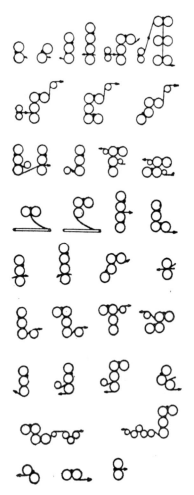

CALENDER GRAIN If rubber or a compound is calendered to form a sheet, orientated internal stresses occur. These give the sheet a high strength and low extensibility in the direction in which it is sheeted, and at right angles to this direction a low strength and high extensibility. The tensions may be removed by heating.

CALGON $(NaPO_3)_6$, sodium hexametaphosphate. Detergent used for removing iron, copper and manganese from textiles before impregnating or coating with rubber (141).

CALIFLUX Group of unsaturated naphthenic hydrocarbons. Extenders and plasticisers (142).

CALOFORT S Calcium carbonate. Surface treated and activated with 3% stearate. Reinforcing filler (143).

CALOPAKE F Precipitated calcium carbonate. White filler (143).

CALOXOL Dispersion of calcium oxide in mineral oil. Prevents development of porosity during continuous vulcanisation of extruded articles in a salt or glass bead bath (409).

CALSIL Calcium silicate. White filler (144).

CALSIL AX Calcium silicate. Reinforcing white filler (145).

CAMP Abbreviation for calcium dimethyl dithiocarbamate.

CAMELBACK Unvulcanised extruded blank for tyre treads. The term is derived from the cross-sectional appearance, which is similar in shape to a camel's back. As with all extrusions, the compounds must have excellent dimensional stability and consistent vulcanisation properties. A mixture of tetramethyl thiuram monosulphide and benzthiazyl disulphide has proven to be an effective accelerator system.

CAMEROON RUBBER Trade name for a wild rubber formerly grown in Cameroon, and obtained primarily from *Kickxia elastica, Ficus Preussii, Carpodinus landolphioides, Carpodinus maxima, Carpodinus uniflora, Clitandra simoni* and *Landolphia comorensis.*

CAMETA Local term for a type of para rubber (natural rubber); between entrefine and Sernamby in quality.

CAOUTCHINO High boiling fraction from the dry distillation of rubber. Discovered by F. K. Himley.

CAOUTCHOUC WHALEBONE Former term for hard rubber.

CAPROLANE Group of commercial nylon fibres, *e.g.* for hoselines, belts (29).

CAPTAN Key name of the US Department of Agriculture for N-trichloromethyl mercapto-4-cyclohexane-1:2-dicarboximide.

CAPTAX TUADS BLEND Mixture of 33·3% 2-mercaptobenzthiazole and 66·7% tetramethyl thiuram disulphide.
Use: accelerator for butyl rubber (51).

CAPUCINE TREE *Northea seychellana.* A *Sapotaceae* (Seychelles) which gives an inferior quality gutta-percha.

CARBAMATE Group of accelerators.
Grades:
P: piperidine pentamethylene dithiocarbamate
PB: lead pentamethylene dithiocarbamate
PC: cyclohexylamine pentamethylene dithiocarbamate
PZ: zinc pentamethylene dithiocarbamate (393).

5-CARBETHOXY RHODANINE Vulcanisation accelerator for SBR. USP 2 721 869 (1955).

CARBOMET S Rumanian produced MPC black.

CARBON BLACKS Consist primarily of spherical particles of 5–500 μ diameter; apart from carbon, contain small quantities of oxygen, hydrogen, nitrogen and sulphur. To *c.* 1910 carbon blacks were used only as pigments although their reinforcing effect was discovered in 1906. In 1912 American manufacturers began to reinforce tyre compounds with channel blacks. The first scientific research was undertaken by Wiegand in 1920. Blacks were classified using a scheme produced in 1943 by the War Production Board of the USA. This indicated the main production processes and the technical properties by letters of the alphabet. According to the method of production, blacks are classified as follows: C, channel black; T, thermal black; F, furnace black. The various types of black differ basically in structure, processing properties and in reinforcement and the physical and chemical properties which they impart to rubber.

Channel blacks. Produced by the partial burning of gaseous hydrocarbons (usually natural gas). Small flames, with a limited supply of air, are allowed to burn against relatively cold metal surfaces. The black deposited is scraped, or sucked, from slowly moving metal rails (channels). The blacks are very active and give vulcanisates with very good strengths. For classification purposes, the following processing properties are indicated:
HPC, hard processing channel
MPC, medium processing channel
EPC, easy processing channel
CC, conductive channel.

Thermal blacks. Produced by thermal decomposition of hydrocarbons in the absence of air. Gases such as methane or carbon monoxide and coal-tar oils are used as raw materials. The particle size is classified as follows:
MT, medium thermal
FT, fine thermal.
Thermal blacks are inactive and improve reinforcement properties only slightly; are primarily used as diluents, and have limited use in synthetic rubbers.

Furnace blacks. Produced by the incomplete combustion of liquid or gaseous hydrocarbons at 1200–1600°C in special furnaces. The black is formed from a large flame and is removed from the burned gases by electrostatic deposition at 60 000–80 000 V by centrifugal force. Depending on the production conditions, various semi-active to highly active types can be made. Classification of these blacks indicates the rubber physical properties, the particle size and the processing:
SRF, semi-reinforcing furnace
HMF, high modulus furnace
GPF, general purpose furnace
FEF, fast extrusion furnace
HAF, high abrasion furnace
ISAF, intermediate super abrasion furnace
SAF, super abrasion furnace
CF, conductive furnace
FF, fine furnace
SCF, semi-conductive furnace.
The following supplementary terms are also used:
CR, channel replacement
ALS, activated low structure
VLS, very low structure
LS, low structure
VLM, very low modulus
LLM, low low modulus
LM, low modulus
HS, high structure
VHS, very high structure.
Channel and furnace blacks account for a large percentage of the black production. Channel blacks contain a higher amount of volatile, chemically bound constituents and consequently have a pH of approx. 4, whereas the furnace blacks have a pH of approx. 9. The German classification is made only on their reinforcing power, distinction being between highly active, active, semi-active and inactive types.
Apart from the above three main groups there are also:

Acetylene blacks. Obtained by the thermal decomposition of acetylene gas (Shawinigan, P blacks) by splitting under pressure, an alternative term is explosion blacks.

Lamp black. Produced by burning liquid distillates (anthracene oil, naphthalene and anthracene residues) in special containers. These blacks are inactive, but differ from the thermal blacks in the nature of their particles: thermal blacks are crystalline whereas lamp blacks have a laminar structure. Compounds containing the blacks have very good processability.

Type	Properties	Trade names
EPC	Large particle size, small surface area, easily processable, good tear and abrasion resistance, high resilience, low heat build-up	Atlantic E 42 Continental AA Croflex 77 Dixiedensed 77 Kosmobile 77 Micronex W-6 Spheron 9 Texas E Witco No. 12 Wyex EPC
MPC	Standard black, medium particle size. Normal processability, good tear and abrasion resistance	Collocarb EPC-20 Carbomet No. 3 GTL Arrow TX and MPC Continental A Dixiedensed HM 566 Micronex Witco No. 1 Atlantic 75, 95, 109 Croflex Th Kosmobile HM S-66 Texas M TX Carbomet S Collocarb MPC-20
HPC	Small particle size, larger surface area, more difficult to process, excellent tear and abrasion resistance	Atlantic 98 Croflex HX Micronex MK II Witco No. 6 Continental F Dixiedensed and S Kosmobile and S Spheron 4
CC	Very small particle size, large surface area, difficult to process, excellent tear and abrasion resistance, gives good electro-conductive properties	Atlantic 156 and 170 Crolac T-10, T-14, T-16 Dixie 5 Dixie Voltex Kosmos and BB Continental R-20 R-30 and R-40 Dixie BB Kosmink Conductex Spheron I, C, N

continues

Type	Properties	Trade names
MT	Coarse particles, small surface area, low modulus, high resilience, accepts high bondings of filler	Atlantic TM Dixitherm M Sterling MT Shell Carbon Croflex TM Kosmotherm M Thermax Velvetex
FT	Smaller particle size than MT, lower modulus, very high resilience, accepts high loadings of filler	Atlantic TF Dixitherm F Miike 20 Sterling FT Croflex TF Kosmotherm F P-33
Acetylene black	High modulus, high heat conductivity, gives electro-conductive rubbers	Shawingan P-Blacks
SAF	Exceptionally small particle size, very difficult to disperse, particularly good abrasion resistance, excellent ageing properties and flex-cracking resistance	Atlantic 150 Croflex 85 Kosmos 85 Statex 160 Aromex SAF Continex SAF Dixie 85 Philblack E Vulcan 9
ISAF	Good abrasion resistance between HAF and SAF, excellent flex-cracking resistance, good ageing properties, good electricity conduction	Aromex 125 Continex ISAF Dixie 70 Philblack I Vulcan 6 Atlantic 135 Croflex 70 Kosmos 70 Statex 125
HAF	Good abrasion resistance, high tensile strength, good flex-cracking resistance, low heat build-up, very good ageing properties, good electricity conduction	Aromex Continex HAF Dixie 60 Philblack O Vulcan 3 Atlantic 130 Croflex 60 Kosmos 60 Statex R

Type	Properties	Trade names
FEF (MAF)	Fast extrusion oil black with medium abrasion resistance, is quickly and easily dispersed, has low viscosity, high resilience, gives smooth extrusion, good dimensional stability, good mould flow, fast curing, good tensile strength, high modulus, good heat conduction, particularly suitable in blends with SRF for heavy tyres, and also for butyl inner tubes	Arovel Continex FEF Dixie 50 Philblack A Sterling SO Atlantic 125 Croflex 50 Kosmos 50 Statex M
FF	Small particle size, similar in its properties to EPC but has a small heat build-up, high dynamic properties, resilience and good ageing properties, fast curing, very low abrasion resistance, good flex-cracking resistance, accepts high loadings of fillers	Statex B High furnace black Sterling 99
GPF	General purpose oil black, combines the high modulus of the HMF type with the resilience of the SRF type, gives smooth extrusion, low heat build-up, good mechanical properties	Atlantic 118 Dixie 35 and 45 Sterling V Arogen GPF Statex G Croflex 35 Kosmos 35 and 45 Ukarb 327 Continex GPF
HMF	High modulus gas black, smooth processing, good physical properties, low volume costs, high resilience, low heat build-up, good-flex cracking resistance, low hysteresis, high load bearing capacity, low shrinkage	Atlantic 120 Continex HMF Dixie 40 Modulex Sterling L and LL Collocarb Croflex 40 Kosmos 40 Statex 93 and 930
SRF	Semi-reinforcing gas black, large particle size, good tensile strength, high modulus and good ageing properties, low heat build-up, good flex-cracking	Atlantic 115 Croflex 20 Essex Gastex NS-Essex

continues

Type	Properties	Trade names
SRF	resistance and reinforcement, high resilience and filler acceptance, low volume costs, low degree of stiffening	Sterling NS, R, S Collocarb SRF Continex SRF Dixie 20 Durnex Kosmos 20 Pelletex Metanex 37 Regal SRF
SRF-HM	Semi-reinforcing gas black with high modulus	Atlantic 115 HM Dixie 20 HM Kosmos 20 HM Continex SRF-HM Croflex 20 HM Furnex H SRF 3
CF	Electrically conducting oil black, small particle size, high resilience and accepts high filler loadings, low heat build-up	Aromex 115 Continex CF Dixie CF Shawinigan Acetylene Black Vulcan C Atlantic CF Croflex CF Kosmos CF Sterling I FB 200 Statex A
SCF	Super conducting gas black	Vulcan SC

Electric-arc blacks. Formed as a subsidiary product during the production of acetylene in an electric arc. The blacks are almost equivalent to the highly active types but have a poorer processability. After treatment, they are differentiated as L-black-T (dry) and L-black-N (wet).

Non-classified blacks:
CK-3, CK-4, equivalent to HAF
Corax B and L, comparable to HMF
Durex O, comparable to SRF
Durex I
Durex 101
Seast

A type of black, which until now had been unclassified, was recently introduced with a low carbon structure:

Neotex 100, oil black with HAF particle size
Neotex 130, oil black with ISAF particle size
Neotex 150, oil black with SAF particle size
Regal 300, oil black with the physical properties of the channel blacks and the cure speed of the furnace blacks
Regal 600, oil black with ISAF particle size.

Type	Particle diameter mμ	EM surface m²/g	BET surface m²/g	Nigrometer index	Oil absorption ml/100 g	Iodine absorption, %	Electrical resistance, ohm/cm³	pH	DPG absorption %
EPC	30	94	100	85	112	16	200	4·3	31
MPC	26	106	120	86	116	17	16	4·9	40
HPC	19–24	105	142	83	119	20	7	4·3	41
CC	8–23	110	227	78	132–160	20	2	4·1	53
MT	473	5·7	6·6	116	39	1·0	9	7·4	
FT	223	16	16	109	51	3·4	120	9·5	
SAF	25	142	134	86	152	16	0·2	9·6	
ISAF	31	121	102–129	90	143	13	0·4	9·1	
HAF	28–45	74–98	65–77	94	126	10	0·5	9·2	10
FEF	80	63	39–43	100	139	5	0·5	9·3	
HMF	41–95	38	34	95–102	81	4	0·5	10·1	10
GPF	98	43	25–32	101	117	4	0·3	9·9	
SRF	80–160	26	22	100–104	85	3	0·9	9·9	10
SRF/HM	130	25	22	103	83	4	0·3	9·8	
CF	31	102		90	143	13	0·4	9·1	
SCF	12		249	90	145	31		8·5	
FF	39		75	92	86	7		9·5	10
CK 3	26–32	94	95	87		14		5·0	20
Corax B	34	58		97				8·7	
Corax L	27	71		94				9·2	
L-Russ-T	20	180		99				8·5	
L-Russ-N	20	180		98				6·8	
Durex 0				108			56	4·3	17
Durex 1				103			99	3·9	23

CARBON TETRACHLORIDE.
CCl$_4$, tetrachloromethane, m.p.
−23°C, b.p. 76·7°C, s.g. 1·59. Colourless, non-flammable liquid. Miscible with practically all organic solvents, immiscible with water.
Uses: solvent for rubber, resins, oils and waxes, also as a flame extinguisher.

CARBO-O-FIL Anthracite powder, s.g. 1·47. Particularly low density filler for all elastomers, especially for accumulators, mats. Non-staining (402).

CARBOPOL Group of synthetic polymers. Soluble in water, or able to swell.
Uses: thickening agents for latex, suspension agents, emulsifiers (14).

CARBOWAX
HO(CH$_2$CH$_2$O)$_x$H, group of polyethylene glycols, m.p. between −65 and 60°C, mol. wt. 200–20 000, index number indicates the mol. wt, with the exception of 20 M (20 000). Soluble in water.
Uses: softener, release agent, and to improve surface finish (146).

CARBOXYL RUBBER Copolymers of butadiene with acrylic, methacrylic or sorbic acids. High tensile strength and good resistance to abrasion.

CARCASE The underlayer of a tyre, consists of rubberised textile layers.

CARCASE The inner layers of textile used to stiffen the inflated tyre.

CARDOLITE Reinforcing filler derived from synthetic resins (147).

CARDAMIDE
Grades:
783: amyl urea
784: N-tert-butyl urea (147).

CARDANOL ETHER Mixture of ethyl ether 3-pentadecenol and less saturated C_{15} phenols. Light brown liquid, b.p. (1 mm) 175–200°C. *Uses:* low temperature softener for SBR (1).

CARIFLEX Trade name for Shell synthetic elastomers (2).

CARIFLEX BR cis-1:4-polybutadiene. 96·5% cis-1:4, 1·9% trans-1:4, 1·6% 1:2 structures (2).
10, 0·2% DPPD, Mooney 45
11, 0·3% Ionol, Mooney 45
25, extended with 50 phr aromatic oil (2).

CARIFLEX IR cis-1:4-polyisoprene with 92% cis-1:4 content.
Grades:
300: contains 0·5% organic acids. Mooney 65
305: easy processing type, free from organic acids, s.g. 0·92. Mooney 55
500: oil extended type with 25% non-staining, naphthenic oil.
All grades contain a non-staining antioxidant (2).

CARIFLEX K-101 New type of polymer which can approach the properties of vulcanised rubber without curing. Loose white crumbs, s.g. 0·94. Solutions which have lower viscosity than usual rubbers are used for pressure and contact adhesives, self-adhesive tapes, impregnation and coating of textiles and paper (2).

CARIFLEX-S Butadiene-styrene copolymer, the number is the same as the ASTM code (2).

CARIFLEX TR Thermoplastic rubber which behaves like a thermoplastic at processing temperatures, but like a vulcanisate at room temperature. Available in a variety of grades with different physical properties.

	TR grades	TR-B grades
s.g.	0·94–1·2	1·03–1·12
Hardness, Shore A	40–85	30–95
Tensile strength kg/cm²	50–204	42–134
Breaking, elongation %	500–1100	370–520
Modulus, 300%, kg/cm²	18–75	25–105
Resilience, %	72–83	

CARNAUBA WAX A hard wax which comes from the needles of the Carnauba palm *Corpernicia cerifera* (Brazil), m.p. 80–86°C, s.g. 0·99. Soluble in benzine, benzene, chloroform and hydrocarbons, insoluble in alcohol. Contains approximately 55% non-saponifiable material.
Uses: in polishes and occasionally in rubber compounds.

CAROB OR LOCUST BEAN FLOUR Obtained from the fruits of *Seratonia siliqua*. Consists primarily of mannan and galactan and contains pectin, lignin and hemicelluloses. Soluble in hot water.
Uses: diluent and thickening agent for latex.

CAROM Trade name for Rumanian produced SBR rubber.

CARPODINUS Lianes and shrub-like climbing plants of the *Apocynaceae* family (Africa).

CASE HARDENING Process used to produce a hard surface on a tough core; alternatively, the appearance of a thin white stripe in the middle of the cross-section of smoked sheet. Caused by drying the sheet too quickly, at too high a temperature, the surface pores become clogged with non-rubber constituents and the diffusion of moisture from the centre is prevented.

CASEIN Phosphor protein from milk. Yellowish powder. Soluble in dilute alkalis. Forms hard, insoluble plastics with formaldehyde.

Uses: improves the mechanical and chemical stability of natural and synthetic latices, thickening agents, dispersing agent, impregnating agent for tyre cords to improve the adhesion of rubber and cord, production of plastics, adhesives. Soluble trade forms are ammonium caseinate and sodium caseinate.

CASLEN Synthetic fibres based on casein.

CASTILLA (Often incorrectly termed castilloa.) Species of the *Moraceae* family which yields rubber (Mexico, and extending to Bolivia and Brazil). Named after the Spanish botanist J. D. del Castillo. The quality of the rubber varies enormously according to the species, *C. elastica* and *C. ulei* being the most important. *C. fallax*, for example, has a latex with a low rubber content, and yields a hard, sticky, non-elastic mass. The latex is highly acidic (pH 4–4·5). Fairly resistant to acids and alkalis, coagulated by alcohol, hot soap solution and resin solution. Dry rubber content 30–40%, resin content 5–15%, particle size 2–3 μ.

C. elastica. Although successfully cultivated in its native Mexico (presumably because of greater yield and longer period of production than the Hevea type), it has never been cultivated on a large scale in Asia. Prior to 1910 attempts were made in Malaya and Indonesia, but yields were small and the trees were susceptible to disease. To make the plantations profitable, cocoa and coffee intercrops were set up in the rubber plantations. Attempts were made in S. America, but proved unprofitable. The trees require almost the same climate and soil as *Hevea*. The chemical composition and rubber content of the trees varies. The resin content is high in young trees, but decreases with the age of the tree. Latex of 12 year old trees contains 7–14% resin (but up to 66% in very young trees), and 26–32% rubber, also protein, tannin, sugars, and an acid, $C_{17}H_{30}O_{10}$.

In the Amazon region the rubber is obtained by either felling the trees or through incisions in them and collecting the milk in holes in the earth, which are lined with leaves. The latex rapidly turns a brown colour and creams easily, so that it can be coagulated by a steady stirring. Coagulation was often achieved by adding the sap of *Ipomoea bona nox* (batata rana), often with the addition of soap. The rubber was dried in the open air and marketed in the form of dirty, black bales (Planhas de caucho). The natural crops were virtually destroyed by uncontrolled cutting.

Lit.: J. v. Wiesner, *Raw Materials of the Plant World*, Vol. 4. Leipzig, 1928.

E. A. Hauser, *Latex*, New York, 1930.

C. O. Weber, *Ber*, 1903, 36, 3108; *Gummi Zeit.*, 1905, 19, 101.

CASTILLO, JUAN DIEGO DEL Spanish botanist, 1744–93. Known for his work in Puerto Rico, participant in Cervantes' expedition to compile a Flora Mexicana. Following Linne's *Jatropha elastica*, the Ule tree was named after him, *Castilla elastica*.

CAST MOULDING Process used to produce thin walled or hollow articles from latex. The latex is poured into a cast, which can be taken apart, water is absorbed in the plaster and a thin rubber layer deposited on the wall. After a specified wall thickness has been reached, the surplus latex is poured away and the cast, together with the rubber layer, dried at 70–100°C. The casting is removed and vulcanised. This process allows the use of high filler contents (*e.g.* up to 300 parts whiting: 100 parts rubber), and the production of articles of widely varying hardnesses.
Uses: include hollow toys, linings, decorative figures.

CATALIN ANTIOXIDANT
CAO-1: 2:6-ditert-butyl-4-methyl phenol
CAO-3: 2:6-ditert-butyl-4-methyl phenol (purified form)
CAO-4: 2:2'-thiobis(4-methyl-6-tert-butyl phenol)
CAO-5: 2:2'-methylene-bis(4-methyl-6-tert-butyl phenol)
CAO-6: identical to CAO-4
CAO-14: identical to CAO-5
CAO-30: 1:1'-thiobis(2-naphthol)

CAO-32: 1:1'-methylenebis(2-naphthol)
CAO-33: 2:6-ditert-butyl-4-methyl phenol.

CATALIN RESIN Group of phenolic resins. Vulcanising agents for butyl rubber. Give heat resistant vulcanisates with high physical properties and low compression set.

CATCH CROP Intermediary growth, *e.g.* of coffee or other species in young rubber plantations.

CATIVO RUBBER Product from the milky sap of the Cativo mangrove (Colombia). Resin-like products, m.p. 50°C.

CAUCHO ANDULLO BLANCO White root rubber. Former trade name for rubber from *Sapium decipiens* (Ecuador).

CAUCHO BLANCO Local name for rubber from *Sapium verum* (Ecuador).

CAUPRENE $C_{32}H_{48}$. Rubbery, polycyclic hydrocarbon. Produced by heating polybutadiene dibromide with zinc dust. The name was coined by Ostromislensky.

CAVEAT Hard rubber discovered by Nelson Goodyear, the younger brother of Charles Goodyear. Consists, for example, of 4 parts sulphur, 12 parts magnesia and 16 parts rubber. Vulcanisation 6–8 hours at 120–150°C. Patented: strictly a rediscovery of the process derived by T. Hancock (1843).

CAVIANA Trade term for wild rubber grown on Caviana, an island in the Amazon. Good, nervy quality.

CAYTUR

4: partial complex of zinc oxide and dibenzthiazyl disulphide
7: 58% cumene diamine, 42% m-phenylene diamine (6).

CAYTUR DA Tertiary amine· Yellow–white crystals, m.p. 95–98°C· s.g. 1·25. Soluble in water, alcohol, esters and ethers. Heat sensitive catalyst for urethane foam with a delayed action. Processing below 38°C, foams at 80–121°C. *Quantity:* 1–2% (6).

CAYTUR-O Tertiary amine. Brown, odourless liquid, s.g. 0·98. Soluble in water, alcohol, hydro-carbons, esters and ethers. Catalyst in the production of urethane foam, somewhat more reactive than conventional amino catalysts. *Quantity:* 1% (6).

CB Abbreviation for carbon black.

CBU Wetting agent for poly-urethanes, based on dichlorobenzidine (148).

CC Abbreviation for conductive channel; a channel black capable of conducting electricity.

CDI Trade name for 3:3′-dichlorodiphenyl diisocyanate (148).

CDPD Abbreviation for cadmium pentamethylene dithio-carbamate.

CEARA RUBBER Natural rubber from *Manihot Glaziovii Müll. Arg.* (Manicoba Ceara). Originally marketed *via* Ceara (Brazil).

CELAFIBRE Acetate rayon yarn (149).

CELITE Diatomaceous earth filler (150).

CELLOBOND Group of reinforcing resin fillers for natural and synthetic rubber; based on phenolic resins and novolaks (151).

CELLOGEN Group of blowing agents.
Grades:
Celogen: p,p′-oxybis(benzene sulphonyl hydrazide)
AZ: azodicarbonamide. Yellow, odourless powder, decomposes at 198°C liberating nitrogen, s.g. 1·63. *Uses:* natural and synthetic rubbers, poly-sulphides, polyethylene and p.v.c. Gives an extremely fine and even cell structure. Non-staining
BSH: benzenesulphonyl hydrazide
OT: p,p′-oxybis(benzo-sulphonyl hydrazide)
RA: p-toluenesulphonyl semicarbazide
TSH: p-toluene sulphonyl hydrazide (23).

CELLOLYN Group of modifying, synthetic resins. *Uses:* diluents, softeners, tackifiers (10).

105

CELLOPREN
Grades:
MO: blend of 85% dinitrosopentamethylene tetramine
with an inactive stabiliser
CC: 85% dinitrosopentamethylene
tetramine with a dispersant
Extra: 70% dinitrosopentamethylene
tetramine with 15% amine
and stabiliser (378).

CELLOSIZE Hydroxyethyl
cellulose. Thickening agent for latex
(112).

CELLOSOLVE Group of sol-
vents, mono- and dialkyl ethers of
ethylene glycol and derivatives.
Grades:
C: ethyl glycol
Methyl-C: methyl glycol
Butyl C: butyl glycol (112).

CELLUFLEX Group of
softeners, *e.g.* for synthetic rubbers,
polyurethanes, polyesters, polyvinyls,
artificial cellulose substances.
Grades:
23: alkyl epoxy stearate
112: cresyl diphenyl phosphate
179-A: tritolyl phosphate (with a
low o- content)
179-C: tritolyl phosphate
179-EG: tritolyl phosphate (electrical
grade)
DBP: dibutyl phthalate
DOP: dioctyl phthalate
FR-2: tri(dichloropropyl)phos-
phate
TPP: triphenyl phosphate
CEF: tri-β-chloroethyl phosphate
(152).

CELLULAR RUBBER Porous
products produced from natural
rubber and latex, with a varying cell
structure, elasticity, density and
hardness. The products are classified
as closed, open and mixed cell,
according to the cell structure. The
size and structure of the cells depends
on the compound and the manner of
production.
Sponge rubber: open cell product
used for household and technical
products (*e.g.* bath sponges, mats,
drop absorbers, filters, cushions).
Was formerly produced by the Lon-
demann process using a two-stage
curve. Today, presses with increasing
steam pressures (0·5–3 atm) are used.
Benzene sulphonhydrazide is suitable blowing agent. Moss rubber:
products with either a closed, open
or mixed cell structure. The cells are
much smaller than those of sponge
rubber. Produced by curing without
pressure or by a back-pressure
process.
Uses: cellular soles, extruded and
moulded seals.
Cellular rubber: closed cells. Produced using the E. Pfleumer process
(DRP 249 777). The compound is
exposed to a pressure of 200 or more
atm of nitrogen, which is absorbed by
the compound. The product is expanded by releasing the gas pressure
after vulcanisation; the gas is retained
in each cell under a slight excess
pressure.
Uses: sealing and insulating
materials.
Foam rubber: open cell products.
Produced by foaming concentrated
latex, causing this to gel and vulcanising the foam.
Uses: include mattresses, cushions.

CELON Polyamide fibre of
nylon 6 (149).

CEMENT General term for a
solution of rubber in hydrocarbons.

CENTRALITE Group of softeners.
Grades:
1: diethyl diphenyl urea
2: dimethyl diphenyl urea
3: methyl ethyl diphenyl urea
4: ethyl phenyl ethyl-o-toluidine
    urea (345).

CEPAR APPARATUS (CEPAR: cure, extrusion, plasticity and recovery). Apparatus for determining vulcanisation properties, extrusion characteristics, plasticity and the elastic recovery of rubber, *e.g.* those properties affecting processing.
USP 2 904 994 (1959).

CEREMUL Group of wax emulsions based on paraffin wax, microwaxes and vaseline, m.p. 49–77°C, pH 7–8·5.
Uses: in processing latex.

CERESINE Hard paraffin, purified mineral wax. White mass, m.p. 65–72°C, s.g. 0·91–0·94. Obtained from the refining of ozocerite and sulphuric acid.
Uses: protective agent against light and groove cracking; primarily usе ᵻ in combination with other softeners to decrease hardness and prevent crystallisation. The products differ widely in their effectiveness because of traces of acid in varying amounts.

CEREX 250 Styrene copolymer. Injection moulding material (5).

CERVANTES, VINCENTE Born 1759, Spain, d. 1829, Mexico. Professor of Botany at Mexico City University. Leader of an expedition to compile a 'Flora Mexicana'; discovered and described the Ule tree, which he named after his associate research worker—Juan del Castillo —*Castilla elastica*. Cervantes made a major contribution to rubber research by describing the production of rubber in Mexico, conducting research on the properties of latex, designing rubber containers, and discovering the use of acetic acid as a coagulant.
Lit.: *Rubb. J.*, 1955, 523.

CETAMOL Group of softeners.
Grades:
AB: butoxyethyl diethyl phosphate
DA: butyl diethyl phosphate
MB: butoxyethyl dimethyl ethyl
    phosphate
P: tripropyl phosphate
Q: tri-p-chloroethyl phosphate.

CETYL MERCAPTAN
$C_{16}H_{33}SH$, hexadecyl mercaptan, m.p. 18°C, b.p. 185–190°C, s.g. 0·847.
Use: peptiser for synthetic rubber.

CETYL PYRIDINE BROMIDE

Wax-like mass. Soluble in alcohol, acetone, chloroform. Catonic, surface active agent.
Uses: reversal of charge of latex (positex, positive latex), emulsion polymerisation.

CEVIAN-N Styrene-acrylonitrile copolymer (435).

CEXO PROCESS Semi-automatic coagulating process for natural latex developed by the Societé des Caoutchouc d'Extreme-Orient (CEXO) (Vietnam). Accounts for all factors which play a part in

coagulation and might influence the quality of the rubber. Coagulation takes place at a constant pH and may be accelerated by increasing the temperature to 35°C. The latex is stirred until flocculation occurs and is then transferred to a coagulating tank. The process facilitates the simultaneous processing of 25 000 l of latex. The rubber is uniform in its technical properties.
Lit.: *Rev. Gen. Caout.*, 1957, 34, 617.

CF Abbreviation for conductive furnace; furnace black which may act as a conductor of electricity.

CFE Abbreviation for chlorotrifluoroethylene.

CHAFFEE CALENDER The first calender, developed by Chaffee (USA) consisted of four steam heated rolls, placed one over the other, the two outside rolls with a diameter of 45 cm and the inner two a diameter of 30 cm. The machine, which weighed 30 tons, was developed for impregnating fabrics and was known as the 'Mammoth'. The top roll had a slower speed than the others. The rubber was passed between the two upper rolls and the fabric between the middle rolls. Modern calenders are based on this design.
Lit.: H. C. Pearson, *Rubber Machinery*, New York, 1915, 93–95.

CHAFFEE'S PROCESSING MACHINE Machine patented by Chaffee (USA) in 1839 which consisted of two steam heated iron rolls in close contact and running at friction speed. By a combined rolling and tearing action, the rubber could be processed into a smooth, soft sheet. The rolls were about 180 cm long and had diameters of 67·5 cm

and 45 cm respectively. This machine was the forerunner of the modern two roll mill.

CHANCEL GRADE Measure for determining the fineness of sulphur and particulate fillers; known as a Chancelometer or sulphurometer (after Chancel, French chemist). Five grams of sieved sulphur or powder are placed in a graduated tube, mixed with a known quantity of water-free ether and shaken for $1\frac{1}{2}$ min. A sediment is allowed to form on a water bath at 15·5°C, until equilibrium is reached. The Chancel scale is read as an expression of fineness. The process is unsuitable for particulate fillers which form gels when mixed with ether (*e.g.* alumina gel); such gels settle with difficulty and the results are often impossible to assess.

CHARPY APPARATUS Impact pendulum used to determine the impact resistance of plastics and electrically insulated materials, according to ASTM D 256-56.

CHATTERTON COMPOUND Cement produced by mixing, *e.g.* gutta-percha, coal-tar oil, resin, colophony.
Uses: adhesion of floor coverings, mats, pram·tyres.

CHC Synthetic polymer based on epichlorhydrin and ethylene oxide, s.g. 27, 26% chlorine. Vulcanisation with 2-mercaptoimidazole.
TN: Hydrin 200 (14).

CHEMIGUM Group of butadiene-acrylo nitrile latices with different monomer ratios and particle sizes from 500–4000 A.

Type	% Total solids	Monomer ratio
200	55	70/30
235CHS	42·5	55/45
236	40	55/45
245B	32·5	67/33
245CHS	42·5	67/33
246	40	67/33
247	42·5	67/33
248	55	67/33

550 carboxylic butadiene-acrylonitrile latex.
Film forming, oil- and solvent-resistant with excellent bonding properties.
Uses: in vulcanised and unvulcanised form for non-woven fabrics, carpet backing, impregnation of textiles, paper, leather, glues, paints, modifier, *e.g.* for asphalt (115).

CHEMIGUM N Group of oil resistant, butadiene-acrylonitrile copolymers, produced catalytically by emulsion polymerisation. The oil resistance and physical properties depend on the acrylonitrile content; increased acrylonitrile content gives an increased oil and solvent resistance, abrasion resistance, tensile strength and hardness; reduction of acrylonitrile content improves elasticity and low temperature properties.

Type	% Acrylonitrile	Mooney	Solubility in Methyl ethyl ketone
N1NS	32–33	115 max	
N3NS	42–43	87–112	65%
N5	42–43	87–112	77% (a)
N6	32–33	40–65	96%
N6B	32–33	40–65	82%
N7	32–33	77–102	60%
N8	32–33	75–102	30%
N300	42–43	57–75	98% (b)
N600	32–33		99%

(a) for cements; (b) without gel (115).

CHEMIGUM SL Group of polyurethanes (115).

CHEMIVIC Acrylonitrile-butadiene copolymers, reinforced with vinyl resin.
Grades:
400: s.g. 1·08, Mooney 40–60. *Uses:* extruded and calendered articles
800: s.g. 1·08, Mooney 78–98. Good stability. *Uses:* extruded and calendered articles (115).

CHEMLOCK Group of bonding agents for metal/rubber composites.
Grades:
301: based on epoxy resins
607: based on a silicone for silicone rubber.
Additional grades: 201, 203, 220, 301, 614.

CHEMOSIL Group of rubber/metal bonding agents produced according to the Chemlock patent.
Grades:
210: primer in a two-part system Chemosil 220/210. Solution of polymers and a dispersion of solids in methyl-butyl ketone glycol ether, s.g. 0·92–0·94, solids content 22–26%, viscosity 31–44 sec, Ford-Becher 3 mm. The resistance to corrosion and heat resistance of the compound is increased by the two-part system, as is the adhesion to some non-ferrous metals and plastics; processing safety is also increased
211: Chemosil 210 with a higher viscosity
220: universal one-part bonding agent. Solution of polymers and a dispersion of solids in xylol

perchloroethylene, s.g. 1·03–1·10, solids content 23–27%, viscosity 94–117 sec, Ford–Becher 3 mm. Bonds NR, SBR and NBR, polychloroprene and butyl rubber to metals, many plastics, wood and glass, regardless of vulcanisation. Vulcanisation temperature 120°C and above. Adhesive strength up to 130 kg/cm². Resistant to corrosion, heat, and corrosive media

510: bonding agent for silicone and Vitone, used with metals, some plastics, glass and textiles. Solution of silanes in methanol; s.g. 0·83, solids content 11·5–12% (382).

CHEMPOR Dinitrosopentamethylene tetramine (428).

CHICLE Chewing gum. Rubber from the latex of *Achras zapota* and *Achras chicle* (*Sapotace*, Mexico, Honduras and Guatemala). The trees grow wild in Mexico, but are cultivated in other tropical countries because of the pleasant tasting fruit. The trees are tapped after 15 years, using V-cuts. Newly cut trees yield 7–11 kg of rubber in the tapping period. The latex is evaporated and formed into cakes. First quality material is called chicle, the lesser quality, chiquibul. In Venezuela, species of *Sapotaceae* known as pendare yield a similar product. Chicle is a white, odourless, crumbly mass, becoming plastic at 50°C. Composition: 40·7% resin, 17·2% rubber hydrocarbon, 9·0% sugar, 8·2% starch and salt, 14·4% water, 10·5% remaining constituents.

CHITTAGONG RUBBER Trade name for rubber from

Willougbeia edulis Roxb. (Burma, Siam, Malaya, Indonesia).

CHLORAFIN Group of odourless and tasteless, chlorinated paraffins. *Uses:* flame retarders and water repellents for rubber and plastics, softeners for vinyl resins (10).

CHLORINATED RUBBER $C_{10}H_{11}Cl_7$. Yellowish powder or flakes, s.g. 1·63–1·66, bulk density 60–420, chlorine content 64–88%. Soluble in benzene hydrocarbons, chloroform, esters, ketones, mineral oils and water. Produced by reacting chlorine with rubber, when, either simultaneously or in succession, hydrogen is substituted by chlorine, chlorine is added to the double bond, and cyclisation and breakdown occur. The precise structure and composition are not known. It is presumed that chlorinated rubber consists of

$$\left[\begin{array}{c} CH_3 \\ | \\ -CH-C-CH-CH_2- \\ |\ \ \ \ |\ \ \ \ | \\ Cl\ \ \ Cl\ \ Cl \end{array} \right]_n$$

with 61·3% chlorine

$$\left[\begin{array}{c} CH_3 \\ | \\ -CH-C-CH-CH- \\ |\ \ \ \ |\ \ \ \ |\ \ \ \ | \\ Cl\ \ \ Cl\ \ Cl\ \ Cl \end{array} \right]_n$$

with 68·2% chlorine.

Production: by introducing chlorine gas into a 4–8% rubber solution, by chlorination of solid rubber under pressure, or by introducing chlorine into stabilised, acidified latex (BP 634 241, 721 512). The chlorinating reaction can be accelerated catalytically. At the beginning of chlorination approx. 50–55% chlorine substitution takes place, then the number of double bonds is quickly reduced as a result of addition. A fast cyclisation reaction occurs in the first phase. Ring formation proceeds to approx. 50% of the theoretical value. The long chain molecules are considerably shortened and become less mobile. Molecular breakdown and the degree of cyclisation determine the viscosity in organic solvents. Smaller quantities of hydrochloric acid are split off under the influence of heat. Stability improves at 60–65% chlorine and above. From solutions, or combined with softeners, chlorinated rubber yields transparent, tough, elastic films (latent elasticity) which are highly resistant to concentrated acids, chemicals and moisture, and flame. During World War II chlorinated rubber was also produced from polyisoprene and butadiene–styrene copolymers, and these were similar in properties to the product obtained from natural rubber.

Uses: corrosion resistant paints in solutions or emulsions, *e.g.* for ships' paints, and printing inks, metal/rubber bonding agents, impregnating agents, electrical insulation, in plastic compounds together with alkyd and diphenyl resins, coumarone–indene resins. Addition to other bonding agents improves resistance to corrosion.

TN: Adeca, Alloprene, Chlorofan, Clortex, Electrogum, Dartex, Pergut, Protex, Tegofan, Tornesit, Parlon (from rubber solutions)
Rulacel (from latex)
Aizen, Detel, Nippon, Paravar, Parlon, Raolin, Para-Stonetex (from solid rubber)
Pliochlor, Parlon-X (from polyisoprene)
Bunalit (from Buna S)
Chlorbuna (from numbered Buna).

Lit.: A. Nielson, *Chlorkautschuk und die übrigen Halogenverbindungen des Kautschuks*, Leipzig 1937. J. Le Bras and A. Delalande, *Les Dérivés chimiques du caoutchouc naturel*, Paris 1950. S. Boström, *Kautschuk-Handbuch*, Vol. 1, Stuttgart 1959, 136 ff.

α-CHLOROACRYLIC ACID

$H_2C=C.Cl.COOH$, 2-chloropropenoic acid, m.p. 65°C. Sublimes and decomposes at 176–180°C. Soluble in water and alcohol.
Use: as an ester, for the production of polyacrylics.

CHLOROBENZENE C_6H_5Cl,

monochlorobenzene, benzene chloride. Colourless liquid, b.p. 131–132°C, s.g. (20°C) 1·107. Miscible with benzene, alcohol, chloroform. Solvent for NR, SR and cements.

CHLOROBUTYL RUBBER

S.g. 0·92, mol. wt. 450 000, unsaturation mol. 1–2%, chlorine content 1·2%, Mooney viscosity ML8/100°C 21 ± 5. Soluble in aliphatic and aromatic hydrocarbons with formation of gel. Antioxidant: 0·2% ditert-butyl methyl phenol, stabiliser: 0·7% calcium stearate. Good storage stability, suitable for the production of non-discolouring articles. Various vulcanisation processes may be used because of the allyl chlorine group; crosslinking can occur either at the double bonds, at chlorine atoms, or at both. Because of the high speed of vulcanisation, the large number of sites of crosslinking and the thermal stability, this type of rubber can be exposed to a long vulcanisation time without reversion. Vulcanisates have a low permanent set, low gas permeability, good resistance to abrasion and excellent heat resistance. Suitable vulcanising agents: zinc oxide, thiurams, primary and secondary amines, sulphur with an accelerator, methylol-phenol resins together with metals or metallic salts.

TN: Chlorbutyl HT 10–66
MD 551.

p-CHLORO-m-CRESOL

6-chloro-3-oxy-1-methylbenzene, 4-chloro-m-cresol. White crystals, with aroma similar to that of cresols, m.p. 67°C. Soluble in alcohol, ether, benzine, benzene, chloroform; as sodium salt, in water. Disinfectant.

Uses: prevention of mould in smoke houses, drying houses and factories on rubber plantations, sometimes used to prevent mould in sheet rubber.

CHLOROFORM

CHLOROFORM $CHCl_3$, trichloromethane. Colourless liquid, m.p. $-63.5°C$, b.p. $61°C$, s.g. 1.484 $(20°C)$. Miscible with ethanol, hydrocarbons and oils.

Uses: solvent for rubbers, resins, plastics, starting material for the production of Teflon.

CHLOROFORM EXTRACT

After extraction with acetone, rubber is extracted with chloroform; bituminous materials are dissolved out and their presence in the rubber can thus be detected.

ASTM D 295-59T
BS 1673: 2: 1954
DIN 53 558.

CHLORONITROETHANE

$$NO_2$$
$$|$$
$$H_3C . C—Cl$$
$$|$$
$$H$$

b.p. 122–128°C, s.g. 1·26. Insoluble in water. Prevents gel formation in self-vulcanising rubber adhesives.

CHLORONITROPROPANE

$CH_3 . CH_2CH(NO_2)Cl$. B.p. 139–143°C. Retarder.

CHLOROPARAFFIN Chlorinated paraffin. Liquid and solid, s.g. 1·14–1·65, chlorine content 40–70%. Soluble in many organic solvents.

Uses: flame and weatherproof impregnations, manufacture of flame proof rubber products, together with antimony trioxide (30 parts Chlorinated Paraffin 70, 15 parts antimonytrioxide). Softener.

CHLOROPRENE

$CH_2=CH—CCl=CH_2$, 2-chloro-1:3-butadiene. Chlorinated butadiene. Volatile, colourless liquid, b.p. 59°C. Produced from acetylene in the presence of copper and ammonium chloride by way of vinyl acetylene. May be easily polymerised into long chains.

CHLOROSULPHONATED POLYETHYLENE

z = ca. 17

'Chemical' rubber. White chips, s.g. 1·12–1·28. Soluble in toluene, benzene, xylene, carbon tetrachloride and mixtures with alcohol, ketones and benzine. Produced by reacting polyethylene with chlorine and sulphur dioxide, thermoplastic elastomer. Variations in the molecular structure of the polyethylene and the quantity of bound chlorine, and sulphur dioxide enable a large variety of products to be obtained. The most widely used is a polyethylene with a mol. wt. of approx. 20 000 with approx. 1·3–1·7% sulphur and 26–29% chlorine. Too low a mol. wt. yields a sticky product with a low tensile strength, while above a certain level, increases in mol. wt. have little effect. The introduction of chlorine helps to prevent the formation of a crystalline structure, while the sulphonyl group gives reactive sites available for vulcanisation. The rapid change of viscosity with increasing temperature makes plasticisation unnecessary. For vulcanisation, the system consists of an organic acid, a metal oxide and an organic accelerator, *e.g.*

Magnesium oxide	20	—
Lead oxide	—	40
Hydrogenated resin		
ester	2·5	5–10
Mercaptobenzthiazole	1	—
Dipentamethylene		
thiuram tetrasulphide	—	1

Resin acids have been shown to be the most effective acids, aliphatic acids the least. Of the metal oxides, magnesium oxide and lead oxide are used and tribasic lead maleate is also used. Thiazoles, thiurams and dithiocarbamates are the most suitable accelerators, whereas the aldehyde condensation products result in scorching and poor physical properties.

Chlorosulphonated polyethylenes may be blended with natural rubber, SBR, nitrile rubber, butyl rubber and neoprene in all proportions. Reinforcement with carbon blacks is unnecessary. The vulcanisates are resistant to ozone, chemicals and heat; have high abrasion resistance and tensile strength; their resistance to heat is good and oil resistance is reasonable. Heat resistance can be improved by adding 1% nickel dibutyl dithiocarbamate and 2% antioxidant.

Uses: include blends with other elastomers to improve their resistance to ozone, tank lining used for corrosive substances, protective coatings, combined sheath and insulation for low voltages.

TN: Hypalon (6), KhSPE (USSR).
Lit.: Du Pont articles, BL 251, 255, 258, 261/63, 267/9, 271, 274, 291, 297, 287, 291, 306, 309, 313, 56–4, 56–10.

CHLOROWAX Group of chlorinated paraffins, s.g. 1·1–1·2, chlorine content 40–73%. Soluble in organic solvents. Flame retardant, water repellent in rubbers, pigments and plastics, softener for chloroprene rubber, plasticiser at processing temperatures. Has a reinforcing effect in SBR, neoprene and nitrile rubber (153).

CHLORYLENE Trade name for trichloroethylene.

CHOLESTEROL
$C_{27}H_{45}OH \cdot H_2O$. Secondary, high alcohol. White flakes, m.p. 148·5°C, b.p. 360°C, with partial decomposition, s.g. (anhydride) 1·052 (monohydrate) 1·03. Soluble in ethers, chloroform, benzene, fats and oils, sparingly soluble in alcohol. Related sterols exist in small quantities in rubber resins (acetone extract) of various plants.

CHONDRILLA Rubber from plants of the *Composita* family; *C. amgibua Fisch.* and *C. pauciflora Lbd.* are important rubber bearers. Rubber exists in the parenchyme cells of the bark and in the twigs and roots; roots 1–1·5% rubber, bark 3·3–5%. The rubber is fairly rich in resin, with twigs yielding 10% resin and the bark 15–20%. The roots are often attacked by insect larvae and the latex flows out and coagulates on the earth. The lumps of rubber which are resinous and contain large quantities of sand, are known as naplyvy (larvae of *Sphenoptera foveola*) or as tsechliki (larvae of *Bradyrrhoa gilveolella*).

CHORO Rubber of inferior quality from *Manihot Glaziovii*; coagulates on the bark and ground.

CHR Synthetic polymer based on epichlorhydrin, s.g. 1·36, 38·4% chlorine. Vulcanisation with 2-mercaptoimidazoline.
TN: Hydrin 100 (14).

CHROME OXIDE Cr_2O_3, chromic oxide, chrome green. Green powder, s.g. 3·5–5·1. Insoluble. Pigment.

CHROME RED $PbCrO_4 . PbO$, basic lead chromate. Red pigment.

CHROME YELLOW $PbCrO_4$, lead chromate. Yellow pigment.

CHRYSILL RUBBER Rubber from *Chrysothamnus nauseosus* (rabbit brush) of the *Composita* (western USA). The bush-like stems contain, on the average, 2–3% pure rubber, in some cases up to 7%. Chewing gum has been obtained from these plants by the Indians in Utah and Cali-

fornia. In 1908, industrial extracting was planned by the USA and achieved during World War II. Only 3-year-old bushes, with an average weight of 2·8 kg, were considered for production. The total produce was estimated at approx. 135 000 t. The rubber may be vulcanised easily, according to the process of Goodspeed and Hall, and yields products of good quality.

CIAGO 2058-P Styrene-butadiene copolymer with 60% bound styrene. Contains a non-staining antioxidant (154).

CIH RUBBER Abbreviation for caoutchouc Indo *Hevea*. Natural rubber with constant properties. Produced by close control of coagulation conditions. Modulus 100% 7.7 ± 0.5 kg/cm^2, Mooney 100 ± 15.

CINNAMALDEHYDE
$C_6H_5CH{=}CHCHO$. Yellow, oily liquid with strong cinnamon-like odour, m.p. $-7.5°C$, b.p. $246°C$, s.g. 1·048–1·052 (25°C). Miscible with oils, alcohol, chloroform and ether. Volatilised by steam.
Uses: odorant for NR, SR and latices.

CINTURATO Brand of tyre produced by Pirelli.

CIRCOSOL Naphthenic oil. Softener and plasticiser for NR and SR, improves processing and resistance to low temperatures.
Grades:
2XH: s.g. 0·9483, Saybolt viscosity 98·9°C, 83 sec for NR and SBR
NS: s.g. 0·9279, Saybolt viscosity 98·9°C, 61 sec for NR, SBR, CR, IIR and oil extended SBR (155).

CIS-4 Trade name for cis-1:4-polybutadiene (130).

CISDENE-100 cis-1:4-polybutadiene with min. 95% cis content (103).

CIS-TRANS ISOMERS

cis-1:4 polyisoprene (natural rubber)

trans-1:4-polyisoprene (gutta-percha)

Stereoisomeric forms which depend on a varying arrangement of the substituent groups at the double bond. A 1:4 addition compound of a diene can appear either in the cis or the trans form. If the substituents are arranged in one direction during chain formation, then a cis form occurs; if they are arranged alternately in two directions, the trans form appears. The isomers are differentiated by their chemical and physical properties. Since spacial distances of relative atoms or atomic groups of the isomers vary in size affinity and stability vary. Natural rubber is a cis-1:4 polymer, whereas gutta-percha has a trans configuration.

CITROFLEX Group of odourless, non-toxic softeners based on citric acid esters; for cellulose and polyvinyl plastics, chlorinated rubber and other polymers.
Use: packaging in the food trade.
Grades:
2: triethyl citrate
4: tri-n-butyl citrate
A2: acetyl triethyl citrate
A4: acetyl tri-n-butyl citrate
A6: acetyl trihexyl citrate
A8: acetyl-tri-2-ethyl hexyl citrate (155).

CK-3 German produced carbon black, comparable to HAF black (156).

CK-4 German produced carbon black, comparable to HAF black (156).

CL Abbreviation for 'Carey Island' clones.

CLASH–BERG TEST Determination of the stiffness and low temperature resistance of flexible polymers, by a torsion method. The variation of torsion angle with the temperature (of a test piece) is determined, using constant torque. ASTM D 1043-51.
Lit.: R. F. Clash and R. M. Berg, *Ind. Eng. Chem.*, 1942, 34, 1218.
Modern Plastics, July 1944.

CLIPPINGS Cuttings. Cuttings of sheet or crêpe, containing air bubbles or impurities.

CLITANDRA ORIENTALIS
C. Arnoldiana. An *Apocynaceae* (tropical Africa) yielding the wild rubber known as Noir du Congo or Cappa. The sap is coagulated by heating or by pouring into boiling water. In W. Africa, acid plant saps are also used, in particular those of the

Bossanga plant (*Costus lucasianus*), the Baobab fruit (*Adamsonia digitata*) and *Tamarindus indica L.*; also the leaves of *Bauhinia reticulata* and *Hibiscus sabdariffa*.

CLONES Selected trees of *Hevea Brasiliensis* with known characteristics, e.g. in relation to production, latex composition, biological and physiological factors, and resistance to disease. Propagation is achieved by bud grafting specially chosen plants or by mother plants selected by cross breeding.

CLOUDS Dull or opaque parts in sheet rubber.

CNS Abbreviation for copper non-staining vulcanisation system, for latex. Shows none of the discoloration or bleeding which frequently occurs in textiles on contact with traces of copper.

Thiazole (MBT,		
ZMBT, CBS)	1·03–1·3	0·2–0·4
Sodium-diethyl		
dithiophosphate	0·4–1·0	0·1–1·5
Activator*	0·2–0·5	0·2–0·5

* Activator: thiourea or substituted thiourea, DOTG, DPG.

COAGULANT WS 46 Polyetherpolyformaldehyde. Brown viscous substances, s.g. 1·09. Soluble in warm water. Coagulant for heat-sensitive natural latex, coagulation temperature 45–46°C (43).

COAGULATEX Mixture of approx. 50% sulphuric acid with 5–6% sodium chloride. Originally used on the plantation as a coagulating agent.

COAGULATING AGENT Coagulant. Substance which causes the coagulation of latex.

COAGULATION Each rubber particle in the latex is surrounded by a protein layer which acts as a stabiliser and prevents self-adhesion. The stability of the latex is determined primarily by the electrokinetic potential of this protective layer. Fresh latex has a pH of approx. 7, the charge on the rubber particles being negative. For proteins in latex, the isoelectric point, at which the negative charge equals the positive charge, occurs at pH 4·8. As the pH of the latex falls, the number of negative charges decreases and the latex becomes unstable; the particles may then approach sufficiently near to each other to adhere in an irreversible coherent coagulum. The coagulation point lies between the two critical electrokinetic potentials (e+ and e−). If the pH lies within this range the latex will coagulate. Outside the range of coagulation two stable zones exist, one negative, the other positive. If sufficient acid is added, the transition through the coagulation range occurs rapidly and a positively charged stable latex is obtained by reverse of charge. The proteins on the latex particles can also be removed by using a stronger, more adsorbent stabilising agent. In concentrated latex, for example, fatty acids are formed by hydrolysis and are saponified by ammonia. These soap particles then partially replace the proteins. At a sufficiently high concentration a complete replacement of proteins takes place. This is used as a practical process for the production of rubber which is poor in protein (so-called G-rubber). The change in potential

116

can also be achieved by replacing the proteins with cationic soap particles (Positex). Coagulation is also exploited as a production process for rubber, by using formic acid, acetic acid or oxalic acid as a 1% (formic acid) solution, or 2% solution. Approx. 4–6 g concentrated formic acid, 8–12 g concentrated oxalic or acetic acids are used per kg rubber. Spontaneous coagulation: if latex is left to stand, spontaneous coagulation occurs after a few hours at a relatively high pH; the causes are as follows:

1. Enzymatic destabilisation resulting in a change in the protective protein layer and partial or total coagulation.
2. Formation of free fatty acids by micro-organisms. The fatty acids partially replace the proteins on the surface of the latex particles and form insoluble magnesium soap particles with the magnesium present. These are precipitated and cause coagulation.

Latices of various clones have varying degrees of stability which are primarily dependent on the $Mg:PO_4$ ratio. The larger the magnesium content, the stronger is the tendency for precoagulation. Use is made of this fact to coagulate the yellow fraction (lutoids, Frey–Wyssling particles) in the production of pale crêpe. Precoagulation may be increased in the plant during the period of leaf change because of changes in tree metabolism. Similarly, during the rainy season, it is more likely to occur because of impurities in the latex, *e.g.* tannins, calcium salts, which are extracted from the bark. Coagulation can be prevented by adding stabilis-ing agents, *e.g.* ammonia, sodium carbonate, caustic soda, formalin.

COAGULUM The sponge-like water-holding rubber mass formed by the addition of coagulants or by spontaneous coagulation.

COALESCENCE Adhesion of latex particles to form larger flocks; first stage of coagulation.

COAL-TAR Neutralised coal-tar with a max. phenol content of 3% is used on the plantations as a protective paint for the tapped surfaces of *Hevea* trees. It prevents fungal diseases on the tapped surface (*e.g. Phytophthora palmivora*) and accelerates bark renewal.

COBALT ALUMINATE Blue inorganic pigment for rubber; s.g. 3·4–3·6.

COBALT CHROME ALUMINATE Bluish green pigment for rubber, s.g. 4·0–4·2.

COCKERILL PROCESS Process for producing rubber by electro-coagulation; based on the migration of the electrically charged latex particles to the anode. The anode is movable and precipitated film is removed continuously by a scraper, washed and dried. Because of the high cost and susceptibility to variation, this process soon became obsolete.
Lit.: Thomas Cockerill, BP 21 441 (1908), 5954 (1910), 5855 (1910), DP 218 927 (1908).

COFILL 11 1:1 blend of resorcinol and Ultrasil VN 3. Used for the RFS bonding system (156).

COHEDUR RL 1:1 solution of resorcinol in a liquid methylol compound (Cohedur A). Used for RFS bonding system, methylol serves as formaldehyde donor in place of hexamethylene tetramine (43).

COHESIVE ENERGY DENSITY CED. The energy necessary to move the polymer molecules in 1 cm³ of liquid sufficiently far apart to prevent them from influencing one another. Used to interpret the swelling behaviour

$$CED = \frac{\Delta H - RT}{V} = \alpha^2$$

where ΔH is the heat of vaporisation, and V is the molecular volume.

COLD CURE Process in which vulcanisation occurs at room temperature. Originally the term applied only to the Parkes process, *i.e.* vulcanisation with carbon disulphide. Other cold processes are the Peachey process, using sulphur dioxide and hydrogen sulphide, and curing with sulphur thiocyanate. Because of the development of latex technology, cold curing processes which are only suitable for thin walled articles have become virtually extinct. Modern ultra-fast accelerators enable latex and dry rubber to be vulcanised at room temperature.

COLD RUBBER Styrene butadiene copolymer polymerised catalytically at 5°C (usual polymerisation temperature for copolymers is 50°C).

COLLOCARB Carbon blacks pelleted with 20% oil (40).

COLLO LIGHT-WEIGHT FOAMS Polyurethane foams.

COLOMBIAN VIRGIN Trade term for the wild rubber of *Sapium biglandulosum* (Brazil) and *S. Thomsinii* (Colombia). Contains approx. 2·5% resin, and 0·6% ash and was formerly used as good quality rubber for the production of rubber solutions.

COLORADO Trade term for wild rubber from the roots of *Actinella Richardsonia* and *Picradenia floribunda utilis*.

COLWYN MEDAL Medal awarded yearly by the Institution of the Rubber Industry Lond., for exceptional scientific or technical achievements in the rubber industry. First award 1929.

COMMINUTED RUBBER Natural rubber produced in crumb form and compressed into compact bales. The trade names for this new type of rubber, which is now sold under technical specifications, include Heveacrumb, Dynat, NATCOM, NAT.

COMMON MILKWEED American term for *Ascelepias syriaca L*, which contains 0·5–4·5% pure rubber in the dried leaves and 0·2–0·5% rubber in the stem.

COMPO Abbreviation for compound; colloquial term for the brown crêpe type rubbers produced from scrap rubber.

COMPRESSION FATIGUE Observed in soft, flexible foam materials during repeated distention and compression. A test piece, at a specified amplitude, frequency and load, is repeatedly compressed to a distended shape, then relieved or sheared alternately on opposite sides. DIN 53 574.

COMPRESSION SET The set which remains after a predetermined time (30 min) after a compressive strain has been removed from a test-piece. Used for evaluating rubbers which may be used under continuous deformation, *e.g.* damping bases, buffers, springs, sealants. Two factors are determined:

1. Constant strain

$$V = \frac{h_0 - h_2}{h_0 - h_1} \cdot 100\%$$

2. Constant stress (28–30 kg/cm^2)

$$V = \frac{h_0 - h_2}{h_0} \cdot 100\%$$

where h_0 is the original thickness of the test piece

h_1 is the thickness in deformed state, and

h_2 is the thickness 30 min after removal of the load.

The test can take place at 20°C (70 h), 70 or 100°C (24 h), and also at very low temperatures (ASTM D 1229). For tests under constant strain, the strain depends on the hardness of the test piece (20–40%).
Specifications: ASTM D 395, D 1229 DIN 53 517
BS 903: 18: 1950.

COMRUB Trade name for crumb rubber made by the Dynat process.

CONAP 1146 Solution of polymers in a graded solvent. Adhesive for metal/urethane assemblies (388).

CONCENTRATED LATEX
Latex fresh from the tree has a rubber content of approx. 33–40%.
To facilitate transport during shipping and since many manufactured articles require a high rubber concentration, the rubber content of latex is increased on the plantations to 60–66%. Only a limited market remains for unconcentrated latex. Concentration can be done by four processes: creaming, centrifuging, electrodecanting and evaporation.
Creaming [Traube's process BP 226 440 (1924)]. As rubber particles have a lower specific gravity than the water phase, they cream when the latex is left to stand. This process occurs slowly but may be considerably accelerated by increasing the temperature and by adding small quantities of colloids such as sodium alginate, tragacanth, Icelandic moss, konnyaku flour or carob seed flour; sodium alginate is the most widely used creaming agent. Latex, preserved with ammonia or with ammonia and sodium pentachlorophenate, is separated until the magnesium ammonium phosphate is removed. It is then placed in tall containers in which it separates into two layers within 4–6 days; into the cream with approx. 60% rubber and 61–65% total solids and into a serum which contains little rubber; the process is continuous. The speed of creaming may be calculated according to Stokes' formula:

$$V = \frac{d^2(D_s - D_k)g}{18\eta}$$

where d is the diameter of particles
D_s is the density of the serum
D_k is the density of the rubber
g is the gravitational constant and
η is the viscosity.

With the addition of colloids the

effective particle diameter is considerably enlarged by the reversible agglomeration of rubber particles and creaming accelerated.

Centrifugation. This process was used as early as 1906 on the Louis d'Or rubber plantation, Tobago, to obtain rubber from Castilla Latex. The larger part of today's production is done by centrifugation based on the W. L. Utermark process [BP 219 635 (1923)]. Fractionation is possible because of the difference in s.g. of the rubber particles (0·91) and serum (1·02). It is carried out in high speed separators, *e.g.* De Laval and Westphalia centrifuges. The value of the g in Stoke's formula is increased approx. 5000 times so that continuous concentration is possible. The latex which has been stabilised with 0·2–0·4% ammonia is separated into two parts of approx. equal volume. The cream fraction has a rubber content of 60% and a max. 62·5% total solids, the skim fraction contains approx. 10% total solids of which 4–6% is rubber. After centrifugation, additional preserving agents are added, *e.g.* gaseous ammonia or sodium pentachlorophenate as a 20% solution. Skim latex, after exhaustion by steam, or neutralisation of the ammonia, contains approx. 4–7% rubber which is directly coagulated or is first subjected to enzymatic decomposition of the proteins. Natural antioxidants are bound firmly to the rubber particles and so remain in the cream, whereas organic acids and other undesirable substances are transferred to the serum.

Electrodecantation (*Electrodialysis*). A process formulated by Pauli and Stamberger [BP 492 030 (1937)]. In which use is made of the migration of negatively charged latex particles in an electric field. Insertion of a permeable membrane, which does not prevent the passage of an electric current, stops the particles reaching, and coagulating on the anodes. Instead, they become concentrated on the membrane. The lighter cream is separated by decanting. Standard production units have approx. 150–200 cells, with 1 cm intervals between the membrane and can be used continuously. Ammonia added as a preservative, increases the charge and therefore the migration speed of the particles. Power consumption is 40–120 W/kg of 60% latex at a potential gradient of 0·8 V/cm. Capacity, starting with 30% latex, amounts to approx. 1 kg of 60% latex/m^2 membrane/h.

Evaporation. By evaporation according to the Revertex process formulated by E. A. Hauser [BP 243 016 (1926)], to obtain concentrates of 70–75% total solids and approx. 66% rubber, with a paste-like consistency. Unlike the other concentration processes this latex retains all the serum constituents, alkalis as preserving agents and soaps as stabilisers. In the original process, the evaporation occurs in a heated rotating drum which has concentric openings on both sides, through which the steam is led off. A small drum rotates inside to ensure thorough mixing of the latex and prevent skin formation. According to another process the heated latex is sprayed under a partial vacuum into a container so that partial evaporation of the water takes place. The latex is taken through the process several times to achieve the required concentration; Revertex is used chiefly in coatings and adhesives. The following quality classifications are

	I	II	III	IV
Solids, min %	61·5	64·0	61·5	64·0
Dry rubber content, min %	60·0	62·0	60·0	62·0
Difference (solids minus DRC) %	2·0	2·0	2·0	2·0
Total alkali, calculated as ammonia, % on wet latex	1·6 min	1·6 min	1·0 max	1·0 max
Viscosity, cp, max, 25°C	50	50	50	50
Sediment, % on wet weight	0·10	0·10	0·10	0·10
Coagulum, % on dry weight	0·08	0·08	0·08	0·08
KOH No., max	0·80	0·80	0·80	0·80
Mech. stability, min, s.	475	475	475	475
Copper, max % on dry weight	0·001	0·001	0·001	0·001
Manganese, max % on dry weight	0·001	0·001	0·001	0·001
Colour (visual)	Not blue or grey coloured			
Smell on neutralisation with boric acid	No foul smell			

used for concentrated latex (ASTM D 1076-59):

Type I, centrifuged natural rubber latex preserved with ammonia or formaldehyde and ammonia

Type II, creamed natural rubber latex preserved with ammonia or formaldehyde and ammonia

Type III, centrifuged natural rubber latex with a low ammonia content and other preserving agents

Type IV, creamed natural rubber latex with a low ammonia content and other preserving agents.

CONDAMINE, CHARLES MARIE DE LA 1701–74. Took part in a French expedition to equatorial America, 1735–45 and sent the French Academy of Science a dark resinous mass, which was called caoutchouc. In 1745 published his article 'Relation Abrégée d'un voyage fait dans l'Interieur de l'Amérique Méridionale', with a section on rubber. Before returning to France he met Fresneau in Cayenne. In 1751 read Fresneau's report to the Academy of Science and mentioned that in 1736 he had sent samples of rubber to Paris, which, however, had never arrived. A section of his accompanying letter was also read out. Condamine had probably never seen a rubber tree, but only the rubber articles made by the Indians; he had merely heard of the trees from others, as from Fresneau.

CONDENSATION A chemical reaction in which two or more molecules combine with the separation of water or simple molecules. The reaction is termed polycondensation when it results in the formation of polymers.

CONGO RUBBER Collective term for various natural rubber types (Africa) from *Kickxia elastica, Ficus nekbudu, F. Bubu* and similar species, such as *Carpodinus congolensis, C. Jumellei, Clitandra zunde, Cl. orientalis, Landolphia Droogmansiana, L. Gentilii, L. Keinei, L. Thollonnii* (root rubber). In the early days of the rubber industry the Congo was second to Brazil in supplying world markets with natural rubber. There were various good quality trade types, such as Kassai rouge, Kassai nor, Lopori, Equateur, Bussira, Ruki, Yakome and Ikelemby sheets; also the more or less impure types, such as mongala, bumba, aruwimi; main trading place was Antwerp. Wild natural rubber is no longer of importance. In 1903 the total output of the Belgian Congo was 6470 tons as opposed to 1747 tons exported from Africa in 1932.

CONSISTENCY Thixotropy. The resistance which a material, subjected to a shearing force (stirring) offers to flow, or permanent deformation. The term is used for materials in which the deformation is not proportional to the shearing force.

CONSISTOMETER Special viscometer used to determine the consistency, *e.g.* Newtonian viscosity, structural viscosity, plasticity of materials such as elastomers, resins, dispersions, emulsions.

CONTINENTAL Trade name for channel blacks (158).

Grade:
A: MPC
AA: EPC
F: HPC
R: CC.

CONTINEX Trade name for furnace blacks (158).

CONTINUOUS FLEXING TEST The resistance to crack formation and its growth may be determined using a standard test piece and applying a continuous flexing strain. Specifications: ASTM D 430-59, D 813-59
BS 903: A10: 1956, BS 903: A11: 1956
DIN 53 522.

CONTINUOUS OSCILLATION TEST Experimental determination of the mechanical properties of products under continuous oscillating strain.
Specification: DIN 50 100.

CONTOGUM Group of modified resin products. Plasticisers, tackifiers (127).

CONTROZON Antiozonant wax, softening point 55°C, s.g. 0·91. Dosage 3–12%.

COPEENBLAK Dispersions of blacks in polyethylene (159).

COPO Trade name for butadiene-styrene copolymers (cold rubber) and latices (160).

COPOLYMER A uniform polymer produced by the polymerisation of chemically distinct monomers which are built into a regulated order, in the final product, *e.g.*

butadiene-styrene (SBR), butadiene-acrylonitrile (nitrile rubber), isobutylene-isoprene (butyl rubber).

COPOLYMERISATION The polymerisation of two distinct monomers to form a single polymer. Polymers are classified according to the arrangement of the monomers in the molecular chain. *Statistical polymers (random polymers)*. The monomer units are arranged statistically in relation to their quantities, *e.g.*

$(-A-B-)_n$	1:1
$(-A-A-B-)_n$	2:1
$(-A-B-C-)_n$	1:1:1

However, the statistical random arrangement cannot occur exactly in a single molecule because of the different polymerisation speeds of the monomers. In the initial stage of polymerisation, molecules are produced with a higher concentration of one component and in the final stage molecules contain more of the other.

Block polymers. The single monomer units follow each other in the molecular chain in longer sequences:

$$[-(A)_x-(B)_x-]_n$$

Graft polymers. The main chains consist of one component and shorter, subsidiary chains for the other component are grafted on to them.

$$\begin{bmatrix} (B)_y & (B)_y & (B)_y \\ | & | & | \\ -(A)_x-(A)_x-(A)_x- \end{bmatrix}_n$$

The polymerisation reaction can be classified as: block polymerisation (pure monomers), solution polymerisation (monomers in suitable solvents), emulsion polymerisation (monomers in an aqueous emulsion), suspension (or pearl) polymerisation (monomers in a non-solvent, usually water, in suspension).

In copolymerisation the intrinsic properties of the pure polymer may be changed. Products with new, independent properties are usually produced, solubility, tendency to crystallise, melting temperature, freezing temperature, mechanical properties being among the properties affected.

COPPER Even minute quantities of copper have a strong catalytic effect on the oxidation of rubber and accelerate its degradation. Not all copper compounds are equally active, active and passive forms being distinguished. The strongest catalytic effect is found in compounds with higher fatty acids (stearates, oleates) and resins, also in salts which are soluble in rubber and form compounds with the fatty acids of the rubber. Inorganic salts have a slight catalytic effect; copper sulphate is approx. 1/5, and copper oxide approx. 1/10 as effective as the stearate, while the sulphide shows only slight activity. Complex compounds of copper, such as those occurring in the phthalocyanine colour pigments, are inactive and these colours may be used in rubber compounding.

Because of its deleterious effect the maximum amount of copper allowed in raw rubber by the RMA is 8 ppm. The quantity of copper which may be added to a rubber article without serious catalytic effect depends on the type of elastomer and the constituents of the compound. Ingredients may be added to the compound as inhibitors which form a complex compound with the copper and render it

123

inactive (*e.g.* salts of ethylene diamine tetra-acetic acid, disalicyl propylene diamine).
Natural rubber contains traces of copper as enzyme constituents. The average content of plantation rubber is 1–5 ppm. The amount in latex is higher and can, depending on clonal origin and the nature of the ground, be up to approx. 12 ppm. A large proportion of this amount, is removed by solution in the serum during coagulation. Copper is a constituent of the polyphenol oxidase (tyrosinase), present in latex. A concentration of tyrosinase, and consequently also of copper is found in the dried strips around the tapped surface, so that the copper content may increase to as much as approx. 20 ppm. Tyrosinase therefore causes oxidation of the phenolic constituents, resulting in darkening of the scraps, probably because of melanine formation. Copper in this rubber is sparingly soluble and the trade types produced from it (brown crêpe) can therefore have a higher copper content level than allowed by the specifications.

COPPER DIETHYL DITHIO-CARBAMATE
$[(C_2H_5)_2N.CS.S—]_2Cu$. Dark brown powder, m.p. 196°C, s.g. 1·70, mol. wt. 360. Accelerator for butadiene-styrene rubber. Gives faster cures than zinc diethyl dithiocarbamate and tetramethyl thiuram disulphide. Decreases compression set, improves abrasion and tear resistance. As a primary accelerator with low sulphur, gives good resistance to heat ageing in compounds.
Uses: particularly for hot air curing, cements, soles, tubing, cable insulation.
TN: + Ethyl cumate (51).

COPPER DIMETHYL DITHIO-CARBAMATE
$[(CH_3)_2N.C(S).S]_2Cu$. Dark brown powder, m.p. above 325°C, s.g. 1·61–1·78, mol. wt. 304. Soluble in acetone, benzene and chloroform, insoluble in water, alcohol and benzine. Ultra-fast accelerator for SBR and IIR. Normally used with thiazole to decrease the tendency to scorch.
TN: Cumate (51)
Robac CuDD (311)
Van Hasselt MTCu (355).

COPPER INHIBITOR
Grades:
50: 50% disalicyl propylene diamine 50% aromatic solvent (6)
65: 65% disalicyl propylene diamine. 35% aromatic solvent
+ X-872, X-872-A: disalicylethylene diamine
+ X-872-L: 80% disalicyl propylene diamine and 20% aromatic solvent. Crystallises at low temperature.

COPPER-2-MERCAPTO BENZTHIAZOLE

Reddish-yellow powder, decomposes above 300°C. Accelerator for SBR, particularly for open steam cures. Gives faster cures than mercaptobenzthiazole (obsolete).
TN: + Copray (51).

CORAL RUBBER Synthetic polymer whose structure and physical properties greatly resemble those of natural rubber. Consists of 93·8% cis-1:4-polyisoprene and a small amount of 3:4 structure, but no trans form. The synthesis occurs on polymerisation of purified isoprene, in a dry, oxygen free atmosphere, at 30–40°C, with 0·1% finely dispersed lithium as catalyst (129).

CORAX Types B and L. High modulus blacks (156).

CORAX PROCESS Process used in Germany for the production of blacks; similar to the furnace process.

CORD Tightly twisted thread made from cotton, rayon, nylon or polyesters; used as the inner, textile lining (carcase) of tyres. Invented in 1888 by J. Moseley (BP 11 804). Tyre cords consist of a dense arrangement of warp threads, held together by a few thin weft threads occurring at approx. 1 cm intervals. This is necessary because in normal textiles the crisscross pattern of the threads results in breakage through friction. Because of the tight twisting the cord has excellent resistance to tensile fatigue, but however, shows a decrease in tensile strength. To ensure adhesion to rubber, the cord must be impregnated with rubber prior to the coating operation: a compound of natural latex with resorcinol-formaldehyde with the addition of vinyl pyridine latex for rayon and polyamide threads is used; isocyanates are used for polyester cord. After impregnation the cord is heated for a few minutes to approx. 140°C, until condensation of the resin occurs on the cord threads. Adhesion to rubber is facilitated by the latex. The quantity of impregnating agent to achieve the best adhesion is 4–8% dry weight. Rayon cord shows 2–12% shrinkage after condensation, according to the tyre. Nylon and polyester cords are stretched while heated after impregnation to improve strength and decrease the tendency to stretch.

CORN RUBBER Former American term (obsolete) for maize oil factice.

CORRICREPE Crumb rubber from Liberia.

COUMARONE-INDENE RESINS Light brown to dark brown, viscous liquids to solids, s.g. 1·08–1·15. Plasticisers, softeners, tackifiers and reinforcing agents for natural and synthetic rubbers with the exception of butyl rubber. The function is dependent on m.p.; resins with m.p. below 50°C are chiefly plasticisers and tackifiers, with m.p. between 50–125°C, they act as softeners and reinforcing agents, and with m.p. above 125°C, as softeners and extenders. Can be mixed directly into a compound or added in masterbatch form, improve processing and electrical insulation properties. The melted resins dissolve rubber and up to 12% sulphur.

COUMARONE RESINS. Benzofuran resins, m.p. 35–170°C. Soluble in benzene, toluene, chlorinated hydrocarbons, ester and ethers. Products of coumarone, polymerised in the presence of acids and resistant to alkalis and acids. *Uses:* softeners, vulcanisation accelerators, in adhesives.

COURLENE Polyethylene yarn (161).

COURPLETA Triacetate fibres (161).

COURPLETY Triacetate fibres (161).

COURTELLE Acrylic fibres (161).

CP 40 Approximate formula $C_{24}H_{43}Cl_7$, chlorinated paraffin. Viscous liquid, s.g. 1·185. Softener and extender for plastics, improves flame resistance (162).

CPD Abbreviation for cadmium pentamethylene dithiocarbamate.

CPV Abbreviation for Centrale Proefstations Vereeniging, Bogor, Java, economic and technological institute of the plantations in Java. Renamed 1958 Balai Penjelidikan Perkebunan Besar BPPB.

CR Term used for chloroprene rubber in the ASTM catalogue D 1418-58T.

CRACK FORMATION Cracks occur in stretched vulcanisates exposed to the weather; they normally lie in the direction of the strain and are presumably caused by ozone. If unstrained vulcanisates are exposed to the weather then fine cracks occur in the rubber in random directions. The formation of cracks depends largely on external conditions such as temperature, ozone concentration, humidity, light intensity and wavelength, and the mechanical strain.

CRANAX Crotonaldehyde-aniline reaction product (5).

CRAZING EFFECT Weather skin, elephant skin. With non-black vulcanisates, a wrinkled surface with a network of small cracks occurs on oxidation caused by weathering and accelerated by light. The surfaces reveal ozone cracks when stretched. Both crack formations can be observed in one sample.

CREEP The change with time of the dimensions of an elastomer or plastic under stress. At room temperature the phenomenon is also termed cold flow. Creep deformation is dependent on the strain, time, temperature, state of vulcanisation, filler and type of rubber, is considerably greater in synthetic rubbers and in loaded compounds.

CREEP STRESS RESISTANCE The maintenance of the static stress induced by a continuously applied force.

CREMONA Polyvinyl alcohol fibres (375).

CREPE RUBBER 1·2–1·5 mm thick rubber sheets, mostly with a rough surface texture and relatively loose structure. Produced by rolling out and washing the coagulum on rolls operating at friction speed. To produce material of an acceptable colour, the latex often has to be fractionally precipitated to remove carotenoids and the Frey-Wyssling particles. According to the clonal origin of the latex, these normally give the crêpe a yellowish to strong yellow colour. The undiluted or slightly diluted latex is mixed with 1/6 to 1/3 of the acid necessary for coagulation, and stirred slowly. After a short period the unstable, yellow fraction separates out and is processed into off-crêpe. The remaining latex is strained and diluted to a rubber content of 16–20%. Approx. 0·6% sodium bisulphite is added to the latex as a bleaching agent, and sometimes it is also necessary to add xylyl mercaptan (0·1%). Coagulation occurs with 1% formic acid (400–500 ml/kg rubber) or with 1% oxalic acid (800–1200 ml/kg). Clonal composition or the latex and the type of crêpe (thin lace crêpe, thick blanket crêpe) determine whether or not addition of xylyl mercaptan, or fractionated coagulation is necessary. After a few hours the coagulum formed is washed thoroughly on ribbed rolls and passed through two or three further pairs of rolls (the last with smooth surfaces) to form thin sheets. As crêpe has a low serum content, preservation is unnecessary. The sheets are dried in hot drying chambers or tunnels at 37°C (2½–4 days), or on drying floors (5–10 days). Higher temperatures have to be avoided as discoloration occurs because of oxidation. The crêpe is packed into bales of approx. 80 kg or, as in Ceylon, rolled to approx. 10 mm thick blankets. First quality crêpe is often packed in paper sacks impregnated with polyethylene or lined with foil.

C-RESIN Low mol. wt. liquid polysulphide containing amines. Pourable mass which hardens (374).

CRESYL DIPHENYL PHOS-PHATE
$(CH_3C_6H_4O)(C_6H_5O)_2PO$. Odourless, colourless liquid, m.p. $-40°C$, b.p. 390°C, flash p. 232°C, s.g. 1·208. Softener and plasticiser for SBR and NBR, improves mastication and processing. Does not affect vulcanisation.
TN: Celluflex 112 (152)
Santiciser 140 (5).

CRI Abbreviation for crystallisation-resistance index.

CRIMEAN SAGHYZ Rubber from *Taraxacum hybernum Stev.* types (*T. gymnamthum, T. megalorhizon, T. allepicum,* Mastikana, Kuljbaba krimska osinja) Mediterranean area and the Crimea. Productive for several years with two vegetation periods per year. Rubber is obtained from the sap tubes of the roots.

CRITICAL TEMPERATURE A term used in the rubber industry for the approx. temperature below which vulcanisation no longer occurs, or occurs only very slowly. Alternatively the temperature below which an accelerator loses its effect.

CRMB Abbreviation for cyclised rubber master batch. Consists of 50:50 cyclised rubber and NR. Good processing and gives a high modulus and high hardness at low density. Gum rubber vulcanisates have a Shore A hardness of 96°.
TN: Cyclite (169, 170)
Cyclatex (168, 114).

CROFLEX Group of reinforcing blacks. Croflex: HPC, 20: SRF, 20HM: SRF-HM, 35: GPR, 40: HNF, 50: FEF(MAF), 60: HAF, 70: ISAF, 77: EPC, 85: SAF, CF: CF, TH: MPC, TF: FT, TM: MT (164).

CROLAC Types T-10, T-14, T-16. Conductive channel blacks (81).

CRONAR Polyethylene terephthalate (6).

CROSSLINKING A term associated with high polymer products indicating the formation of a

127

network, normally with a three-dimensional structure, by intermolecular bridges consisting of molecular chains with reactive groups. The main chains in crosslinked products may be separated only by thermal degradation or highly reactive agents. The properties of polymers are affected considerably by the degree of crosslinking, *i.e.* by the number of crosslinks. In elastomers such crosslinking is termed vulcanisation, even when sulphur does not cause the crosslinks. Solubility, softening and plasticity are normally decreased by crosslinking.

The properties of crosslinked materials depend on the type of the main chain and the number of crosslinks. To achieve elasticity the number of crosslinks must be small.

CROSS, ROBERT In 1875 brought seedlings of the Ule tree from Brazil to England. These were sent to India in 1876. In May, 1876 was once again sent by the India Office to Brazil and in November returned with 1080 *Hevea* seeds. Of these seeds only 3% germinated. A hundred of these were later sent to Ceylon as the number of plants was increased by means of seedlings. While Cross was in Brazil, H. Wickham returned to England in 1876 with 70 000 seeds.

CROTONIC ACID
CH_3—CH_2=$CH.COOH$, β-methylacrylic acid. Monoclinic needles, m.p. 72°C, b.p. 185–189°C, s.g. 1·018. Soluble in alcohol. Softener for synthetic rubbers used in production of copolymers with vinyl acetate.

CROYDAX 2-mercapto-benzthiazole surface treated with oil (120).

CRF Abbreviation for channel replacement furnace black. Substitute for channel black.

CRYLOR Formerly Fibre D. Polyacrylonitrile fibre (165).

CRYPTOSTEGIA Plants of the *Asclepiadaceae* family; once sources of good quality wild rubber (Madagascar), particularly *C. grandiflora* and *C. Madagascariensis.*

CRYSTALLISATION First order transition. In most polymers crystallisation occurs at low temperature as a result of orientation of the molecules. It is characterised by changes in hardness and modulus. Crystalline elastomers are stiff, though not necessarily brittle. The brittle condition (glass transition) occurs only at much lower temperatures.

Natural and synthetic rubbers are generally considered amorphous, but under certain conditions, *e.g.* low temperatures, compression of unloaded materials, or during stretching, signs of crystallisation may appear.

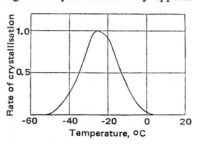

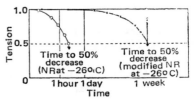

Crystallisation (orientation) of natural rubber occurs during storage at low temperature, causes decrease in elasticity and increase in stiffness. The rate of crystallisation depends on the temperature and load and reaches its maximum rate at $-26°C$. At an elongation of 100%, the rate of crystallisation is increased 10 times, at 200% elongation, approx. 100 times. Crystallisation may be significantly retarded by the simple modification of latex or solid rubber, *e.g.* with thiols, and modified natural rubber may be produced which scarcely crystallises. The time taken for the tension in a stretched test piece to decrease, at a given temperature and elongation, to 50% its original value is the crystallisation half-life and is a measure of the degree of resistance to crystallisation. The relation of the half-life of the modified rubber to that of non-modified rubber is a measure of the retardation of crystallisation.
Lit.: *Trans. IRI*, 1954, 30, 144.
J. Pol. Sci., 1959, 36, 77.
ASTM D 832-59.

CRYSTEX Sulphur insoluble in rubber at normal processing temperatures (guaranteed 85% insoluble). Metastable form, transformed into soluble sulphur (during vulcanisation).
Use: for vulcanised compounds in which sulphur blooming is undesirable (166).

CSN Czechoslovakian standard.

CUDC Abbreviation for copper diethyl dithiocarbamate.

CUMAKABALLI RUBBER Bartaballi. Rubber-like product used in Guyana as a cheap substitute for balata.

CUMAR Group of p-coumarone/p-indene resins.
Uses: extenders, processing aids, tackifiers for metal/rubber bondings and cements (29).

CUMARIN

$$C_6H_4 \begin{array}{c} O—CO \\ | \\ CH=CH \end{array}$$

1:2-benzopyrone, coumarin. Orthorhombic crystals, m.p. 68–70°C, b.p. 297–299°C, s.g. 0·94. Soluble in alcohol, chloroform, ether and oils.
Uses: odorant for rubber compounds.

CUMBT Abbreviation for copper dimercaptobenzthiazolate.

CUMD Abbreviation for copper dimethyl dithiocarbamate.

CUMENE HYDROPEROXIDE Dimethyl benzoyl hydroperoxide. Colourless oily liquid, b.p. 53°C, s.g. 1·062. Soluble in acetone, alcohol and hydrocarbons. Catalyst in the emulsion polymerisation of styrene butadiene rubber at 5°C (cold rubber), hardening agent for polyester resins (383).

CUP LUMPS After the cups collecting the latex from the tree have been emptied, the tree usually continues to drip sap slightly. The latex coagulates because of the effect of bacteria and is taken out of the cups, on the next tapping, as small lumps. These so-called cup lumps are usually processed to thin brown crêpe. Frequently this rubber is classified according to its colour before being traded. Fresh cup lumps normally give Brown 1X.

CURACAP Group of accelerators:
Cadmium PD: cadmium pentamethylene dithiocarbamate
Lead PD: lead pentamethylene dithiocarbamate
MBT: 2-mercaptobenzthiazole
MBTS: 2,2′-dibenzthiazyl disulphide
PPD: piperidine pentamethylene dithiocarbamate
Zinc MBT: zinc-2-mercaptobenzthiazole
Zinc PD: zinc pentamethylene dithiocarbamate (241).

CURADE Blend of organic salts of various amines. Activator (100).

CURE Vulcanisation or the conditions under which vulcanisation takes place.

CURE BLEND Group of master batches produced by coagulation of an SBR latex containing a non-discolouring antioxidant with a 50% phr accelerator:
MS: tetramethyl thiuram monosulphide
MT: tetramethyl thiuram disulphide
MZ: zinc dimethyl dithiocarbamate (31).

CURING BAG An inflatable bag made of a heat resistant compound (usually butyl rubber), used for the vulcanisation of tyres. During vulcanisation the curing bag is kept under pressure with steam, air or water, thereby shaping the tyres in the mould.

CUROMETER Apparatus for determining the vulcanisation curve

of elastomers by the changes in shear modulus. Developed by RAPRA.
Lit.: *Trans. IRI*, 1962, 37, 206.

CUSSON'S DUROMETER Instrument (obsolete) for determining the hardness of rubber. Used a great deal in Britain around 1920.

CV Often incorrectly termed CV cure. Abbreviation for continuous vulcanisation.

CV RUBBER Constant viscosity NR (about 65 Mooney). Stabilised by the addition of hydroxylamine hydrochloride. Abbreviation not to be confused with continuous vulcanisation.

CWH Abbreviation for Chemische Werke Hüls.

CYANACRYL Fast curing acrylic elastomer. Remains flexible at low temperatures (21).

CYANAFLEX Diphenylamine-acetone reaction product (21):
50: 50% with 50% carbon black
100: normal type.

CYANAPRENE 4590 Polyurethane. Shore A 85–87° (21).

CYANOGUANIDINE
$NH_2.C(:NH).NH.CN$, m.p. 205°C.
Accelerator.
TN: Ultracen (with zinc oxide) (167).

CYANOX Group of antioxidants (21):
8: octylated diphenylamine, s.g. 0·98
53: alkylated bisphenol. White powder, s.g. 1·74

B: diphenylamine. Soluble in benzine, benzene and acetone, m.p. 78–85°C, s.g. 0·98
DH: N,N'-bis(1:4-dimethyl pentyl)-p-phenylene diamine
IP: N-isopropyl-N,N'-phenyl-p-phenylene diamine
LF: alkylated phenol. Brown, viscous liquid, s.g. 1·06.

CYCLATEX A material produced from latex and consisting of equal parts natural latex and natural rubber cyclised in latex form. Brown sheets, s.g. 0·95. Reinforcing agent for natural rubber. Gives products with a high modulus and high hardness; improves tear resistance, reduces the swell and gives a good surface finish. May be mixed with rubber in the normal way, may also be used as a processing aid for neoprene compounds (168, 114).

CYCLE Complete operation of a moulding press from the time of closing to the next time of closing.

CYCLISATION The ring formation in a chemical compound. In rubber, denotes transformation into an isomeric product with C_{10} rings of the same chemical formula $(C_5H_8)_x$ but with a higher degree of saturation and different chemical and physical properties.

CYCLISED RUBBER By heating rubber and reacting it with various chemicals it may be isomerised to cyclic products which may be soft, hard, non-elastic or thermoplastic according to the degree of cyclisation. The chemical formula is the same as in natural rubber

$(C_5H_8)_x$, but there is a lesser degree of unsaturation. Softening p. 90–120°C, s.g. 0·992, mol. wt. 2000–10 000. Soluble in aromatic and aliphatic hydrocarbons, solubility depending on the degree of cyclisation and the method of production. The first cyclisation process was introduced by J. G. Leonhardi in 1781, using sulphuric acid. Cyclisation proceeds as follows:

Since cyclisation can occur at any site in the rubber molecule, single C_5 units may be enclosed between the rings.

1. *Cyclisation with an acid.* Products made by this process are called thermoprenes. Cyclisation is achieved by using sulphuric acid and organic derivatives of the general formula $R.SO_2.X$, sulphonic acid and sulphonyl chloride, by phenols in the presence of sulphuric acid or phosphoric acid, or by using trichloracetic acid, etc.; p-toluene sulphonic acid and toluene sulphonyl chloride are most widely used. Rubber is mixed

with approximately 10% reactants on a mill and heated for 1–4 h at 125–145°C. The properties depend on the degree of cyclisation and production conditions. The products can be tough masses similar to gutta-percha or hard, crumbly masses. Rubber cyclised with phenol is much more difficult to break, has a low mol. wt., is readily soluble and gives highly concentrated solutions with a low viscosity. Cyclisation can also be carried out in latex stabilised with cationic or non-ionic wetting agents. Concentrated sulphuric acid is used and the product is obtained in the form of a fine powder (m.p. 130°C) by pouring the latex into hot water.
2. *Cyclisation by halides.* The reaction of dissolved rubber with tin tetrachloride, titanium tetrachloride, aluminium trichloride, antimony pentachloride, and other chlorides yields an amorphous, colourless powder, s.g. 0·98; soluble in most rubber solvents. Rubber may be cyclised in solid form with chlorostannic acid ($H_2SnCl_6.6H_2O$) in 2–5 h at 130–150°C. Products of this type are known as Pliolite and Pliofilm.
3. *Thermal cyclisation.* Heating a rubber solution for several hours at 250–300°C gives approx. 40% yield, the rest being lost as volatile decomposition products. The product is a yellow powder, mol. wt. approx. 2500, insoluble in acetone and alcohol, soluble in many solvents.
4. *Reduction by dehydrohalogenation.* Heating rubber hydrochloride or hydrobromide with zinc dust gives a cyclised rubber with a mol. wt. approx. 2500–15 000.
Uses: Reinforcing agent for rubber, *e.g.* in shoe sole compounds, bonding agent for printing inks, lacquers and paints, metal/rubber bonds (Vulcalock process); golf balls, impregnations, insulation, mechanical goods. Cyclised latex acts as a reinforcing agent for latex compounds used for dipped articles and foam rubber.
TN: Alpex 450J
CRMB
Cyclatex
Cyclite
Fenolac
Plastoprene
Pliofilm
Pliolite
Thermoprene

CYCLITE Mixture of equal parts of natural and cyclised rubber. Identical with Cyclatex (169, 170).

CYCLOCURE Accelerator of undisclosed composition (obsolete) (5).

CYCLODECATRIENE-(1, 5, 9) Trimer of butadiene. Obtained by polymerisation of butadiene by using a Ziegler catalyst under suitable conditions.

CYCLOHEXANE C_6H_{12}, hexahydrobenzene, hexamethylene. Colourless liquid, m.p. 6·5°C, b.p. 80·7°C, s.g. 0·77–0·78. Miscible with alcohols, hydrocarbons, chlorinated hydrocarbons, amines. Solvent for natural and synthetic rubbers with the exception of NBR, and for adjusting the viscosity of cements.

CYCLOHEXANOL $C_6H_{11}OH$, hexalin, hexahydrophenol. Hygroscopic crystals with camphor-like odour, m.p. 23–25°C, b.p. 161°C, s.g. 0·96. Miscible with alcohol, aromatic and aliphatic hydrocarbons. Solvent for NR and SR with the exception of NBR.

CYCLOHEXANONE $C_6H_{10}O$. Oily liquid, m.p. $-45°C$, b.p. $155·6°C$, s.g. $0·948$. Miscible with alcohol and most organic solvents. Solvent for NR, synthetic rubbers and vinyl polymers.

CYCLOHEXYLAMINE ACE-TATE White crystalline powder, s.g. $1·12$. Soluble in water and alcohol. *Uses:* coagulant for latex dipped articles. Has no effect on properties, ageing or odour.
TN: Coagulant CHA (43).

N-CYCLOHEXYL-2-BENZ-THIAZOLE SULPHENAMIDE

Pale grey, non-hygroscopic powder, m.p. $95-100°C$, s.g. $1·27-1·31$, mol. wt. 254. Soluble in aromatic hydrocarbons, chlorinated hydrocarbons, carbon tetrachloride, acetone and ethyl acetate, sparingly soluble in benzine and alcohol, insoluble in water and diluted alkalis. Effective accelerator with a delayed action, and processing safety in compounds containing reinforcing blacks and silicate fillers. Extremely active at high temperatures. Vulcanisation temperature $135-150°C$, 5% zinc oxide and 1% stearic acid are necessary, boosting is not usually required. In blends with tetramethyl thiuram disulphide, suitable for rapid vulcanisation at high temperatures, *e.g.* in continuous vulcanisation of cables, etc. Gives a high modulus, high tensile strength and elasticity, good resistance to abrasion, good ageing properties and resistance to flex cracking. Suitable for natural rubber, synthetic rubbers (SBR and NBR) and reclaim.

Quantity: $0·5-1\%$ with $3·5-2·5\%$ sulphur, or $0·5\%$ with $3·5-2·5\%$ sulphur and $0·3\%$ diphenyl guanidine. Highly loaded compounds $1·5\%$ with $3-4\%$ sulphur and $0-0·5\%$ diphenyl guanidine.
Uses: highly loaded compounds, compounds with HAF blacks, tyres, inner tubes, conveyor belts, hoses, cable insulation, footwear, mechanical goods. Because of its bitter taste it is unsuitable for products which come into contact with food.
TN: Accel CZ (16)
Accelerator A
CBS
Conac S (6)
Cydac (21)
Delac S (23)
Furbac (22)
Nocceler CZ (274)
PSANNSD-70 (70% master batch with polyisobutylene) (282)
Rhodifax 16 (20)
Santocure (5)
Vulcafor HBS (60)
Vulkacit CZ (43)
Santocure DT (with diphenyl guanidine and di-2-benzthiazole disulphide) (60)
Santocure RF-1 (with 2-(2:4-dinitrophenyl thio)benzthiazole (60)
Vulkacit FZ (with basic accelerator) (43).
Lit.: J. Org. Chem., 1949, 14, 921-34.
USP 2 191 657, 2 354 427, 2 383 793, 2 419 283, 2 495 085, 2 191 656.
BP 655 668, 517 451.

CYCLOHEXYLETHYLAMINE
$(CH_2)_5CH.NH.C_2H_5$. Colourless liquid, b.p. $165°C$, s.g. $0·85$. Soluble in benzene, carbon tetrachloride, methylene chloride, alcohol, acetone, ethyl acetate, insoluble in water. Mild basic accelerator but used primarily as a booster. Effective in blends with

dithiocarbamates. Suitable in combination with zinc ethyl phenyl dithiocarbamate for dipped articles, impregnations, textile coatings, and self-vulcanising compounds and adhesives. Used in benzene solution with zinc ethyl phenyl dithiocarbamate for curing dipped goods; zinc oxide is necessary. Suitable for natural rubber, SBR and nitrile rubber.

Quantity: secondary, 0·5%; self-vulcanising compounds, 1·5–2%; sulphur, 1·8–3·5%; synthetic latices 1% with 2·5% colloidal sulphur.
TN: Accelerator HX (408)
Vulkacit HX (43).

CYCLOHEXYL ETHYL-AMMONIUM CYCLOHEXYL ETHYL DITHIOCARBAMATE

$$H_5C_2 \diagdown \atop H_{11}C_6 \diagup N-CS-S-N \diagup \atop \diagdown \atop C_6H_{11}^{C_2H_5}$$

Yellowish crystalline powder, m.p. 92°C, s.g. 1·08. Soluble in benzene, carbon tetrachloride, methylene chloride, alcohol, ethyl acetate, acetone, water. Ultra-accelerator; chiefly used as a mixed accelerator with zinc ethyl phenyl dithiocarbamate and piperidine pentamethylene dithiocarbamate. Has a limited storage stability (approx. 5 months), gives a non-toxic, odourless and non-staining vulcanisate. It is a slow accelerator without a metal oxide and only develops its full power in the presence of zinc oxide; fatty acids are unnecessary. Gives a high modulus and good ageing properties. Onset of cure is retarded by benzoic acid, phthalic anhydride, N-nitrosodiphenylamine and 2-mercaptobenzimidazole. Zinc dimethyl dithio-

carbamate and mercaptobenzthiazole act as boosters. Because of the fast speeds of boosted combinations, low processing temperatures are necessary and the accelerators should be added at the end of the mixing cycle. A two-part procedure, one compound without sulphur and one without an accelerator, is also recommended. Vulcanisation at room temperature takes a few days, at increased temperatures occurs rapidly.
Quantity: 0·5–0·8% with 0–0·5% piperidine pentamethylene dithiocarbamate and 2–3% sulphur.
Uses: alone or as a secondary accelerator, *e.g.* for dipped articles, self-vulcanising solution, textile coatings, impregnations, latex articles; less suitable for press cures. Suitable for natural rubber, SBR and butadiene, acrylonitrile copolymers.
TN: Accelerator 774 (408).
Vulkacit 774 (43).

N-CYCLOHEXYL-p-TOLUENE SULPHONAMIDE
$CH_3.C_6H_4.SO_2.NH.C_6H_{11}$, m.p. 56°C, b.p. 350°C, s.g. 1·125.
TN: Santiciser 1-H (5).

CYCLOHEXYL STEARATE
$C_{17}H_{35}COOC_6H_{11}$, m.p. 28–29°C. Softener.

CYCLOLAC S.g. 1·01–1·04. Thermoplastic, shock resistant terpolymer of styrene, butadiene and acrylonitrile. Has good electrical properties, dimensional stability, chemical and heat resistance. Used in blends with synthetic rubber (172).

CYCLONOX Group of cyclohexanone peroxides (171).

CYCOLOY 800 Polymer mixture: acrylonitrile-butadiene-styrene. Polycarbonate (172).

CYCOVIN KA Acrylonitrile-butadiene-styrene copolymer (172).

CYMAC
201: methylstyrene-acrylonitrile copolymer
400: polymethylstyrene.
Thermoplastic material (21).

CYNACRYL Fast curing acrylic elastomer. Remains flexible at low temperatures (21).

CYNAFLEX Diphenylamine-acetone reaction product (21):
50: 50% with 50% carbon black
100: normal type.

CYURAM
DS: tetramethyl thiuram disulphide
MS: tetramethyl thiuram monosulphide (21).

CYZATE
B: zinc dibutyl dithiocarbamate. 50% dispersion
E: zinc diethyl dithiocarbamate. 50% dispersion (21).

CYZONE IP N-isopropyl-N-phenyl-p-phenylene diamine (21).

D

DACCO Extender and reclaiming oil.

DA COSTA PROCESS Coagulation process for natural rubber. Used infrequently since the beginning of the plantation industry; as in the original Brazilian methods for smoking rubber, wood smoke was introduced into the latex. BP 19 730 (1910).

DACRON Formerly known as Fibre V and Amilar. Synthetic polyester fibre produced by the reaction of dimethyl terephthalate and ethylene glycol, s.g. 1·38. Soluble in o-chlorophenol and hot m-cresol. Spun and stretched in a melt-spin process. In production since 1953 (6).

DAHOMEY Natural rubber from *Ficus elastica*. Originates from Dahomey and came on the market in root-like form.

DAHQ Abbreviation for ditert-amyl hydroquinone.

DAIFLON Polytrifluoro-chloroethylene (173).

DAKES' MACHINE Machine used for the continuous foaming of latex. A cylindrical chamber, constructed with concentric ribs in ring formation, contains a fast running rotor with similar ribs. Latex and air, brought together in an axial direction, are mixed under increased pressure and on leaving the chamber form a homogeneous foam.

DALPAC Group of antioxidants based on 2:6-ditert-butyl-4-methyl phenol (10).

DALTOCEL Polyester. Starting material for the production of polyurethane foams (60).

DALTOFLEX 1 Solid millable polyurethane rubber (60).

DAMBONITE
$C_6H_6(OH)_4(OCH_3)_2$, dimethylinositol, mesoinositol-1:3-dimethyl ether, m.p. 195°C. Optically inactive. Was found in Gabon rubber (*Ladolphia* and *Ficus spp.*) and the sap of *Castilla elastica, Dyera Lowii* and *Dyera costulata.*

DAMBOSE $C_6H_6(OH)_6.2H_2O$, optically inactive inositol, m.p. (water-free) 224°C, s.g. 1·524. Obtained from hydrolysis of bornesite (d-methylinositol).

DAMMAR Pale yellow dried resin obtained by tapping trunks of *Shorea wiesneri* and other *Shorea* and *Hopea* species (Sumatra, Malaya and Siam). M.p. approx. 180°C, softening p. approx. 90–100°C. Soluble in petroleum hydrocarbons, turpentine oil and chloroform, sparingly soluble in alcohol and ether. *Uses:* adhesive for paints and medical tapes, tackifier in rubber compounds, aids adhesion of talc in coating rubber bales. In Indonesia the colophony obtained from the resin of *Pinus sylvestris* is also known as Dammar.

DAMPING Loss of energy during constant loading and unloading operations; reduction of the amplitude of oscillation as a function of time. The damping loss represents the loss of mechanical energy, which appears as heat due to internal friction. Amplitude is given by:

$$A = A_0 \cdot \exp(-\beta t)$$

where A_0 is the amplitude at time $t = 0$, t is the time, β is the Damping

Constant for linear vibrations given by

$$\beta = \frac{k}{2m}$$

k is the coefficient of friction and m is the vibrating mass. Damping is directly proportional to β and depends on the type of elastomer, the compound and the form of the oscillating strain. It plays a role in all oscillating rubber systems, *e.g.* tyres, springs, damping mountings. Lit.: DIN 53 513.

DANTUNG Local term for *Payena stipularis Burck* (*P. sumatrana Miq.*) (Sumatra and Celebes); also the gutta-percha which it yields.

DAPICHO Zapis. Porous, corklike, elastic solid derived from the coagulation of sap from various tree species. Occurs in bogs of S. American jungles. Used locally as cork, *e.g.* bottle stoppers.

DAPM Abbreviation for p,p'-diaminodiphenylmethane.

DAPON M Trade name for prepolymer of diallyl isophthalate (184).

DAREX Group of butadiene-styrene copolymers.
Types:
40:	50:50 SBR 1502 and styrene resin coprecipitated from latex
43G:	85% styrene content. Reinforcing agent for natural and synthetic rubber
620L:	latex 42%, solids/monomer ratio 50:50
622L:	latex 42%

632L: latex 52%, butadiene/ styrene ratio 10:90. Reinforcing agent for natural and synthetic latices.

Group of softeners:
DBP: dibutyl phthalate
DIBA: diisobutyl adipate
DIDA: didecyl adipate
DIDP: didecyl phthalate
DIOA: dioctyl adipate
DIOP: diisooctyl phthalate
DOP: di-2-ethyl hexyl phthalate
DOS: dioctyl sebacate (174).

DARTEX Chlorinated rubber (175).

DARTEX PROCESS Process for bonding metals to rubber using chlorinated rubber.

DARVAN Group of wetting and dispersing agents for latex:
1: sodium salt of a polyalkyl naphthalene sulphonic acid. Yellow powder, pH (1% aqueous solution) 8–10·5. *Uses:* general, especially for blacks
2: sodium salt of a naphthalenesulphonic acid condensate with larger alkyl chains. Dark brown powder, pH (1% aqueous solution) 7–8·5. Selective dispersing agent, especially for zinc oxide, alumina, sulphur and tellurium
3: sodium salt of a naphthalenesulphonic acid condensate with an inorganic material. Pale grey brown powder, especially for sulphur
7: polyelectrolyte in aqueous solution. Colourless liquid, s.g. 1·16, pH 9·5–10·5, viscosity 75 cp at 25°C. Useful over a wide pH range (both acid and alkaline). Non-discolouring and suitable for highly loaded aqueous dispersion. Shows little tendency to foam.

DAXAD Polymerised salts of alkyl arylsulphonic acids. Dispersing agents, wetting agents, emulsifiers, stabilisers:
11: the polymeric sodium salt of an alkyl naphthalene sulphonic aid. *Uses:* stabiliser in emulsion polymerisation of butadiene styrene copolymers
11-KLS: as 11 but potassium salt
21: monocalcium salt of a polymeric alkyl aryl sulphonic acid
23: dispersing agent for antioxidants in SBR, and polymerisation endstop for SBR (174).

DBA Abbreviation for dibenzylamine.

D–B–A Mixture of dibenzylamine and monobenzylamine. Activator for dibutyl xanthogen disulphide and zinc dibutyl xanthate. Accelerator at room temperature (231).

DBF Abbreviation for dibutyl fumarate.

DBK Abbreviation for 'Dyaberiyakande' clones.

DBM Abbreviation for dibutyl maleate.

DBMC Abbreviation for ditert-butyl-m-cresol.

DBP-60 Russian produced butadiene-piperylene copolymer in latex form.

DBP VALUE Dosage of dibutyl phthalate (in cm^3) taken up by 1 g of black. Absorption method for determining the structure indices of blacks.

DBPC Abbreviation for 2:6-ditert-butyl-p-cresol (176).

DBPPD Abbreviation for N,N'-disec-butyl-p-phenylene diamine.

DBR Trade name for dibenzoyl resorcinol (61).

DBRM Abbreviation for Deli-Batavia Rubber culture Maatschappij clones (Sumatra).

DBUD Abbreviation for dibutyl-ammonium dibutyl dithiocarbamate.

DC- Group of polymethyl siloxanes.
Uses: primarily as liquid dielectrics, lubricants, release agents, antifoam agents (7).

DC ANTIFOAM A Silicone compound.
Uses: antifoam agent, e.g. in aqueous solutions, emulsions, latex (7).

DCHP Abbreviation for dicyclohexyl phthalate.

DCP Abbreviation for dicapryl phthalate.

D.D.A. Antioxidant; secondary aromatic amine. **D.D.A.-EM:** aqueous emulsion of D.D.A. (43).

DDDM Abbreviation for 2:2'-dihydroxy-5:5'-dichloro diphenylmethane.

DD LACQUER DD: Desmodur and Desmophen. Lacquer and lacquer dye based on reaction products of diisocyanates and polyesters.

DDMA Abbreviation for decamethylene dimethyl azodicarboxylate.

DDP Abbreviation for didecyl phthalate.

DE Abbreviation for Defo elasticity.

DEBS Abbreviation for N-diethyl benzthiazyl sulphenamide.

DECAHYDRONAPHTHALENE $C_{10}H_{18}$, decalin, m.p. $-50°C$, b.p. 188°C, s.g. 0·89. Solvent for resins, waxes, fats, oils, miscible with normal and chlorinated hydrocarbons, alcohols, esters, acetone.

DECAMETHYLENE DIMETHYL AZODICARBOXYLATE DDMA. Sulphur-free vulcanising agent for rubber.
Uses: in compounds, to test the influence of the various ingredients on heat ageing.
Lit.: *Polym. Sci.*, 1953, **11**, 83.
Rubb. Chem. Techn., 1954, **27**, 459.

DECOMPOSITION Destruction of thick walled rubber articles by internal heat build-up caused by dynamic strain.

DEFO ELASTICITY DE. The recovery of an elastomer or compound, after 60% compression of a test piece (10 mm in height and 10 mm diameter) at 80°C, during a recovery period of 30 sec. DIN 53 514.

DEFO HARDNESS DH. The force which will compress a test piece (10 mm in height and 10 mm in diameter) to 4 mm in 30 sec at 80°C. A measure of the plasticity of elastomers and compounds as determined by the Defo plastimeter. DIN 53 514.

DEFO-PLASTIMETER Defo meter, deformation measuring apparatus. Apparatus working on the principle of the parallel plate plastimeter, after Baader. The plasticity (Defo-hardness, DH) is expressed as the weight in g which compresses a test piece (10 mm in height and 10 mm in diameter) prewarmed to 80°C, to 4 mm in 30 sec. The apparatus also enables the elastic recovery, 30 sec after unloading, to be expressed as a percentage.

DEGREE OF BRANCHING Expressed as the ratio of the degrees of polymerisation determined by the osmotic pressure and the dilute solution viscosity:

$$V = \frac{PG_{osm}}{PG(\eta)}$$

DELAC
3C: sulphenamide
A: diphenyl guanidine acetate
J: N-nitrosodiphenylamine
O: diphenyl guanidine oxalate
P: diphenyl guanidine phthalate
S: N-cyclohexyl benzthiazyl-2-sulphenamide (23).

DELALOC MACHINE Early form of dynamometer; hand operated.

DELAYED ACTION ACCELERATOR Accelerator giving a scorch curve which rises only after a long time, then rises rapidly. This effect is desirable for good processing safety. It is achieved by taking a normal accelerator, *e.g.* MBT, and blocking the active group with a compound so that the accelerator becomes ineffective. At the vulcanisation temperature the compound splits off so that the normal activity of the accelerator is regained and activation is achieved by the group which has been dissociated.

DELRIN Polyoxymethylene obtained by polymerisation of anhydrous formaldehyde. Hard, inert thermoplastic material.

DELTA-A FUNCTION Parameter used for the comparative assessment of the reinforcing effect of fillers.

DE MATTIA TESTER OF TENSILE FATIGUE Continuous flexing machine used to investigate the formation of cracks on bending and the growth of cracks in rubber test pieces. The strip-like test piece of 152 × 25 × 6·35 mm has a mould groove in the middle, across the breadth. The test pieces are fastened on one side to an adjustable stationary clamp and on the other, to a clamp moved by an eccentric drive along the plane of the middle line towards the opposite pair of clamps at 300 ± 10 flexes/min. The crack formation can be determined by comparison with a standard specimen and expressed as resistance to crack formation (common logarithm of kilocycles).
The cracking can also be reported as the number of cycles needed to reach a specified length of crack between 2 and 20 mm. The crack growth is expressed as the number of cycles needed to increase the crack length from 2 to 4, 4 to 8, and 8 to 12 mm.

Specification: ASTM D 813-59, D 430-59
BS 903: A10: A11: 1956
DIN 53 522.

DENAX Diphenyl guanidine.

DENAX DPG Diphenyl guanidine (428).

DENIER Unit of thread size; weight in g of 9000 m length of thread.

DENKA (JAPANESE PRO-DUCED) POLYCHLOROPRENE RUBBERS

Grade	Equivalent to
A90	Neoprene AD, Perbunan C 320, C 330, Butaclor MA40
M30	Neoprene WM 1, Perbunan C 211, Butaclor MC 31
M40	Neoprene W, Perbunan C 210, Butaclor MC 30
M120	Neoprene WHV, Butaclor MH 30, Perbunane 230
PM40	Neoprene GNA, Butaclor SC 21
S40	Neoprene WRT, Perbunan C 110, Butaclor MC 10

DEODORANTS ND AND O Deodorants for rubber compounds (6).

DEPANOL Group of solvents based on terpene hydrocarbons (217).

DEPOL Mixture of higher terpene hydrocarbons. Solvents for rubbers, plastics and resins (217).

DERIPHAT Group of amphoteric, surface active materials; salts of N-alkyl-β-aminopropionic acid and N-alkyl-β-iminodipropionic acid. *Uses:* emulsifiers for emulsion polymerisation (55).

DERRY PROCESS Process used during the first years of plantation management for obtaining rubber by coagulation, using wood smoke. Although this process has no practical significance, it is interesting historically as the first attempt at a continuous processing method. A rubber belt dipped into a latex container and carried a thin layer of the latex into a smoke chamber.
R. Derry BP 6 858 (1911).

DESERT MILKWEED Local term (California) for *Asclepias subulata*. The dried stems contain 0·5–6·5% pure rubber.

DESMODUR Group of isocyanates.
15: 1:5-naphthylene diisocyanate
AP: TH phenol adduct
H: 1:6-hexamethylene diisocyanate
R: triphenyl methane triisocyanate and Desmodur R
T: tolylene diisocyanate. Mixture of T-2:4 and T-2:6
TH: polyester resin based on tolylene diisocyanate and polyalcohol. 75% solution in ethyl acetate
THN: as TH, but free from tolylene diisocyanate. 60% solution in toluene ethyl acetate 1:1
TT: dimer of 2:4 tolylene diisocyanate
44: 4:4'-diphenyl methane diisocyanate (43).

DESMODUR R 20% solution of triphenyl methane triisocyanate,

140

CH(C$_6$H$_5$—N=C=O)$_3$ in methylene chloride; s.g. 1·32. Can be diluted with benzene, toluene, ethyl acetate and trichloroethylene. Nontoxic, incombustible. Gives oil resistant bonds of natural and synthetic rubbers with metals, and good adhesion between rubber and textiles; also for self-vulcanising adhesives and lacquers for rubber articles (43).

DESMOLIN Polyisocyanate for water free impregnations and coatings (43).

DESMOPHEN Group of polyesters which react with diisocyanates to form polyaddition products used for paints, adhesives, impregnants and, in general, corrosion resistant products.
1200: polyester of butylene glycol, adipic acid and a triol (3:3:1). Mol. wt. approx. 1200-2000
2000: polyester of ethylene glycol and adipic acid. Mol. wt. approx. 2000, hydroxyl No. 50-60 (43).

DETERGENT A-40 Detergent E-40, Stanvawet 0. Mixture of 40% alkyl aryl sulphonate with sodium sulphate.
Uses: wetting agent, emulsifier in talc coating for rubber bales, the removal of impurities from inferior quality rubber (slabs, scraps) in which the initial material is soaked for 1–2 days in a 0·2% solution with the addition of 0·1% trisodium phosphate before being sheeted (178).

DETERMINATION OF DESTRUCTION In the Martens tester, spheres of 30 mm φ rotate in a V-shaped groove under a given load. The accelerated destruction of materials is caused by internal friction and the high temperatures produced. The number of revolutions of the driving wheel is measured up to the time of destruction of the spheres.

DEV Abbreviation for 'Devon' clones.

DE VRIES, PROF. OTTO 1881–1948. 1915–1930, director of the Rubber Research Establishment, Java. Known for his extensive research into the production and quality of natural rubber. His book, *Estate Rubber, its Preparation, Properties and Testing*, 1920, remains a standard text on the subject.

DEWANIL WDS 65% aqueous dispersion of paraffin wax for latex compounds. Has no effect on thickening or coagulation (32).

DFFDA Russian produced di-β-naphthyl-p-phenylene diamine.

DGM-O Group of Russian produced blacks. The code number gives the specific surface area in m^2/g.
80: active anthracenic black.
    Specific surface area 88 m^2/g
105: properties resemble those of channel blacks.

DH Abbreviation for Defo hardness.

DHS Abbreviation for dihexyl sebacate.

DI Group of accelerators:
IV: zinc dimethyl dithiocarbamate
VII: zinc diethyl dithiocarbamate (180).

DIACELLER
DM: dibenzthiazyl disulphide
K: reaction product of formaldehyde and an amine
M: mercaptobenzthiazole (392).

DIACETONE ALCOHOL
$(CH_3)_2 . C(OH) . CH_2CO . CH_3$,
4-hydroxy-4-methyl-2-pentanone.
Inflammable liquid, m.p. $-47°C$, b.p. 168°C, s.g. 0·931–0·94. Miscible with water, alcohol and other solvents. Solvent for cellulose acetate, cellulose nitrate, resins, oils, fats, waxes, and chlorinated rubber.
TN: Tyranton
Pyranton A.

DIAK Group of vulcanising agents for fluoroelastomers (6):
1: hexamethylene diamine carbamate
2: ethylene diamine carbamate
3: (formerly LD-214) dicinnamylidene-1:6-hexane diamine
4: alicyclic amine salt, m.p. 145–155°C, s.g. 1·23.

DIALLYL PHTHALATE
$C_6H_4(COOCH_2CH=CH_2)_2$. Colourless, oily liquid, m.p. $-70°C$, b.p. (4 mm) 156–175°C, flash p. 166°C, s.g. 1·12. Soluble in aromatics, alcohol, ketones, and esters. Polymerises with peroxide as catalyst.
Uses: in polymeric form as plasticiser for polyester resins, adhesives.

4:4′-DIAMINODIPHENYL-AMINE
$NH_2 . C_6H_4 . NH . C_6H_4NH_2$.
m.p. 158°C. Antioxidant. Discolours slightly.
TN: Oxynone (obsolete) (5).

4:4-DIAMINODIPHENYL-METHANE
$NH_2 . C_6H_4 . CH_2 . C_6H_4 NH_2$,
4:4′-methylene dianiline. Crystallises from water and benzene, m.p. 91·5°C, b.p. 398°C, s.g. 1·12, technical product is in the form of brown, waxlike lumps, m.p. 73–89°C. Soluble in acetone, ethylene dichloride, benzene, alcohol, ether, insoluble in water. Antioxidant, accelerator and activator for NR, SBR, neoprene and neoprene latices. Strongly activates vulcanisation, particularly in compounds containing thiazoles. Decreases processing safety; this can be partially overcome by suitable accelerator combinations. Causes brown staining and can also discolour vulcanised articles. Slight antioxidant action in unvulcanised and vulcanised compounds. Improves the flex-cracking resistance of tyre carcases and transmission belts, also improves abrasion resistance, stiffens unvulcanised NR compounds. Antifrosting agent in NR. Tendency to bloom at over 0·5%.
Quantity: NR, for normal uses 0·25–0·35%, antifrosting agent 0·3–0·5%, SBR 0·5% neoprene GN 0·5–1%.
TN: NA-11 (6)
Tonox (23)
Tonox D (23)
Robac 44 (311).

2:4-DIAMINOTOLUENE
$C_6H_3(CH_3)(NH_2)_2$. m.p. 99°C.
Antioxidant.
TN: Neozone B (6).

2:5-DITERT-AMYL HYDRO-QUINONE

DAH, DAHQ. Yellowish-brown powder, m.p. 165°C, s.g. 1·05. Antioxidant for unvulcanised rubber, also for unsaturated resins and oils. Stiffens unvulcanised compounds and reduces cold flow. Does not stain or migrate.
Uses: tackifier, *e.g.* for unvulcanised films.
Quantity: 0·5–4%.
TN: Antage DAH (16)
Inhibitor DAHQ (193)
Santovar A (5).

2:4-DITERT-AMYL PHENOL
$(C_5H_{11})_2C_6H_3OH$. Liquid, s.g. 0·92. General purpose antioxidant.
TN: Diamylphenol (176)
110 (4).

DIAMYL PHTHALATE
$C_6H_4(COOC_5H_{11})_2$. Yellowish liquid, m.p. −55°C, b.p. 342°C, flash p. 171°C, s.g. 1·02. Plasticiser for rubber, p.v.c., nitro and acetyl celluloses and other plastics.

DIANISIDINE DIISOCYANATE
3:3′-dimethoxy-4:4′-diphenyl diisocyanate. Dark-grey powder, m.p. 121°C, s.g. 1·20. Crosslinking agent for urethane elastomers with hydroxyl end groups. Used in adhesives to assist the adhesion of vinyl elastomers and nitrile rubber to synthetic textiles.
TN: DADI (148).

DIATOMACEOUS EARTH
Kieselguhr, tripoli, bergmehl. Fine particle size, low density, grey to reddish powder, 70–90% amorphous silicic acid, with additional constituents such as iron, phosphorus, sulphur, calcium and magnesium. Consists of microscopic structures of single diatoms (*Bacillariophyta*). Obtained from fresh water deposits; organic substances and iron $(FeCl_3)$ are removed by heating with ammonium or sodium chloride.
Uses: filler, filters, heat insulations.

DIAZOAMINOBENZENE
$C_6H_5N=N.NH.C_6H_5$. Brown powder, m.p. 90–95°C, decomposition temperature 150°C, in rubber compounds approx. 100°C, s.g. 1·17. Highly soluble in rubber. Blowing agents safe to process at temperatures up to 95°C. Gives a uniform cell structure and develops a high blowing pressure in compounds so that the plasticity of the compound is of little importance: a great hardness range is possible. Has no effect on vulcanisation or ageing, causes an orange yellow fluorescent discoloration of compounds. Can be blended with sodium bicarbonate.
Quantity: in soft compounds 1–4%, hard compounds 2–5%, in blends with sodium bicarbonate 0·5–1·5%, compounds highly loaded with carbon blacks up to 20%.
TN: Porofor DB (43)
Unicel (6)
Vulcacel A (60)
Vulcacel AN (60).

DIAZOETHER Initiator for emulsion copolymerisation of butadiene and styrene, particularly at low temperatures. Use was discovered in the USA in 1944 by W. B. Reynolds; 2-(4-methoxy benzene diazomercapto)naphthalene was previously used. Diazoether was later replaced by hydroperoxide/iron pyrophosphate mixtures.
TN: MDN.

DIAZOTHIOETHER

RN: NSR', DTE. Group of polymerisation catalysts used in the development period of SBR cold rubber. Produced by coupling diazotised aromatic amines with aliphatic or aromatic mercaptans. Give a rapid polymerisation reaction at low temperature (3½ h at 5°C, conversion 60%). Later replaced in technical production by sugar-containing and sugar-free hydroxyperoxide/iron pyrophosphate systems. The diazothioethers which were widely used were: 2-(4-methoxy-benzene diazomercapto)naphthaline (MDN), produced from diazotised p-anisidine and β-naphthyl mercaptan; 1-(2:4-dimethyl benzene diazomercapto)-naphthalene (α-XDN), diazotised 2:4-dimethyl aniline, and α-naphthyl mercaptan.

DIBA Abbreviation for di-iso-butyl-adipate.

DIBENZAMIDODIPHENYL DISULPHIDE

Odourless, pale yellow crystalline powder, m.p. 136–143°C, s.g. 1·35. Peptiser for natural rubber, butadiene-styrene copolymers, and reclaim. Active above 115°C. Has a strongly plasticising effect and enables considerable reduction of masticating time and power. Has no influence on vulcanisation of natural rubber but in SBR increases the tendency to scorch, has no effect on ageing properties. Sulphur has a strongly retarding effect and 0·2% completely prevents plasticisation. Channel blacks also act as retarders, but are somewhat weaker than sulphur. In natural rubber, zinc oxide, calcium carbonate and furnace blacks have only a slight effect, if any. In SBR, furnace blacks and zinc oxide have a slightly

retarding effect. In natural rubber, diphenyl guanidine and mercaptobenzthiazole cause additional softening while other accelerators, *e.g.* benzthiazyl disulphide, tetramethyl thiuram disulphide, sulphenamides retard slightly, as do most antioxidants. In SBR the plasticisation is completely halted or strongly retarded by mercaptobenzthiazole, benzthiazyl disulphide, tetramethyl thiuram monosulphide, tetramethyl thiuram disulphide, phenyl aminomethyl-2-benzthiazyl sulphide and diphenyl guanidine. Antioxidants have little or no effect.

Quantity: natural rubber 0·50–0·5%, SBR 0·5–3%, reclaim 1–2%, blends of SBR, NR and reclaim 0·05–3%.

TN: Pepton 22 (21, 22) Peptazin BAFD.

p,p'-DIBENZOYL QUINONE DIOXIME

dibenzoyl-p-quinone dioxime, p-quinone dioxime dibenzoyl ester. Brown powder, decomposes at temperatures above 200°C, insoluble in acetone, benzene, ethylene dichloride and water. Good storage stability. Very fast sulphurless vulcanising agent for sulphur-free vulcanisates. Activated by oxidising agents and gives vulcanisates with good heat resistance. Maximum physical properties are achieved by combination with sulphur. Similar to p-quinone dioxime but has a more pronounced delayed action and gives greater processing safety because of the benzoyl groups. Effective in natural rubber, butyl rubber and SBR; gives fast

cures and imparts a high modulus. Particularly suitable for butyl rubber in combination with lead tetraoxide (Pb$_3$O$_4$). Acidic ingredients including fatty acids and channel blacks, should be avoided as they increase the tendency to ignite. Carbon blacks should be used to obtain good physical properties. Phthalic anhydride p,p′-aminodiphenylmethane, N-nitrosodiphenylamine, thiazoles, thiurams and dithiocarbamates retard and therefore increase the processing safety. Sulphur, also has a slight retarding effect.
Quantity: 6% with 6% lead tetraoxide.
Uses: for fast vulcanising butyl compounds with a high modulus, air bags made from butyl rubber.
TN: Accélérateur 113 (23)
Actor DQ (16)
BSAM (85)
Butalene BD (19)
Dibenzo GMF 113 (mixed with 67% clay) (23)
DPQD
PPD-70 (70% with 30% polyisobutylene) (282)
PDSD-715 (65% with 6·5% sulphur and 28·5% polyisobutylene) (282)
Rhenocure BQ (42)
Vulnoc DGM (274).

DIBENZTHIAZYL DIMETHYL THIOUREA

$$\left[\begin{array}{c} \text{C—S—CH}_2\text{—NH—} \end{array} \right]_2 \text{CO}$$

1:3-bis(2-benzthiazyl mercaptomethyl)urea. Yellowish powder, m.p. 220°C, s.g. 1·29–1·38. Hardly soluble in any solvent. Accelerator for natural rubber, SBR, NRB and latices; needs zinc oxide and fatty acid. Can be used alone but works better in combination with an activator. Confers good processing and vulcanisa-

tion properties. Suitable for compounds with a high loading of retarding filler, *e.g.* clays.
Quantity: natural rubber 0·45–1·4% with 1–4% sulphur, boosted with 0·1–1% cyclohexyl benzthiazyl sulphenamide, 0·125% tetramethyl thiuram disulphide, or 0·25–0·5% diphenyl guanidine; SBR 0·8–1·5% with 1·5–4% sulphur, boosted with 0·6% cyclohexylbenzthiazyl sulphenamide or 0·5% diphenyl guanidine.
Uses: footwear, tyres, inner tubes, insulations, mechanical goods, latex and in continuous vulcanisation.
TN: El-Sixty (5).

2:2′-DIBENZTHIAZYL DISULPHIDE

2:2′-dithiobis(benzthiazole), benzthiazyl disulphide (MBTS). Pale yellow non-hygroscopic powder, m.p. 160–179°C (pure product, 179°C), s.g. 1·45–1·50. Soluble in carbon disulphide, benzene hydrocarbons, chlorinated solvents, sparingly soluble in alcohol and acetone, insoluble in water and dilute alkalis; various trade products show some variation in solubility. General purpose accelerator of the mercapto group with a delayed action, has excellent storage stability; medium fast with excellent processing safety and little tendency to cause scorching. Sufficiently active for cures above 143°C with a marked delayed action and with a long vulcanisation plateau. Unvulcanised compounds may be safely stored. Used instead of mercaptobenzthiazole in compounds requiring a relatively high degree of processing safety. In combination with activated dithiocarbamates, effective in preventing scorch. Boosted by thiurams, dithio-carbamates, aldehyde-amines, alkaline reclaims and basic ingredients. Vulcanisation temperatures 135–160°C. Non-staining non-discolouring. Plasticises natural rubber and neoprene in the absence of sulphur and fillers. Has a retarding effect during the processing and vulcanisation of neoprene. When used with ultra-accelerators, improves processing safety; requires 5% zinc oxide and 1% stearic acid with 1·5–3% sulphur, also can be activated by lead oxide. Products show good ageing properties.

Quantity:

	Primary %	Secondary %
Natural rubber:		
MBTS	1–2	0·5–1·0
Zinc dimethyl dithiocarbamate	—	0·5–0·2
Sulphur	3–2	2·0–1·0
Tellurium	—	0·0–0·5
SBR:		
MBTS	1–3	0·25–0·75
Cyclohexyl benzthiazyl sulphenamide	—	0·6–0·3
Sulphur	1–3	1·0–3·0
NBR:		
MBTS	1–2	0·75
Cyclohexyl benzthiazyl sulphenamide	—	0·3
Sulphur	0·2–2	0·2–2·0
Reclaim:		
MBTS	0·8–1·2	—
Sulphur	1–3	—
Butyl rubber:		
MBTS	0·25–1	—

Latex: has little influence on stability and prevulcanisation. 1–2% with 1–2% sulphur and 0·1–2% potassium hydroxide.
Neoprene: 0·25–1% as retarder.
Boosting: 0·05–0·3% dithiocarbamate or thiuram or 0·1–0·5% guanidine (DPG or DOTG), dibutyl-ammonium oleate or butyraldehyde-aniline condensation product. Reclaim compounds with too high an alkalinity tend to be 'scorchy'.
Uses: include mechanical goods, tyres, conveyor belts, insulation, light-coloured articles, frictioned materials, footwear (with DPG, but without stearic acid), in butadiene-acrylonitrile copolymers, foam rubber. Vulcanisates have a characteristic smell and bitter taste and are unsuitable for medical articles, or products which come into contact with food.
TN: Accel DM (16)
Accélérateur MBTS
Accélérateur rapide GR
Accélérateur rapide GS (19)
Accélérateur rapide 201 (20)
Altax (51)
Ancatax (22)
Belger DM
Diaceller DM
Diosit
Duracap MBTS (241)
Eveite DM (59)
MBTS
Mercapto S
Nocceler M (274)
PAD-60 (60% with 40% polyiso-butylene) (282)
Poly-Zole MBTS
Rapidex KY
Rodform Altax (51)
Soxinol DM
Thiofide (5)
Vulcafor MBTS (60)
Vulcaid MBTS (3)
Vulkacit DM (43)
Vulkacit DM/C
Vulkator DX (364)

Wobezit DM
Blends with hexamethylene tetramine and diphenyl guanidine:
Ureka White F (5)
Vulkacit F (43)
Vulcafor FN (60)
Blends with tetramethyl thiuram disulphide:
Vulcafor DAU (60)
Vulcafor DAUF (60)
Blends with undisclosed material:
Vulcafor DA (60).

DIBENZYLAMINE

Pale yellow to brown liquid, b.p. (10 mm) 168–172°C, s.g. 1·03. Soluble in acetone, benzene and chlorinated hydrocarbons, practically insoluble in water. Amine activator for rubber and latex, particularly for use with zinc butyl xanthate and butyl xanthogen disulphide. With these ingredients it forms ultra-fast accelerator combinations particularly suitable for low temperatures. Vulcanisation times range from 24 h at room temperature to 20 minutes at 100°C. May be retarded by acid compounding ingredients, *e.g.* clays and channel blacks. Mixtures with xanthates and carbon disulphide behave like thiazoles and thiurams. Needs zinc oxide and sulphur in the usual quantities, fatty acids are unnecessary and retard. Accelerator and dibenzylamine should be mixed in a separate part of the compound without sulphur; this should only be added to the part with sulphur when the complete compound is to be used. Dibenzylamine may also be used alone in the compound if the accelerator is added when the compound is warmed for further processing. Carbon disulphide can be added by dipping or painting the compounds containing dibenzylamine so that scorching may be avoided.
Uses: textile coatings, impregnations, gum stocks, fancy goods, bathing caps, bathing shoes, cold curing adhesives and other items made from natural rubber and SBR.
Quantity: dibenzylamine 2%, xanthate 2%, tetramethyl thiuram disulphide 0–0·25%, sulphur 2%.
TN: DBA (23)
Vulcaid 28 (3).

DIBENZYL ETHER
Colourless liquid, m.p. 3·5°C, b.p. 295°C, s.g. 1·043. Soluble in alcohol. Softener; improves the elasticity of butadiene-styrene and butadiene-acrylonitrile copolymers. Increases elasticity and resistance to extremely low temperatures, while tensile strength, modulus and hardness are lowered. Also suitable for latices in emulsion form. Used in friction mixes in combination with resins.
Quantity: softener 5–10%; for higher elasticity and resistance to low temperatures 20–60%.

DIBENZYL SEBACATE
$(CH_2)_8(COOCH_2C_6H_5)_2$. Pale yellow liquid with light, pleasant odour, m.p. 28°C, b.p. (4 mm) 265°C, flash p. 236°C, s.g. 1·05. Softener for NR, SR and vinyl polymers; improves low temperature resistance. Has no effect on vulcanisation.
TN: Harflex 90 (210)
Morflex 250 (264).

DIBUTOXYETHOXYETHYL ADIPATE
$[C_4H_9(OC_2H_4)_2COO(CH_2)_2]_2$, b.p. (3 mm) 260°C, s.g. 1·025. Softener.
TN: TP-95 (208).

DIBUTOXYETHYL ADIPATE
$(C_4H_9OC_2H_4OCOCH_2CH_2)_2$, DBEA, m.p. $-34°C$, b.p. (4 mm) 200–220°C, s.g. 0·997. Softener.
TN: Adipol BCA
Staflex DBEA.

DIBUTOXYETHYL PHTHALATE
$C_6H_4(COOC_2H_4OC_4H_9)_2$, m.p. $-50°C$, b.p. (4 mm) 210–223°C, flash p. 208°C, s.g. 1·063. Softener. *TN:* Kronisol (184).

DIBUTOXYETHYL SEBACATE
$(CH_2)_8(COOC_2H_4OC_4H_9)_2$, m.p. $-10°C$, b.p. (10 mm) 255°C, s.g. 0·964–0·970. Softener. *TN:* Staflex DBES.

DIBUTYL ADIPATE
$C_4H_9OCO(CH_2)_4COOC_4H_9$, b.p. (17 mm) 168°C, s.g. 0·961. Insoluble in water. Softener.

DIBUTYLAMINE
$(CH_3CH_2CH_2CH_2)_2NH$. Colourless liquid, m.p. $-62°C$, b.p. 160°C, s.g. 0·767. Soluble in alcohol, benzine, benzene, chloroform, sparingly soluble in water, miscible with hydrocarbons.
Use: activator for a few accelerators.

DIBUTYLAMMONIUM DIBUTYL DITHIOCARBAMATE

$$C_4H_9\underset{C_4H_9}{\diagdown}N-\overset{\overset{\textstyle S}{\|}}{C}-S-N\underset{\diagdown C_4H_9}{\overset{\diagup C_4H_9}{}}$$

Light brown powder, m.p. 45–47°C, s.g. 1·20. Accelerator.
TN: Robac DBUD (311).

DIBUTYLAMMONIUM OLEATE
$(C_4H_9)_2.NH_2.O.CO.C_{17}H_{33}$.

Yellow brown to dark brown liquid, s.g. 0·87. Activator for thiazoles and thiurams. Gives good processing safety, increases the vulcanisation range and flattens the vulcanisation curve. Effective also as softener and processing aid. The vulcanisates have good ageing properties (particularly in compounds containing thiazoles and thiurams). Decreases the modulus and heat build-up. Stearic and other fatty acids are unnecessary. Improves the dispersion of fillers. Suitable for natural rubber, synthetic elastomers and latices.
Quantity: 0·25–1%.
TN: Actogen (31)
Barak (6)
Activator 1102 (22).

2:6-DITERT-BUTYL-α-DIMETHYLAMINO-p-CRESOL
White to yellowish powder, b.p. (40 mm) 179°C, m.p. 95°C, s.g. 0·97. Non-staining and non-discolouring antioxidant for natural and synthetic rubbers and latices, improves heat resistance.
TN: Ethyl antioxidant 703 (88).

DIBUTYL FUMARATE DBF.
Used as comonomer in polymerisation reactions.

2:5-DITERT-BUTYL HYDROQUINONE

White powder, m.p. 215°C, s.g. 1·095. Antioxidant for unvulcanised rubbers and compounds, unsaturated resins and oils. Stiffens compounds and reduces cold flow. Non-staining. Polymerisation inhibitor. *Quantity:* 0·5–4%. *Uses:* notably for adhesives, unvulcanised sheeting. *TN:* Santovar O (5).

3:5-DITERT-BUTYL-4-HYDROXYBENZYL ETHER

White powder, m.p. 135–136°C. General purpose antioxidant. *TN:* Ionox 201 (2).

DIBUTYL MALEATE DBM.
Unsaturated diester. *Uses:* comonomer in polymerisation reactions.

2:6-DITERT-BUTYL-α-METHOXY-p-CRESOL White to yellowish powder, m.p. 101°C, s.g. 1·073. Non-staining and non-discolouring antioxidant for NR, SR and latices. *TN:* Ethyl antioxidant 762 (88).

2:6-DITERT-BUTYL-4-METHYL PHENOL

2,6-ditert-butyl-p-cresol, 4-methyl-2:6-ditert-butyl phenol, butylised hydroxytoluene, BHT, DBPC. White

to yellowish crystals, m.p. 70°C, b.p. 265°C, s.g. 1·048. Soluble in alcohol, benzene, methyl ethyl ketone and toluene, insoluble in water. Non-staining and non-discolouring antioxidant for natural and synthetic rubbers and latices; protects against normal oxidation and heat ageing, improves colour retention. *Quantity:* 0·01–3%. *Uses:* include light-coloured articles, articles used in food and drink industry, pressure sensitive adhesive tapes, insulation, coatings, gum compounds, impregnations.
TN: Amoco 533 (78)
Antioxidant AC-1
BHT (15, 5, 176)
Catalin Antioxidant CAO-1, CAO-3 (15)
Dalpac 4 (10)
DBC
DBPC (176)
Deenax (131)
DTHP
Gasoline AO-29 (6)
Ionol (2)
Kerobit TBK (32)
Nonox TBC (60)
Oxygard (246)
Sustane BHT (352)
Tenamene 3 (193)
Tenox BHT (193)
Topauoloc (60)
Wytox BHT (237).

DITERT-BUTYL PEROXIDE
$(CH_3)_3C.O.O.C(CH_3)_3$,   DTBP. Inflammable liquid, m.p. −40°C, b.p. (284 mm) 80°C, s.g. 0·794. Soluble in most organic solvents, in monomers and intermediary polymers. *Uses:* polymerisation catalyst, catalyst for vulcanisation of vinyl silicone rubber. *TN:* Trigonox B (171). X-1960 (112).

149

2:6-DITERT-BUTYL PHENOL

$$(CH_3)_3C-\underset{\underset{}{\bigcirc}}{\overset{\overset{OH}{|}}{}}-C(CH_3)_3$$

Light straw-coloured solid, m.p. 37°C, s.g. 0·91. General purpose antioxidant.
TN: Ethyl antioxidant 701 (88).

DIBUTYL-p-PHENYLENE DIAMINE

$$-\underset{}{\bigcirc}-\left[NH-\overset{\overset{CH_3}{|}}{CH}-C_2H_5 \right]_2$$

N,N'-disec-butyl-p-phenylene di-amine. Red liquid, b.p. 295–300°C. Antioxidant and antiozonant in natural and synthetic rubbers.
Quantity: up to 2%, higher concentrations cause dermatitis.
TN: Amoco 532 (72)
Gasoline AD-22 (6)
Tenamene 2 (193)
Topanol M (60).

DIBUTYL PHTHALATE

$C_6H_4(COO.C_4H_9)_2$, phthalic acid dibutyl ester, DBP. Colourless liquid, m.p. −35°C, b.p. 339°C, s.g. 1·05. Soluble in most organic solvents, insoluble in water. Softener and extender, particularly for chlorinated rubber, butadiene-styrene and butadiene-acrylonitrile copolymers. Gives marked increase in elongation at break, though tensile strength, modulus and hardness are decreased, improves low temperature resistance.
Quantity: softener 5–10% for low temperature flexibility and high elongations 20–60%. Used also as a plasticiser, *e.g.* for pvc nitrocellulose, resins, lacquers, plastics or as a fat free lubricant and for antifoam agents.

TN: Celluflex DBP (152)
Darex DBP (174)
Dutch Boy NLA–10 (191)
Harflex 140 (210)
Morflex 140 (264)
PX 104 (305)
Staflex DBP (330)
Vestinol C (189)
Witcizer 300 (215).

DIBUTYL SEBACATE

Colourless, odourless liquid, m.p. −8°C, b.p. 345°C, s.g. 0·93. Miscible with ketones, aliphatic and aromatic hydrocarbons, immiscible with water. Softener, *e.g.* natural rubber, nitrile rubber, polyvinyl acetate. Improves flexibility at low temperatures, suitable for articles used in the food industry.
TN: Harflex 40 (210)
Monoplex DBS (33)
Morflex 240 (264)
PX-404 (305).

DIBUTYL SUCCINATE

$C_4H_9OCO(CH_2)_2COOC_4H_9$, m.p. −19°C, b.p. 255°C, s.g. 0·97. Softener.

DIBUTYL TARTRATE

$C_4H_9OCO(CHOH)_2COOC_4H_9$, m.p. 20–22°C, b.p. 292–312°C, flash p. 170°C, s.g. 1·093. Softener.
TN: Kesscoflex (132)
Witcizer (215).

1:3-DIBUTYL THIOCARBA-MIDE
White powder, m.p. 60°C, s.g. 1·06. Accelerator for neoprene, antiozonant for natural rubber, SBR, NBR and neoprene.
TN: Pennzone B (120).

DIBUTYL XANTHOGEN DISULPHIDE

$C_4H_9O.CS.S—S.CS.OC_4H_9$, butyl xanthogen disulphide. Brown liquid, s.g. 1·15–1·17. Soluble in acetone, benzene, ethylene dichloride. Accelerator for natural rubber, SBR and their latices, reclaim and acrylonitrile-butadiene copolymers. Boosted by dibenzylamine, monobenzylamine or other amines, gives effective ultra-fast accelerator system for low temperature cures. A satisfactory state of vulcanisation is obtained overnight at room temperature, or in 20–30 min at 100°C. Channel blacks, clays, and acidic compounding ingredients retard the vulcanisation. An amine combination can be further activated by adding thiazoles, thiurams and dithiocarbamates. The product is non-staining and non-discolouring. Zinc oxide and sulphur are necessary in the usual quantities, fatty acids retard and are unnecessary. Processing should be in two parts, one with and one without sulphur, and the blending of the two should take place immediately before use. Similarly, the amine can be added to the compound first, while the other accelerator is added when the compound is warmed on the mill.

Uses: impregnations, textile coatings, gum stocks, cements, adhesives, latex articles.

TN: CPB (23).

DICALITE Types RC, RF, RN, RO, RW; diatomaceous earth. Clay filler (181).

DICAPRYL ADIPATE
$C_8H_{17}OCO(CH_2)_4COOC_8H_{17}$, m.p. − 15°C, b.p. (4 mm) 213–217°C, s.g. (20°C) 0·9135. Insoluble in water. Softener.

TN: Harflex 280 (210).

DICAPRYL PHTHALATE
$C_6H_4(COOC_8H_{17})_2$, DCP, m.p.

− 60°C, b.p. (15 mm) 227–234°C, flash p. 201°C, s.g. (25°C) 0·965. Pale yellow liquid with pleasant odour. Softener for NR and SR, particularly for NBR and vinyl polymers; improves low temperature properties.

TN: Elastex 80-P (107)
Harflex 180 (210)
Monoplex DCP (33)
Morflex 120 (264).

DICAPRYL SEBACATE
$(CH_2)_8(COOC_8H_{17})_2$, b.p. (4 mm) 231–235°C, s.g. (20°C) 0·9136. Softener.

TN: Harflex 80 (210).

3:3-DICHLOROBENZIDINE

White, needle-like crystals or greyish powder, m.p. 132°C, s.g. 1·25. Soluble in alcohol, benzene and glacial acetic acid, insoluble in water.

Uses: crosslinking agent for urethane rubber.

2:4-DICHLOROBENZOYL PEROXIDE Polymerisation catalyst, crosslinking agent for polymers, particularly for silicone rubber.

TN: Perkadox PDS 40 (40% in silicone oil) (171).

DICHLOROBUTADIENE Polymers or copolymers of 2:3-dichlorobutadiene with 1:2 dichlorobutadiene or chlorovinyl acetylene. Show good adhesion to NR and synthetic elastomers and to metals, used as a basis for metal to rubber bonding agents.

TN: Loxit 3000.

3:3-DICHLORO-4:4'-DI-AMINODIPHENYL METHANE

4:4'-methylenebis(2-chloroaniline), methylenebis-o-chloroaniline. Pale brown powder, m.p. 100–105°C, s.g. 1·39. Soluble in warm methyl ethyl ketone, ethyl acetate, aromatic hydrocarbons. Crosslinking agent for liquid urethane elastomers, crosslinking occurs within 5–7 min at 120°C.
Quantity: 10–20%.
TN: MOCA (6).

DICHLOROPHENE

bis(5-cloro-2-hydroxyphenyl)methane, 2:2'-methylene-bis-(4-chlorophenol), 2:2'-dihydroxy-5:5'-dichlorodiphenylmethane. Crystalline, m.p. 177°C. Soluble in alcohol, ether, methanol, isopropyl alcohol, petroleum ether, sparingly soluble in toluene, insoluble in water.
Uses: bactericide for rubber articles.
TN: Didroxane
Di-phenthane 70
G-4
Parabis
Preventol G-D
Tentiatol.

2:4-DICHLOROPHENOXY-ACETIC ACID

Colourless crystals, m.p. 138°C. Readily soluble in alcohol, sparingly soluble in water. Fertiliser used in 1% concentration with a natural oil bearer (*e.g.* 5 parts palm oil and 3 parts mineral oil) to stimulate latex production. The material is placed above or below the tapping incision, on the lightly scratched bark. Latex output is increased by approx. 40–70%, although the dry rubber content of the latex decreases slightly. Higher concentrations, placed at the base of the trunk, are used to kill old rubber trees. The output of the trees increases by 100–150% for a few weeks, before they die.

DICINNAMYLIDENE-1:6-HEXANE DIAMINE

Brown powder with cinnamon-like smell, m.p. 82–88°C, s.g. 1·09.
Uses: vulcanising agent for fluoroelastomers.
Quantity: 2–3·5%. Vulcanisation cure: moulded under pressure for 30 min at 149°C, followed by 24 h post-cure in hot air at 204°C.
TN: Diak No. 3 (previously LD-214) (6).
Lit.: *Du Pont Report* 59-4.
Viton Bull., 2, 8.

DI-o-CRESOL MONO-SULPHIDE

Grey powder, m.p. above 110°C, s.g. 1·27. Soluble in alcohol and acetone, partially soluble in aromatic

152

solvents. Antioxidant, stains slightly; protects against heat and normal oxidation. Also effective as copper inhibitor.

TN: Nonox CC (60).
Santowhite CM (5).

DICUMYL PEROXIDE
$[C_6H_5C(CH_3)_2O]_2$. Vulcanising agent for rubber articles requiring good heat resistance.

TN: Di-Cup (10)
Perkadox BC 40 (40%) (171)
Perkadox SB (95%) (171).

DI-CUP Dicumyl peroxide. Type 40C, 40% dicumyl peroxide absorbed on calcium carbonate. White powder, s.g. 1·53.

Uses: vulcanising and crosslinking agent for natural rubber, SBR, NBR, CR, polyethylene, Hypalon, silicone rubber; polymerisation catalyst for vinyl monomers and polyester resins. Gives peroxide crosslinked vulcanisates with C–C bonds having good physical properties, good resistance to normal and heat ageing and low compression set. Non-staining, may be used in white, translucent and transparent compounds. Unsuitable for vulcanisation in hot air. May be used for press and open steam cures (10).

DICYCLOHEXYL PHTHAL-
ATE $C_6H_4(COOC_6H_{11})_2$, DCHP, m.p. 58–65°C, b.p. (5 mm) 212–218°C, flash p. 207°C, s.g. 1·148. Softener.

TN: Elastex DCHP
KP-201 (184)
Vestinol HX (189).

DIELECTRIC STRENGTH
Measure of the highest voltage that rubbers and plastics can withstand.

The test is carried out between two plate-like electrodes, applying an alternating voltage (50 Hz), which is increased by a regulator until the sample fails. The results are given in kV/mm. The dielectric strength depends on the thickness of the test piece. DIN 53 596.

DIENE RUBBER Trade name for stereospecific rubbers (129).

DI-ESTEREX N Mixture of 2-(2:4-dinitrophenyl thio)benzthiazole and diphenyl guanidine acetate. Accelerator (23).

DIETHANOLAMINE
$(CH_2CH_2OH)_2NH$, 2:2'-dihydroxy-diethylamine, 2:2'-iminodiethanol. Colourless, viscous liquid with a weak ammoniacal odour, b.p. 269°C, s.g. (30°C) 1·088. Miscible with water, methyl alcohol and acetone. Strong base. Activator in compounds with white reinforcing fillers, processing aid, neutralises the acidity of clays.

Uses: in SBR, NBR, CR and latices.

DIETHOXYETHYL ADIPATE
$C_2H_5OC_2H_4COO(CH_2)_4$
$. COOC_2H_4OC_2H_5$,
m.p. −70°C, b.p. (4 mm) 165°C, s.g. (25°C) 1·036. Plasticiser.

DIETHOXYETHYL PHTHAL-
ATE $C_6H_4(COOC_2H_4OC_2H_5)$, m.p. 31°C, b.p. (4 mm) 188–215°C, flash p. 186°C, s.g. (20°C) 1·12.

TN: Ethox (184).

DIETHYL ADIPATE
$C_2H_5OCO(CH_2)_4COOC_2H_5$, m.p. −14°C, b.p. 245°C, s.g. 1·002. Insoluble in water. Plasticiser.

DIETHYLAMMONIUM DIETHYL DITHIOCARBAMATE

$(C_2H_5)_2N.CS.S.NH_2(C_2H_5)_2$. White crystalline powder with ammoniacal odour, m.p. approx. 80°C, s.g. 1·11. Soluble in benzene and water, sparingly soluble in chloroform, solubility in rubber is greater than 5%. Ultra-accelerator for latex compounds, activator for zinc isopropylxanthate, accelerator for butadiene-styrene copolymers. Limited shelf life. Slight tendency to discolour vulcanisates.

TN: Superaccélérateur 3010 (20)
Vulcafor DDCN (60).

DIETHYL-2-BENZTHIAZOLE SULPHENAMIDE

Dark brown liquid with amine-like odour, s.g. 1·17. Soluble in benzene, carbon tetrachloride, alcohol, ethyl acetate, acetone, methylene chloride, petroleum ether, insoluble in water. Effective accelerator with delayed action for natural and synthetic rubbers. Derived from mercaptobenzthiazole, with the active group blocked by diethylamine, so that the accelerator is ineffective at processing temperatures, at vulcanisation temperature cleavage occurs in two components, both of which are reactive. Gives good heat resistance in low sulphur mixes. Used in combination with tetramethyl thiuram disulphide for fast cures at high temperatures, *e.g.* continuous vulcanisation of cables.

Quantity: 0·3–1·2% with 0·1–0·4% secondary accelerator and 2–2·8% sulphur.

Uses: tyres, hose lines, mechanical goods, latex compounds of Revertex.

Unsuitable for articles which come into contact with food.

TN: Accélérateur rapide (400)
Accélérateur rapide A, R, Z (19)
Accelerator AZ (408)
Vulkacit AZ (43).

DI-2-ETHYL BUTYL AZELATE

$C_6H_{13}OCOC_7H_{14}COOC_6H_{13}$, m.p. −65°C, b.p. (5 mm) 230°C, s.g. (20°C) 0·934. Insoluble in water. Plasticiser.

TN: Plastolein 9050 DOZ (196).

DI-2-ETHYL BUTYL PHTHALATE

$C_6H_4(COOC_6H_{13})_2$, m.p. −50°C, b.p. 350°C, flash p. 193°C, s.g. (20°C) 1·01. Plasticiser.

DIETHYL CARBANILIDE

$C_2H_5(C_6H_5)N.CO$
$.N.(C_6H_5)C_2H_5$,
diethyl diphenyl urea. Crystalline powder, m.p. 79°C. Antioxidant.

DIETHYL DIPHENYL THIURAM DISULPHIDE

Accelerator activated by piperidine.
TN: Accélérateur rapide TE (19).

DIETHYLENE GLYCOL

$CH_2CH_2O.CH_2CH_2OH$. Colourless, odourless hygroscopic liquid, m.p. −10°C, b.p. 244–245°C, s.g. (20°C) 1·118. Miscible with water, alcohol, acetone, ether. Insoluble in benzene and carbon tetrachloride. Activator, plasticiser and softener for natural rubber, synthetic rubbers and latex. Activator in mixes containing silicate fillers; increases the hardness and improves the water resistance of highly filled compounds.

154

DIETHYLENE GLYCOL DIBENZOATE
$C_6H_5.CO(OCH_2CH_2)_2O$
$.COC_6H_5$.
m.p. 16–28°C, b.p. (5 mm) 230–242°C, s.g. 1·18. Plasticiser.
TN: Benzoflex 2–45
Plastoflex DGB (45).

DIETHYLENE GLYCOL DIPELARGONATE
$(C_8H_{17}COOC_2H_4)_2O$, m.p. − 30°C, b.p. (5 mm) 229°C, s.g. (20°C) 0·966. Plasticiser.
TN: Morpel X-931 (264)
Plastoleine 9055 (196).

DIETHYLENE GLYCOL DIPROPIONATE
$(CH_3CH_2COOCH_2CH_2)_2O$, b.p. (4 mm) 110–118°C, s.g. (20°C) 1·066. Solubility in water 3·6%. Plasticiser.

DIETHYLENE GLYCOL DIRICINOLEATE
$(C._{17}H_{32}(OH)COOCH_2CH_2)_2O$, mlp. − 15°C, s.g. (25°C) 0·953. Plasticiser.

DIETHYLENE GLYCOL DISTEARATE
$(C_{17}H_{35}COOC_2H_4)_2O$, m.p. 38°C, s.g. (25°C) 0·96. Plasticiser.
TN: Morpel X-932 (264).

DIETHYLENE GLYCOL MONOLAURATE
$C_{11}H_{23}COOC_2H_4OC_2H_4OH$.
Oily liquid, m.p. 9–18°C, s.g. (25°C) 0·95. Soluble in alcohol, benzene, toluene, chlorinated hydrocarbons, acetone, ethyl acetate, insoluble in water. Plasticiser, dispersing agent for NR, SR and latices. Effective in latex as an emulsifier and a stabiliser.
TN: Morpel X-870 (264)
Glaurin
Diglykollaurat.

DIETHYLENE GLYCOL MONO-OLEATE Reddish, oily liquid, s.g. 0·93. Softener and plasticiser for natural and synthetic rubbers, with the exception of butyl rubber, stabiliser and emulsifier in latices.

DIETHYLENE GLYCOL MONORICINOLEATE
$C_{17}H_{32}(OH)COO(CH_2CH_2O)_2H$.
Yellowish to amber liquid, m.p. − 60°C, s.g. (25°C) 0·927. Softener and plasticiser for NR, SR and latices with the exception of butyl rubber. Has no effect on vulcanisation.

DIETHYLENE GLYCOL MONOSTEARATE Waxy white mass, m.p. 51–54°C, s.g. 0·96, acid No. 96–104, saponification No. 160–164. Plasticiser for natural rubber, synthetic rubber and latices.

DIETHYLENE TRIAMINE
$NH_2(CH_2)_2NH.(CH_2)_2NH_2$.
Colourless liquid, m.p. − 39°C, b.p. 207°C, s.g. 0·96. Miscible with water. Activator in emulsion polymerisation.
TN: DETA.

DI(2-ETHYL HEXYL)ADIPATE
$C_4H_8[COOCH_2CH(C_2H_5)C_4H_9]_2$,
m.p. − 75°C, b.p. (10 mm) 232°C, flash p. 196°C, s.g. (20°C) 0·927. Insoluble in water. Low temperature plasticiser for NR, SR and vinyl resins.
TN: Adipol 2 EH
Cabflex Di-OA
Elastex 60-A
Flexol A-26
Morflex 310
PX-238
Staflex DOA.

155

DI(2-ETHYL HEXYL)AZELATE

$C_8H_{17}OCOC_7H_{14}COOC_8H_{17}$, m.p. $-65°C$, b.p. (5 mm) $237°C$, s.g. (20°C) 0·918. Insoluble in water. Plasticiser for cellulosic plastics, p.v.c., and SR.
TN: Morflex 410
Staflex DOZ
Plastolein 9058 DOZ.

DI(2-ETHYL HEXYL)HEXAHY-DROPHTHALATE

$C_6H_{10}(COOCH_2$ $.CH(C_2H_5)C_4H_9)_2$, m.p. $-54°C$, b.p. (10 mm) $232°C$, flash p. $218°C$, s.g. (20°C) 0·959. Plasticiser.
TN: Flexol CC-55.

DI(2-ETHYL HEXYL) PHTHALATE

$C_6H_4(COOC_8H_{17})_2$, octoyl. Clear oily liquid, b.p. $386°C$, flash p. $210°C$, s.g. 0·98. Plasticiser for NR, SR, vinyl resins and cellulose esters, improves low temperature flexibility.
TN: Celluflex DOP
Darex DOP
Dutch Boy NL A-20
Elastex 28-P
Flexol DOP
Morflex 110
PX-138
Staflex DOP
Witcizer 312.

DI(2-ETHYL HEXYL) SEBACATE DOS.

Clear, light-coloured liquid, b.p. (5 mm) $250°C$, s.g. 0·917. Processing aid and low temperature plasticiser for NR, SR and vinyl resins.
TN: RC Plasticiser DOS.

DIETHYL PHTHALATE

$C_6H_4(COO.C_2H_5)_2$, phthalic acid diethyl ester. Colourless, odourless liquid, b.p. $198°C$, s.g. 1·12. Miscible with water, benzine oil, alcohol, ether and many organic solvents. Plasticiser for NR and in particular for SR, does not affect vulcanisation, improves flexibility at low temperatures.

DIETHYL SEBACATE

$C_2H_5COO(CH_2)_8COOC_2H_5$. Colourless liquid, m.p. 1·3, b.p. $308°C$, s.g. 0·965. Miscible with alcohol, ether and many other solvents. Plasticiser.

DIETHYL SUCCINATE

$C_2H_5COO(CH_2)_2COOC_2H_5$, m.p. $-22°C$, b.p. $218°C$, s.g. (25°C) 1·048. Plasticiser.

DIETHYL TARTRATE

$C_2H_5COO(CHOH)_2COOC_2H_5$, m.p. $17°C$, b.p. $280°C$, s.g. (25°C) 1·19. Plasticiser.

1:3-DIETHYL THIOCARBA-MIDE

White powder, m.p. $72·5°C$, s.g. 1·098. Accelerator for neoprene, antiozonant for NR, SBR, NBR and neoprene.
TN: Pennzone E
Robac DETU (311).

DIETHYL THIOCARBAMYL 2-MERCAPTOBENZTHIAZOLE

2-benzthiazyl-N,N-diethylthiocarbamyl sulphide. Pale yellow powder, m.p. $69°C$, s.g. 1·27. Primary accelerator for NR, SBR, and NBR, gives quick cures which are scorch free. Secondary accelerator and activator with delayed action for NR and SBR. Non-staining and non-discolouring.
TN: Ethylac (4).

DIHEXYL ADIPATE
$C_6H_{13}OCO(CH_2)_4COOC_6H_{13}$, m.p. −8°C, b.p. (8 mm) 204–208°C, s.g. (20°C) 0·925–0·938. Softener.

DIHEXYL PHTHALATE
$C_6H_4(COOC_6H_{13})_2$, b.p. (5 mm) 210–220°C, s.g. (20°C) 1·007. Softener.
TN: Di-HP, Harflex 160 (210) Morflex 160 (264).

DIHEXYL SEBACATE
$C_6H_{13}OCO(CH_2)_8COOC_6H_{13}$, DHS, b.p. (2 mm) 184°C, s.g. (20°C) 0·920. Softener.
TN: Harflex 60 (210).

DI-HP Abbreviation for dihexyl phthalate.

DIHYDROABIETYL ALCOHOL
$C_{19}H_{31}CH_2OH$, m.p. 32–33°C, s.g. (25°C) 1·008. Insoluble in water. Plasticiser, *e.g.* for cellulose nitrates, ethyl celluloses, vinyl chloride and to a certain extent, polymethyl methacrylate, and vinyl chloride acetate. *TN:* Abitol (10).

DIHYDROPERFLUORO ALKYL ACRYLATES
Polymers and copolymers (with butadiene) up to the hexyl group give vulcanisable rubber-like products with good resistance to hydrocarbon solvents. Butadiene increases the flexibility at low temperatures.

1:2-DIHYDRO-2:2:4-TRI-METHYL-6-PHENYL QUINO-LINE

Antioxidant and protective agent against flex cracking. m.p. approx. 80°C, s.g. 1·11. Particularly suitable for articles used under dynamic strain. Causes heavy discoloration and is only suitable for dark-coloured articles.
Quantity: 1–2%, in extreme cases up to 4%. Blends with N.N′-diphenyl-p-phenylene diamine show improved resistance to flex cracking but bloom at 2% and above.
Quantity: 1–2%. The blends can also be used in latex.
TN: Santoflex B (5)
Santoflex BX (15% diphenyl-p-phenylene diamine), s.g. 1·12 (5)
Santoflex 35 (35% diphenyl-p-phenylene diamine), s.g. 1·12 (5)

2:2′-DIHYDROXY-4:4′-DI-METHOXY BENZOPHENONE
White, crystalline powder, m.p. 130°C. Physical protective against light ageing with a high ultra-violet absorption capacity.
Uses: in urethane foam materials.
TN: Uvinul D-49 (85).

4:4′-DIHYDROXYDIPHENYL
$HO.C_6H_4.C_6H_4.OH$. White to pale yellow powder, m.p. 260°C, s.g. 1·22. Soluble in alcohol, acetone, ethyl acetate, slightly soluble in benzene and methylene chloride, insoluble in carbon tetrachloride, benzine and water. Antioxidant against normal ageing for hot and cold vulcanised rubbers, also protects against copper and manganese. Suitable for latex compounds. Non-staining and suitable for white and light-coloured articles. Recommended in blends with di-β-phenyl-p-phenylene diamine.
Quantity: 0·75–1·5% for high temperature vulcanisation, 0·4–0·8% for sulphur chloride vulcanisation.
TN: Antioxidant DOD (43).

DIHYDROXYPHENYL RUBBER $[C_5H_8(C_6H_4OH)_2]_n$.
Amorphous powder. Soluble in alkalis, acetone, esters and pyridine, produced by heating rubber dibromide with phenol at 150°C. May be coupled with diazoamines to form pigments.

DIISOBUTYL ADIPATE $[(CH_2)_2COOCH_2CH(CH_3)_2]_2$, DIBA, m.p. $-20°C$, b.p. (4 mm) 135–147°C, flash p. 160°C, s.g. 0·95. Insoluble in water. Softener.
TN: Cabflex Di-BA (139)
Darex DIBA (174).

DIISOBUTYL AZELATE Clear, oily, light brown liquid, s.g. 0·936. Low temperature plasticiser for SBR, NBR, CR, IIR and vinyl resins.
TN: Plasticiser 4300.

DIISOBUTYL PHTHALATE $C_6H_4[COOCH_2CH(CH_3)_2]_2$. DIBP. Colourless, odourless liquid, m.p.$-50°C$, b.p. 327°C, flash p. 174°C, s.g. 1·04. Softener for synthetic rubbers, particularly NBR, natural rubber, vinyl polymers, cellulose plastics; increases plasticity and decreases the mixing time.

DIISODECYL ADIPATE $C_{10}H_{21}OCO(CH_2)_4COOC_{10}H_{21}$. Pale, oily liquid, m.p. $-72°C$, b.p. (4 mm) 242°C, flash p. 219–227°C, s.g. (25°C) 0·918. Softener for natural rubber, synthetic rubbers and vinyl resins.
TN: Adipol XX
Cabflex DDA (139)
Darex DIDA (174)
Elastex 20-A (107)
Morflex 330 (264)
PX-220 (305).

DIISODECYL PHTHALATE $C_6H_4(COOC_{10}H_{21})_2$. Clear liquid, m.p. $-35°C$, b.p. (4 mm) 249–256°C, s.g. 0·967. Softener for natural and synthetic rubbers and vinyl resins.
Uses: insulation for wire and cables.
TN: Cabflex DDP (139)
Darex DIDP (174)
Elastex 90-P (107)
Morflex 130 (264)
PX-120 (305).

DIISO-OCTYL AZELATE $C_8H_{17}OCOC_7H_{14}COOC_8H_{17}$, m.p. $-65°C$, b.p. 225–244°C (4 mm), s.g. 0·918 (20°C). Insoluble in water. Softener.
TN: Cabflex Di-OZ (139)
Plastolein 9057 DIOZ (196).

DIISO-OCTYL FUMARATE DIO. Unsaturated diester.
Use: comonomer in polymerisation reactions.

DIISO-OCTYL MALEATE DIOM. Unsaturated diester.
Use: Comonomer in polymerisation reactions.

DIISO-OCTYL PHTHALATE $C_6H_4(COOC_8H_{17})_2$, m.p. $-45°C$, b.p. (5 mm) 239°C, flash p. 210°C, s.g. 0·981. Softener.
TN: Cabflex Di-OP (139)
Darex DIOP
Dutch Boy NL A-30 (191)
Elastex 10-p (107)
Harflex 120 (210)
Hercoflex 100 (240)
PX-108 (305)
Staflex DIOP (330)
Witcizer 313 (215).

DIISO-OCTYL SEBACATE

$C_8H_{17}OCO(CH_2)_8COOC_8H_{17}$, m.p. $-58°C$, flash p. 246°C, s.g. 0·916. Softener.
TN: Monoplex DIOS (33).

N,N'-DIISOPROPYL BENZ-THIAZYL-2-SULPHENAMIDE

2-mercaptobenthiazyl diisopropyl sulphenamide. Pale brown flakes, m.p. 55–60°C, s.g. 1·23. Accelerator with marked delayed action. In compounds loaded with time furnace blacks (SAF) it enables higher processing temperatures to be used and a wider range of vulcanisation; the vulcanisation temperature should be above 138°C, 3–5% zinc oxide and the normal quantity of stearic acid are necessary. Vulcanisates have good ageing properties.
TN: DIBS (21)
Dipac (120).

N,N'-DIISOPROPYL-p-PHENYLENE DIAMINE

$CH(CH_3)_2-NH-\langle\bigcirc\rangle-NH-CH(CH_3)_2$

Reddish-brown fluid. General purpose antioxidant.
TN: Tenamene 5 (193).

DIISOPROPYL XANTHOGEN DISULPHIDE

$(CH_3)_2CHO.CS.S.COCH(CH_3)_2$, bisisopropyl xanthogen, Diproxide, Dixie. The only solid xanthogen disulphide with a relatively high m.p. Effective modifier in emulsion polymerisation of SBR and NBR; used for the first time in 1942 in the production of Buna S 3. Also acts as a powerful vulcanisation accelerator. Produced by the oxidation of potassium isopropyl xanthate with potassium persulphate.

DILATANCE Antithixotropy. The rise in viscosity of a material on the application of an increased shearing force.

DILATANT LIQUID Pseudoplastic liquid. A liquid in which the apparent viscosity or consistency increases suddenly when the shearing (stirring) force is increased.

DILATOMETER Apparatus for measuring change in volume of solids.

DILAURYLTHIODIPRO-PIONATE White powder, m.p. 37–40°C. Antioxidant and stabiliser for polyolefins (polyethylene, polypropylene). Softener for elastomers.
TN: Topanol CH (60).

DILEX Sulpholignin. Dispersing agent for rubber compounds (183).

DIMENSIONAL STABILITY Capacity of rubber or plastic article to maintain its original shape and size.

DI-p-METHOXYDIPHENYL-AMINE

$(CH_3.O.C_6H_4)_2NH$, 4:4'-dimethoxydiphenylamine. Greyish black, granular material, m.p. 103°C, s.g. 1·25. Antioxidant. Effective against normal ageing, heat and flex-cracking. Has no effect on plasticity or vulcanisation. Discolours in sunlight.
Quantity: 1–2%, or in smaller amounts if used with other antioxidants to improve flex-cracking resistance. Does not bloom below 1·5%.
TN: Thermoflex A (blend) (6)
Vulcaflex A (blend) (60).

DIMETHOXYETHOXYETHYL ADIPATE

$[CH_3(OC_2H_4)_2COO(CH_2)_2]_2$, b.p. (3 mm) 200°C, s.g. (25°C) 1·104. Softener.
TN: TP-98.

DIMETHOXYETHYL PHTHALATE

$C_6H_4(COOC_2H_4OCH_3)_2$.　　Clear liquid, m.p. −40°C, b.p. 310°C, s.g. 1·17. Softener. Benzine and oil resistant, compatible with butadiene-acrylonitrile copolymers and neoprene. Used in synthetic rubbers in the aircraft industry.
TN: Methox.

DIMETHYLAMINE

$NH(CH_3)_2$. Gaseous at normal temperatures, b.p. 7°C, s.g. 0·680 (4°C). Readily soluble in water, gives a strong alkaline reaction, soluble in alcohol and ether. Accelerator.

DIMETHYLAMINOMETHYLENE DIMETHYL DITHIO-CARBAMATE

$(CH_3)_2N.CS.S.CH_2.N(CH_3)_2$. Ultra-fast accelerator.
TN: Robac DAMD (311).

DIMETHYLAMMONIUM DIMETHYL DITHIOCARBA-MATE

White crystals which decompose on contact with air. Extremely active ultra-fast accelerator in the presence of zinc oxide, 'scorchy'. Vulcanisation temperature 141°C.
Quantity: 0·5%.
TN: Naugatex 112 (23)
Naugatex 535 (38% aqueous solution) (23)
Ultra DMC (obsolete)
Vulcafor DDD (60).

DIMETHYL AZOISOBUTYR-ATE

Polymerisation catalyst and blowing agent for plastics.

α,α′-DIMETHYL BENZYL HYDROPEROXIDE

$C_6H_5C(CH_3)_2OOH$.　　Colourless liquid. Miscible with alcohol, acetone, esters, hydrocarbons and chlorinated hydrocarbons.
Use: polymerisation catalyst in redox systems.

2-DI-(α-METHYL BENZYL)-4-METHYL PHENOL

Russian produced heat and light stabiliser for NR, SBR and PB-vulcanisates.
TN: Alkofen MBF.

2:3-DIMETHYL-1:3-BUTA-DIENE

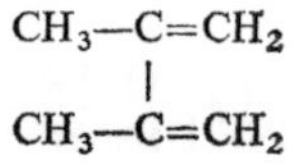

$$CH_3-C=CH_2$$
$$|$$
$$CH_3-C=CH_2$$

Diisopropenyl. Liquid, m.p. −76°C, b.p. 69°C, s.g. (20°C) 0·727. Diene hydrocarbon, used during World War I as a basis for methyl rubber. Produced by reduction of acetone to pinacol with subsequent distillation (over aluminium) under pressure, at 429–470°C.

N-1:3-DIMETHYL BUTYL-N′-PHENYL-p-PHENYLENE DIAMINE

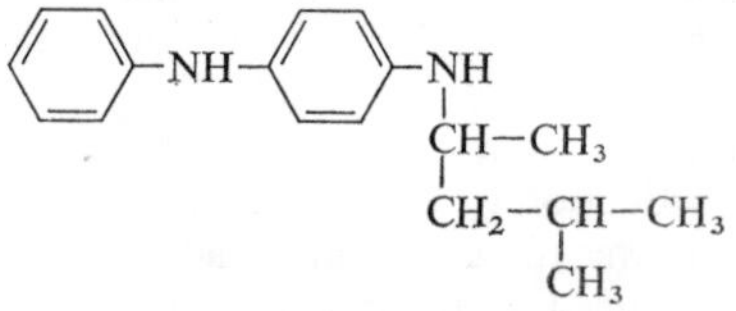

Brown lumps (flakes), m.p. 40–50°C, s.g. 1·01–1·07. Soluble in ethanol,

acetone, methylene chloride, carbon tetrachloride and benzene. General purpose antioxidant and fatigue protector, especially against ozone and subdynamic conditions. Used in NR and SR.
Dosage: 0·4–3%.
TN: Antioxidant 4020 (43)
Antozite 67 (51)
Flexzone 7-L (23)
Santoflex 13 (5)
Wingstay 300 (115).

DIMETHYL CYCLOHEXYL-AMMONIUM DIBUTYL DITHIOCARBAMATE

$(C_4H_9)_2N.CS.S.HN(CH_3)_2.C_6H_{11}$. Brown liquid, s.g. 0·96. Ultra-fast accelerator for cures at room temperature or slightly above. Vulcanisation speed can be accelerated by adding mercaptobenzthiazole.
Uses: primarily for fast vulcanising cements, coatings.

	Accelerator %	MBT %	Sulphur %
Solid rubber NR cements and adhesives SBR cements and adhesives	2 3 5	– 1 1	2–4 2–4 1·5–3

TN: RZ 50-A (50% solution)
RZ 50-B (for latex)
RZ 100 (5).

DIMETHYL DI(1-METHYL PROPYL)-p-PHENYLENE DIAMINE

$$\left[\begin{array}{c} CH_3-CH-CH-N- \\ \quad\quad | \quad\ | \\ \quad\quad CH_3\ \ CH_3 \end{array} \right]_2 \!\!\bigcirc$$

Reddish-brown liquid, s.g. 0·933. Antiozonant for natural rubber and SBR, particularly useful under static conditions.
TN: Eastozone 32 (193).

DIMETHYL DINITROSO-TEREPHTHALAMIDE

Colourless, yellow, crystalline powder, m.p. 118°C; upon decomposition is used as a blowing agent, s.g. for vulcanisable elastomers, p.v.c., plastisols, epoxy resins; produces 99% nitrogen. Has a good processing safety, gives a fine uniform cell structure. Since the product has a slight tendency to explode it is used in a 70:30 mixture with mineral oil.
Quantity: for elastomers 3·5–7% of 70:30 mixture, for p.v.c. and plastisols up to 25%.
TN: Nitrosan (6).

3:3'-DIMETHYL-4:4'-DIPHENYL DIISOCYANATE

Bitolyl diisocyanate. White to yellowish flakes, m.p. 69–71°C, s.g. 1·197.
Uses: isocyanate component in urethane elastomers.
TN: TODI (148).

DIMETHYL DIPHENYL THIURAM DISULPHIDE

White, crystalline, non-hygroscopic powder, m.p. (pure substance) 206°C (trade product) approx. 175°C, s.g. 1·33. Soluble in benzene and methylene chloride, sparingly soluble in carbon disulphide, alcohol, ethyl

acetate, acetone, insoluble in benzine and water. 8·79% sulphur is available for vulcanisation. Secondary accelerator in blends with tetramethyl thiuram disulphide and zinc ethyl phenyl dithiocarbamate, vulcanising agent. Has excellent processing safety with a delayed action and gives fast cures at relatively low temperatures; vulcanisates possess good ageing properties.
Quantity: primary 0·3–0·5% with 2·2–2·5% sulphur, secondary 0·1–0·2% with 0·15–0·25 tetramethyl thiuram disulphide and 2·2–2·4% sulphur.
Uses: articles which come into contact with food; light-coloured and coloured articles, fast vulcanising moulded and dipped articles, coatings and impregnations.
TN: Accelerator J (408)
Robac J (311)
Vulkacit J (43).

DIMETHYLFORMAMIDE

$$H-C-N \begin{smallmatrix} CH_3 \\ \\ CH_3 \end{smallmatrix}$$
$$\| \\ O$$

Colourless liquid, m.p. −61°C, b.p. 153°C, s.g. 0·945. Miscible with water, ethanol, ether, ketones and benzene.
Uses: solvent for plastics, polyacrylonitriles, pvc, etc.

2:5-DIMETHYLHEXANE-2:5-DIHYDROXYPEROXIDE
White, crystalline powder, catalyst for polymerisation at high temperatures.
TN: USP-611 (354).

DIMETHYL ISOBUTYL CARBINYL PHTHALATE
$C_6H_4[COOC(CH_3)_2$
　　　　　　$.CH_2CH(CH_3)_2]_2$,

b.p. 190–195°C, s.g. (25°C) 0·983. Softener.
TN: Plastoflex 520 (45).

2-(2:6-DIMETHYL-4-MOR-PHOLINTHIO)BENZTHIAZOLE

Brown flakes, m.p. 84°C, s.g. 1·23–1·29. Delayed action accelerator.
TN: Santocure 26 (5).

DIMETHYL OXYETHYL PHTHALATE
Colourless liquid, b.p. (4 mm) 190–210°C, viscosity (20°C) 58 cp, s.g. 1·168–1·174. Softener for SBR, NBR, CR and vinyl polymers. Has no effect on vulcanisation. Vulcanisates have a good tensile strength and elongation at break with a low hardness; improves surface appearance.

2-DIMETHYL-4-OXYMETHYL-1:3-DIOXALANE

Odourless and tasteless liquid, b.p. (10 mm) 82–83°C, viscosity (20°C) 11 cp, mol. wt. 132, s.g. 1·064. High vapour pressure. Highly miscible with water and with many organic solvents.
Uses: solvent, *e.g.* for plastics, resins, softeners, sulphur.
TN: Imnulit (technical quality)
Solketal (chemically pure) (384).

DIMETHYL PHTHALATE
$C_6H_4(COOCH_3)_2$, phthalic acid

dimethyl ester. Oily, colourless liquid with aromatic odour, m.p. 5·5°C, b.p. 284°C, s.g. 1·19. Soluble in water, sparingly soluble in benzine and oils. *Uses:* Softener in rubber solutions, solvent and softener for cellulose acetate and cellulose acetate-butyrate.

DIMETHYL POLYSILOXANE

$$\left[\begin{array}{cc} CH_3 & CH_3 \\ | & | \\ -Si-O-Si- \\ | & | \\ CH_3 & CH_3 \end{array}\right]_x$$

Colourless plastic material, s.g. 0·98. Readily soluble in benzine, benzene, carbon tetrachloride, ethyl acetate, insoluble in alcohol and acetone. Silicone rubber has excellent resistance to temperatures between −55°C and 200°C, excellent electrical properties and also good physical properties which are not affected by the temperature. Peroxides are used as crosslinking agents. Compounds are vulcanised by moulding in hot air or in steam, postcuring is in hot air without pressure. The compounds must be stored before vulcanisation, to achieve the best properties.
TN: Silopren B (43).

DIMETHYL SEBACATE
$CH_3OCO(CH_2)_8COOCH_3$. Clear liquid, m.p. 24°C, b.p. 294°C, s.g. (25°C) 0·990. Softener.

DIMETHYL SULPHOXIDE

$$O=S\begin{array}{c} CH_3 \\ CH_3 \end{array}$$

DMSO. Selective solvent for Orlon, nitrocellulose, cellulose acetate, resins, aromatic and unsaturated hydrocarbons and compounds containing sulphur.

DIN Abbreviation for Deutsche Industrie Norm.

DI-β-NAPHTHYL-p-PHENY-LENE DIAMINE

$$C_{10}H_7-NH-\!\!\!\langle\bigcirc\rangle\!\!\!-NH-C_{10}H_7$$

Pale grey powder which darkens on contact with air, m.p. 230–235°C, s.g. 1·20. Soluble in acetone, sparingly soluble in benzene, chloroform and carbon disulphide, insoluble in water, alcohol and benzine. One of the most effective and economic antioxidants for protection against normal oxidation, heat ageing and the effects of copper and manganese. Can be used in relatively small quantities; increases the effectiveness of other antioxidants when added in small quantities. Prevents gel formation during thermal degradation of SBR. Causes a slight discoloration. Activates accelerators of the mercapto type.
Quantity: 0·2–0·75%, 0·1–0·2% mixed with other antioxidants (*e.g.* 0·8% dioxydiphenyl).
Uses: include light-coloured articles, rubber bands, coated textiles, latex articles (in combination with other antioxidants), cable compounds. Unsuitable for articles which come into contact with food.
TN: Aceto DIPP (25)
Age-Rite White (51)
Antioxidant DNP (43)
Antage F (16)
Antigen F
Antioxidant 123 (22)
Nocrac White (274)
Nonox CI (60)
Santowhite CI (5).

163

2:4-DINITROPHENYL DI-METHYL DITHIOCARBAMATE

Pale yellow, crystalline powder, m.p. 140–150°C, s.g. 1·57. Soluble in chlorinated hydrocarbons, sparingly soluble in acetone and benzene, insoluble in water and benzine. Fast accelerator with good storage properties at normal temperatures, decomposes on heating in contact with air. Compounds and vulcanisates can cause dermatitis. Gives a medium vulcanisation range, good plateau effect and good processing safety; needs zinc oxide and sulphur in normal quantities, fatty acids are unnecessary but can be added. Activated by basic materials. Primary or secondary accelerator with aldehyde-amines, excellent secondary accelerator with thiazoles. Causes yellow discoloration and is unsuitable for white or light-coloured articles. Vulcanisates can cause slight staining on light-coloured articles.

Quantity: natural rubber 0·75–0·9% with 1·75–3·5% sulphur, SBR 1·25% with 2% sulphur, or 0·15% with 1·5% benzthiazyl disulphide and 3% sulphur.

Uses: tyre treads, carcases, conveyor belts, mechanical goods.

TN: Safex (23).

2-(2:4-DINITROPHENYL)-THIO BENZTHIAZOLE

Yellow to orange yellow powder, m.p. 140–145°C,. s.g 1·53–1·61.

Soluble in benzene, chloroform, ethyl acetate, acetone and methylene chloride, sparingly soluble in alcohol, carbon tetrachloride and dilute alkalis, almost insoluble in water, insoluble in dilute acids. Relatively slow accelerator, boosted by basic accelerators. Guanidines are the most suitable boosters and give compounds with good processing safety and fast vulcanisation; zinc oxide and fatty acid are necessary. Gives a product with a high modulus and good tensile strength. Without an activator may be used for thick articles which should be vulcanised very slowly. The commercial product is normally a mixture with guanidines and basic accelerators.

Quantity: 0·4–0·5% with 0·6–0·8% diphenyl guanidine and 2–4·5% sulphur.

Uses: mechanical goods, sponge rubber, soles, heels, tyre treads. Suitable for hot air vulcanisation, *e.g.* rubber shoes, unsuitable for goods which come into contact with food.

TN: Ureka Base (5)
Mixtures with diphenyl guanidine:
Di-esterex (23)
Ureka, Ureka HR (57)
Vulcaid 333 (3)
Vulcaid 555 (3)
Vulkacit U (43)
Mixtures with diphenyl guanidine acetate:
Di-esterex N (23)
Mixtures with diphenyl guanidine phthalate:
Ureka B (5)
Ureka DD (5)
Mixtures with diphenyl guanidine phthalate, diphenyl guanidine acetate and mercaptobenzthiazole:
Ureka White (5).

p-DINITROSOBENZENE

Pale brown powder, s.g. 1·57. Vulcanising agent for unsaturated elastomers, particularly for isobutylene-isoprene copolymer, together with sulphur and a thiuram accelerator. Gives a fast cure with a high modulus; improves heat resistance. Stiffens unvulcanised compounds. The commercial product is a 25% concentration stabilised with an inorganic base.
Uses: inner tubes, neoprene latex.
TN: Vulcafor BDN (60).

N:4-DINITROSO-N-METHYL ANILINE

Extremely reactive material which can cause otherwise inert materials to react, *e.g.* reacts with blacks and other fillers in rubber to form chemical bonds between the polymer and the filler. Acts as a promoter in butyl rubber and improves the crosslinking properties of polymers with a low degree of unsaturation, increases the elasticity, modulus, abrasion resistance, Mooney viscosity, but decreases the flow of unvulcanised compounds, improves the blending of SBR with natural rubber. Must be incorporated in a Banbury mixer at 150°C, before the other compounding ingredients are added.
Quantity: 0·5–1·5%.
TN: Elastopar (5).

DINITROSOPENTAMETHYLENE TETRAMINE

$$\begin{array}{ccc} CH_2{-}N{-}CH_2 \\ |\quad\ \cdot|\quad\ | \\ O{=}N{-}N\quad H_2C\quad N{-}N{=}O \\ |\quad\ \ |\quad\ | \\ CH_2{-}N{-}CH_2 \end{array}$$

Yellow powder with light, characteristic odour, decomposition temperature 190–200°C, in the presence of an activator 80–180°C, s.g. 1·45. Soluble in dimethyl formamide. Blowing agent for NR, synthetic rubbers, pvc and polyethylene. Develops a high gas pressure so the influence of plasticity on the uniformity of cell structure is decreased. Stearic acid and sometimes salicylic acid, urea or melamine diethylene glycol are necessary as activators. Slightly activates acidic accelerators. Has no effect on ageing properties. Non-staining, does not bloom below 15%.
Quantity: for open cells 1–5%, for closed cells 1–15%, secondary with sodium bicarbonate 0·5–1·5%, 1·3% activator.
TN: Aceto DNPT (40%, 80%, 100%) (25)
Cellopren MO (85%) (378)
Celloprenon CC (85%) (378)
Celloprenon Extra (70%) (378)
Isopor (414)
Nitrocel (414)
Opex 40 (40% with 60% inert filler) (237)
Opex PL-80 (80% with 20% ester softener) (237)
Opex 100 (237)
Porofor DNO/N (43)
Unicel ND (40%) (164)
Unicel NDX (80%) (6)
Unicel 100 (6)
Vulcacel BN
Vulcacel B 40 (40%) (127).

DINONYL ADIPATE

$(CH_2)_4(COOC_9H_{19})_2$, m.p. 165°C, flash p. 232°C, s.g. (20°C) 0·9168. Insoluble in water. Softener.
TN: Morflex DNA (264)
PX-209 (305).

DINONYL NAPHTHALENE

$C_{10}H_6(C_9H_{19})_2$, b.p. 200–270°C (20 mm), s.g. (20°C) 0·946. Softener.
TN: Morpel X-928 (264).

DINOPOL Group of softeners.

DIOA Abbreviation for diisooctyl adipate.

DIOCTYL ADIPATE
$C_8H_{17}OCO(CH_2)_4COOC_8H_{17}$. Pale yellow liquid, m.p. $-70°C$, b.p. 205–220°C (4 mm), s.g. (20°C) 0·928. Insoluble in water. Softener for natural rubber, synthetic rubbers, vinyl and cellulosic plastics.
TN: Adipol 10 A
Cabflex Di-OA (139)
Darex DIOA (174)
DOA
DIOA
Good-Rite GP 230 (14)
Harflex 220 (210)
Morflex 300 (264)
Polycizer 332 (127)
PX-208 (305)
Staflex DOA and DIOA (330).

DIOCTYL-p-PHENYLENE DIAMINE
B.p. approx. 400°C, s.g. 0·912. Antioxidant. Primarily used as an antiozonant for dynamic and static conditions. The product migrates to the rubber surface over a relatively long period, thus protects against ozone attack, particularly recommended for SBR.
TN: Eastozone 30 (193)
OZO-88 (281)
Tenamine 30
UOP 88 and 288 (352).

DIOCTYL PHTHALATE
$C_6H_4(COOCH_2CH(C_2H_5)C_4H_9)_2$. Colourless liquid, b.p. 220–248°C, s.g. 0·983–0·989. Insoluble in water, miscible with many organic solvents. Softener for SBR, NBR, CR, polyvinyls, nitrocellulose, cellulose acetobutyrate. Has no effect on vulcanisation. Gives vulcanisates with good ageing properties, resistance to low temperatures, high strengths and good electrical properties. Particularly suitable for electrical insulations.
TN: Harflex 150 (210)
Morflex X-926 (264)
Polycizer 162 (127)
Vestinol AH (189).

DIOCTYL SEBACATE
$C_8H_{17}OCO(CH_2)_8COOC_8H_{17}$, DOS. Yellowish liquid with faint odour, m.p. $-55°C$, b.p. 248°C (4 mm), s.g. (20°C) 1·449. Softener for plastics, natural and synthetic rubbers, particularly for CR and vinyl polymers.
TN: Darex DOS (174)
Dutch Boy NLC-20 (191)
Harflex 50 (210)
Monoplex DOS (33)
Morflex 210 (264)
PX-438 (305)
Staflex DOS (330).

DIOF Abbreviation for diisooctyl fumarate.

DIOL 80 Cycloparaffin. Softener (41).

DIOLEFIN Aliphatic hydrocarbon with two double bonds, *e.g.*
$$CH_2=CH-CH=CH_2$$

DIOLIN A Octadecanediol diacetate. Softener and release agent.

DIOM Abbreviation for diisooctyl maleate.

DIOP Abbreviation for diisooctyl phthalate.

DIORID Synthetic fibre produced from a copolymer of polyvinylidene chloride and acrylonitrile (32).

DIOS Abbreviation for diisooctyl sebacate.

DIOSITE Dibenzthiazyl disulphide (391).

DIOX 7 Tert-butylisopropylbenzene hydroperoxide. Initiator in emulsion polymerisation (185).

DIPENTAMETHYLENE THIURAM DISULPHIDE
$H_{10}C_5=N.CS.S.S.CS.N=C_5H_{10}$.
Yellowish odourless powder, m.p. 110–112°C, s.g. 1·39. Soluble in alcohol, acetone, benzene, insoluble in water. Accelerator with 10% sulphur available; acts therefore as a vulcanising agent, also as an activator for thiazoles; zinc oxide is necessary. Sulphurless compounds tend to produce products with excellent ageing properties and good heat resistance.
Quantity: 0·25–0·35% with 2–2·5% sulphur, 3–4% for sulphurless vulcanisation.
Uses: articles which come into contact with food.
TN: Robac PTD and PTD 86 (311). PTD 86 (has added sulphur and can be used instead of tetramethyl thiuram disulphide).

DIPENTAMETHYLENE THIURAM HEXASULPHIDE

$$\left[\begin{array}{c} CH_2-CH_2 \\ CH_2 \quad\quad N-C- \\ CH_2-CH_2 \quad \overset{\parallel}{S} \end{array} \right]_2 S_6$$

Buff powder, m.p. 104–118°C, s.g. 1·48–1·52. Very active accelerator, even in the absence of sulphur, for Hypalon.
TN: Accelerator 4P (22)
Robac Thiuram P-25 (311)
Sulfads (51)
Tetrone (6).

DIPENTAMETHYLENE THIURAM MONOSULPHIDE
$H_{10}C_5=N.CS.S.CS.N.=C_5H_{10}$.

Yellow, crystalline powder with characteristic odour, m.p. 98–102°C, s.g. 1·38. Soluble in alcohol, acetone and benzene, insoluble in water. Extremely safe accelerator with marked delayed action. Suitable for natural rubber, butadiene-styrene and butadiene-acrylonitrile copolymers, isobutylene-isoprene copolymers, neoprene. May be used in press and open steam cures; zinc oxide is necessary. Non-staining. Activates thiazoles. Compounds with a minimal sulphur content have extremely good ageing properties.
Quantity: 0·3% with 2–5% sulphur, compounds with low sulphur content 2–4% with 0·5% sulphur.
TN: Robac PTM (311).

DIPENTAMETHYLENE THIURAM TETRASULPHIDE

$$\left[\begin{array}{c} H_2 \quad H_2 \\ C-C \\ H_2C \quad\quad\quad N-C-S-S- \\ C-C \quad\quad \overset{\parallel}{S} \\ H_2 \quad H_2 \end{array} \right]_2$$

Pale grey to colourless powder, m.p. 110–117°C, s.g. 1·41–1·48. Soluble in chlorinated hydrocarbons, carbon disulphide, sparingly soluble in solutions of hydrocarbons and ketones, insoluble in water. Ultra-fast accelerator. Extremely active, 25% sulphur available for vulcanisation. Gives high modulus. Extra sulphur is not added when good ageing properties are required. Has relatively good processing safety at normal processing temperature and without sulphur; scorches when sulphur is present. The simultaneous use of a thiazole accelerator is advisable. Active at temperatures above 110°C. Vulcanisation temperature: sulphurless NR compounds 141°C, latex 100–110°C, nitrile rubber 150°C. To

167

improve processing safety litharge, lead oleate or N-nitrosodiphenylamine may be used as a retarder, but only in compounds in which the colour is unimportant. Recommended as a secondary accelerator with tetramethyl thiuram disulphide for sulphurless compounds with good ageing resistance. Used with zinc mercaptobenzthiazol it is suitable for foam rubber, when used with dibenzothiazyl disulphide, vulcanisates with very low compression set may be produced. Non-staining and non-discolouring. Thiuram polysulphides generally give better results than normal sulphur systems in NBR.

Quantity: natural rubber and SBR 1–2·5% or 1–5·1% with 0·75–1·25% tetramethyl thiuram monosulphide and without sulphur; latex 0·5–3% with 1–1·5% zinc mercaptobenzthiazole and 0–1·5% sulphur, nitrile rubber 1·5–3% with 1·5–0·5% dibenzthiazyl disulphide; butyl rubber 1–2% without sulphur; Hypalon 20 1–3% without sulphur, or 0·25–1% with 1–0·5% dibenzthiazyl disulphide.

Uses: natural rubber, latex, nitrile rubber, SBR, reclaim, butyl rubber.

TN: Accelerator 4 P (22)
Kuracap PTD (241)
PTetD 70 (70% with 30% polyisobutylene) (232)
Robac Thiuram P 25 (311)
Sulfads (51)
Tetrone A (6)
Van Hasselt DPTT (355).

DIPENTENE

CH$_3$

C
CH$_2$ CH$_3$

Kautschin, d,1-limonen, cinene. Colourless liquid, b.p. 176–178°C, s.g. 0·84. Miscible with alcohol. Occurs in turpentine oil, and is one of the main constituents obtained during the dry distillation of rubber, and is formed when polymerising isoprene.

Uses: solvent for resins and waxes, softener and dispersing agent for NR and SR.

DIPENTENE 122 Terpene hydrocarbon. Solvent and reclaiming oil (10).

DIPHENOXYETHYL DIGLYCOLATE
(C$_6$H$_5$OC$_2$H$_4$OCOCH$_2$)$_2$O, m.p. 70–75°C. Softener.
TN: Pycal 60 (96).

DIPHENOXYETHYL FUMARATE
(C$_6$H$_5$OC$_2$H$_4$OCO)$_2$C$_2$H$_2$, m.p. 125°C, s.g. (25°C) 1·08. Plasticiser for plastics.
TN: Pycal 62 (97).

DIPHENYLAMINE-ACETONE CONDENSATION PRODUCTS
Antioxidants for natural rubber and synthetic elastomers.
Products formed at a low temperature: yellowish powder, m.p. 85–95°C, s.g. approx. 1·10. Soluble in acetone and ethylene dichloride, sparingly soluble in benzene, insoluble in water. Has a negligible effect on vulcanisation. Causes a yellowish discoloration in sunlight, does not bloom. Gives good protection against normal oxidation, heat and flex-cracking.
Quantity: 0·5–1%, light-coloured articles 0·25%, inner tubes 2–3%.
Uses: tyres, inner tubes, footwear,

insulations and light-coloured articles.

TN: Aminox (23)
Naugatex 505 A (aqueous dispersion) (23)
Naugatex 503 E (50% paste) (23)
Nonox B (60).
Products formed at a high temperature: dark-brown liquids, s.g. approx. 1·1. Soluble in acetone, benzene, ethylene dichloride, insoluble in water. Strongly retard vulcanisation. Cause dark discoloration in sunlight and stain articles which come into contact with them. Migration occurs during vulcanisation. Gives good protection against heat, normal oxidation and flex cracking.

Quantity: for the optimum effect 1–1·5% but good protection is obtained with 0·25%.

Uses: tyres, tubes, mechanical goods and articles in which discoloration or staining is unimportant.

TN: BLE (23)
BLE 25 (low viscosity) (23)
Naugatex 519 (aqueous dispersion) (23)
Nonox BL (60)
Santoflex DPA (5).

DIPHENYL CARBAMYL DIMETHYL DITHIOCARBAMATE

$(C_6H_5)_2N.CO.S.CS.N(CH_3)_2$.
Vulcanisation accelerator.
TN: ONC (obsolete) (23).
Vulcaid 666S (3).

DIPHENYL CRESYL PHOSPHATE

Colourless liquid, b.p. above 245°C (20 mm), s.g. 1·21. Miscible with most organic solvents, insoluble in water. Softener and extender for butadiene-styrene and butadiene-acrylonitrile copolymers. Increased loadings result in increased elongation at break with simul-

taneous decreases in hardness and tensile strength.
Quantity: 10–20%.

4:4′-DIPHENYL DISULPHONYL AZIDE

White powder, m.p. 142–145°C, s.g. 1·51. Soluble in acetone, sparingly soluble in benzene and toluene, insoluble in water. Blowing agent which produces nitrogen, for NR, SR, thermoplastics and thermosetting materials. Also effective as a crosslinking agent in NR, SBR and CR.
TN: Nitropore CL-100 (237).

DIPHENYL ETHYLENE DIAMINE

$(C_6H_5NHCH_2)_2$. Pale brown powder, m.p. 60–75°C, b.p. 228°C, s.g. 1·14–1·21. Slightly discoloring antioxidant which gives complete protection against heat and normal oxidation. Retards flex-cracking, plasticises compounds during processing.
TN: Nonox DED (60)
Stabilite (100).

DIPHENYL GUANIDINE

White, crystalline, odourless powder, m.p. 145–148°C, s.g. 1·1–1·2. Soluble in alcohol, benzene hydrocarbons, chlorinated solvents, carbon disulphide, insoluble in water. Safe accelerator for general use, gives a reasonably fast cure. Vulcanisation temperature 135–160°C, disperses easily, good processing safety because of the high critical temperature. Causes slight discoloration and cannot be used in light articles except as a booster. Decreases the stability of latex. Does not give good ageing properties and is therefore seldom used as the sole accelerator; blends with DPG need a primary accelerator which is also an antioxidant. Strongly boosts accelerators of the mercapto type.

Quantity: primary 1–2% with 3·5–2·5% sulphur and 3–5% zinc oxide, secondary 0·1–0·25% with 1–0·75% dibenzthiazyl disulphide, 2·5% sulphur and 3–5% zinc oxide.

Uses: as the sole accelerator in thick walled articles which must be vulcanised slowly, *e.g.* solid rubber tyres and in mountings designed to cushion shock effects, booster for mercapto accelerators, for footwear.

TN: Accelerator 12 (6)
Accelerator D (408)
Accelerator X 28
Accélérateur D (19)
Accélérateur DPG
Ancazide DPG
Belger DPG (22)
Denax
DPG
Dynamine
Eveite D (59)
Helastene F
Naccomine
Noceller D (274)
Nurac
Phénaldine
P (DPG) D-65 (65% with 35% polyisobutylene) (282)
Rapidex KA
Skeed X
Soxinol D
Vulcafor DPG (60)

Vulcaid ND (252)
Vulcogene ND
Vulkacit D (43)
X-28
XLO
Xylite.

DIPHENYL GUANIDINE ACETATE
$[(C_6H_5NH)_2C=NH]HOOC.CH_3$.
Booster for thiazole accelerators with a delayed action. Rarely used.
TN: Delac A (23).

DIPHENYL GUANIDINE DIBENZYL DITHIOCARBAMATE

$[(C_6H_5)_2H—C—SH.NC.(NH.C_6H_5)_2]$
$\quad\quad\quad\quad ||$
$\quad\quad\quad\quad S$

Ultra-fast accelerator. Powder with aromatic odour, m.p. 140°C, s.g. 1·0. Vulcanises at 115°C.
Uses: in adhesives.
TN: W 29 (366).

DIPHENYL GUANIDINE DIBUTYL DITHIOCARBAMATE
White powder, m.p. 123°C, s.g. 1·14. Ultra-accelerator.
TN: Ultex (100).

DIPHENYL GUANIDINE OXALATE
$[(C_6H_5NH)_2C=NH]_2H_2C_2O_4$.
Booster for thiazole accelerators with a delayed action. Rarely used.
TN: Delac O (25).

DIPHENYL GUANIDINE PHTHALATE
$[(C_6H_5NH)_2C=NH]_2$
$.C_6H_4(COOH)_2$.
White powder, m.p. 178°C, s.g. 1·21. Soluble in alcohol, insoluble in benzene, toluene and water. Booster

with delayed action for thiazole accelerators. May be used in natural rubber, synthetic elastomers and latices, most effective in natural rubber. Gives improved processing safety compared to boosting with pure guanidines. Causes slight discoloration, does not bloom and is suitable for light-coloured articles (with the exception of pure white). Used alone it is an extremely slow accelerator.
TN: Guantal (5)
Delac P (23).

4:4'-DIPHENYL METHANE DIISOCYANATE

OCN—⟨benzene⟩—CH₂—⟨benzene⟩—NCO

di-p-isocyanate phenyl methane, methylene bis(4-phenyl isocyanate). Yellow solid, m.p. 40–41°C, b.p. (0·1 mm) 256–158°C, s.g. 1·20. Soluble in aromatic and chlorinated aromatic hydrocarbons, esters, ethers, nitrobenzene. Added to elastomers and cements to improve the adhesion of materials such as textiles, glass, wood, leather and plastics to elastomers. Activates the vulcanisation of rubber compounds, has no effect on ageing, causes discoloration. Crosslinking agent for polyurethanes at room temperature.

Quantity: 15–30 phr in cements. Cements gel when left to stand. The block is removed when heated at temperatures above 145°C so that the diisocyanate becomes reactive again.

Uses: in latex and aqueous systems for adhesive compounds for synthetic fibres, bonding agent and vulcanising agent for some elastomers. (Hylene MP.)

TN: Desmodour 44 (43)
Hylene M (MDI) (6)
Hylene M-50 (MDI-50) (50% solution in o-dichlorobenzene) (6)
Mondur M (263)
Multrathane M (263)
Nacconate 300 (263)
Vulcafor VCC (solution in Xylene) (60)
XDI (148).

DIPHENYL-p-PHENYLENE DIAMINE

⟨benzene⟩—NH—⟨benzene⟩—NH—⟨benzene⟩

Grey powder, m.p. 138–153°C, s.g. 1·21. Soluble in acetone, benzene and ethylene dichloride, insoluble in water. Antioxidant for natural rubber and butadiene–styrene copolymers.
Quantity: natural rubber 0·35%, SBR 0·7%.
Uses: articles in which a maximum resistance to flex cracking is essential, *e.g.* tyre treads, cable sheathings, soles, mechanical goods. Protects against heat, normal oxidation, copper and manganese. Unsuitable for light-coloured articles, tends to bloom.
TN: Agerite DPPD (51)
Agerite XPX (51)
Antigen P
Antioxygéne D.I.P. (19)
DPPD (5)
Hoboken (14)
JZF (23)
Nonox DPPD (60)
Permanay 18 (20)
Reaction products of this material with a complex diarylamine ketone aldehyde:
BLE Powder (23)
Flexamine (23)
Naugatex 219 (aqueous dispersion) and 279 (BLE-Paste) (23)

Santoflex 75, mixture with 25%
6-dodecyl-1:2-dihydro-2:2:4-
trimethyl quinoline (5).

DIPHENYL PHTHALATE
$C_6H_4(COOC_6H_5)_2$. White powder,
m.p. 69°C, b.p. 250–257°C (14 mm),
s.g. (25°C) 1·28. Softener.

DIPHENYL PICRYL HYDRA-ZYL DPPH.
Small quantities prevent the formation of carbon gel in rubber compounds during mixing.

DIPHENYL PROPYLENE DIAMINE
$CH_3.CH(NH.C_6H_5)$
$.CH_2.NH.C_6H_5$.
Antioxidant for latex.
TN: Stabilite L (100).

DIPHENYL SULPHON-3:3'-DISULPHONYL HYDRAZIDE

$$NH_2-NH-SO_2-\overset{O}{\underset{O}{\overset{\|}{\underset{\|}{S}}}}-SO_2-NH-NH_2$$

Blowing agent, which also influences cure above the active temperature, 148°C.
TN: Porofor D-33 (43).

DIPHENYL XYLENYL PHOSPHATE
Colourless, viscous liquid, b.p. 238–251°C (5 mm), s.g. 1·18. Miscible with most solvents. Softener and extender, in particular for NBR and CR.
Quantity: 10–30%.
TN: Disflamoll XDP (43).

DIPPD
Abbreviation for N,N′-di(2-octyl)-p-phenylene diamine.

DIPPED ARTICLES
Thin walled articles produced by dipping formers in vulcanisable rubber solution or latex compounds, *e.g.* gloves, dummies, prophylactics, balloons, catheters, medical articles, meteorological balloons.

DIPPING
Process in which a rubber solution or latex compound is deposited on a mould by dipping. After the water or solvent has been evaporated, a film remains, which is vulcanised hot or cold. The dipping procedure is repeated a number of times according to the thickness of the film required.

DIP PROCESS
Process of the US Rubber Reclaiming Co. for reclaiming scrap rubber, according to USP 2 415 449. At the beginning of thermal mechanical degradation in the presence of oxygen, vulcanised natural and synthetic rubber show a considerable increase in plasticity, which shortly afterwards decreases rapidly due to cyclisation. The plasticity increases again on further heating (*see* diagram). The

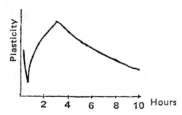

172

process may be employed to plasticise dry, fibre and metal-free ground scrap rubber (approx. 25 mesh US Standard) in 3–5 min., using a recalimator, *e.g.* a type of extruder, under increasing screw pressure at 188–215°C.

DIPROPYLENE GLYCOL DIBENZOATE

$C_6H_5CO(OC_3H_6)_2OCOC_6H_5$, m.p. −12 to −35°C, b.p. 225–235°C (5 mm), s.g. (25°C) 1·129. Insoluble in water. Softener.
TN: Benzoflex 988
Plastoflex MGB (45)
Flexol 77-G (146).

DIPROPYL PHTHALATE

$C_6H_4(COOC_3H_7)_2$, DPP, b.p. 129–132°C (1 mm), s.g. (25°C) 1·071. Softener.

DISALICYL ETHYLENE DIAMINE

Copper and manganese inhibitor, used particularly for natural rubber, to a lesser extent for SBR and neoprene. Does not act as an antioxidant. Boosts thiazoles and thiurams. Suitable for latex.
Quantity: 0·25–0·5%.
TN: Copper Inhibitor X-872-A (50% plus 25% cumarone resin and 25% stearic acid) (6).

DISALICYL PROPYLENE DIAMINE

Brown liquid, s.g. 1·07. Active constituent of copper inhibitors, prevents catalytic oxidation through copper. Has no effect as a conventional antioxidant and must always be used in combination with antioxidants. Activates thiazoles and thiurams and causes slight discoloration. Suitable for all elastomers and latices.
Quantity: natural rubber 0·3–1·5%, SBR 0·3–1·75%, neoprene 0·3–1·25%, latices 0·3–1·75%.
TN: Copper Inhibitor X-872-L (80% solution)
Copper Inhibitor 65 (65% solution)
Copper Inhibitor 50 (50% solution) (6).

DISEASES *Hevea Brasiliensis* can be attacked by a number of plant diseases. The appearance of specific diseases is dependent on the environment. Similarly, different clones reveal varying disease resistance. Some diseases show specific symptoms, but with others, a decline in production and changes in metabolism are the first signs of deviation from normal; advance symptoms become apparent later. Where leaf diseases are concerned, growth and production may be prevented by the effects of disrupted assimilation. Diseases of the trunk and roots can result in the destruction of the tube root vessels.

Root diseases. Parasites enter the living root tissue and cause a blockage in the vessel and destruction of the tissue.
White root mould: *Fomes lignosus* (*Leptoporus lignosus*).
Red root mould: *Ganoderma pseudoferreum.*
Brown root mould: *Fomes noxius.*
Root flange disease: *Ustulina vulgaris.*
Root putrefaction: *Sphaerostilbe repens.*

Diseases of the tapped surface and the trunk.
Vein cancer, bark rot: *Phytophthora palmivora.*
Patchy cancer: *Phytophora palmivora and Pythium complectens.*
Mould putrefaction: *Cerastomella fimbriata.*
Pink disease: *Corticium salmonicolor.*
Bark brownness: physiological disease caused by too intensive tapping.
Prevention of latex flow. Caused by mechanical damage to the cambium or the sclerenchyme bark (infiltration of stone cells).
Leaf and branch diseases.
Mildew: *Oidium heveae.*
Leaf fall: *Phytophthora palmivora Helminthosphorium Hevea.*
S. American leaf disease: *Dothidella Ulei* (exclusive to S. America, catastrophic in its effects).
Dying of leaves: physiological disease caused by lack of mineral substances, or by burning due to the heat of the sun in very young plants.
Dying of twigs: primarily caused by lack of mineral substances, secondarily by the appearance of *Botryodiplodia Theobromae, Gloeosporium albo-rubrum, Phyllosticta spp., Phomopsis Heveae.*
Pests. Attack by pests, termites, boring beetles, larvae, lice, locusts, leaf-eating insects, snails, epiphytes, and higher parasitic plants are only destructive to a minor degree in *Hevea.*
Lit.: 'Zeikten en Plagen van Hevea brasiliensis in Indonesie', Bogor, 1955.
Roger, 'Phytopathologie des Pays Chauds', Paris, 1951–53, 2 vols.
A. Sharples, 'Diseases and Pests of the Rubber Tree', London, 1936.
M. H. Lanford, 'Hevea diseases of the Amazon valley', *Boll. Techn. Inst. Agron. Norte (Bélém)* 1953, 27, 1–29.

R. N. Hilton, 'Maladies of Hevea in Malaya', Kuala Lumpur, RRIM, 1959.

DISODIUM METHYLENE DINAPHTHALENE SULPHONATE Yellow powder, s.g. 1·62. Soluble in water. Anionic, surface active agent. Dispersing agent and stabiliser for latex. Non-discoloring, non-foaming. Has no effect on vulcanisation. Prevents coagulation of natural latex during mixing, processing, and storage.
TN: Dispersol LN (33% solution) (60).

DISPERSED RUBBER Synthetic dispersions similar to latex are difficult to produce, and are obtained by being produced from reclaim. The dispersing agent is mixed with the reclaim and slowly replaced by a protective colloid solution (*e.g.* casein). First a water in rubber emulsion is formed which turns into a rubber in water emulsion at a certain concentration. Bentonite, colloidal clays and Wilkinite, for example, are particularly useful dispersing agents capable of absorbing large quantities of water. Besides casein, synthetic wetting agents (alkyl or alkyl aryl sulphonates) may also be used. Films of such dispersions are fairly water sensitive, soaps can also be formed by the direct addition of fatty acids and alkalis to the rubber. Resin soaps may also be used but cause a frothy dispersion with a high viscosity. Potassium carbonate, in the presence of potassium silicates or volatile alkalis, is also suitable. Compared with natural latex, synthetic dispersions are more viscous and have better chemical and mechanical stability. Films produced from such dispersions, however, do not have such good physical properties. Dispersions are unsuitable for dipped articles but especially suitable for cements, adhesives, carpet underlays.

DISPERSITE Aqueous dispersions of natural rubber, synthetic rubbers, reclaim, vinyl resins and other plastics (23).

DISPERSOL Group of surface active materials. Stabilisers and sensitising agents for latex:
L, LNT, LR, disodium methylene dinaphthalene sulphonate
T, sodium alkyl naphthalene sulphonate (60).

DITHIOBISBENZANILIDE
Yellow powder, m.p. 135°C, s.g. 1·35. Soluble in chloroform, particularly soluble in acetone, benzene and trichlorethylene. Active above 115°C. *Dosage:* Nr, Polyisoprene 0·05–0·5%, SBR, NBR 0·5–3·0%. *TN:* Peptisant 10 (20).

DITHIOCARBAMATES Salts of unstable or hypothetical disubstituted dithiocarbamic acids. Ultrafast accelerators. The materials may be alkali salts, amino salts, heavy metal salts (zinc, lead, copper, bismuth). Dithiocarbamates may be oxidised to the thiuram disulphide, an accelerator. They are produced by the reaction of secondary amines with carbon disulphide and sodium hydroxide. The sodium salt is soluble in water, and other dithiocarbamates may be precipitated from it. Secondary aliphatic amines react with carbon disulphide to form dithiocarbamic acids.
The amines of chief interest are dimethylamine and diethylamine, but piperidine, dibutylamine and dibenzylamine are also used. The zinc salts are the most active and most widely used accelerators of this type. Sodium, selenium, bismuth and copper salts are only used to a small extent. Their esters give accelerators with a delayed action.

$$2R_2NH + CS_2$$
$$\rightarrow R_2NC\!\!-\!\!S(R_2NH_2)$$
$$\underset{S}{\overset{\|}{}}$$

Estimation: when treated with dilute acids the dithiocarbamates are precipitated as the salts of the reactant acid, with the formation of carbon disulphide. To an aqueous solution of the accelerator, a quantity of N. HCl or sulphuric acid is added, sufficient to complete decomposition of the accelerator. The mixture is warmed on a water bath until the clouding caused by carbon disulphide has disappeared. The excess acid is then titrated with potassium hydroxide using methyl red or methyl orange as indicator; 2 mol. acid equal 1 mol. accelerator. The carbon disulphide formed during the reaction may be absorbed by alcoholic potassium hydroxide when potassium xanthate is formed quantitatively. This is titrated with N/10 iodine solution. The dithiocarbamate is calculated from the quantity of iodine used.

$$CS_2 + C_2H_5OH + KOH \rightarrow$$

$$\underset{OC_2H_5}{\overset{SK}{C}\!\!=\!\!S} + H_2O$$

$$2\,\underset{OC_2H_5}{\overset{SK}{C}\!\!=\!\!S} + I_2 \longrightarrow \underset{OC_2H_5}{\overset{S}{\overset{\|}{C}}\!\!-\!\!S\!\!-\!\!S\!\!-\!\!\underset{OC_2H_5}{\overset{S}{\overset{\|}{C}}}} + 2KI$$

175

DITHIOPHOSPHATES

Group of accelerators similar in structure to thiuram disulphides. Act as accelerators and sulphur donors at relatively low temperatures and yield thiourea or diphenyl thiourea to activate rapid curing at 100°C; particularly useful because the vulcanisates do not cause brown staining in the presence of traces of copper, e.g. in textiles.

DITHIOPYROMUCIC ACID

The zinc and lead salts of this acid have been used occasionally as accelerators.

$$\left[\begin{array}{c} CH-CH \\ \| \quad \| \\ CH \quad C-CS.S- \\ \diagdown O \diagup \end{array} \right]_2 \text{ Pb or Zn}$$

DITHIURAM Pentamethylene thiuram disulphide (393).

DI-o-TOLYL ETHYLENE DIAMINE

$CH_3.C_6H_4.NH.CH_2.CH_2$
$.NH.C_6H_4.CH_3.$

Brownish violet powder, s.g. 1·12. Slightly discoloring antioxidant for NR, SR and latex.
TN: Stabilite Alba (100).

DI-o-TOLYL GUANIDINE

White, odourless powder, m.p. 170–174°C, s.g. 1·1–1·2. Readily soluble in chlorinated solvents, soluble in acetone, alcohol, ethyl acetate, methylene chloride, slightly soluble in benzene, insoluble in carbon tetrachloride, benzine and water. Safe, slow vulcanising, general purpose accelerator. Vulcanisation temperature 125–140°C, disperses easily, the relatively high critical temperature ensures good processing safety. In latex has no influence on prevulcanisation, but decreases the stability. Causes slight discoloration and is not recommended for white articles except as a booster. Gives good ageing properties, and may be used therefore as an antioxidant. Has a strong synergistic effect in combination with thiazoles and other acid accelerators.

Quantity: primary accelerator 0·75–2% with 3·5–2·5% sulphur and 3–5% zinc oxide, secondary 0·1–0·5% with 1–0·75% dibenzthiazyl disulphide, 2·75–2·5% sulphur and 3–5% zinc oxide.

Uses: for a large variety of products with the exception of white articles and footwear. Normally used as a secondary accelerator, is seldom used independently.

TN: Accélérateur DT (19)
Accélérateur DOTG (20)
Accelerator 18 (6)
Ancazide DOTG (22)
DOTG
Eveite DOTG (59)
Vulcafor DOTG (60)
Vulkacit DOTG (43)
PDOTGD 65 (65% with 35% polyisobutylene) (282).

DI-o-TOLYL GUANIDINE DIPYROCATECHOL BORATE

Greyish brown powder, m.p. 165°C, s.g. 1·15–1·27. Fast accelerator for neoprene, medium processing safety which may be improved by addition of sodium acetate. Causes slight discoloration in light articles, but restrains the yellowish discoloration of vulcanisates. Boosts thiazoles and thiurams.

Quantity: 0·25–1%. In natural and synthetic rubbers, in concentrations of 0·5–2% it acts as an antioxidant, protecting against normal ageing and offering slight protection against heat ageing and flexcracking. Blooms in concentrations over 2%.

TN: Permalux (6).

DI-o-TOLYL THIOUREA
$CH_3.C_6H_4.NH.CS.NH.C_6CH_3$.
White, crystalline powder, m.p. 144–148°C, s.g. 1·25. Accelerator (obsolete).

Quantity: 1–3%.
TN: + A-22 (5)
+ Accelerator 17 (6)
DOTTU
DOTT
Eveite DOT.

DITRIMETHYL HEXYL SEBACATE Odourless, colourless liquid, s.g. 0·91. Miscible with ketones and aromatic hydrocarbons. Softener.

DIVINYL BENZENE DVB.
Clear liquid, m.p. − 67°C, b.p. 199°C, s.g. 0·93. May be easily polymerised, used for crosslinking and gel formation in the production of SBR.

DIXIE Trade name for a group of reinforcing blacks:

5: CC. Conductive channel black
20: SRF. Semireinforcing furnace black, Type 20 NS non-staining
20: HM, SRF-HM. Semireinforcing black with high modulus
35: GPF. General purpose furnace black

40: HMF. High modulus furnace black
45: GPF
50: FEF (MAF). Furnace black with good abrasion resistance, extrudes quickly and easily
60: HAF
70: ISAF
85: SAF
BB: CC
CF: CF
R-1: MPC. Pelleted
Voltex: pelletised, conductive channel black (164).

DIXIEDENSED Trade name for a group of pelleted channel blacks. 77: EPC, HM: MPC, S: HPC, S-66: MPC (164).

DIXITHERM Thermal black. F: FT, M: MT (164).

DIXYLYL DISULPHIDE

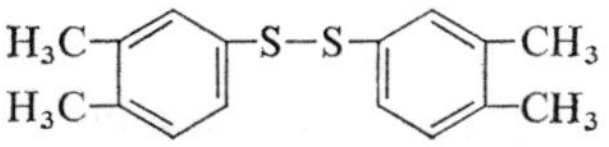

Dark brown oil, s.g. 1·02. Reclaiming agent for natural and synthetic rubber. Reduces reclaiming time and gives reclaims with a high plasticity, smooth surface and high tackiness. Also effective as a processing aid for very tough or scorched neoprene compounds. Disperses easily and can also be added as an aqueous emulsion.

Quantity: 0·15–1·5% with reclaiming oil and softeners; for the reclaiming of scorched neoprene compounds 2–3%.
TN: RR-10 (6).

DKG Abbreviation for Deutsche Kautschuk Gesellschaft.

DMASK Russian produced accelerator; dimethylamine mixed with 2-mercaptobenzthiazole.

DMBS Abbreviation for N-dimethyl benzthiazyl sulphenamide.

DMFPK Russian produced dimethyl phenyl-p-cresol.

DMHPPD Abbreviation for N,N'-di(3-methyl heptyl)p-phenylene diamine.

DMS Abbreviation for dimethyl sebacate.

DMSO Abbreviation for dimethyl sulphoxide.

DNA Abbreviation for Deutscher Normen Ausschuss.

DNODP Trade name for octyl decyl phthalate (5).

DNPPD Abbreviation for dinaphthyl-p-phenylene diamine.

DOB Local term for rubber from *Ficus Vogelli* (tropical W. Africa). The product is rich in resin; contains 50% rubber and 40% resins.

DOCUMENTATION A decimal classification scheme, known as the RABRM–RAPRA or Dawson code is used for the documentation of rubber literature. It is based on the reference publication 'Rubber Abstracts' (published by RAPRA) and the 'Documentation Analytique' of the Revue General du Caoutchouc (compiled by the Institut Francais du Caoutchouc). The original edition was produced by T. R. Dawson for the Library and Intelligence Section of RABRM and published in 1937 (*J. Rubb. Res.*, 1937, **6**, 67–132). After a few subsequent additions an improved code was published in 1942 (*J. Rubb. Res.*, 1942, **11**, 23–65). In 1952–53, RABRM, IFC and the Royal Society collaborated in producing an improved classification based on the 1937 code. The scheme is subdivided as follows:

Main tabulation. The code comprises 10 main groups, symbolised by the numbers 0–9, each group being divided decimally into subgroups. The numbers have no absolute value (decimal classification). The main tabulation gives a systematic division of materials, auxiliary products and machines.

Subsidiary tabulation A. Intended for tabulation of production methods, properties, test methods, processing and use of materials, auxiliary materials and end-products.

Subsidiary tabulation AZ. Tabulation of the use of rubber and its derivates. Grouping is according to products, not materials. Thus, for example, literature on 'Shoe soles made from crêpe, latex or rubber' appears under the group classification 3626 (Products of raw rubber), 2626 (Products of unvulcanised latex) or 3666 (Products of vulcanised rubber), but not in group 66F24 which includes general remarks about soles.

Main tabulation:

0　General, the rubber industry as a whole

1　Latex and rubber producing plants, plantations, cultivation, collection

2　Natural latex

2S　Synthetic latex

3　Raw rubber

3G　Gutta-percha and similar products

3N　Reclaim of natural rubber

3S Synthetic rubbers and similar products
4   Compounding ingredients
5   Fibres and textiles
6   Vulcanised rubber. 66 Uses, rubber products
7   Processes and materials
8   Machines and apparatus
9   Organisation, Economy.

Additional tabulation A:
1   Sources, origins, preparation, production, raw materials for manufactured goods
2   Types, general description, incidental characteristics, history
3   Quantities, properties, uses
4   Analysis and testing
5   Treatment, processing methods, construction, protection, control
6   Uses, transformations, accessories, treatment of materials.

Additional tabulation AZ:
A   Tyres, inner tubes and similar products used in the motor trade
B   Drive belts, conveyor belts and transmission
C   Cables and electrical insulation
F   Footwear
G   Games, toys, sport goods
H   Hose, tubing
M   Mechanical goods in general, not otherwise specified
R   Roads, floor coverings
S   Medicinal and dental articles made from soft rubber
T   Rubberised textiles and similar rubber articles, impregnated articles, rubber bands and fibres.
X   Various products of soft rubber (incl. foam rubber)
Z   Hard rubber; ebonite, vulcanite.

By combining the three tables the following combinations are possible:
Main tabulation + A. Method of production, characteristics, methods of testing, etc. of materials of the main tabulation.

Main tabulation + 5 (A) + 4 (main tabulation). Treatment of material with an added substance (5 (A): treatment, 4: added substance).
Main tabulation + 626 + AZ. End-product, unvulcanised, in a particular field of use.
Main tabulation + 666 + AZ. End-product, vulcanised, in a particular field of use.
Main tabulation + 626 + AZ + A. Method or production, properties, etc. of an end-product, the use of the material (main tabulation) in a particular field: unvulcanised.
Main tabulation + 666 + AZ + A. As above, but vulcanised end-product.
$-66\ AZ$–A. Method of production, properties, etc. of an end-product without giving the type of material from which it is made.
The places in the main tabulation in which a coupling with the secondary tabulation A or AZ is possible are termed (A) and (AZ).
This system is not based on the international decimal classification (DC) which is insufficient for this specialised field.

DOCUMENTATION ANA-LYTIQUE Reference section of the French rubber journal *Revue Général du Caoutchouc*, published by the Institut Francais du Caoutchouc. The articles in this journal, which are not French publications, are taken from *Rubber Abstracts* and translated. Appears monthly. Only one side of the page is printed on.

DODECANOLACTAM

$$HN–(CH_2)_{11}–C=O$$

Produced from cyclododecatriene, a

trimer of butadiene. May be polymerised to polydodecanolactam (polyamide 12), which has similar properties to Perlon.

6-DODECYL-1:2-DIHYDRO-2:2:4-TRIMETHYL QUINOLINE

Dark, viscous liquid, s.g. 0·93. Antioxidant for natural and synthetic rubbers. Gives excellent protection against flex cracking and normal oxidation. Causes a strong, dark stain and is suitable only for black articles.
Quantity: 2–3%.
TN: Santoflex DD (5).

n-DODECYL MERCAPTAN
$C_{12}H_{25}SH$, DDM, dodecanethiol, tert-dodecyl mercaptan, lauryl mercaptan. Oily, colourless liquid with mercaptan odour, b.p. 86°C (technical product), s.g. 0·84. Soluble in methanol, ether, acetone, benzene, ethyl acetate. Mixture of various isomers. Modifiers in emulsion polymerisation of SBR and NBR for controlling mol. wt.

DODECYL TRIMETHYL-AMMONIUM CHLORIDE
Coagulant, sensitiser and processing aid in natural rubber latex, blends with SBR latex and foam rubber compounds. Non-staining and non-discolouring. Should be added as a 10% aqueous solution.

DODIGEN 226 Quaternary ammonium compound, mixture of alkyl dimethyl benzylammonium chlorides. Aqueous solution with 50% active material. Cationic wetting agent and emulsifier, effective antimicrobe agent in rubberised textiles (217).

DOF Abbreviation for di-2-ethyl hexyl fumarate.

DOG FACTICE Group of factice types (385).

DOLAN AKRYL Previously KSF Kelheim. Staple fibre of polyacrylonitrile; in production since 1952 (186).

DOLOMITE $CaCO_3.MgCO_3$, calcium-magnesium carbonates. Inorganic filler.

DOM Abbreviation for di-2-ethyl hexyl maleate.

DOMEI PROCESS Process for the production of propellants by destructive distillation of rubber. Used industrially during the Japanese occupation of Indonesia and Malaya in World War II.

DOP Abbreviation for di-2-ethyl hexyl phthalate.

DOPTAX Accelerator consisting of a mixture of 2-mercaptobenzthiazole, diphenyl guanidine and a small quantity of selenium and phenyl-β-naphthylamine (167).

DOPTAX AZ Mixture of 2-mercaptobenzthiazole and 15% cyanoguanidine (167).

DOS Abbreviation for dioctyl sebacate.

DOTG Abbreviation for di-o-tolyl guanidine.

DOTT Abbreviation for di-o-tolyl thiourea.

DOTTU Abbreviation for di-o-tolyl thiourea.

DOW Group of softeners:
77: tri-p-tert-butyl phenyl phosphate
276-V-2. 276-V-9: poly-α-methyl styrene (61)
P1: diphenyl tert-butyl phenyl phosphate
P2: monophenyl tert-butyl phenyl phosphate
P3: diphenyl-p-chlorophenyl phosphate
P4: monophenyl di-o-chlorophenyl phosphate
P5: diphenyl o-xenyl phosphate
P6: monophenyl di-o-xenyl phosphate
P7: tri(p-tert-butyl)phosphate.

DOWANOL Group of softeners, solvents, etc.:
1: phenyl glycol
2: propylene glycol phenyl ether
3: ethylene glycol-p-sec-butyl phenyl ether
4: ethylene glycol-p-tert-butyl phenyl ether
5B: dipropylene glycol phenyl ether
33B: propylene glycol-o-chlorophenyl ether
43B: propylene glycol-p-sec-butyl phenyl ether
44B: propylene glycol-o-sec-butyl ether
93B: blend of mono-, di- and tripropylene glycol methyl ether (61).

DOW CORNING 199 Silicone glycol copolymer. Pale yellow liquid, s.g. 1·08. Soluble in polyglycols, amines and toluene diisocyanate. Used for improving cell structure in polyurethane foams (7).

DOWICIDE Preservative for latex, NR and SR.
A: sodium-o-phenyl phenate
G: sodium pentachlorphenate (61).

DOW LATEX Group of copolymer latices:
700: p.v.c. latex with 50% total solids, s.g. (latex) 1·14, s.g. (dry material) 1·41, pH 8, viscosity 20, particle size 1600 Å. Anionic emulsifier. *Uses:* textile impregnation
744B: vinyl chloride/vinylidene chloride latex with 50% total solids, s.g. (latex) 1·20, s.g. (dry material) 1·48, pH 8, viscosity 20, particle size 2000 Å. *Uses:* textile and paper impregnation, paints.
2582: vinyl toluene/butadiene latex with 41% total solids, s.g. (latex) 0·98, s.g. (dry material) 0·9555, pH 10·5, viscosity 25, particle size 1200 Å. Anionic emulsifier. *Uses:* acid resistant bonding agents, adhesives, paper impregnation, dipping agent.
2647: acrylic latex with 47% total solids, s.g. (latex) 1·07, s.g. (dry material) 1·15, pH 8·5, viscosity 80–90. Non-ionic emulsifier. *Uses:* pigments, paper, textiles, building materials (61).

DOW PROCESS Process used for the production of styrene. Alkylation of benzene with ethylene produces ethyl benzene, which is converted by catalytic dehydrogenation to styrene.

DOWTHERMS Oily liquids used as heat transfer agents at high temperatures:

Dowtherm A: eutectic mixture of diphenyl and diphenyl oxide, m.p. 12°C, b.p. 258°C, stable up to 285°C
Dowtherm B: stabilised o-dichlorobenzene, for temperatures between 150–260°C (61).

DP Abbreviation for degree of polymerisation.

DP Group of softeners (187).

DP Butadiene-piperylene copolymers, Russian produced latex, Types 25 and 50.

DPDB 6169 Polyethylene copolymer, s.g. 0·931, Shore A Hardness 27. Extremely tough and flexible, has good resistance to corrosion and is easily processable. Has little resistance to chlorinated hydrocarbons (112).

DPG Abbreviation for diphenyl guanidine.

DPGA Abbreviation for diphenyl guanidine acetate.

DPG NUMBER Diphenyl guanidine absorption. A measure of the influence of a filler on vulcanisation; depends on the chemical nature of the surface. As absorption increases, retardation of vulcanisation increases at a parallel level.

DPGO Abbreviation for diphenyl guanidine oxalate.

DPGP Abbreviation for diphenyl guanidine phthalate.

DPPD Abbreviation for diphenyl-p-phenylene diamine.

DPPH Abbreviation for diphenyl picryl hydrazyl.

DPQD Abbreviation for dibenzoyl-p-quinone dioxime.

DPR Abbreviation for depolymerised rubber; natural rubber liquidised by a thermochemical process. Produced in varying grades of viscosity, both vulcanised and unvulcanised. Softener, for natural and synthetic rubbers, additive for resins, waxes, plastics, adhesives, rubber band compounds. Unvulcanised DPR is used for articles produced by a casting process, *e.g.* rolls.
TN: Lorival (251).
Lit.: *India Rubb. World*, 1954, **130**, 211–3.

DPR SYNTHETIC N-27 Liquid butadiene-acrylonitrile copolymer with 27% solid material.
Uses: softener and modifying agent for plastics and synthetic resins (256).

DPTD Abbreviation for dipentylmethylene thiuram disulphide.

DPTM Abbreviation for dipentamethylene thiuram monosulphide.

DPTT Abbreviation for dipentamethylene thiuram tetrasulphide.

DRALON (Previously Bayer acrylic fibre.) Staple fibre of polyacrylonitrile. In production since 1953 (43).

DRAPEX Group of softeners for NR, SR and neoprene (188).

DRAWINELLA Acetate fibre (359).

DR CLONES Abbreviation for *Dothidella* resistant clones.

DRESINATE Group of emulsifiers produced from modified resin acid.
Uses: in polymerisation of synthetic rubber, give latices with little foam. Stabilisers and dispersing agents for latex (10).

DRILLAX Benzthiazyl-2-diethyl sulphenamide.

DRV Abbreviation for Deutscher Reifenhändler Verband.

DRY RUBBER CONTENT Determined by coagulating latex with 1 % ferric acid at pH 4·8–5·0, sheeting the coagulum and drying it at 70°C. Expressed as a percentage of the total latex. Strictly, this figure represents the rubber hydrocarbon content plus the small quantity of protein, resin, minerals, and so forth, present in the rubber, in practice the percentage of solids remaining after drying is accepted as the rubber content of the latex.

DSV Abbreviation for dilute solution viscosity.

DTBP Abbreviation for (1) di-tert-butyl peroxide, (2) ditert-butyl-p-cresol.

DTET Abbreviation (French) for tetra ethyl thiuram disulphide (disulfure de tétra éthyl thiurame).

DTMT Abbreviation (French) for tetramethyl thiuram disulphide (disulfure de tétra méthyl thiurame).

DU CROS, SIR ARTHUR PHILIP 1871–1955. President of the Dunlop Rubber Co., pioneer of the tyre industry.

DULAS 557 Trade name for piperidine pentamethylene dithiocarbamate.

DUMB-BELL (T-test piece.) Dumb-bell shaped test piece used in determining tensile strength and modulus of rubber. ASTM D 412-62 T, DIN 53 504.

DUNLOCRUMB Trade name for crumb rubber made by the *Hevea* crumb process (from Dunlop Malayan Estates).

DUNLOP, JOHN BOYD 1840–1921. Scottish veterinary surgeon and inventor, pioneer of pneumatic tyres. In 1887, constructed a pneumatic tyre for his son's tricycle; the discovery was patented the next year, and production was taken up through the Pneumatic Tyre and Booth Cycle Agency in Belfast, in association with William Harvey Du Cros, in 1890. Dunlop handed the patent to Du Cros for a small sum and received 1500 shares in the company. Difficulties arose when it was discovered that the principle of air-filled tyres had been discovered by Thompson in 1846; however, on the basis of the 1888 patent it was possible to maintain the position of the company. The company was sold in 1896 to E. T. Hooley, who invested £5 million in its development, and Dunlop took no further part in its development.
Lit.: Jean MacClintock, *History of the Pneumatic Tyre*, 1923.

DUNLOPILLO Foam rubber produced by the Dunlop Rubber Co.

BP 332 525 (1930). Synonym for foam rubber.

DUNLOP PENDULUM A striking pendulum used to determine the resilience of rubber.
Specification: BS 903: A8: 1963.

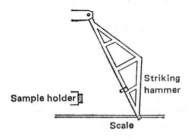

Sample holder

Striking hammer

Scale

DUNLOP TRIPSOMETER
Apparatus for determining the resilience of rubber. The striking pendulum consists of a steel plate 41·9 cm in diameter and 1·43 cm thick with a striking sphere of 4 mm.
Specification: BS 903: A8: 1963.

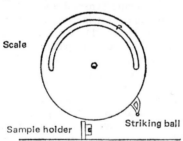

Scale

Sample holder   Striking ball

DU PONT FLEXOMETER
Machine for investigating the flex cracking of rubber, and the ply separation in test pieces, comprised of textile layers bonded by rubber. The test pieces, which are plied to textiles to prevent their stretching, measure 25·4 × 101·6 mm. Twenty-one test pieces are put together by bonding them onto an endless band.

The apparatus consists in principle of four discs of 76·2 mm diameter, over which the band runs in a double V form. The motivating disc runs at 860 rpm and drives the band at 95 c/min. The discs are under a tension of 7·7 kg (171 lb). The results are expressed in cycles, the crack formation compared with a standard scale.
Specification: ASTM 430-59.

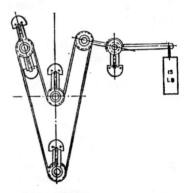

DUPRENE Original term for polychloroprene; the name was changed to neoprene in 1939. Sometimes used also for Type GR-M (now GN).

DURADENE 75:25 stereo-specific SBR with a low heat build-up. Produced by solvent polymerisation with lithium as a catalyst (129).

DURANIT Butadiene-styrene copolymer with a high styrene content (the reverse of Buna), s.g. 1·02–1·04. Reinforcing filler, for rubber. Facilitates processing and reduces mixing times; improves flexibility and decreases calender shrinkage. When mixing a compound with Duranit fillers and other compounding ingredients, excluding softeners, the Duranit is added to the rubber at

60 or 70°C. High friction causes temperature to rise to 120–150°C.
Type 15S, styrene content 84%; Type 30, 70%; Type 40, 60%; Type B, 60%.
Uses: insulations, floor coverings, jointings, soles.
Quantity: 10%.

DURAX N-cyclohexyl-2-benzthiazyl sulphenamide (51).

DUREX O Comparable to SRF black (156).

DUREZ Group of phenol-formaldehyde resins. Primarily used as softeners for SBR, CR, NBR and vinyl polymers, also as reinforcing agents and auxiliary vulcanising agents for NBR. Improve heat and tear resistance, electrical properties, oil and chemical resistance, and give increased hardnesses. Also used as bonding agents for nitrile rubber/metals and as modifying agents for latex (162).

DUROMETER Instrument for measuring the hardness of rubber. The resistance of the rubber to penetration by a blunt indentor is determined by compression of a spring with fixed characteristics. The Hardness 0 is characterised by penetration depth of 2·54 mm and a spring loading of 56 g, Hardness 100 corresponds to zero penetration and a spring loading of 822 g. DIN 53 505.

DUROXON Group of synthetic waxes made by the Fischer–Tropsch process.
Uses: fillers, lubricants, also for improving surface appearance (190).

DUTCH BOY Trade name for lead compounds used as stabilisers for vinyl resins, and for various softeners:
DS-207: dibasic lead stearate: lubricant
Dyphos: dibasic lead phosphite
Dythal: dibasic lead phthalate
Normasal: n-lead salicylate. Scorch-free activator
Plumb-O-Sil: mixture of lead silicate with silica gel
Tribase: tribasic lead sulphate
NLA-10: dibutyl phthalate
NLA-20: di-2-ethyl hexyl phthalate
NLA-30: diiso-octyl phthalate
NLC-20: dioctyl sebacate (191).

DUTREX Group of aromatic petroleum hydrocarbons. Dark, odourless liquids, s.g. 0·991–1·032. Softeners, plasticisers, processing aids and extenders for NR, SBR, CR, NBR and p.v.c.:
6: aromatic resin. Lubricant
7: heavy oil. Softener for natural rubber and SBR
15: modified plasticiser with an asphalt basis. Used for natural and synthetic rubbers
20: highly aromatic resin. Softener for vinyl chloride and acrylonitrile rubber
25: oil with a high aromatic content. Softener for p.v.c.
31: paraffinic oil. Softener for butyl rubber
44: cresol base. Softener for acrylonitrile.
In SBR and natural rubber for general use, Dutrex 6, 7, 6H (SPX-97), 20, 15E, 15W; for light-coloured compounds, Dutrex 32 and 39; for butyl rubber, Dutrex 31 and 32; for nitrile rubber, Dutrex 21 and 25, for neoprene W types, Dutrex 6 and 20 (2).

DUTRAL Trade name for ethylene propylene rubber (5a).

DV Abbreviation for direct vulcanisation; process by which an unvulcanised sole compound may be vulcanised directly onto the upper part of the shoe.

DVCHB-70 Russian produced butadiene-vinylidene chloride copolymer/latex.

DVM SOFTNESS NUMBER (DVM = Deutscher Verband für die Materialprüfung der Technik). The softness number of rubber is expressed as the change in penetration of a polished hardened steel sphere under different loads (initial compared with total load). The load 1000 g, correlation with Shore:

Shore Hardness	DVM No.	Shore Hardness	DVM No.
30	115	70	40
35	102	75	33
40	90	80	26
45	80	85	19
50	70·5	90	12
55	62·5	95	6·5
60	55	100	0
65	47		

DYERA Various types (*Dyera* family), *Apocynaceae* (Borneo and Malaya); chiefly *D. Lowii, D. costulata* and *D. laxiflora,* produce Jelutong. The trees attain a height of up to 45 m and trunk circumference of 7–8 m.

DYLEX LATEX Group of styrene-butadiene copolymer latices with a monomer ratio between 90:10 and 50:50.
Uses include impregnation of paper, emulsion paints and bonding agents (176).

DYNAGEN Plastic made from polyisobutylene (192).

DYNAGEN ZP-139 Propylene oxide rubber; copolymer of propylene oxide and unsaturated epoxides. Gives good ageing and ozone resistance, good physical properties (31).

DYNAMINE Trade name for diphenyl guanidine.

DYNAMOMETER A machine used to determine the tensile strength, elongation at break and modulus of elastomers and other materials. Also an apparatus used to determine the power of a machine or some other force, may be used to determine the power of a drive belt.

DYNAMOMETER PB (PB = Pierre Breuil) Breuil–Cillard dynamometer. Machine for determining the tensile and elongation properties of rubbers, textiles and other materials.

DYNAT Natural rubber produced in crumb form by mechanical cutting or by extruding the wet coagulum, and compressed into compact bales of approx. 34 kg after drying. Packed in polyethylene, transported in palettes of 1 ton.
Type WF, produced from latex
CL, produced from cup lumps
TL, produced from dry film of tree lace
S, produced from skim latex (386).

DYNEL Synthetic fibre of a copolymer of 60 % vinyl chloride and 40 % acrylonitrile, highly resistant to corrosion and chemicals. Properties lie between those of pure acrylonitrile fibres, such as Orlon and Acrilan, and those of p.v.c. fibres.

Uses: include insulating materials and fillers. The copolymer is also used as a corrosion resistant coating, *e.g.* in shipbuilding, machine parts, cars (112).

DYNSTAT APPARATUS Apparatus used to determine the flex-resistance and impact strength of plastics.

DYPHOS Dibasic lead phosphite. White powder, s.g. 6·94. Stabiliser for chlorinated rubber, vinyl polymers and copolymers, chlorinated hydrocarbons (191).

DYTHAL Dibasic lead phthalate. White, crystalline powder, s.g. 4·6. Heat and light stabiliser for vinyl polymers and copolymers (191).

E

E 244 Aqueous emulsion of a butadiene-acrylonitrile-styrene polymer. Softener for p.v.c. (43).

E 342 Silicone rubber in three different hardness grades. Specification: DTD 818.
Uses: with gaskets, O-rings (60).

EASTOZONE Group of ozone antioxidants.
Grades:
30: dioctyl-p-phenylene diamine
31: di-3-(5-methyl heptyl)-p-phenylene diamine
32: dimethyl-di(1-methyl propyl)-p-phenylene diamine (139).

ECODIENE Abietic-maleic acid resin (252).

EDF CRYSTALS Condensation product of ethylene diamine and formaldehyde. Aldehyde-amine accelerator (194).

EEA Abbreviation for ethylene-ethyl acetate copolymers.

EFA Abbreviation for ethyl chloride-formaldehyde-ammonia condensate.

EJF Reaction product of 2-mercaptobenzthiazole formaldehyde and p-toluidine (120).

EKAGOMME Polysulphide rubber, identical to Thiokol A (252).

EKANDA RUBBER Wild rubber from *Raphionacme utilis* (Angola). Obtained by pressing the chopped, beet-like nodules.

ELA Mixture of mono- and diphosphate esters of long-chain alcohols. Light-brown liquid, s.g. 0·99. Processing aid and lubricant for elastomers and plastics, prevents sticking to rolls and calender rolls and improves extrusion without influencing vulcanisation in natural rubber, SBR and neoprene. Causes slight retardation in butyl rubber above 0·5%.
Quantity: 0·25–2% mixed into the compound shortly before the mixing cycle or at the onset of sticking (6).

ELAPRIM Butadiene-acrylonitrile copolymer.
Uses: include tubes, gaskets, antistatic conveyor belts, linings (59).

ELASTEX Group of softeners for natural rubber, synthetic rubber and vinyl polymers.

Grades:

10-P:	diiso-octyl phthalate
18-P:	iso-octyl isodecyl phthalate
20-A:	diiso-decyl adipate
28-P:	dioctyl phthalate
36-R:	polyester with low mol. wt.
37-R:	polyester with high mol. wt.
40-P:	butyl isodecyl phthalate
48-P:	butyl octyl phthalate
50-B:	butyl cyclohexyl phthalate
60-A:	dioctyl adipate
82-E:	mixture of n-octyl and n-decyl phthalates
90-P:	diiso decyl phthalate
DCHP:	dicyclohexyl phthalate (107).

ELASTICITY According to Hooke's law, strain and stress are proportional within a certain limit for homogenous, isotropic bodies. For extension:

$$\varepsilon = \alpha \cdot \sigma$$

where

$$\frac{\Delta l}{l} = \frac{\sigma}{E}$$

ε is the extension

σ is the tensile stress

α is the degree of extension, elasticity coefficient, proportion factor, relation of strain to stress $(= \varepsilon/\sigma)$, and

l is the length.

The reciprocal value of the degree of extension $E = 1/\alpha$ is called the elasticity modulus, stretch modulus or Young's modulus, and is one of the main properties used to characterise the elasticity of a material.

Hooke's law is also valid in an analogous form for shearing strain:

$$\gamma = \beta \cdot \tau$$

where

$$\text{shear strain} = \frac{\tau}{G}$$

where

γ is the slippage

β is the degree of shearing

τ is the shearing stress, and

G is the shearing modulus.

Between the elasticity modulus and the shearing modulus is the relation

$$E = 2G(1 + v)$$

where v is Poisson's ratio.

On compression the elastic resistance is expressed by the compressibility coefficient:

$$K = \frac{dV}{dP}$$

The valid range of these equations is defined as the corresponding limits of ε and γ; beyond these limits, small plastic deformations may be observed together with the elasticity phenomenon. Taken in its strictest sense, Hooke's law is valid only for ideal, elastic bodies. If rubber is stretched slowly then, after releasing the tension, a deformation remains which only recovers slowly; this is called the elastic after-effect. After an elastic deformation the stress necessary to maintain the deformation decreases with time. This gradual reduction is termed relaxation. Valid for a viscous body according to Maxwell:

$$\frac{d\tau}{dt} = G \qquad \frac{dy}{dt} = \frac{\tau}{\lambda}$$

From this the relaxation time may be deduced:

$$\tau = \tau_0 \cdot \exp\left(-\frac{t}{2}\right)$$

where
  τ_0 is the original stress
  τ is the remaining stress after time t, and
  λ is the relaxation time, *i.e.* the time necessary for the reduction of the stress to $1/e$ of its original value τ_0.

ELASTIC PROPERTIES Determination of hysteresis loss, using the relationship between energy regained on unloading test pieces of soft rubber, and the work of loading. DIN 53 511.
1. Determined from the tensile strain with a fixed elongation. DIN 53 511 Bl. 1.
2. Determined from the tensile stress with a fixed loading. DIN 53 511 Bl. 2.
3. Determined using the compression strain with a fixed amount of compression. DIN 53 511 Bl. 3.
4. Determined using the compression strain with a fixed loading. DIN 53 511 Bl. 4.

ELASTINE Oleic acid factice. Produced using sulphur chloride and diluted with copal resin.

ELASTOMER S Copolymer of 100 parts butadiene, 37 parts styrene and 112 parts isopropyl ketone (61).

ELASTOMERS Terminology adopted in 1939 by H. L. Fisher for natural and synthetic vulcanisable products, which reveal elastic properties after crosslinking, can be stretched to at least double their length at room temperature and, on removal of the tension, quickly return to their original length.

ELASTOMETER Modification of the durometer. Measures the recovery of an indentor pressed into a fixed depth of the test piece.

ELASTOMETER PB (PB = Pierre Breuil.) Simple elastometer of an older type. Comprised of a graduated glass tube in which a steel ball falls down onto a rubber plate. The height of rebound is measured against a scale on the side of the tube.

ELASTOTHANE 455 Urethane elastomer, suitable for metal/rubber components (208).

ELECTRICAL RESISTIVITY Specific electrical resistance, referred to a 10 mm cube between two parallel surfaces.
Volume resistivity: the electrical resistance of the interior of a test piece, or of a finished rubber article of any shape.
Surface resistivity: the electrical resistance of a liquid, blooms or impurities on the rubber surface; normally only of interest when the inner resistance of the rubber test piece is considerably greater than that of the impurity.
Insulation resistivity: electrical resistance between any two electrodes situated in, or on a test piece, or in a finished rubber article.
Specification: DIN 53 596.

ELECTRODEPOSITION Production of rubber goods from latex by electrocoagulation, the rubber particles being deposited either on a mould, which acts as anode, or on a permeable, non-conducting

189

former, which lies in a **D.C.** field between anode and cathode.
Uses: for articles to be covered with a rubber coating, for thin walled articles.

ELECTROFINES Group of chlorinated paraffins. L 75, liquid, 75% chlorine, 45, 50 (389).

ELECTROGUM French produced chlorinated rubber.

ELEPHANT FOOT The normally thickened base of a grafted tree of the *Hevea* type.

ELONGATION The extension of a test piece under a given tensile stress. Expressed as a percentage of the original length.

ELONGATION AT BREAK Maximal stretching capacity of a test piece at the moment of break, determined by the original length of the test piece and expressed in percentage.

ELP TYRE Abbreviation for extra low pressure tyre.

EL SIXTY Dibenzthiazyl dimethyl thiourea (5).

ELVACET Emulsion of polyvinyl acetate. Extender and modifying agent for latex (6).

ELVAX Ethylene-vinyl acetate copolymer. Tough, flexible, has close affinity to paraffin.
Uses: modifying agent for paraffins used, *e.g.* for impregnation; on addition of 10–30%, the paraffins become totally plastic (6).

EMERY Group of softeners.

Grades:
2301: methyl oleate
2302: n-propyl oleate
2221: glycerol mono-oleate (196).

EMKA RUBBER (After Van der Mark and Kremer.) Produce of the small plantations in existence at the start of the rubber industry. After the rubber coagulum had been sheeted, it was folded, blown up as balls, dried in the open, cut, and dusted with talc.

EMPYREUMA The acetone soluble portion of carbon blacks.

EM SURFACE AREA The surface area of fillers. Determined by using the electron microscope.

EMULPHOR Group of emulsifiers and dispersing agents; polyethylene ethers with a high content of, *e.g.* fatty alcohols, fatty acids, which are soluble in water. Types O, ON, and ELA are particularly suitable for latex (32, 85).

EMULSION Aqueous polyvinyl acetate dispersion.
Grades:
NP: non-plasticised
P: plasticised (59).

EMULSION 1073 Polystyrene emulsion. Plasticised with 25% butadiene (189).

EMULSION PAINTS Highly pigmented, plasticised synthetic latex compounds which require almost no addition of solvent to improve film formation, or of silicone oil to prevent bubbles. Protective agents are necessary to prevent rusting, *e.g.* sodium benzoate, ethyl aniline phosphate, and a thickening agent may also be required. *Hevea* latex is unsuitable as a basis for emulsion paints.

EMULSION POLYMERISA- TION Process in which monomers are catalytically polymerised as an aqueous emulsion; additional modifiers or catalysts may be necessary. This emulsion is then precipitated as a finely dispersed latex. By adding short stops the reaction may be halted at the required level of conversion. Monomers which have not been converted may be separated from the latex by distillation. The latex may be used in this form, after stabilisers have been added if necessary, or the polymer may be removed by coagulation, and dried.

EMULSOGEN Alkyl polyglycol ether, Types O and ON. Non-ionic dispersing agents, emulsifiers and stabilisers (217).

EMULTEX A Lecithin concentrate, s.g. 1·03, viscosity 4790 cs at 21·1°C, 1830 at 37·8°C, 650 at 60°C, pH (2% aqueous dispersion) 6·5–7·5. Stabiliser for latex emulsion colours (197).

EMULVIN Emulsifiers, stabilisers and crosslinking agents for latex.
Grades:
S: polyether thioether. Brown, wax-like mass, s.g. 1·19. Soluble in water. Improves stability and structure of latex foam without influencing the physical or ageing properties. *Quantity:* 7 parts of a 15% solution to 100 parts rubber
W: aromatic polyglycol ether. Yellowish brown, viscous liquid, s.g. 1·13. Stabiliser for latex, emulsifier (43).

ENDOR Activated zinc pentachlorothiophenate. Grey green, odourless powder, s.g. 2·40. Peptiser. Effective at high and low temperatures, has no effect on vulcanisation or on ageing and physical properties. Concentrations over 0·25% cause green discoloration in light-coloured articles. Sulphur and accelerators can retard the peptising action. *Quantity:* natural rubber 0·1–0·4%, SBR 1–4%.

ENERGY AT BREAK If the tensile stress of rubber is plotted against the strain at the break point, the area beneath the curve indicates the total energy of the rubber. The graph enables the energy at break of compounds to be assessed in relation to the filler which they contain, the curves obtained being characteristic of the type of filler. Fillers with a curve above the base line (basic compound) show reinforcing properties. The curves may be used for comparing fillers in terms of a Delta-A function.

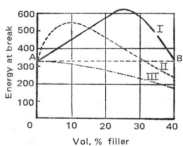

Vol. % filler

I   Black
II   Zinc oxide
III Barium sulphate.

ENJAY 3509 Trade name for ethylene-propylene terpolymer (131).

ENJAY BUTYL Trade name for various types of butyl rubber.

191

Type	Former classification	Mol. % unsaturation	Mooney	Properties and Uses
035 or (065)	GR-I-35	0·6–0·1	38–49	Best ozone and chemical resistance. Cables and insulation compounds
150, 165*	GR-I-50	1·0–1·4	41–49	Proofings, mouldings, hose
215, 265*	GR-I-15	1·5–2·0	41–49	Cures more rapidly than the 100 series. Air hose, mouldings
217, 267*	GR-I-17	1·5–2·0	60–70	Like 215, but with higher Mooney
218, 268*	GR-I-18	1·5–2·0	70–80	Standard type. Air hose, extrusions and mouldings, tyre inner liners, adhesives, corrosion resistant coatings
325, 365*	GR-I-25	2·1–2·5	40–50	Cures more rapidly than the 200 series. Heat resistant products such as belting and steam hose; mouldings
HT series			50–60	Chlorinated butyl rubber, cures more rapidly compatible with NR, SBR, CR, IIR; heat resistant to 205°C.

* Non-discolouring.

All types may be obtained also with a non-staining antioxidant. The wrapping normally used is polyethylene film, as this disperses easily in the compound above 120°C. For roll compounds at low temperatures, and for the production of adhesives, delivered in impregnated cartons (131).

ENJAY BUTYL CO. (New York.) Associated company of the Esso Standard Oil Co., Baton Rouge, La, and the Humble Oil and Refining Co., Baytown, Texas. Created to take over the production of butyl rubber from the American Government, after it had become a private enterprise.

EN RUBBER Methyl rubber produced in Leverkusen, 1914–1918.

EPC Abbreviation for easy processing channel black.

E/P COPOLYMER Abbreviation for ethylene-propylene copolymer.

EPDM (EPT) abbreviation for ethylene-propylene terpolymer, according to ASTM D 1418.

EPDR-60 Ethylene-propylene terpolymer (14).

EPICHLORHYDRIN RUBBER Polymer based on epichlorhydrin (CHR), or epichlorhydrin and ethylene oxide (CHC). Because of the high chlorine content, the polymers show high oil resistance and are non-flammable.
TN: Hydrin 100 (CHR)
Hydrin 200 (CHC) (14).

E P I C H L O R H Y D R I N E 1-chloro-2:3-epoxy propane, chloromethyl oxirane, γ-chloropropylene oxide. Colourless liquid, m.p. $-25.6°C$, b.p. $117.9°C$, s.g. 1.181. Miscible with alcohol, ether, chloroform, ethylene trichloride, carbon tetrachloride, immiscible with petroleum hydrocarbons.
Uses: starting material for epoxy resins, solvent for natural and synthetic resins and cellulosic plastics.

EPM EPR. Abbreviation for ethylene-propylene rubber according to ASTM D 1418.

EPOLENE Low mol. wt. polyethylenes.
Uses: release agents, prevent tackiness in unvulcanised compounds, compatible with waxes (193).

EPOXY RESIN Ethoxylin resin, condensation product of epichlorhydrin with polyalcohols or polyphenols. Stable, liquid to thermoplastic resin, which may be hardened with polyamines, dicarbonic acid anhydrides. By esterification with unsaturated acids, modified products are produced which form air drying films. Hardened products, s.g. 0.98–1.3, are resistant to solvents, acids and alkalis, have good mechanical properties, are resistant to flexing and temperatures up to 150°C. May be diluted with fillers (quartz powder).
Uses: cast products, sealing compounds, jointings, coatings, paints, bonding agents for metal/rubber components.
TN: Araldite (94).
Epikote, Epon (2).

EPR Abbreviation for ethylene-propylene rubber.

EPSYN Group of EPDM polymers with ethylidene norbornen as third monomer (160).

EPT Abbreviation for ethylene-propylene terpolymers, mostly with dicyclopentadiene as the third monomer.

EPTAC Group of accelerators (6):
1: zinc dimethyl dithiocarbamate
2: 50% zinc dimethyl dithiocarbamate and 50% tetramethyl thiuram disulphide
4: zinc dibutyl dithiocarbamate.

EPTD Abbreviation for diethyl diphenyl thiuram disulphide.

EPTM Abbreviation for diethyl diphenyl thiuram monosulphide.

EQUATORIAL RUBBER Natural rubber from the Congo, reasonable in quality, probably from different *Landolphia* types. Commercially produced in the form of small balls.

ERRATHENE Irradiated polyethylene. In an oxygen-free atmosphere resists temperatures up to 200°C (198).

ERWASOLE Reclaims formerly produced in Germany, using the digestion process.

ERYTHRENE Former term for butadiene.

ESTABEX Group of secondary plasticisers, also light and heat stabilisers based on esters, for chlorinated rubber, p.v.c. and other chlorinated products. Increase the adhesion of chlorinated rubber coatings to metals (171).

ESTANE Trade name for polyurethane (14).

ESTAX Group of softeners based on various fatty acid esters.

ESTER AMINE
R.COOC$_2$H$_4$N(CH$_3$)$_2$. Produced from dimethylamine, ethylene oxide and paraffinic acid with 12 C atoms. *Uses:* emulsifier for the polymerisation of Buna.

ESTEREX Dinitrophenyl benzthiazole sulphenamide (120).

ESTERSIL Hydrophobic, colloidal silica whose surface active groups are esterified with primary alcohols. By substitution of the hydrophilic hydroxyl groups (OH) and the alkoxy groups (OR), hydrophobic and organophilic properties are achieved on the surface of the particles.
Uses: primarily as a reinforcing filler, dusting agent, release agent.

ESTYNOX Group of softeners and stabilisers, *e.g.* for chlorinated rubber, p.v.c., cellulose nitrate, ethyl cellulose.
Grades:
308: epoxyacetoxy stearate (triglyceride)
366: isobutyl epoxyacetoxy stearate (fatty acid ester)
403: butyl epoxy stearate (monofatty acid ester) (9).

ETA EXTRACTION Azeotropic mixture of 70 vol. % ethanol and 30 vol % toluene for the extraction of vulcanisable rubber.

ETHANITE Former term for thioplasts produced in Germany by Bayer; trade name for Belgium produced thioplasts.

ETHOXYDIHYDROTRI-METHYL QUINOLINE

6-ethoxy-1:2-dihydro-2:2:4-trimethyl quinoline. Dark, viscous liquid, s.g. 1·04. Soluble in benzene, benzine, carbon tetrachloride and alcohol, partially soluble in acetone and ether. Antioxidant, mainly effective against ozone and flex cracking, more effective in SBR than in natural rubber. Causes discoloration and staining, unsuitable for light-coloured articles.
Quantity: 1–2%, especially for dynamically stressed articles, into which the compounding of waxes is not desired.
TN: Polyflex (246)
Santoflex AW (5)
Nocrac AW (274).

ETHYL ANTIOXIDANT Group of antioxidants (88):

701: 2:6-ditert-butyl phenol
702: 3:4′-methylene bis(2:6-ditert-butyl phenol)
703: 2:6-ditert-butyl-α-dimethyl-amino-p-cresol
712: 4:4′-bis(2:6-ditert-butyl phenol)
720: 4:4′-methylene bis (2-methyl-6-tert-butyl phenol)
733: mixture of 75% 2:6-ditert-butyl phenol, 10–15% 2:4:6,-tritert-butyl phenol and 15–10% o-tert-butyl phenol
736: 4:4′-thiobis(2-methyl-6-tert-butyl phenol)
762: 2:6-ditert-butyl-α-methoxy-p-cresol.

ETHYL CHLORIDE-FORM-ALDEHYDE-AMMONIA REACTION PRODUCT Dark brown, viscous liquid, s.g. 1·1–1·2. Soluble in water and acetone, insoluble in benzene and chlorinated hydrocarbons. Accelerator. Vulcanises quickly with a relatively low critical temperature, stiffens unvulcanised mixtures. Can be retarded by means of channel black, alumina, silica or other acidic materials. Activation by thiazole, thiuram, guanidine and heptaldehyde-aniline reaction products; needs zinc oxide and sulphur in normal quantities, fatty acids are not essential but can be added as weak retarders. Causes slight discoloration in white articles, but can be used in coloured wares with the exception of light pastel shades; discoloration does not occur in sunlight. Modulus and hardness are increased, elongation at break is decreased.
Uses: include hot air and press vulcanisates, footwear, impregna-tion, sponge rubber, mechanical goods.
Quantity: natural rubber 0·5–3% with 1·5–3% sulphur, SBR 0·25% with 0·75% benzthiazyl disulphide and 3% sulphur.
TN: Trimene Base (23)
Trimene with stearic acid (23)
Vulcafor EFA (6).

ETHYLENE CHLORHYDRIN $CH_2Cl.CH_2OH$ 2-chloroethanol, m.p. $-67°C$, b.p. 127–130°C, s.g. 1·195 (20°C). Colourless liquid. Miscible with water and alcohol. Solvent for NR and SR.

ETHYLENE CHLORIDE $CH_2Cl—CH_2Cl$ Ethylene dichloride, dichloroethane, chloroethylene. Colourless liquid with chloroform-like smell, m.p. $-36°C$, b.p. 84°C, s.g. 1·26. Miscible with ethanol and ether, sparingly soluble in water.
Use: solvent for rubber, resins, bitumens, fats and oils.

ETHYLENE DIAMINE $NH_2CH_2CH_2NH_2$, 1:2-diamino-ethane. Colourless, viscous, strongly alkaline liquid, m.p. 8·5°C, b.p. 116–117°C, s.g. 0·895. Soluble in water (by formation of a hydrate), and alcohol.
Uses: stabiliser for latex, antigelation agent for cements based on NR and SR, solvent for sulphur and casein, emulsifier.

ETHYLENE DIAMINE CARBAMATE White, hygroscopic powder with ammonia-like smell, m.p. 145–155°C, s.g. 1·37. Soluble in water, insoluble in non-polar solvents.
Use: vulcanising agent for fluoro-elastomers.
TN: DIAK No. 2 (6).

ETHYLENE DIAMINE TETRA-ACETIC ACID EDTA; as di-, tri- or tetra-sodium salt. White powder. Soluble in water, pH (1 % solution): 4·5 (2Na), 9·3 (3Na), 11·8 (4Na). Reacts with metal ions to form non-ionising chelates.

Uses: stabiliser for latex, partly or totally as a substitute for ammonia; copper and manganese inhibitor in rubber and latex mixtures, especially for spreading and impregnation, removes copper and manganese from textiles for use with rubber.

TN: Versene
Sequestrene
Complexon
Trilene B, etc.

ETHYLENE DIAMINE TETRAISOPROPANOL
$[(CH_3—CHOH—CH_2)_2N—CH_2]_2$
Colourless, viscous liquid, b.p. 230°C (6 mm).

Use: crosslinking agent for polyurethanes (prepolymers).

TN: Quadrol.

ETHYLENE DICHLORIDE
$CH_2Cl.CH_2Cl$ 1:2-dichloroethane. Colourless, oily liquid, m.p. $-36°C$, b.p. 83–85°C, s.g. 1·257 (20°C). Miscible with alcohol, ether and chloroform. Solvent for natural and synthetic rubber, and in particular for cements.

ETHYLENE GLYCOL RICINOLEATE
$C_{17}H_{32}(OH)COOCH_2CH_2OH$, m.p. $-21°C$, s.g. 0·97 (20°C). Softener.

TN: Flexricin 15 (9).

ETHYLENE POLYAMINE
Yellow to reddish brown liquid, s.g. 0·99. Soluble in water, benzene, alcohol, carbon tetrachloride, methyl chloride, ethyl acetate and acetone. Strong basic semi-ultra accelerator. Recommended in combination with di-o-tolyl guanidine for compounds containing white factice or retarding fillers (leather and cork powders), and with latex mixtures; activation with zinc oxide increases the tendency to scorch, fatty acids retard the commencement of vulcanisation. Processing safety may be improved by adding magnesium oxide. Has a strong activating effect on dithiocarbamates, thiurams and thiazoles. Vulcanisates have extremely high modulus and high elasticity.

Quantity: erasers, 1–1·5 % with 5–6 % sulphur, 1 % di-o-tolyl guanidine and 25 % magnesium oxide; latex mixture for impregnation, 1 % with 2 % sulphur and 5 % zinc oxide.

TN: Vulkacit TR (43).

ETHYLENE PROPYLENE RUBBER EPR devised by G. Natta and produced commercially by Montecatini. Prepared by solution polymerisation of molar proportions of ethylene and propylene. Completely saturated polymer, crosslinked by peroxides. Types which can be vulcanised with sulphur may be produced by incorporation of conjugated dienes (EPT: ethylene-propylene terpolymer). Highly resistant to oxygen and ozone, good electrical properties, dynamic qualities similar to SBR. Immiscible with SBR and NR, miscible with butyl rubber. Can be oil extended to a considerable extent. Has potential in tyre construction.

TN: AP Rubber
C-23 (59).

ETHYL HEXYL OCTYL PHENYL PHOSPHITE Brown fluid, s.g. 0·94–0·95. Antioxidant.

ETHYLIDENE ANILINE
(Formerly known as EA.) Acetalde-hyde-aniline condensation product.

ETHYL MERCURIC CHLOR-IDE Was suggested as a bactericide for latex in concentrations of 0·01–0·1% (to latex) together with 0·2–0·75% ammonia (BP 499 578). Has disadvantage of being highly poisonous.

ETHYL-β-NAPHTHYLAMINE

Antioxidant for latex, with no effect on stability and vulcanisation, is protective against normal ageing and heat, discolours through the effect of light.
Quantity: 1–2%.
TN: + Neozone L (55% water dispersion) (6).

ETHYLON Polyethylene film (199).

ETHYL PHTHALYL ETHYL GLYCOLATE
Colourless, odourless liquid, b.p. 190°C (5 mm), s.g. 1·18. Softener and plasticiser for SBR, NBR, and vinyl resins; improves mastication and processing, without effect on vulcanisation.
TN: Santiciser E-15 (5).

α-ETHYL-β-PROPYLACRO-LEIN-ANILINE CONDENSA-TION PRODUCT
Yellow to brown liquid, s.g. 0·99. Soluble in benzene, carbon tetrachloride, methyl chloride, alcohol, ethyl acetate, acetone, insoluble in water. Extremely active, versatile semi-ultra accelerator for solid rubber and latex, also for compounds with reclaim content. Produces high modulus, excellent tensile strength and elasticity. Activator for dithiocarbamates, thiazoles and thiurams. Activation by mercaptobenzthiazole, dibenzthiazyl disulphide, tetramethyl thiuram disulphide and zinc oxide. Causes discoloration, unsuitable for light-coloured articles. Retarded by alumina and carbon blacks. Vulcanisation temperature above 134°C.
Uses: solid rubber tyres, inner tubes, conveyor belts, driving belts, mechanical goods, compounds with a high reclaim content.
TN: Vulkacit 567 and 567 Extra (43) Phenex (100).

ETHYL SILICATE BAYER 40
Condensed polyethyl silicate with approx. 40% SiO_2. Water-white liquid, s.g. 1·02. Soluble in benzene, benzine, carbon tetrachloride, ethyl acetate, alcohol, acetone, methyl chloride, insoluble in water.
Use: adhesive for silicone rubber/metal bonds (43).

ETHYL-p-TOLUENE SUL-PHONAMIDE
$CH_3C_6H_4SO_2NHC_2H_5$, m.p. 58°C, b.p. 340°C, flash p. 188°C, s.g. 1·17. Softener.
TN: Santiciser 3 (5)
Santiciser 9 (mixture of the o- and p- forms) (5).

ETRTO Abbreviation for European Tyre and Rim Technical Organisation. Founded in 1964 to standardise the production of tyres and rims for lorries.

ETU Abbreviation for ethylene thiourea (2-mercaptoimidazoline).

EUCASIN Water-soluble ammonium casein.

EUKANOL W 40% aqueous emulsion of a butadiene-acrylonitrile-styrene terpolymer. Plasticiser for p.v.c. (43).

EUPHORBIA Branch of the *Euphorbiaceae* family, which includes species varying from small plants to trees and is world-wide in distribution; 44 types secrete rubber. The most important of these is *E. Tirucalli* (tropical Africa), a bush or tree with latex rich in resin, but produces a low quality rubber. Also known as broom euphorbia because of its closely interwoven branch structure. The dry mass contains approx. 15% rubber and 75% resin.

EUPHORBIACEAE Plant family of world-wide distribution, certain species of which are rubber producing: *Hevea, Euphorbia, Micranda, Manihot, Jatropha, Sapium.*

EUROPRENE Italian produced butadiene-styrene rubber. Numbers of individual types are identical to those of the ASTM code for SBR.
Grades:
SS: SBR with a high styrene content
N: butadiene acrylonitrile rubber
cis: cis-1:4-polybutadiene (200).

EVA Abbreviation for ethylene vinyl acetate polymers.

EVEITE Group of accelerators (59).
Grades:
4 MT: tetramethyl thiuram disulphide
101: butyraldehyde-aniline condensation product

202: dinitrophenyl phenyl thiazyl sulphide and diphenyl guanidine
303: dinitrophenyl phenyl thiazyl sulphide
A: aldehyde-amine
D: diphenyl guanidine
DM: dibenzthiazyl disulphide
DOT: di-o-tolyl thiourea
DOTG: di-o-tolyl guanidine
F: undisclosed compound
L: sodium diethyl dithio-carbamate
M: mercaptobenzthiazole
MST: tetramethyl thiuram monosulphide
P: zinc ethyl phenyl dithiocarbamate
T: tetra ethyl thiuram disulphide
TC: thiocarbanilide
UR: hexamethylene tetramine
Z: zinc diethyl dithio-carbamate.

EV SYSTEM Abbreviation for efficient vulcanising system. Vulcanisation system with a plateau curing curve, good resistance to heat ageing and a low compression set.

EX-B 501-4 One part bonding agent for bonds of nitrile and chloroprene rubbers to metals, many hard plastics, *e.g.* polyamides, wood and glass fibre. Other elastomers do not bond satisfactorily. The bonding strength is independent of the compound, method of vulcanisation or temperature. Has a bond strength of up to 130 kg/cm^2, resists rusting, solvents, and temperatures up to 150°C. Composition: solution of organic polymers and dispersion of solid materials in methyl ethyl ketone-xylene blend (2:1), s.g. 0·94, solid content 22–26% (382).

EXON Group of vinyl polymers (372).

EXPANSION COEFFICIENT Cubic expansion coefficient of amorphous and crystalline natural rubber is linearly related to temperature and decreases slightly with decreasing temperature. With amorphous rubber it is 0·000 67 per degree at 25°C. The temperature coefficient per degree C is 0·000 000 7. With the crystalline form it is approx. 15% lower than for the equivalent amorphous form. Below −72°C both forms have a coefficient of 0·0002 per degree C.

EXTENDER 600 Modified vegetable oil copolymer. Extender and softener (210).

EXTRUDER Machine for continuously forming long lengths of rubber or plastics compounds by forcing the compound through a die. Feed is usually by means of a screw but ram extruders are sometimes used (*e.g.* in the extrusion of p.t.f.e. coagulated dispersion polymer). Special extruders include among their uses cable covering (T-head), combination extrusion of tyre tread and sidewall (different compounds). The traditional 'hot feed' rubber extruder has a short barrel (length: diameter ratio, say 6:1), 'cold feed' extruders may have a ratio of up to, say, 10:1, and plastics extruders as high as 16:1. The feed area, the various zones of the barrel, the head, and the die should be equipped with devices for adequate temperature control.

EXTRUSION Continuous method of shaping profiles by means of an extruder. The compound is forced by a screw through an opening (die) which is the correct shape of the profile.

Uses: mainly for hose-lines, tyre treads, seals.

F

F-3 Russian produced polytrifluorochloroethylene.

F-50 The temperature at which a test piece, under specified conditions reaches a hardness exactly half way between 100 and the hardness of the test piece at room temperature. The value depends on the crystallisation occurring and is a means of indicating the resistance of the vulcanisate to low temperatures.

FACTICE Polymerised product of unsaturated vegetable oils with sulphur or sulphur monochloride. On reaction with sulphur at 130–150°C a light brown to dark brown factice is produced. The reaction takes 6–8 h. Vulcanisation accelerators may be used to decrease the time. Specific gravity may be reduced by extending with mineral oils. The reaction product with 15–25% sulphur monochloride yields a white factice. The reaction is exothermic and takes place, as with vulcanisation, only in the presence of sufficient double bonds. Linseed oil, rape-seed oil, soya oil, castor oil, sesame oil, train oil are among the suitable raw materials. Normal factices are only slightly soluble, but can be made soluble by the addition of organic bases during production, increasing the pH to approx. 9·5. A simultaneous reaction with added lyophobic materials considerably increases resistance to swelling. Vulcanisation of the oils is interrupted and the materials form a solid gel which softens above 80°C and may easily be dispersed into rubber mixes. On vulcanisation of such compounds the factices also vulcanise. Factice has excellent resistance to ozone, improves dimensional stability and rigidity of unsupported articles in hot air cures, improves extrusion and calendering, gives smooth finishes with a good, soft drape, improves ageing properties, acts as a softener and a good electrical insulator. Increased loadings of factice decrease hardness and abrasion resistance. Factice is necessary in the production of erasers and is frequently used in cable compounds.

Sulphur chloride (white) factice. Basis: vegetable oils. Reaction at 30–50°C is strongly exothermic, with hydrochloric acid being liberated. To reduce the speed of reaction, the oil or sulphur monochloride is diluted with benzine which also reduces the temperature through evaporation. The factice may be stabilised by adding calcium oxide or magnesia; it is a white to yellowish, crumbly mass, with an odour of sulphur monochloride.

Sulphur (dark) factice. Basis: vegetable or animal oils. Sulphur is added at 140–160°C, 10–25% according to the type of oil and the required hardness. The factice may be extruded, *e.g.* with paraffin, mineral oil or bitumen. Sulphur factice does not affect vulcanisation, but white factice normally has a retarding effect because of the liberation of hydrochloric acid during vulcanisation and highly basic compounds are therefore necessary.

FACTOPRENE Special type of factice used for neoprene; improves oil resistance and resistance to compression set; also the dimensional stability of compounds (114).

FADE-O-METER Apparatus for investigating the accelerated oxidation of rubbers and plastics by light and the light fastness of these materials. An electric arc is used which produces a spectrum similar to that of the sun. On discharge, practically no ozone is formed.

FAKAU Abbreviation for Fachnormenausschuss Kautschukindustrie Frankfurt am Main, Zeppelin-Allee 69.

FARADAYINE A low boiling fraction, discovered by Himley in 1835, from the products of the dry distillation of rubber.
Lit.: F. K. Himley, Diss, *De Caoutchouc ejusque destillationis siccae productis et ex his de caoutchino, novo corpore ex hydrogenis et carboneo composito,* Göttingen, 1835. Ann., 1838, **27**, 40.

FATIGUE CRACKING In rubber subjected to dynamic strain cracks appear where the strain is greatest. These cracks are orientated in the direction of strain and finally cause a complete break. The resistance to such cracking depends on the type of compound, temperature, oxygen and ozone concentration, frequency and amplitude of the mechanical strain and on the shape of the article. Cracking may be reduced by using antioxidants.

FBRAM Abbreviation for Federation of British Rubber and Allied Manufacturers.

FBRMA Abbreviation for the Federation of British Rubber Manufacturers Association.

F-C-B Acetaldehyde-aniline condensation product. Accelerator. Gives products a high surface glaze. *Quantity:* 1–2%.

F CLONES Abbreviation for 'Fordlandia' clones.

FED Abbreviation for formaldehyde-ethylene diamine condensate.

FEF Abbreviation for fast-extruding furnace black.

FERRIC FERROCYANIDE $Fe_4[Fe(CN)_6]_3$. Paris, Berlin, Prussian, Chinese or mineral blue. Blue pigment for rubber articles. Insoluble in water, alcohol and dilute acids.

FERRIC OXIDE Fe_2O_3. Pigment, variable in colour, but normally red to brown.

FERRO Group of stabilisers for vinyl resins, based on mixtures of fatty acid compounds of barium, cadmium and zinc and on organic phosphorus compounds of sodium and barium (203).

FERROUS IODIDE $FeI_2.4H_2O$. Almost black solid, sometimes reddish green liquid. Extremely corrosive, decomposes in air and on contact with light.

Uses: catalyst for the production of certain types of polyacrylic rubber.

FEUERLAND MACHINE H 50 Apparatus used to determine fatigue properties at temperatures of -20 to $70°C$.

FF Abbreviation for fine furnace black.

FIBRAVYL (Formerly Phofibre, Fibrovyl.) P.V.C./rayon yarn. Fr. P. 913 164, 913 927.

FIBRE V Synthetic fibre, condensation polymer of ethylene glycol and terephthalic acid (6).

FIBRO Viscous rayon yarn (161).

FIBROCETA Acetate rayon yarn (161).

FIBROLANE Protein fibre (161).

FICUS Member of the *Moraceae* family widely grown in the tropics and subtropics. In the early days of the rubber industry a considerable quantity of rubber was obtained from the various species. Originally only *F. Elastica* was cultivated for its rubber yield, but was soon supplanted by *Hevea* type. In 1956 only a single plantation in New Guinea was known to exist.

FICUS CALLOSA Pangsar, Ilatilat. A tree attaining up to 30 m (Java), whose sap contains a highly active proteolytic enzyme and has rapid coagulating effect on latex from *Hevea Brasiliensis*. With smaller quantities, rubber with a low nitrogen content and a heat sensitive latex may be produced.

FICUS ELASTICA Assam rubber tree.

Discovered in 1810 by Roxburgh, native to India and South Asia. A tree which grows to approx. 30 m and has countless aerial roots and leather-like leaves. In Bengal is called Kusnir, in Sumatra, Rambung and Kadjai, and in Java, Pohon karet. The annual yield of 5-year-old trees is approx. 25–50 g and increases to 2–2·5 kg for trees approx. 15 years old. Plantation cultivation had been tried in Java as early as 1864 by the Dutch Forestry Commission, and in 1872 the Anglo-Java Plantation Co. planted approx. 30 ha on the Pamanoekan Tjiassan, which was worked until 1930, it was the oldest rubber plantation in the world. The plantation was tapped regularly between 1886–1895, and later at irregular intervals, particularly when the price of rubber was high. Later, in Assam, India, Sumatra, Java and Malaya smaller plantations were laid out. However, since the rubber yield rapidly decreases through regular tapping, the plantations were given up before World War I. The greatest quantity of rubber came into trade from wild groups of trees or from neglected plantations and was mostly collected by the local people. The sap of *F. elastica* has always been used locally for the production of torches or for making baskets and containers watertight. The latex contains 37–38% rubber, 2·5% resin, 0·4% protein, 0·4% ash, particle size 1–5 μ, somewhat unstable, but may be stabilised by adding ammonia. The latex is white but discolours when left to stand. It does not coagulate easily with acids or alkalis. Complete coagulation occurs with alcohol, ether and acetone.

FIELD OF ACTIVITY The space

occupied by a dissolved molecular chain in a solution of polymers.

$$V = \left(\frac{nl}{2}\right)^2 \pi d$$

where

  n is the degree of polymerisation
  l the length of the structural unit
  d the diameter of the structural unit.

FILLERS Mostly particulate

constituents of a compound; added in large quantities to the rubber, they improve the physical properties or lower the volume cost. According to their degree of effectiveness, they are graded as reinforcing or non-reinforcing (active or non-active). The first group includes the carbon blacks, fine, light-coloured fillers and reinforcing synthetic resins; second includes diluent materials, coloured pigments, white pigments, also fillers which influence vulcanisation and act as activators, *e.g.* zinc oxide, magnesium oxide, magnesium carbonate, litharge, red lead, and calcium hydroxide.

White reinforcing fillers. The development of synthetic rubber types which can only be satisfactorily used with reinforcing fillers, and of coloured articles to meet market demands, has enabled the development of light-coloured, reinforcing fillers to make rapid progress since 1945. To characterise the large number of fillers according to their properties and method of production, the following classification was devised: (according to F. Endter and H. Weslinning, *Kautschuk u. Gummi*, 1956, **5**, 130 WT).

1. Characterisation by method of production:

PP: pyrogenic process (formation of the solid particles by molecular separation in the gas phase, *e.g.* hydrolysis of halogenous silanes in super heated steam

TP: thermal process (from solid materials)

WP: aqueous process (precipitation from solution)

MN: modified natural product (product treated afterwards)

N: natural product.

2. Chemical constituents: with some materials the chemical formulae are given (*e.g.* SiO_2), while with non-stoichiometric materials the symbols for the main constituents are given together, thus a dash Al/SiO_2. Constituents present in small amounts which have no influence on the effects of the filler are not given.

3. Reinforcing power:

LR: low reinforcing

MR: medium reinforcing

HR: highly reinforcing

SR: super reinforcing

ER: extremely reinforcing.

Examples: $PP-SiO_2$-ER, pyrogenically produced silica dioxide with an extremely reinforcing action.

Classification	Trade product
$PP-SiO_2$-ER	Aerosil
$PP-Al_2O_3$-ER	Aluminium oxide P
$TP-Al_2O_3$-SR	AOK I
$WP-SiO_2$-SR	Ultrasil VN 3
$WP-SiO_2$-HR	Durosil
$WP-Al/SiO_2$-HR	AS 7
$WP-Ca/SiO_2$-MR	Calsil
WP-Al/OH-MR (and LR)	Tegs
$WP-CaCO_3$-MR	M 1057

Classification	Trade product
$MN-CaCO_3$-LR	surface-treated champagne whiting
$N-CaCO_3$-LR	champagne whiting
N-Kaolin-LR	kaolin (china, dixie, Windsor clay, mikrolin)

FILLER SURFACE The surface area of a filler gives a good indication of its behaviour in compounds, as does the mean diameter of the particles. Surface area and particle size may be determined either by a 50 000–75 000 electron microscope magnification (EM surface), or by the hydrogen-absorption isotherm (BET surface). The surface area is given as m^2/g. The relation of the EM surface to the considerably larger BET surface is termed the index of roughness, and gives an indication of the influence of the filler on vulcanisation.

FINGOMAINTY Trade term for wild rubber from *Landolphia hispidula* (Madagascar).

FINGOTRA Trade term for wild rubber from *Landolphia crassipes, L. grammifera, L. Madagascariensis*, lianes of the *Apocynaceae* family (Madagascar). Was valued as reasonably good rubber.

FIRESTONE FLEXOMETER Apparatus for determining the dynamic compression fatigue of rubber. The test piece, in the form of a pyramid (base 54 × 28·6 mm, top surface 50·8 × 25·4 mm, height 38·1 mm), is placed under a constant unilateral load with an oscillating, rotating movement of 800 c/min. The time taken to obtain a predetermined deformation is measured and shown automatically. ASTM D 623-58.

FIRESTONE PLASTIMETER Extrusion plastimeter in which a sample compound, volume 5·45 cm^3, at a fixed temperature, is forced by a ram through a nozzle. The extrusion time (between 2–60 sec) is taken as the plasticity index. The ram is driven by an air pressure of 0·7 kg/cm^2 (10 lb/in^2) for soft compound, 1·4 kg/cm^2 (20 lb/in^2) for extrusion compound and 2·1 kg/cm^2 (30 lb/in^2) for master batches. There is practically no correlation between the results obtained and the parallel plate instruments and the Mooney values. The results, however, agree to a reasonable extent, with the practical extrusion and calendering characteristics.

FIRI Abbreviation for Fellow of the Institution of the Rubber Industry. Fellowship awarded for outstanding achievements in the field of rubber.

FISH EYES (Cats' eyes, jargon.) Undissolved pieces of rubber in rubber solutions.

FIXING PROCESS Production of thick walled articles from latex compounds by dipping. The dipping form is first placed in a 10–30% solution of a coagulating agent, *e.g.* lactic acid, formic acid, acetic acid, acetates, calcium nitrate, calcium chloride, in alcohol. After the solvent has evaporated the former is dipped in the latex compound. With an even layer of coagulant on the former, a uniform product is produced with an even thickness of film. The thickness of the film can be varied by altering the length of the dipping time and the concentration of the coagulant. BP 252 674, 1926.

FIXING SALT Cyclohexylamino acetate. Coagulating agent for dipped latex articles (43).

FLAME RESISTANCE Resistance of rubber and plastics to the effects of fire; applies primarily to electrical goods. May be determined with an incandescent rod apparatus, according to Schramm/Zebrowski. DIN 7705, 53 459, VDE 0302.

FLECTOL Group of antioxidants.
Grades:
A, B, H: polytrimethyl dihydroquinoline
White: undisclosed material (5).

FLETCHER-GENT APPARATUS Apparatus used to determine the dynamic properties of rubber; uses forced oscillations.

FLEXAMINE Mixture of a complex diaryl amine-ketone condensation product and 35% N,N'-diphenyl-p-phenylene diamine. Brown powder, m.p. 75–90°C, s.g. 1·10. Soluble in acetone, benzene, ethylene dichloride, insoluble in water and benzine. Antioxidant, has an activating effect on vulcanisation. Vulcanisates stain light-coloured articles on contact. Causes a brown discoloration of vulcanisates in light, and migrates during vulcanisation. With an excess (over 1%) the N,N'-diphenyl-p-phenylene diamine blooms. Gives excellent protection against heat, oxidation, flexing fatigue, copper and manganese.

Quantity: for optimum effect 1·5–2%, for articles in which lack of bloom is important less than 1%. Gives a good protection at 0·2%.
Uses: natural rubber, SBR, reclaim, tyres, mechanical goods, cable sheathings, soles, insulations, compounds with reclaim, compounds which could become contaminated by copper, and in compounds requiring maximum protection against heat and flex cracking (23).

FLEXCRACKING Formation of cracks in rubber articles and the extension of existing cracks through dynamic fatigue caused by repeated bending. The characterisation of crack formation and extension is carried out by using the equipment listed in the entry, Flex Resistance.

FLEX CRACKING RESIST-ANCE The logarithm of the number of kilocycles necessary to achieve a certain degree of crack formation, may be determined using the De Mattia machine.

FLEXOL Group of softeners for natural and synthetic rubber, vinyl resins, and plastics (146).

FLEXOMETER Apparatus for determining the change of properties in rubber under dynamic flexing strain.

FLEXON (Formerly Barum-Flon.) Czechoslovakian produced polytetrafluoroethylene.

FLEXRICIN Group of softeners based on ricinoleic acid.
Grades:
9:     propylene glycol ricinoleate
13:    glycerine monoricinoleate
16:    isobutyl ricinoleate

25:    ethylene glycol ricinoleate
61:    modified methyl acetyl ricinoleate
62:    modified methoxyethyl acetyl ricinoleate
66:    isobutyl ricinoleate
P-1:   methyl ricinoleate
P-1C: methoxyethyl ricinoleate
P-3:   butyl ricinoleate
P-4:   methyl acetyl ricinoleate
P-4C: methoxyacetyl ricinoleate
P-6:   n-butyl acetyl ricinoleate
P-8:   glycerine triacetyl ricinoleate (9).

FLEX RESISTANCE The resistance of rubber test pieces to crack formation and the extension of existing tears produced by dynamic fatigue is tested by means of an alternating bending strain. Apparatus: De Mattia, Du Pont, Ross, and Scott flex machines.
Specifications: ASTM D 430-59, D 813-59, D 1052-55
DIN 53 522
ISO recommendation 172, 173.

FLEXURAL FATIGUE If rubber is subjected to constant changes of stress, cracks form on the surface and normally extend in the direction of tensile stress. The determination of stability against cracking usually consists of repeated bending of the test piece under specifically fixed conditions, using either De Mattia or Ross apparatus. This gives comparative results suitable for laboratory research.

FLEXZONE Group of antiozonants for dynamic applications.
Grades:
3-C: N-isopropyl-N'-phenyl-p-phenylene diamine
5-L: N-sec-butyl-N'-phenyl-p-phenylene diamine

6-H: N-phenyl-N′-cyclohexyl-p-
phenylene diamine (23)
7-L: N-1:3-dimethyl butyl-N′-
phenyl-p-phenylene diamine
68: N-phenyl-N′-cyclohexyl-p-
phenylene diamine.

FLOCCULATION Agglomeration of latex particles to larger flakes without the formation of a coherent coagulu, the first stage of coagulation.

FLOSBRENE Group of fluid styrene-butadiene copolymers, 25% bound styrene manufactured by emulsion polymerisation (103).

Type	Viscosity (poises 25°C)	Mol. wt.
VLV	400	2 000
LV	2 000	3 000
MV	4 500	5 000
HV	8 000	15 000

Lit.: *Rubb. World*, May 1968, 83.

FLUAVIL Probably $C_{20}H_{32}O$. Amorphous part of gutta-percha resin, softening p. approx. 100°C. Soluble in alcohol.

FLUIDITY Reciprocal of viscosity.

$$\Phi = \frac{1}{\eta}$$

FLUON 170 Polytetrafluorethylene powder, lubricant for rubber compounds and thermoplastics (60).

FLUOREL Copolymer of vinylidene fluoride and hexafluoropropylene. Fluoroelastomer, s.g. 1·85, Shore A Hardness 40. Vulcanisable with organic amines, *e.g.* hexamethylene diamine carbamate, per-

oxides, or by radiation. Vulcanisates are resistant to corrosive chemicals, lubricants and solvents, have a low permanent set and withstand temperatures of up to 205°C continuously.
Uses: in aeroplane structures (1).

FLUOROCARBON RUBBER Linear copolymer of vinylidene fluoride and hexafluoropropylene or fluorotrichloroethylene, with approx. 65% fluorine. White, translucent mass, s.g. 1·85. Soluble in low mol. wt. ketones. In comparison with other polymers has a low mol. wt. Has good physical and rubber-like properties up to 200°C, resistant to oils and solvent at high temperatures. May be vulcanised to rubber-like products by peroxides, polyamines (*e.g.* hexamethylene diamine carbamate), or beta and gamma radiation. Compounds must contain fillers (reinforcing silicates or blacks) and an acceptor (metal oxide) to absorb traces of free acids.
Uses: include heat-resistant, chemically and mechanically stable products, O-rings, gaskets, tubes and diaphragms.
TN: Viton
Kel-F
Fluorel
SKF (USSR).

FLUOROELASTOMERS Fluorine containing polymers and copolymers with an excellent resistance to heat and cold, ozone, oxygen, solvents and chemicals; have good electrical properties and low water absorption. Soluble in esters and ketones. Special products which are primarily used in the aircraft industry. Production:
1. *Copolymers of fluorolefins.* Polymerisation of vinylidene fluoride with

hexafluoropropylene (Viton A) or chlorotrifluoroethylene (Kel-F) with halogenised peroxides as a catalyst in an aqueous dispersion at 20–60°C.
2. *Polyfluorolefins.* A product of this group is polyfluorobutyl acrylate (Poly FBA) obtained by emulsion polymerisation of 1:1-dihydroperfluorobutyl acrylate.
3. *Butadiene copolymers.* Copolymers of butadiene with approx. 30% 1:1-dichloro-2-difluoroethylene or 1-chloro-2:2-difluoroethylene. The products may be used either in solid form or as latex. Normal rubber processing machines are used. Reinforcement may be achieved by the use of blacks or reinforcing silicates; vulcanisation by polyamines, peroxides or isocyanates.
Uses: include impregnations and coatings.

FLUOROPREN 2-fluoro-1:3-butadiene polymer. Has similar properties to SBR.

FLUOROTHENE Trade name for polychlorotrifluoroethylene.
Grades:
FYTD: melting viscosity at 230°C, 5 megapoise
FYTH: melting viscosity at 230°C, 15 megapoise
FYTS: melting viscosity at 230°C, 50 megapoise (137).

FOAMREX L Antifoam agent for latex compounds (178).

FOAM RUBBER Porous rubber articles produced by foaming vulcanisable latex compounds, gelling the foam and curing. The first method used for the direct processing of latex was initiated by P. Schidrowitz and H. A. Goldsbrough [BP 1111 (1914), DRP 321 092, USP 1 156 184]. The latex was mixed with a blowing agent (ammonium carbonate) which, during coagulation with acetic acid, evolved a gas and caused the latex to foam. Mechanical foaming by stirring was introduced in 1927, and in 1928–29 the gelling process was developed by Dunlop, the first foam products coming on the market in 1930. Of the various production processes the Dunlop process is the most widely used [USP 1 852 477 (1932)]. In principle it consists of mixing a 60% latex with an accelerator, sulphur, an antioxidant, 1% soap, and ingredients such as mineral oil, pigments, fillers, then foaming mechanically, after a maturing period of at least 24 h, to 5–12 times the original volume. At the end of the process zinc oxide and a gelling agent are added. The liquid foam is poured into moulds and coagulates in 5–10 min. Curing is in open steam or by high frequency heating. The vulcanisate is washed, water removed by centrifuging, and the product dried. With thick articles a honeycomb structure is formed by a cored mould so that the maximum thickness of the walls does not exceed approx. 2·5 cm.
Foaming agents: potassium soaps of oleic acid, castor oil or resins are preferred. Many synthetic wetting agents have an undesirable effect on the coagulation.
Accelerators: zinc salts, *e.g.* zinc diethyl dithiocarbamate and zinc mercaptobenzthiazole, piperidine pentamethylene dithiocarbamate, mercaptobenzthiazole, sodium dibutyl dithiocarbamate, are preferred.
Gelling agents: include sodium silicofluoride, alkaline fluoritanates, nitropropane, and ammonium persulphate and trioxymethylene.
The latex may be foamed batchwise

in a mixer, using wire beaters, or continuously, using foaming machines, *e.g.* the B. Oakes machine. The first attempts to use the latter were made in 1936 (BP 471 899) and such machines have been used industrially since *ca.* 1947.

When producing foam rubber it is important that coagulation takes place without the foam disintegrating or the cell structure being destroyed. Depending on the compound used and the processing method, the products shrink up to 12% after vulcanisation.

Another frequently used process is the Talalay process [USP 2 432 353 (1947)], in which hydrogen peroxide and a catalyst are added to the compound. Because of the delayed action in the peroxide decomposition the liquid compound can be poured into the moulds and foamed in closed moulds by the liberated oxygen. The foam is frozen to break the cell walls, coagulated by introducing carbon dioxide, and vulcanised after thawing.

The foam has a very high compression modulus, very low compression set, high resilience, may be washed and has bactericidal properties (*e.g.* for hospital beds). Its uses also include upholstery, mattresses, underlays, packing. After tyre rubber it is the type of rubber most widely used in industry.

Synthetic latices: SBR latex is difficult to process alone and is usually used in blends with natural latex. SBR latex gives products with very poor physical properties, poor resistance to heat and wet and has extremely poor foam stability. Neoprene latices give high densities, poor low temperature resistance and shrinkage up to 25%. Their advantages lie in their flame resistance and resistance to ozone and exposure to sun-rays. For special applications foam materials are also produced from nitrile and vinyl latices and from polyurethanes.

FORD VISCOSITY The time taken for a liquid to flow out of an orifice under its own hydrostatic pressure; measured by using the Ford–Becher viscometer.

FORMIC ACID H.COOH, m.p. 8·4°C, b.p. 100·7°C, s.g. 1·226. Most commonly used coagulating agent for natural rubber. Formic acid is added, as a 1% solution, to fresh latex, thinned down to 15–20%. For coagulation at a pH of 4·5–4·8, 4–5 g of concentrated acid is used per kg of rubber.

FORMOPON Sodium sulphoxylate formaldehyde (38).

FORTAFIL A 70 Precipitated aluminium silicate, s.g. 2·1. White reinforcing filler which imparts high hardness and wear resistance (60).

FORTISAN Acetate silk (152).

FOURCROY, ANTHOINE FRANCOIS DE 1735–1809, French chemist. In 1791 discovered the preserving effect of alkali on latex and in 1804 suggested the transportation of preserved latex to France.
Lit.: *Leçons élémentaires d'Histoire naturelle et de Chimie*, Paris, 1782.

FPT Abbreviation for formaldehyde-p-toluidine condensate.

FRANSIL Ultra-white silicic acid. Reinforcing filler.

FREE SULPHUR The free elementary sulphur extracted from rubber by acetone.

FREEZE GRINDING NR, SR, polystyrene, vinyl plastics, acrylic resins and many other materials which cannot be ground by the conventional process, may be cooled with liquid nitrogen according to the process of the Linde Air Products Co., USA, and thus made brittle and crushable. The use of nitrogen also prevents infiltration of atmospheric oxygen.

FREEZING POINT METHOD Process for concentrating SBR latex. Developed in Germany during World War II. The latex is gelled by cooling, after being foamed mechanically to seven times its original volume; when cold it separates into a granular form of concentrate and a clear serum which is removed. On warming the semi-solid concentrate to room temperature liquid latex is obtained.

FRESNEAU, FRANCOIS (Seigneur de la Gataudiere) 1703–1770. French engineer in Cayenne. First took an interest in the rubber articles brought from Para by the Portuguese and Indians. Led an expedition into the jungles of Guiana and in 1747 discovered various rubber-yielding trees, among them *Hevea Brasiliensis*. The earlier announcement of the discovery in Mexico was probably concerned with *Castilla elastica*. Studied the methods of rubber tapping and processing in the provinces of Quito and Esmeralda. In 1751 told La Condamine of his discoveries, who read out his 'Memoire' (written 1747) to the French Academy. Fresneau was aware of the potential of rubber and attempted the production of shoes, tubes and water flasks, sending a few samples to the French Government, who received them with interest. Also suggested that rubber costumes should be made for divers. After his return to France, 1749, undertook research on dissolving rubber. He found the ideal solvent to be turpentine, which was later independently discovered by Herissan and Macquer. In his writings also gives a fairly accurate description of the *Hevea* tree.

FREY-WYSSLING PARTICLES Spherical non-rubber particles, approx. 2μ in size, reddish brown, yellow or almost colourless. Occur either singly or in groups in natural latex; colour is brought out by carotinoids. Discovered by Frey-Wyssling (*Arch. Rubb. Cult.*, 1929, **13**, 392), they form part of the so-called yellow fraction of the latex and can be removed by fractional coagulation, *e.g.* as in the production of pale crêpe.

FRICTIONING Impregnation of a fabric with a thin rubber layer by friction, using a calender. For example, with a three roll calender, the middle roll runs at a faster speed than the two outer rolls. The rubber compound is pushed into the upper nip between the rolls while the material runs, at this point through the lower nip, the rubber compound being forced into the textile structure.

FROTHY SHEETS Sheets which contain porous or scarred patches caused by fermentation during coagulation.

FR-S Group of butadiene-styrene copolymers and latices; cold and hot polymerised and oil extended types (129).

F RUBBER Raw rubber type produced by flocculation of a latex preserved with formaldehyde. The fresh latex is mixed with formaldehyde, purified by sedimentation and flocculated with acid. The crumbs are then washed with water and coalesced to larger particles by steam at approx. 65°C. The product is then rolled out into sheets and dried. F rubber requires less mastication than sheet rubber and gives a harder product.
Uses: in particular for shoe sole compounds.

FT Abbreviation for fine thermal black.

FT- (200, 300.) Group of waxes; aliphatic hydrocarbons with a high mol. wt.
Uses: lubricants, softeners, prevention of tack (190, 205).

FUNTUMIA *Kickxia.* Branch of the *Apocynaceae* family (Africa). *F. elastica* was planted in large quantities in Africa at the beginning of the rubber planting era. By 1912 approx. 6000 ha were estimated in Africa.

FUNTUMIA ELASTICA (*Kickxia elastica*), African rubber tree (*Apocynaceae*). Produce the silk rubber originally traded in Lagos. The trees grow wild, up to 30 m high, and occur abundantly throughout tropical Africa (Sierra Leone, Congo, Liberia, Nigeria, Cameroon, Uganda, the Ivory Coast), sometimes in groups of up to 500. To obtain the sap, spine-like cuts were made leading into a main cut. The rubber was usually obtained by evaporation of the water, and kneaded into bales for the market. Because of the small size of the rubber particles, the latex is much more difficult to coagulate than that of other plants. In Cameroon a plantation culture was attempted, but because of the mediocre yield and the long intervals necessary between tapping, the attempt was soon abandoned. During World War II, regardless of the high cost, production was restarted on existing plantations.

FURAC Accelerator (obsolete).
Grades:
2: zinc salt of dithiopyromucic acid
3: lead salt of dithiopyromucic acid (206).

FURATONE Group of furfural derivatives.
Grades:

1444:	reinforcing resin for acrylonitrile polymers (1, 147)
NC-1006:	furfuraldehyde derivative. s.g. 1·15
NC-1008:	resin-like, thermoplastic material, s.g. 1·25. Processing aid and extender for synthetic rubber
NV-1012:	low viscosity liquid, 20–50 cp (25°C), b.p. 210°C, s.g. 1·110. Antiozonant for neoprene and butyl rubber compounds.

FURNAL R 300 Rumanian produced gas black, Type FF. Produced from methane.

FURNEX SRF black (3).

FURNEX H SRF gas black with a high modulus (159).

FURNEX HV High modulus, smooth extruding, carbon black (159).

FVRS Abbreviation for Feuilles à vulcanisation rapid contenant de serum de centrifugation. A specialised rubber type developed by the IRVC in Laikhé, Vietnam, by combining normal latex with skim latex (centrifugal serum). Properties lie between those of normal smoked sheet and skim rubber.

	RSS 1	FVRS
Nitrogen %	0·37	0·83
NH$_2$ Index	0·072	0·230
Mixture w. 50% EPC, acceleration with MBT/DPG		
Module 300 % kg/cm^2	90	110
Breaking strength kg/cm^2	280	270
Elongation at break %	650	600
Shore Hardness	65	72

FX CLONES F clones. Group of clones, grown in S. America (US Experimental Station, Turrialba, Costa Rica) from Indonesian and Malayan *Hevea* hybrids resistant to *Dothidella*. Still at the experimental stage, the clones are primarily a cross between hybrids of *H. Brasiliensis* and *H. benthamiana*, but sometimes also selections from the Fordlandia clones.

G

GABUN Natural rubber of *Landolphia* and *Ficus* type (native to the Congo). Came into trade as bales and so-called tongues. The trade terms for the different qualities were loango, ogoway, gabun, mayumb, batanga, kamerun, and bata. Hard, good quality gabun bales were formerly used for the production of mechanical goods and impregnated textiles.

GALAPREN Polyurethane foam (207).

GALEX Trade name for dehydroabietic acid (208).

GALLIE-PORRITT TESTER Apparatus used to determine the grit content of powder, washed by a jet of water through a sieve fastened to a funnel.

GAMBIA Trade name for natural rubber from various *Landolphia* and *Ficus* types (Sudan, Senegambia and the Senegal River area). Was marketed in bales of varying quality.

GASOLIN ANTIOXIDANT Group of antioxidants (6):
22: N,N'-disec-butyl-p-phenylene diamine
26: phenylene diamine isomers. Reddish brown fluid, s.g. 0·9
29: 2:6-ditert-4-methyl phenol.

GBOGBOI Local term for a wild rubber from Sierra Leone, probably *Funtumia elastica*.

GEBAGEN RESINS Group of pure hydrocarbon resins produced by polymerisation of unsaturated hydrocarbons.
Types:
J and J-T: neutral, unsaponifiable indene resins, which differ in softening p. and solubility
P: slightly polymerised styrene
ML-60: copolymer of cyclopentadiene, indene and linseed oil.
Uses: improve the dispersion of sulphur and the structure of vulcanisates of NR and SR, also the processability of SR; compatible with chlorinated and cyclised rubbers (43).

GEER ANTIOXIDANT TEST Test for investigating the accelerated ageing properties of rubber according to G. Geer. The test pieces are aged in an oven with circulating air, usually at 76°C, for 7, 14, and 28 days, after which the tensile strength may be determined. For synthetic rubber higher temperatures are used. This ageing test shows a relatively good correlation with normal ageing at room temperature. To avoid error, only test pieces of a similar nature may be tested together. ASTM D 573-53, BS 903: A19: 1956, DIN 53 508.
Lit.: *India Rubb. World*, 1916, **55**, 127.

GEER OVEN An oven with circulating air; used for accelerated ageing of rubber test pieces, usually at 70°C. The physical properties are determined after a specified length of time.

GE FIBRE Synthetic fibre; copolymer of 70% acrylonitrile, 25% ethyl acrylate and 5% ethyl methacrylate (189).

GEHMAN TEST Method used to determine the cold resistance of elastomers. Using a torsion wire, the torsion angle of strip test pieces (approx. 38 × 3 mm) is determined as a function of temperature. From the torsion/temperature curve, the temperatures T_2, T_5, T_{10}, and T_{100} are ascertained, at which the torsion modulus has 2, 5, 10 and 100 times the value of the modulus at 23°C. ASTM D 1053-61.

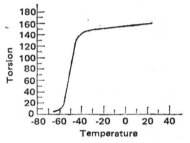

Temperature dependence of the torsion of natural rubber

GELITEX Heat sensitive, concentrated (60%) latex used particularly for the production of foam rubber. Contains all the chemicals necessary for the production of foam rubber, with the exception of the curing system. Developed by the IRCV in Laikhé (Vietnam).

GEL RUBBER The benzene insoluble part of rubber.

GENAPOL Group of wetting agents, dispersing agents, emulsifiers and stabilisers.
Grades:
LRO: modified lauryl sulphonate. Anionic

LRT: lauryl sulphonate-triethanol-
amine salt. Anionic
X: group of alkyl polyglycol-
ethers. Non-ionic, wetting and
dispersing agents for carbon
blacks (217).

GENITRON Group of blowing
agents (209).
Grades:
AB: azoisobutyramide
AC: azodicarboxylacid
         diamide
AZDN: azoisobutyronitrile
BH: benzyl hydrazone
BSH: benzene sulphonhydrazide
CHDN: azocyclohexyl nitrile
OB: p,p'-oxybisbenzene
         sulphonyl hydrazide
N: nitrourea
SA: salicyl aldazine.

GEN-TAC LATEX Latex of a
vinyl-pyridine copolymer. Improves
the adhesion of rubber to rayon,
nylon and other synthetic fibres (31).

GENTHANE S Millable polyurethane, polyaddition product of a
polyester of adipic acid, ethylene
glycol, and propylene glycol with
diisocyanates. Vulcanisable with peroxides. Has excellent resistance to
abrasion, resistant to oils, solvents,
ozone, and oxygen. Stable up to
150°C.
Uses: include solid rubber tyres,
corrosion resistant coverings, hose
lines and in the aircraft industry (31).

GENTRO Trade name for SBR
rubber (31).

GENTROJET Trade name for
SBR rubber (31).

GEON Group of plastics and
latices based on p.v.c. or blends of
vinyl resins with butadiene-acrylonitrile polymers. Many types may be
vulcanised. The products are resistant
to ozone and solvents and have good
physical properties.
Uses: include insulating materials,
protective coverings, impregnations
(14).

GEPOLIT Vinyl chloride-styrene
copolymer. Processing aid for p.v.c.
compounds, particularly for floor
coverings (189).

GERKE PROCESS Heat treatment of black master batches before
vulcanisation.

GERMAN RUBBER SOCIETY
E. V., Frankfurt-am-Main, Zeppelin-
allee 69. Scientific society of chemists
and engineers specialising in rubber.
Founded 1926.

G. E. RUBBER Silicone rubber
(198).

GE SILICONE RUBBER Group
of silicone rubber types and compounds (198).

GE SILICONES Group of silicone oils.
Uses: release agents (198).

GETAH Rambung, karet.
Malayan term for rubber.

GETAH DUJAN Getah putih,
white rubber. Local term for gutta-
percha from *Palaquium treubii* (Banka
Island: Indonesia). Less important
than gutta-percha from *P. oblongifolium.*

GETAH MARAU Local term for gutta-percha from *Payena Bawun* (New Guinea). Sub-standard product.

GETAH MELABUAI Local term for a product, similar to jelutong, and obtained from *Alstonia grandifolia* (Indonesia, Sumatra).

GETAH MUNTAH Local term (Borneo) for the raw gutta-percha obtained from wild trees. Frequently adulterated, *e.g.* with wood flour, sago flour, and clay.

GETAH NALU Local term for the sub-standard gutta-percha from *Payena mentzelii* and *Syderoxylon kernbachianum* (New Guinea).

GETAH SUNTIK Local term for gutta-percha from the *Payena leerii* (S. Sumatra). The term suntik (Malayan: to sting) has arisen because the trees have been tapped by 'stinging' them with long poles with iron spikes at the end. In N. Sumatra the same product is called ringi. The gutta-percha has a yellow colour, a high resin content, and is excellent in quality.

GETREN 4 Water soluble release agent based on an organic modified polysiloxane. Used at a dilution of 1–3%. In contrast to the use of silicone oil emulsion, the release agent can be removed by washing with cold water (436).

GEUNS, JAN VAN Dutch chemist, 1799–1865. Was producing rubber goods in 1828, from 1833 with the addition of sulphur; presumably was influenced by an article by F. Lüdersdorff (*Tech. u. Ökon. Chem.*, 1832, **15**, 359). Between 1833–1837 he probably discovered that, to produce better properties, the rubber should be heated after mixing with sulphur. The production at that time of fire-hoses, which were elastic even in the cold, confirms that the vulcanisation process discovered by Goodyear in 1839, was already known and had been put to use.

GHEZ PROCESS Process for reclaiming old rubber. The crumbled, fibrous material is replasticised in a bath of molten asphalt, the fibres being carbonised.

GILSONITE Asphalt, m.p. 110–160°C, s.g. 1·15. Extender and tackifier.

GLASS TRANSITION γ transition, α anomaly, second order transition, vitrification. The change of a polymer from a viscous or elastic nature to a brittle, glass-like condition. The transition is characterised by the temperature at which a sudden change or discontinuity occurs in the primary thermodynamic properties (specific heat, thermal coefficients of expansion). The transition takes place at a relatively low temperature and is caused by prevention of the rotation of the complex molecules; it is similar to the freezing of a liquid. Some glass transition temperatures are:

natural rubber	$-73°C$
NR vulcanisates (with 4% bound sulphur)	$-60°C$
SBR	$-61°C$
polyisobutylene	$-74°C$
polybutadiene	$-85°C$
polymethacrylate	$3°C$
polystyrene	$81°C$
hard rubber (32% bound sulphur)	$80°C$

In vulcanisates the transition temperature increases with the increase in percentage of bound sulphur. ASTM D 832-59.

GLAURIN Trade name for diethylene glycol monolaurate (30).

GLIMMER (Biotite, muscovite). Monoclinic, hydrated sodium, potassium and aluminium silicates, s.g. 2·6–3·0.
Uses: finely pulverised, as a dusting agent, *e.g.* in tubing and mechanical goods. Or suitable for articles in which a high quality surface finish is required.

GLOXIL Light-coloured, semi-reinforcing silicate filler for NR and SR, s.g. 2·6. Contains 85% SiO_2, 10% Al_2O_3, quartz:kaolin 3:1. Suitable for translucent products (387).

GLYCERINE
$CH_2OH.CHOH.CH_2OH$, 1:2:3-propanetriol, trihydroxypropane, glycerol. Syrup-like, hygroscopic, sweet tasting liquid, b.p. 290°C, s.g. 1·265. Miscible with alcohol and water. Solidifies to ortho-rhombic crystals when left standing a long period at 0°C.
Uses: processing aid for rubber and latex, dispersing agent, softener, release agent, gives a smooth surface, surface detackifiers, protects against oxygen and ozone, antifreezing agent in emulsion polymerisation below freezing p., extracts thiophenols in reclaiming processes.

GLYCERINE DIACETATE
$C_3H_5(OH)(OCOCH_3)_2$, diacetine. Colourless liquid, m.p. −30°C, b.p. (11 mm) 142–153°C, s.g. 1·186.

Soluble in water and alcohol. Softener, solvent.

GLYCERINE MONOACETATE
$CH_3COOCH_2CHOH.CH_2OH$, monoacetine, acetine. Colourless, hygroscopic liquid, m.p. −30°C, b.p. (30 mm) 160–178°C, s.g. (25°C) 1·190. Soluble in water and alcohol. Softener.

GLYCERINE MONOLAURATE
$C_{11}H_{23}COOCH_2CHOH.CH_2OH$, m.p. 26–28°C, s.g. (25°C) 0·970. Softener.
TN: Morpel X-574 (264).

GLYCERINE MONO-OLEATE
$(C_{17}H_{33}COO)C_3H_5(OH)_2$. Yellow paste, m.p. 18°C, s.g. (25°C) 0·945. Insoluble in water. Softener.
TN: Emery 2221 (196)
Morpel X-570 (264).

GLYCERINE MONORICIN-OLEATE
$C_{17}H_{33}COOC_3H_5(OH)_2$. Reddish, oily liquid, m.p. −50°C, flash p. 265°C, s.g. (25°C) 0·981. Soluble in alcohols. Softener.
TN: Flexricin 13 (9).

GLYCERINE MONOSALI-CYLATE
$CH_2OH.CHOH.CH_2OCO$
$.C_6H_4OH.$
Softener, antioxidant.
TN: Glyceryl Salicylate E
Glycosal.

GLYCERINE MONOSTEAR-ATE
$C_{17}H_{35}COOCH_2CHOH.CH_2OH$, monostearine. White, wax-like solid, m.p. 55–58°C, s.g. 0·97. Soluble in hot organic solvents. Softener.
TN: Morpel X-710 (164).

GLYCERINE TRIACETATE

$CH_3.COOCH_2CH(OCOCH_3)$
　　　　　　　$.CH_2OCOCH_3,$

triacetine, enzactine. Colourless liquid, m.p. $-78°C$, s.g. $258–260°C$, s.g. $1·16$. Soluble in alcohol, ether, chloroform, benzene. Softener, solvent.

GLYCERINE TRIACETYL RICINOLEATE

$[C_{17}H_{32}(OCOCH_3)COO]_3C_3H_5$, m.p. $-35°C$, flash p. $290°C$, s.g. $(25°C)$ $0·965$. Clear, yellowish liquid with slight aroma. Softener for natural and synthetic rubber and latices. Increases cold resistance, extensibility and release properties of NBR.

TN: Flexricin P-8 (9).

GLYCERINE TRIBUTYRATE

$C_3H_5(OCOC_3H_7)_3$, tributyrine, butyrine. Colourless, oily liquid, m.p. $-75°C$, b.p. $305–310°C$, s.g. $1·035$. Soluble in alcohol and ether. Softener.

GLYCERINE TRIPROPIONATE

$C_3H_5(OCOC_2H_5)_3$, tripropionine, m.p. $-50°C$, b.p. (20 mm) $177–182°C$, s.g. $(20°C)$ $1·078$. Solubility in water $0·3\%$. Softener.

GLYOXAL

$$O=C-H$$
$$|$$
$$O=C-H$$

Biformal, oxaldehyde, ethanedial. White prisms, m.p. $15°C$, b.p. $50·4°C$, s.g. $1·14$. Soluble in water-free solvents. Easily polymerised.

Uses: Synthesis of plastics and synthetic fibres, vulcanisation retarder.

GOLD RUBBER

Thin, transparent rubber sheets with embedded aluminium powder. Produced by evaporation of latex and painting the surface with a suspension of aluminium powder in a solution of crêpe rubber in benzene; vulcanised by sulphur chloride, light reflected through the aluminium powder imparts a 'gold' shimmer. Not produced commercially.

GOLF BALL

Since 1848 compact balls of unvulcanised guttapercha have been used, eventually with the addition of some rubber. The modern golf ball was invented by C. Haskell, 1898 (BP 17 554, USP 622 834), and consists of a core bound with a thin, stretched rubber band. The length of the unstretched rubber is approx. 25 m; the stretched rubber around the core is about 150 m long. The ball is then covered with gutta-percha or balata. Recently nylon, vinyl acetate–vinyl chloride polymers or polyethylene have been used as the cover, sometimes being diluted with butyl rubber, styrene/butadiene polymers with a high styrene content, or gutta-percha (BP 494 031/1936, USP 2 722 264/1955). The surface of the ball has small indentations, whose shape and depth control the flight characteristics and distance. The weight is fixed at 45·93 g, and the diameter at 41·20 mm (min) in Britain, and 42·71 mm in the USA.

GOMA, LA

Spanish journal devoted to rubber. First appeared in 1929, published by Guerrero, Barcelona.

GOODRICH FLEXOMETER

Apparatus for determining heat build-up and the appearance of fatigue. The test piece is subjected alternately to preloading and to compression caused by an eccentric. The dynamic, alternating strain occurs at 1800 c/min. The development of heat is measured by a thermoelement; thus increase in temperature, static and dynamic compression and set can be determined. ASTM D 623-45, Method A.

GOODRICH PLASTIMETER
A compression plastimeter with a constant load. A cylindrical test piece 1 cm in height and 1 cm² area is compressed between two plates of 1 cm² area under a fixed load for a specified time, and is then allowed to recover for the same length of time. The plasticity is expressed as the ratio of compressibility and elastic recovery.

$$P = \frac{h_0 - h_2}{h_0 + h_1}$$

where

h_0 is the original height of the test piece

h_1 is the height after compression, and

h_2 is the height 30 sec after removal of the pressure.

GOOD-RITE Group of products of the Goodrich Chemical Co.
Grades:
Erie: thiazyl disulphide compound. Dark brown liquid. Produced by oxidation of a compound of 85% 4:5-dimethyl-2-mercapto-thiazole and 15% 4-ethyl-2-mercaptothiazole. Accelerator with delayed action (obsolete)
GP-233: dioctyl adipate
GP-235: octyl decyl adipate
GP-236: didecyl adipate

GP-261: dioctyl phthalate
GP-265: octyl decyl phthalate
GP-266: didecyl phthalate
Resins: group of styrene-butadiene copolymers with a high styrene content. Uses: fillers, reinforcing and stiffening agents for articles which are extremely hard but of low density
Vultrol: N-nitrosodiphenylamine (14).

GOODYEAR, CHARLES 1800–1860, b. New Haven, Conn., USA. Originally worked in his father's hardware shop. From 1831, after long being aware of the poor quality of existing rubber, which became tacky in hot weather and hard in the cold, he devoted himself, despite the lack of funds and bad health, to improving the quality of rubber, since he was convinced of its value. In 1838 he learned a method of drying rubber articles by dusting with sulphur from Nathaniel Haywood (solarisation of rubber with sulphur, USP 1090, 1839) and patented it. A government order for impregnated post-bags was a failure, however. After constant failures he tried the effect of heat on rubber/sulphur compounds. The production of these compounds was made possible by a machine, invented by Hancock, for mixing rubber. In 1839 he discovered that on prolonged heating the original plastic mass became elastic. He patented the discovery, called metalisation in 1841. The term vulcanisation was coined by William Brockedon, in England, and accepted by Goodyear. In 1837 he patented his acid gas process (USP 240): the treatment of rubber with nitric acid or nitric acid vapour, but soon discovered that the change was superficial. Goodyear also used basic lead carbonate as an accelerator (USP 3663, 1844).

GOODYEAR-HEALY PENDULUM Apparatus for determining resilience of rubber, according to ASTM D 1054.

$$R = \frac{1 - \cos \text{ rebound angle}}{1 - \cos \text{ angle of drop}} \times 100$$

GORDON PLASTICATOR Mastication machine used for plasticising rubber; in principle a large extruder in which the rubber is plasticised between a spiral screw and the walls of the machine, at 150°C or above, and then ejected from the machine as a tube. Once the tubes have been split open, the sheets are ready for processing. The degree of plasticity may be controlled by altering the speed of the screw. Screw diameter 380–500 mm, driving power up to 500 hp, capacity up to 5000 kg/h.

GOST USSR standards.

GOUGH–JOULE EFFECT The increase in modulus (stiffness) of strained rubber which occurs on increasing the temperature. If a freely hanging rubber strip is subjected to a stress under load, it retracts when heated; unstressed rubber, on the other hand, stretches on heating. At constant deformation, the tension of the rubber increases. This effect was discovered by Gough, in 1805, with raw natural rubber (*Mem. Lit. Phil. Soc.*, Manch., 2nd series, 1805, 1, 288) by holding rubber between his lips he discovered that it grows warm on stretching and cold on retracting. This observation was confirmed by Joule (*Trans. Roy. Soc., London,*

1859, 149, 91) by taking exact measurements with vulcanised rubber. The effect can be demonstrated with W. B. Wiegand's pendulum. The construction consists of a pendulum, loaded both above and below the suspension point and fastened to a stand by a piece of stretched rubber (100–200%). As the pendulum swings outwards, the rubber band is stretched further. When heated by a specially designed source of light rays, the rubber band contracts and the pendulum swings to zero and beyond, when the band immediately stretches again. On swinging out again, the band once more comes into the line of the light ray, contracts because of the warmth and the process is repeated. In the middle position of the pendulum, the light ray is screened and the rubber band is

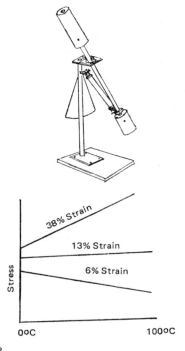

cooled by an air stream. In principle, the pendulum is a periodic working machine producing heat on the basis of the Carnot cycle. The extent of the effect depends on the degree of elongation. At minimal elongation, the effect is zero.

The effect can play a part in the construction of mechanical goods. Rubber springs become stiffer in warm weather; the rubber parts around rotating axles grow warm through friction, and contract, possibly causing a blockage.

GPF Abbreviation for general purpose furnace.

GR-A Abbreviation for government rubber acrylonitrile. Former term (USA) for nitrile rubber.

GRAFT POLYMERS Copolymer comprised of a homologous main chain of one component and a subsidiary chain of a second.

```
A—A—A—A—A—A—A—A—
    |              |
B—B—B          B—B—B
```

Graft polymers can be produced in latex form from natural rubber, *e.g.* by the addition of acrylonitrile or styrene to give subsidiary chains, polymerisation depending on peroxide catalysts. Graft polymers have good processing properties. The graft polymers of natural rubber are called Heveaplus.

GRANULAR POLYMERISATION Simple polymerisation process for the production of polyacrylates. The monomer is polymerised with water (approx. 8:1) and a catalyst in a mixer and is precipitated as a pure, granular polymer.

GRAPHTHOL COLOUR PIGMENTS Type Vulco. Group of colour pigments for use in rubber; free of ionic copper and manganese. *Quantity:* 1 % (421).

GRASSELERATOR Group of accelerators originally produced by the Grasselli Chemical Co., Rubber Service Department, New York; manufacture has been taken over by du Pont, and the trade name is obsolete.

Grades:
101: acetaldehyde-ammonia
102: hexamethylene tetramine
327: undisclosed compound
552: piperidine pentamethylene dithiocarbamate (Accelerator 552)
808: butyraldehyde-aniline condensation product (Accelerator 808)
833: butyraldehyde-butylamine condensation product (Accelerator 833).

GREASY SHEET Tacky sheet rubber; feels either fatty or damp to the touch, due to hygroscopic constituents of the serum, and is the result of insufficient washing or rinsing.

GREEN BOOK Colloquial term for a booklet produced by the Rubber Manufacturers Association, New York, describing the types and forms of packing for natural rubber. Since July, 1962, it has appeared under the title, *International Standards of Quality and Packing for Natural Rubber Grades.* Formerly published as *Type Descriptions and Packing Specifications for Natural Rubber Grades used in International Trade.*

GREEN COMPOUND Unvulcanised compound.

GREEN TYRE Built tyres which have not been vulcanised.

G REVERTEX Graft polymer of natural rubber with polymethylmethacrylate in latex form with 60% total solids.
Grades:
10:   10% methacrylate
K 40: 40% methacrylate.

GR-I Abbreviation for government rubber isobutylene. Former term (USA) for butyl rubber.

GRILAMID Swiss polyamide 12.

GR-M Abbreviation for government rubber monovinylacetylene. Former term (USA) for neoprene.

GRP Abbreviation for glass reinforced plastics.

GR-P Abbreviation for government rubber polysulphide. Term (USA) for a thiokol rubber produced from a mixture of ethylene and propylene dichlorides. Production was started in accordance with the Baruch Committee plan at the beginning of World War II, but not continued.

GR-S Abbreviation for government rubber styrene; later the term SBR (styrene-butadiene rubber) was used. Most of the types are produced from various combinations of catalysts, modifiers, stabilisers, emulsifiers, reaction temperatures and styrene:butadiene ratios. GR-S, with 78% butadiene and 22% styrene was the most important synthetic rubber product of the Allies during World War II.

GR-S-X Production of GR-S was formerly controlled by the US Government; test samples produced under widely differing conditions were denoted by the letter X, and a number. The test resulted in improved types of GR-S and latex.

GT-25 Light brown flakes, m.p. 66°C, s.g. 1·30. Peptiser and activator for neoprene GT.

GUANIDINE

$$HN{=}C\Big\langle {}^{NH_2}_{NH_2}$$

Iminourea. White crystals, m.p. 50°C. Soluble in alcohol and water. Strongly basic.
Uses: production of accelerators and plastics, stabiliser for p.v.c. resins.

GUAYULE Rubber from *Parthenium argentatum*.

GUMA Resinous Guayule rubber modified with maleic acid.

GUMMI German term for vulcanised natural and synthetic rubber. Term originates from the mistaken idea that rubber comes from a plant gum, *e.g.* Gum arabic.

GUMMIDUR Active zinc oxide of fine particle size.

GUMMI ELASTICUM Former term for rubber.

GUMMI OPTIMUM The earliest term used for rubber; recorded by P. M. d'Anghiera, 1530.

GUTHRIE MACHINE Semiautomatic machine with four (Cadet

type) or five (Aristo type) pairs of rolls; used to roll out sheet rubber. The roll pairs have the same nip setting, but run at increasing speed to account for the elongation of the coagulum. The rolls in each pair run at an even speed, and both compress and give a smooth finish to the coagulum. The last pair, which is ribbed, impresses the characteristic pattern.

GUTTA Abbreviation for gutta-percha.

GUTTA-PERCHA $(C_5H_8)_n$, from the Malayan getah pertja (getah: sticky plant sap, rubber; pertja: blocks, lumps). Hard, thermoplastic isomer of rubber with a trans structure; polymorphous and exists in a stable, α, and unstable, β form, with a crystalline structure in its normal state, softening p. 60–100°C. Soluble in aromatic solvents and chlorinated hydrocarbons when cold, in paraffinic hydrocarbons when warm. Contains 70–80% rubber hydrocarbon and 20–30% resin (alban and fluavil). Obtained from the sap of trees of the *Sapotaceae* family (E. Asia). The main sources are: *Palaquium acuminatum Burck* (Sumatra), *Pal. Clarkeanum King et Gamble* (Malaya), *Pal. gutta Burck* (*Isonandra gutta Hooker*) (Singapore) extinct, *Pal. gutta Burck* (*var. sessiliflora*) (Borneo), *Pal. borneense Burck* (Borneo), *Pal. leiocarpum Boerl.* (Borneo), *Pal. oblongifolium* (Borneo), *Payena Havilandi King et Gamble* (Malaya), *Payena Leerii Kurz* (Burma, Malaya, Sumatra), *Payena stipularis Burck* (Sumatra, Celebes).

A number of other species yield gutta-percha or similar products of a poorer quality. Most gutta-percha on the market is from *Pal. oblongifolium* and *borneense*.

The first sample of gutta-percha was brought to Europe by the English traveller, John Tradescant, in the middle of the seventeenth century, and is mentioned in his son's book (J. Tradescant, 'A Collection of Rarities Preserved at Lambeth Bridge near London', London, 1656) as a type of wood—'mazer wood'—which became rubbery in warm water. The product was forgotten as a curiosity, but rediscovered, independently, by J. d'Almeida and W. Montgomerie in 1843. Samples sent to the Royal Society of Arts in London caused great excitement. In 1845 W. Siemens suggested its use as an insulating material for telegraph cables. In 1847 the first underwater cable was produced and laid as an experiment two years later by W. Breit at Folkstone. By 1845 Lagrenee had large quantities of gutta-percha at his disposal in Paris and in 1846 Alexandre, Carbriol and Duclos took out the first patent for its use. Gutta-percha was originally obtained from trees, growing wild, by felling them. *Isonandra gutta*, which grew in large quantities in Singapore, was virtually extinct by 1857. The various types were often differentiated by naming them after their port of trade, *e.g.* Pahang, Banca, Bulongan, Sarawak, Padang, Siak, Bagangulie, Sundie, getah taban merah, Bandjar putih. In 1885 the Dutch Government founded the first gutta-percha plantation in Java, with approx. 120 ha, which increased considerably within 10 years. *Pal. oblongifolium* and *obovatum* were used for planting, as well as other hybrids of the species.

The Tjipetir and Selbourne plantations in Java and Malaya are the only cultivated gutta-percha plantations left today. The plants are kept in bush form by periodic pruning. Gutta-percha is extracted from the leaves after picking. The first attempt at extraction was made by Jungfleisch in 1888. In 1892, Rigole suggested the use of carbon disulphide (a method which he patented). Serullas (1896) used toluene as an extracting agent and then precipitated the gutta-percha with acetone. The varying degrees of solubility of gutta-percha in warm and cold petroleum ether were used by Obach. This type of gutta is green in colour because of the chlorophyl extracted at the same time (Gutta verde). The Ledeboer process was an attempt to separate gutta-percha mechanically. The gutta-percha content of the leaves is 1·8–2·3%, young twigs contain 0·7% and the bark approx. 1·2%. It is obtained today by mechanical crushing of the leaves and extraction with hot water. Today, gutta-percha has been superseded by plastics in the cable industry, apart from repair of existing cables; it is used in the production of golf balls and for temporary teeth fillings.

GUTTA-PERCHA, ABYSSINIAN Substance similar to gutta-percha, but of poorer quality; of unknown origin. Tacky, acidic, becomes rubbery on heating.

GUTTA TABAN CHAIA Local term for gutta-percha from *Palaquium polyanthum* (Burma). Low quality product with approx. 50% gutta-percha and 45% resin.

H

H-50D Aniline-acetaldehyde reaction product (6).

HA Abbreviation for heptaldehyde-aniline condensate.

HAF Abbreviation for high abrasion furnace.

HAFTMITTEL 7110 Colourless, viscous fluid, s.g. 1·23, pH approx. 7·5. An adhesive obtained by thermal decomposition of formaldehyde. By mixing with resorcinol and reinforcing silica, the adhesive may be used, for example, in treating fabrics or cords. Dosage for optimum adhesion: 2·5% Haftmittel 7110, 2·5% resorcinol, 20–30% Vulkasil S (43).

HAIRLOCK PROCESS Process used to bind loose fibres, *e.g.* coconut, sisal, horsehair, with latex, into rubbery cushions and lightweight material which keeps its shape.

HAKUENKA Group of Japanese produced activated calcium carbonates. Reinforcing fillers.

HALANE Dichlorodimethylhydantoin. Vulcanisation retarder (299).

HALAR Copolymer of ethylene and chlortrifluorethylene (29).

HA LATEX Latex with a high ammonia content (0·8%).

HALOWAX Group of chlorinated naphthalenes (137).

HANCOCK, THOMAS 1786–1865. A blacksmith by trade; he was introduced to Mackintosh and became interested in the use of rubber for waterproof materials, elastic threads and balls. Noticed that freshly cut sheets of rubber adhere to one another when pressed together. Discovered the process of mastication in 1824 and constructed the first masticator, which consisted of a roll, covered with spikes, which rotated in a chamber. The original idea of this machine was to cut the rubber into small pieces, but it produced a dough-like mass because of the heat which was developed. He recognised the importance of this in the shaping of rubber. After experimenting with vulcanised rubber, he modified Goodyear's process independently, and patented vulcanisation in 1843 [BP 9952 (1843)]. Vulcanisation was achieved by mixing rubber with certain ingredients and sulphur and heating in an autoclave; alternatively the rubber was dipped in molten sulphur. He discovered that prolonged heating in molten sulphur resulted in dark ebonite, and called the process vulcanisation after the god of fire and the smithy art, Vulcanus. An inquiry, later conducted in the USA, recognised Goodyear as the discoverer of vulcanisation.
Lit.: *Personal Narrative of the Origin and Progress of the Caoutchouc or India Rubber Manufacture in England*, London, 1857.

HANCOCK MEDAL Awarded by the Institution of the Rubber Industry, London, for exceptional work in the rubber field, not necessarily of a technical nature.

HANCORNIA SPECIOSA Mangabeira tree. An *Apocynaceae* (widely spread in Brazil). Gives poor quality rubber known as Mangabeira.

HANGKANG Djongkang, angso. Local term for *Palaquium leiocarpum* Boerl and the guttapercha obtained from it (Borneo). Marketed in large bales, dark-brown on the surface, and grey to reddish inside; the product is hard and brittle. Contains approx. 39% water, 0.3% impurities, $14–15\%$ pure guttapercha and up to 50% resin. Is frequently used to adulterate guttapercha of a higher quality.

HARCRUMB Trade name for rubber powder (169).

HARDNESS The relative resistance of rubber to the insertion of a blunt object; depth of indentation is determined using a spherical or conical indenter. Usually expressed as depth of indentation or as an arbitrary unit. Determination of hardness is related to measurement of the modulus of elasticity under slight compression, usually,

$$H = k \frac{L^x}{D^y}$$

where
L is the load
D is the diameter of the indentor, and
k is a constant, involving the modulus of elasticity.

For a cylindrical indentor

$$x = y = 1$$

For a spherical indentor with an indentation of less than $0.8D$

$$x = 0.75$$

and

$$y = 0.5$$

A number of methods and instruments have been evolved for the assessment of hardness and are enumerated in the following standards:

Soft rubber. ASTM Hardness ASTM D 314-58. BS Hardness (identical with International Rubber Hardness) BS 903: 19: 1950, BS 903: A7: 1957, ASTM D 1415-56T, Wallace measure of hardness. Pusey-Jones plastimeter ASTM D 531-56. Shore Hardness A (identical with Rex Hardness, general durometer hardness) ASTM D 676, BS 903: 20: 1950, DIN 53 505. Softness DIN 53 503. Instrument with a constant indentation (after Buist and Kennedy) BP 617 465.

Hard rubber. Rockwell Hardness ASTM D 530-60 T. Shore Hardness D ASTM D 1484-59, DIN 53 505.

The following points help ascertain the hardness of a compound based on rubber, fillers and softeners:

For 100 parts polymer	Base hardness (gum stock)
NR, cold SBR	40
Hot SBR	37
Oil-extended cold SBR (25% oil)	31
Oil-extended cold SBR (37·5% oil)	26
Butyl	35
NBR, neoprene	44

Fillers, Softeners,	Change of hardness
Channel blacks, HAF, FEF	+ 0·5 per part
ISAF	+ 0·5 per part + 2
SAF	+ 0·5 per part + 4
SRF	+ 0·33 per part
Thermal blacks, clays	+ 0·25 per part
Whiting	+ 1 per 7 parts
Factice, mineral rubber	− 0·2 per part
Liquid softener	− 0·5 per part

According to Philback and the Philprene formulary.

HARD RUBBER Ebonite, vulcanite, whalebone caoutchouc. Dark, hard, tough vulcanisate with a high sulphur content and viscoelastic (thermoplastic) properties above 80°C. The first patent was taken out by Thomas Hancock (1843), who produced the first hard product by heating rubber in molten sulphur. Nelson Goodyear claimed the priority for himself in 1851. According to his process rubber is mixed with sulphur. Most probably Goodyear already knew of hard rubber in 1831. Hard rubber was originally conceived as a substitute for ebony, hence the name ebonite. It has a sulphur content of 25–47%, whereas soft rubber has a max. of 5%. Vulcanisates with a sulphur content of 5–25% are inferior in quality. The formula $(C_5H_8S)_n$ with a theoretical sulphur content of 32% was postulated. Additional sulphur can probably enter into the molecules by substitution of hydrogen atoms, when hydrogen is liberated as hydrogen sulphide. The properties of hard rubber improve with increase in sulphur content.

Hard rubber can be produced from rubber/sulphur mixtures alone or from a normal compound, *e.g.* from natural rubber, SBR, nitrile rubber, and neoprene, but not from butyl rubber or thiokol; it can also be made from latex, for use as coverings and linings by spraying and dipping. According to the compound, the vulcanisation period is from 15–30 min to 15 h at 100–165°C. Rubber may be replaced in large quantities by reclaim. Ebonite dust is particularly useful as a filler. Vulcanisation can be carried out in a press, in hot air, or in hot water. The formation of hard rubber is exothermic; the heat liberated for a 68:32 compound is approx. 300 cal/g. Certain precautions are necessary when vulcanising thick articles, because of the evolution of hydrogen sulphide gas and, possibly, high heat build-up.

Properties: tensile strength 600–800 kg/cm^2, elongation at break 3–8%, elasticity modulus 3×10^4 kg/cm^2, s.g. 1·13–1·18, hardness, Shore A 93°, specific heat 0·30–0·35 cal/g/°C, coefficient of linear expansion 0·8 $\times 10^{-5}$/°C, water absorption max. 0·25%, water diffusion $1·5 \times 10^{-8}$ g/cm^2/h/mm Hg. The most important attributes of hard rubber are its resistance to chemicals and its excellent electrical insulation properties, but it has been supplanted in many, hitherto important applications by cheaper plastics. It withstands temperatures up to 60°C, gives good thermal insulation. The usual rubber solvents cause swelling. The low level of water absorption may be decreased further by using purified rubber, which has a low protein content. Hard rubber may be machined with water or drilling oils as cooling agents.

Uses: include corrosion resistant coverings, battery cases, electrical insulation, tie layers in rubber/metal bonds, hard mechanical goods, combs.

HARFLEX Group of softeners and plasticisers for synthetic elastomers, vinyl resins and other plastics:
40: dibutyl sebacate
50: dioctyl sebacate
60: dihexyl sebacate
80: dicapryl sebacate
90: dibenzyl sebacate
120: diiso-octyl phthalate
140: dibutyl phthalate
150: di-n-octyl phthalate
160: dihexyl phthalate
180: dicapryl phthalate
220: dioctyl adipate
280: dicapryl adipate
300: polyester. Brown liquid, s.g. 1·089 (20°C)
325: polyester. Brown liquid, s.g. 1·097 (25°C)
330: polyester. Brown liquid, s.g. 1·077 (25°C)
375: polyester. Brown liquid, s.g. 1·065 (25°C)
500: polyester, b.p. 238–246°C, s.g. 0·930 (25°C) (210).

HARRIES, CARL DIETRICH 1866–1923. Graduated 1890 at Berlin University; assistant of A. W. von Hoffmann and Emil Fischer. Studied the effects of nitrous gases and ozone on rubber and prepared nitrosites and the ozonides. The method of ozonolysis which he developed remains the best method of determining the basic structure of, for example, the rubber molecule, terpenes, fatty acids, vitamins. His work on the production of dienes, their polymerisation and the structure of polymers hastened the synthesis of rubber. He also discovered the catalytic effect of alkali metals on the polymerisation of butadiene.

HARTOLAN Group of wool wax alcohols. Softeners and emulsifiers (211).

HARUB Trade name for crumb rubber made by the Hevea crumb process.

HARVEX LSF Acryl alkyl amine. Ultra-accelerator with slight delayed action (39).

HAUSER, ERNST A. b. Vienna, 1897–1956. 1932–35 chief chemist at the Semperit Austrian/American Rubber Co. in Vienna; from 1935 professor of chemical technology at the Massachusetts Institute of Technology. His work included the invention of Revertex, studies of mol. wt. distribution and its correlation with physical properties, and several books, among them 'Latex' (1927 Germany; 1930 Britain). Editor of the first German rubber technology handbook (*Handbuch der Gesamten Kautschuktechnologie*, Berlin, 1935, 2 vols.).

HB- Group of hydrocarbon softeners (5).

HC 5 Yellowish-brown resin. Accelerator of undisclosed constitution (obsolete).

HCPC Group of resin emulsion and solutions used as coatings, paints, impregnating agents, tackifiers, neutralisers, flame retardants, extenders and modifiers (212).

HD POLYETHYLENE Abbreviation for high density polyethylene.

HEAT RESISTANCE The ability to withstand permanent changes in physical properties over prolonged periods at high temperatures. The degree of heat resistance depends on the elastomer, the ingredients in the compound, and the conditions in which the article is used. The heat degradation in NR and SR increases logarithmically with temperature and is also dependent on the atmospheric oxygen concentration. To determine heat resistance, the tensile strength, hardness and other properties are measured before and after heat is applied. Test conditions should be as close as possible to those in which the articles will be used. Determination according to: ASTM D 454, D 574, D 865, BS 903: 13: 1950, DIN 53 508.

HEAT SENSITISING By adding certain chemicals, latex compounds may be made heat sensitive.
1. Kaysam process: by adding weakly coagulating electrolytes (ammonium salts, sodium silicofluoride) and zinc oxide or carbonate, compounds are obtained which remain stable at room temperature but which coagulate rapidly at 50–60°C. Used in the casting process. USP 2 153 184 (1938).
2. Polyvinyl methyl ether: soluble in water at room temperature but becomes insoluble at 35–40°C and is precipitated; by adding this to latex compounds which have been treated with formaldehyde, compounds are obtained which remain stable at room temperature for a long period, but which coagulate spontaneously at 35°C. Used in casting and dipping processes. DP 869 861 (1940).
3. Trypsin process: concentrated latex, the proteins of which are decomposed by trypsin, is made sensitive to heat by adding zinc oxide to the compound. Compounds are stable at room temperature for 1–2 h, but coagulate within a few minutes at approx. 50°C. Used in the casting process.
4. 2-Mercaptobenzimidazole: 0·5–1% addition gives a compound stable for a few hours at room temperature. Coagulates rapidly. Used in the casting process.

HECTORITE Hydrothermally modified ash of the montmorillonite group, similar to bentonite but with magnesium instead of aluminium in the structure (magnesium bentonite). Has strongly thixotropic properties. *Uses:* thickening agent, stabiliser, dispersing agent for aqueous systems and latices.

HELASTINE F Trade name for diphenyl guanidine.

HELIOZONE Blend of petroleum waxes, softening p. 73°C, s.g. 0·90. Protective agent against light for natural rubber, SBR, NBR, neoprene and latices under static conditions. Does not affect ageing properties, heat ageing, flex cracking; improves the surface appearance of vulcanisates.
Quantity: 0·5–2% (6).

HEPTAC Mixture of 2-mercaptobenzthiazole and hexamethylene tetramine (407).

HEPTALDEHYDE-ANILINE ($C_6H_5N=CH-C_6H_{13})_x$; condensation product. Dark-brown liquid, s.g. 0·94. Soluble in acetone, benzene and chlorinated hydrocarbons, slightly soluble in benzine, insoluble in water. Ultra-accelerator. Gives safe processing. May be retarded by channel blacks, clays and acidic softeners. Activated by basic substances such as DPG, DOTG and ammonia. Good secondary accelerator with thiazoles and thiurams. Zinc oxide and sulphur are necessary in usual quantities, however, it vulcanises gum stocks with sulphur alone. Fatty acids, though not essential may be used. Has a plasticising effect on NR, gives a high modulus. Causes slight discoloration in light-coloured articles which can be masked by using strong white pigments.
Quantity: 0·12–0·35% and 2·25–3%

sulphur, hot air vulcanisation 0·4% with 1% DPG and 1–5% sulphur. *Uses:* gum stocks, hot air vulcanisates (boots), air hose, white side walls for tyres.
TN: A 20 (5)
Heptene base (23)
Vulcaid 444 (3).

n-HEPTANE $CH_3(CH_2)_5CH_3$. Volatile liquid, b.p. 98·4°C, s.g. 0·684 (20°C). Soluble in alcohol and chloroform. Solvent for NR and SR, with the exception of NBR.

HEPTENE Blend of one part Heptene base (heptaldehyde-aniline) and four parts mineral oil (23).

HEPTYL DIPHENYLAMINE Mixture of mono- and diheptyl diphenylamine. Brown liquid, s.g. 0·97. Soluble in alcohol, chloroform, benzene and carbon disulphide, insoluble in water. General purpose antioxidant for natural and synthetic rubber, particularly effective in neoprene. Non-discolouring.
TN: Agerite Stalite (51)
Agerite Gel (blend with a wax, m.p. approx. 65°C, s.g. 0·97) (51).

HERANDRA Local term for a wild rubber of the *Mascarenhasia* type (Madagascar).

HERAX Group of Czechoslovakian produced accelerators (428):
N: mixture of 2 parts tetramethyl thiuram disulphide and 1 part 2-mercaptobenzthiazole. For butyl rubber compounds
UTS: activated hexamethylene teramine. Basic secondary accelerator for use with MBT, CBS and MBTS. Can be used with DPG, but has somewhat lower activity.

HERCLOR Homo- and co-polymers of epichlorhydrin, fully saturated polymers with excellent resistance to ageing and oxidation, and with resistance to hydrocarbon solvents.
H: polymer of epichlorhydrin
C: co-polymer with ethylene oxide. (Hercules Powder Co.)

HERCOFLEX Group of softeners.
Grades:
150: octyl decyl phthalate
200: diiso-octyl phthalate
250: 50:50 mixture of dioctyl phthalate and dioctyl decyl phthalate
290: octyl decyl adipate
600: pentaerythritol fatty acid ester, s.g. 0·994
610: pentaerythritol fatty acid ester, s.g. 0·997
900: polyester, b.p. 197°C (4 mm), s.g. 1·23 (10).

HERCOLYN Hydrogenated methyl abietate. Light-brown liquid with slight smell of turpentine, s.g. 1·02. Softener and tackifier for natural rubber, reclaim, SBR, polychloroprene and latices, and butyl rubber (10).

HERECOL Trade name for a group of modified butadiene copolymers (213).

HERMAT Group of accelerators (428).
Grades:
Cu:　　　copper dimethyl dithiocarbamate
FEDK:　zinc ethyl phenyl dithiocarbamate
TMT:　　tetramethyl thiuram disulphide
ZDK:　　zinc diethyl dithiocarbamate
ZDM:　　zinc dimethyl dithiocarbamate
ZnMBT: zinc-2-mercaptobenzthiazole.

HERNANDEZ, FRANCISCO In *Rerum Medicarum Novae Hispaniae Thesaurus*, written in 1570, Rome, 1649, pp. 50–1, he describes the Mexican rubber tree (ule tree) (*Castilla elastica*).

HERREIRA, ANTONIO 1549–1625. Described Columbus's journey, mentioning a game played with rubber balls by the natives of Haiti. Also refers to trees in Mexico which, on incision, gave a sap from which an elastic substance could be produced.

HERRON Group of softeners and processing aids based on petroleum products.
Grades:
-H-T: aromatic hydrocarbons
Plas.: naphthenic hydrocarbons
SPO: processing oil
Wax: (also −6) paraffin wax (89).

HETEROPOLYMER Polymer consisting of a mixture of various polymer types, *e.g.* two or more co-polymers, or copolymers with monopolymers.

HEVEA Branch of the Euphorbiaceae family, the name is derived from the local term, Hhévé. Grows in the province of Esmeraldas (Ecuador) and was described by La Condamine in 1745 in his report on the rubber tree. The term was adopted by Aublet in his description of nomenclature for *Hevea Peruviana* (*H. Guianensis*).

HEVEA ACID An acid discovered by G. S. Whitby in the acetone extract of raw rubber, and later identified by him as impure stearic acid.
Lit.: *Trans. IRI*, 1925, **1**, 12.
India Rubb. J., 1925, **70**, 382.
J. Chem. Soc., 1926, 1448.

HEVEA BENTHAMIANA Species with a white latex (N.W. Amazon valley, upper tributary of the Orinoco). Crossed with *H. Brasiliensis* to cultivate disease resistant clones (*Dothidella ulei*).

HEVEA BRASILIENSIS MULL. ARG. Discovered by Fresneau in 1747 in French Guiana. *H. Guaianensis*, discovered by Aublet in 1762 was at first thought to be identical with the *H. Brasiliensis*, but was later termed *Jatropha elastica* by Linee. *Hevea* grows in a warm, humid climate, 10° North and South, up to 600 m above sea-level and needs a yearly rainfall of at least 170–250 mm. Occurs naturally on the banks of the Amazon and its tributaries, at a density of 8–10 trees per ha. There are two types:
1. Var. *latifolia*. Has broad leaves, a white bark and little branch growth on the trunk (seringueira branca). Grows close to rivers.
2. Var. *angustifolia*. Has narrower leaves and a thick, soft, dark bark (seringueira preta), the trunk is fairly heavily intersected with branches and reaches a greater height (30 m). Grows on slightly higher ground.
The second variety yields a better rubber, has a higher production rate, and was brought to the east by Wickham in 1867. The Brazilian wild rubber was obtained between January and June. In the tapping areas (seringal) which stretched up to 100 km², every 100 trees were linked together by paths (estrada). Two paths were tapped by each worker on alternate days, a tapping system which has been found successful even on the plantations. To tap the tree an incision was made into the bark. The so-called Para method was used to produce rubber. A spade-shaped piece of wood was covered with latex and this was dried with smoke. This was repeated until the single layers formed bales of 10–50 kg weight. The smoke was produced by burning wood and a type of palm kernel (*Attalea speciosa, Maximiliana regia, Orbiguya Martiana*). The bales were a dark-brown colour, amber towards the middle, and composed of millimetre-thick layers. The rubber was called 'para fin' or 'boracha fina'. The types which had not been smoked so well were called 'para entrefin' or 'boracha entrefina'. The scraps coagulated on the trees and together with other wastes were kneaded, unsmoked, into bales; this product being known as Sernamby or Nigroheads. The annual yield was approx. 2–4 kg for approx. 100–120 tappings.
Plantation industry. The significance of *Hevea* for the plantation industry was first mentioned in 1869 by James Collins, the curator of the Museum of the Pharmaceutical Society, London. In 1870 Sir Clements Marckham of the India Office, Sir Joseph Hooker, director of the Botanical Gardens at Kew, and Sir Dietrich Brandis, director of the Forest Department, showed interest in bringing *Hevea* to the east.
1873: under the authority of the India Office, Collins ordered several hundred seeds to be brought to

England; these were transported by a Mr Farris. Twelve of the seeds germinated at Kew Gardens, six were propagated by shoots and six were transported to Calcutta, but without success.

1875: Wickham sent several seeds to Kew; none survived.

1875: Kew received some seeds *via* the India Office from an unknown source. Of these, 378 were sent to Calcutta; none survived.

1876: 70 000 seeds were received from Wickham, approx. 2700 grew to seedlings; of these 1919 were sent to Ceylon, 18 to Java (in 1876) and four more to Java (1877). Some sent to Singapore and Perak in 1876 did not arrive.

1876: 100 plants were sent from an unknown source to Ceylon.

1876: 1090 plants were brought by R. Cross; 400 were kept at Kew, of which only 3 % survived, 680 went to William Bull, of which 14 survived. 100 plants were sent to Ceylon.

1877: 100 plants were sent to Ceylon.

1877: 100 plants were sent to India and Burma.

1877: Ceylon sent 22 plants to Singapore, which bore seeds for the first time in 1881; these were distributed in Malaya and Borneo.

1893: 90 000 seeds were distributed in Ceylon. The trees in Ceylon flowered for the first time in 1881; in 1882 36 seeds were obtained; 1883, 260 seeds; 1884, over 1000 seeds. In 1886 and 1889 seeds were sent to Buitenzorg (Java), in 1887 to the Fiji Islands, 1888 to Borneo and German E. Africa. 1901 to Sumatra and the Gold Coast. The trees existing today in the east are the result of the relatively few plants sent out and cultivated there. Initially only the seeds of wild rubber trees were cultivated and the resulting trees had a low average yearly production, and varied enormously. Around 1918 vegetative propagation from specially chosen mother plants began, and was augmented by artificial pollination, both to increase the production rate and to select characteristics, such as resistance to disease and to wind breakage, good bark regeneration, and white latex. The former production rate of 300–600 kg per year (180–250 trees per ha.) was increased to 1000–2000 kg. Some modern clones produce at the high annual rate of 3000 kg. *Flower biology. Hevea Brasiliensis* is, like many other *Euphorbiaceae*, self-fertilising, *e.g.* the same tree carries both male and female flowers. Blossom time occurs at the same time as the leaf change. The stem of the bloom is a panicle, on which the female flower occurs at the end, while the more copious male flowers are found closer to the branch of the panicle. Cross-fertilisation carried out by insects is essential for the growth of the fruit.

HEVEACRUMB NR produced in granular form by a mechanico-chemical process, using castor oil as a separating agent for the damp granules. The dried rubber is compressed into compact bales. Properties as RSS (Process of the Rubber Research Institute of Malaysia).

HEVEA GUIANENSIS *H. Peruviana.* The most widely spread, and probably the oldest species of *Hevea*, which was tapped particularly in Colombia for its wild rubber yield. Described by Aublet in 1775.

HEVEA PLUS Graft polymer of natural rubber with methyl methacrylate or styrene, or blends of rubber with polymerised monomer produced from latex.

Grades:

M-G: graft polymer of latex and monomeric methyl methacrylate, polymerised with butyl peroxide and tetraethylene pentamine

M-M: blend of rubber with polymethyl methacrylate, obtained by common coagulation of the latices

S-G: graft polymer with styrene.

The index number shows the amount of other polymers present, *e.g.* M-G 30, M-G 49.

With a polymer content of less than 50%, the products are relatively hard sheets which soften at 70°C and can be processed and vulcanised like normal rubber. Products with a higher polymer content fall into crumb form.

Production: concentrated latex is diluted 1:1 with an aqueous dispersion of monomer methyl methacrylate (the stabiliser having been washed out), together with 0·2% (on the rubber content) tert-butyl hydroperoxide. Finally, 0·2% tetraethylene pentamine is added as a 10% solution. After polymerisation (for approx. 2 h) the polymer is coagulated by adding 3 parts v/v 0·1% formic acid.

The self-reinforcing graft polymers are exceptional in their physical properties at a high hardness level (90° BS), their good flow properties while being shaped, and their excellent electrical properties. They have the resilience of rubber plus the hardness of ebonite or hard plastics. Because of the omission of fillers, stiff, translucent products in various colours can be made from these materials. M-G 23, for example, as a gum compound has similar properties to an HAF tread compound, but has a considerably better tear resistance (De Mattia test). M-G latices containing 4–10% polymethyl methacrylate are particularly suitable for foam rubber, because they allow considerable compression without loss of stiffness or load capacity.

Lit.: *NRPRA Bull.* (1).

HEVEA SEED OIL Pale yellow drying oil contained in the kernels of *Hevea Brasiliensis* at approx. 38–47%; s.g. 0·918, iodine No. 127, saponification value 192, Hehner No. 95·1, RM No. 0. Not a commercial crop, as the seed has to be processed while still fresh, stored seeds containing up to 25% free fatty acids.

Uses: possible substitute for linseed oil, softener for rubber.

Lit.: *Analyst,* **56,** 1931, 61–2.

HEVEA SPRUCEANA A species found in the Amazon swamps; has a low rubber content of poor quality and a watery latex. Hybrids with *Hevea Brasiliensis* are highly disease resistant.

HEVEENE Liquid hydrocarbon, b.p. 315°C. Causes a fairly heavy sediment in the dry distillation of raw rubber. Name given by A. Bouchardat.

HEVEIN The protein fraction of *Hevea* latex; characterised by 5% sulphur content and low mol. wt., approx. 10 000. Soluble in water at all levels of pH, isoelectric point 4·5. Sulphur is present as cystine. Contains 15 amino acids, but not methionic or cysteine. Probably has no influence on the colloidal properties of latex.

Lit.: Archer, *Biochem. J.*, 1960, **75**, 236.

HEXACHLOROBENZENE

C_6Cl_6. Colourless prisms, m.p. 227°C, b.p. 326°C, s.g. 2·0. During the production of the gamma isomer used as an insecticide, large quantities of alpha and beta isomers accumulate. Up to 10% of these can be used as diluents for natural and synthetic rubbers; in hard compounds up to 20%. The alpha isomer is more compatible with rubber than the beta isomer. BP 745 096.

HEXACHLOROCYCLO-PENTADIENE Substitute for sulphur as a rubber vulcanising agent. For soft articles 1–5% is sufficient, but the quantity may be increased up to 200% to give products of increasing hardness and stiffness. BP 725 634, 732 948, USP 2 732 362, 2 732 409 (1956).

HEXACHLORONAPH-THALENE $C_{10}H_2Cl_6$, m.p. 136–139°C. Antioxidant.

TN: Halowax No. 1014 (137).

HEXACHLOROPHENE

2:2′-methylene bis-3:4:6-trichlorophenol, 2:2′-dihydroxy-3:3′:5:5′:6:-6′-hexachlorodiphenylmethane. Crystalline powder, m.p. 164–165°C. Soluble in alcohol, acetone, ether, chloroform, propylene glycol, vegetable oils and dilute alkalis, insoluble in water.

Uses: preservative for rubber and latex (potassium salt), phenol coefficient of the monopotassium salt 125.

TN: G-11
AT-7

Hexosan
Gamophen
Exofene.

HEXALIN Trade name for cyclohexanol (6).

HEXAMETHYLENE AMMONIUM-HEXAMETHYLENE DITHIOCARBAMATE

$(CH_2)_6N.CS.S—H_2N(CH_2)_6$. Colourless powder. Ultra-fast accelerator (obsolete).

TN: + Latac (6).

HEXAMETHYLENE DIAMINE CARBAMATE HMDA carbamate, HMDAC. Vulcanising agent for fluorocarbon polymers and polyacryl rubber.

TN: DIAK No. 1 (6).

1:6-HEXAMETHYLENE DIISOCYANATE

$OCN(CH_2)_6NCO$, 1:6-hexane diisocyanate, m.p. −67°C, b.p. 127°C (10 mm), s.g. 1·046 (20°C).

Uses: production of polyurethane (Vulcaprene); little used because it is highly toxic and irritates the mucosa.

TN: Desmodur H (43)
Mondur HX (263).

HEXAMETHYLENE TETRAMINE

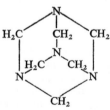

White, rhombic crystals or slightly hygroscopic powder, sublimes at 270°C without melting, s.g. 1·3. Soluble in water, chloroform, alcohol, sparingly soluble accelerator for slow cures, used primarily to boost mercapto and thiuram accelerators. The vulcanisates have a characteristic odour. Zinc oxide is necessary. Gives a low modulus and unsatisfactory ageing properties. An antioxidant is necessary.

Quantity: primary (for thick walled articles) 0·6–0·8% with 3–4% sulphur, secondary, 0·1–0·35% with 0·25–1% mercapto or thiuram accelerator and 2–3·5% sulphur.

TN: Accelerator H (6408)
Aceto HMT (25)
Aminoform
Eveite UR (59)
Formaldehyde-ammonia
Formin
+ Grasselerator 102
Hexa
Hexamine
Hexamine tetramine (215)
Nocceler H (274)
Urotropin
Vulkacit H (43)
White salt.

HEXAMETHYL PHOSPHORIC ACID TRIAMIDE Clear liquid. Ultra-violet stabiliser in vinyl plastics.
TN: HPT.

n-HEXANE　$CH_3(CH_2)_4CH_3$. Colourless, volatile liquid, b.p. 69°C, s.g. 0·66 (20°C). Miscible with alcohol and chloroform. Solvent for NR and SR with the exception of NBR.

1:2:6-HEXANETRIOL
$HO.CH_2CH(OH)CH_2.CH_2$
$.CH_2CH_2.OH.$
Viscous liquid, b.p. 178°C (5 mm), s.g. 1·106. Miscible with alcohol and water, trihydric alcohol.
Uses: polyester, polyurethane elastomers, softener.

HEXAPLAS Softener for butadiene-acrylonitrile and isobutylene-isoprene copolymers, neoprene, vinyl polymers and copolymers.
Grades:
BUT: polybutylene adipate. Plasticiser for nitrile and chloroprene rubbers
LMV: modified polypropylene adipate. Plasticiser for nitrile, chloroprene and butyl rubbers
PPA: polypropylene adipate
PPL: modified polypropylene adipate. Plasticiser for nitrile, chloroprene and butyl rubbers, and for vinyl chloride polymers
PPS: modified polypropylene adipate. Pale brown liquid, s.g. 1·05. Soluble in esters, ketones, aromatic hydrocarbons and chlorinated hydrocarbons. Non-extractable, does not tend to migrate.
Quantity: up to 30% (60).

HF LATEX Abbreviation for hydrazine-formaldehyde latex. Reaction of both components in latex gives vulcanisates with a high modulus and improved tear resistance.

	HF latex 4–8%
60% NR latex, 0·2% NH_3	10 000
KOH, 20% solution	30–60
Hydrazine hydrate	215–430
Formaldehyde	311–622

Five minutes after addition of the constituents, the latex is brought to pH 9·5 by adding potassium hydroxide. The latex may be stored for an unlimited period.

HIBITITE Acid derivative of a reaction product of a diaryl substituted thiourea and an aldehyde-amine. Acidic retarder (5).

HIBITITE L Reaction product of mercaptobenzthiazole and ethylene oxide (5).

HI-FAX Polyethylene produced by the low pressure process (10).

HIGH FREQUENCY HEATING The heating of materials through dielectric losses in a high frequency field; the material is placed between two electrodes and thus absorbs energy.

HIKATOL Japanese produced thioplasts.

HI-SIL Group of precipitated, hydrated silicon dioxides, s.g. 1·95. Reinforcing white filler (134).

	101	233	X303
Particle size, micron	0·03	0·022	0·022
BET surface, m²/g	119	150	160
Refractive index	1·44	1·46	1·45
Oil absorption, g/100 g	160	170	248
pH, 5% suspension	9·0	7·3	4·5
SiO₂	85%	87%	88%

HMDA Abbreviation for hexamethylene diamine carbamate.

HMDAC Abbreviation for hexamethylene diamine carbamate.

HMDA CARBAMATE Abbreviation for hexamethylene diamine carbamate.

HMF Abbreviation for high modulus furnace.

HMT Abbreviation for hexamethylene tetramine.

HOBOKEN N,N′diphenyl-p-phenylene diamine (14).

HOCK, LUTHER 1890. Professor of physical and colloidal chemistry, casting, research into rubber. Recipient of the Carl Dietrich Harries Plaque, 1955.

HOCK'S ELASTICITY TESTER Old-type, simple, falling-ball apparatus used to determine the resilience of rubber. A ball falls onto a test piece held at an angle, and rebounds at an angle to the vertical. The angle made by the ball to the horizontal is taken as a measure of the resilience.

HOEKSTRA PLASTOMETER A parallel plate apparatus for determining the plasticity of very small test pieces. The test piece, compressed to 1 mm thickness by the plastimeter, is loaded with 10 kg for 30 sec in a closed chamber between two 12 mm diameter plates, steam-heated to a temperature of 100°C. The compression is measured on a scale graduated to 1/100 mm (216).

HOFMANN, PROF. FRITZ
1866–1956. From 1897 worked in the Elberfelder colour industry, from 1906 on research into polymerisable diolefins. He first attempted the synthesis of rubber from isoprene, but later concentrated on butadiene whose ability to polymerise had been discovered by Kondakow in 1900. In 1909 his attempts at hot polymerisation succeeded, resulting in the so-called Nor rubber. During World War I the production of methyl rubber (H and W) was started under his direction, at approx. 150 tons per month. In 1918 he became head of the Coal Research Institute of the Kaiser Wilhelm Gesellschaft in Breslau. Although methyl rubber was only suitable for hard rubber and production was stopped after the war, his research laid the foundation for further development in the field of synthetic rubber, *e.g.* in 1926, to the technical production of butadiene and to the erection of the first Buna producing factory at Schkopau.

HOLOFOL PROCESS Process invented by W. Opavsky for producing articles from unvulcanised, calendered sheets. Both sides of the sheet are partially vulcanised and it is treated with a swelling agent. The sheet splits as the swelling agent evaporates after heating. The hollow body is then fully vulcanised.
Lit.: *India Rubb. World*, 1954, 130, 213.
BP 687 629, 807 632.

HOLZRICHTER, HERMANN 1910. Conducted research into continuous polymerisation, redox catalysts (amine process), and preparation of synthetic rubber.

HOOKE, ROBERT Physicist, 1635–1703. Professor of Geometry at Gresham College, London. Secretary of the Royal Society.

HORNA Group of masterbatches and granulated pigments for the rubber industry (424).

HOSTAFLON Group of fluoro polymers (217).
Grades:
C: polytrifluorochloroethylene
TF: polytetrafluoroethylene.

HOSTALEN Polyethylene produced by the low pressure process (217).

HOSTAPAL Group of alkyl aryl polyglycol ethers. Non-ionic emulsifiers (217).

HOSTAPHAT L 317 Phosphoric acid esters. Antistatic agents (217).

HOSTAPON Group of fatty acid condensation products. Anionic active emulsifiers, wetting agents and cell regulators for the production of sponge rubber. Foam stabilisers for foam rubber.

HOSTAPUR Alkyl polyglycol ether, Types WX and CX. Non-ionic wetting agent (217).

HOT MIXING PROCESS Process which increases the speed of the rotors in internal mixing, raising the temperature to 150–180°C, a higher degree of plasticity being achieved with a lower degree of degradation. Facilitates the use of fillers.

HOUDRY PLANT Plant for producing butadiene and butylene by catalytic dehydration of butane. The C_4 fraction contains approx.

12% butadiene and 25% butylene. The butadiene is isolated in the copper-ammonium acetate plant.

HPC Abbreviation for hard processing channel.

HPT Abbreviation for hexamethyl phosphoric acid triamide.

HP TYRE Abbreviation for high pressure tyre.

HRH SYSTEM Hi-Sil. Abbreviation for the RFS system; hexamethylene tetramine resorcinol and precipitated silica.

HS 160 Hard, cellular product made from Hycar 1001, nylon and phenolic resin. Filling material for propellers.

HT-1 Heat resistant polyamide fibre.
Uses: include electrically insulating materials, reinforcement of heat resistant hose, special-purpose belts, conveyor belts (6).

H TENSILAC Types 40 and 41. Accelerator of undisclosed composition (obsolete) (6).

HTP Abbreviation for high temperature polymerisation.

HTV Abbreviation for high temperature vulcanisation.

HUAYULE *Parthenium argentatum*, Guayule rubber.

HUCEL Blowing agent of undisclosed composition. White, odourless powder, stable up to 100°C, s.g. 1·47. Non-discolouring. Suitable for natural and synthetic rubbers.

Quantity: soft compounds 1–3%, hard compounds 1–5%. Optimum effect is achieved at 140°C with a delayed action accelerator. For cellular hard rubber, it should be used in combination with sodium carbonate (114).

HUEMEGA Local term for *Micranda minor* (Peru). The sap was often mixed with that of *Hevea Brasiliensis*.

HULE MACHO *Hule Colorado*, local term used on the east coast of S. America for *Castilla tuna Hemsel* and the rubber obtained from it.

HUTENEX Flexible, compounded factice. Processing aid for extruding and calendering (114).

HUTEX 50% aqueous dispersion of a high quality factice for use in latex.
Uses: gloves, balloons, foam rubber (114).

HX HPC black (40).

HYAMINE 2389 Quaternary ammonium chloride. Sensitising agent for latex (33).

HYCAR Group of butadiene-acrylonitrile copolymers (14).
Grades:
1000 × 88: 45% acrylonitrile, s.g. 1·00, Mooney 65–95. Excellent oil resistance. Age-Rite Stalite is used as the stabiliser
1000 × 132: approx. 45% acrylonitrile, s.g. 1·00, Mooney 45–65. Excellent resistance to oils and solvents. *Uses:* transformer gaskets, oil hose

1001: (formerly Hycar OR 15) approx. 40% acrylonitrile, s.g. 1·00, Mooney 80–110. Excellent resistance to oils and solvents. Age-Rite Stalite is used as the antioxidant

1002: (formerly Hycar OR 25) approx. 31% acrylonitrile, s.g. 0·98, Mooney 75–110. Good oil resistance, excellent water resistance. Age-Rite Stalite used as the stabiliser

1011: (formerly Hycar OR 15 EP, now obsolete) approx. 40% acrylonitrile. Mooney 85–105. Easily processed

1012: (formerly Hycar OR 25 EP, now obsolete) approx. 31% acrylonitrile. Mooney 85–105. Easily processed

1013: approx. 26% acrylonitrile. Mooney 70–95. Easily processed (obsolete)

1014: (formerly Hycar OR 45 EP) approx. 21% acrylonitrile, s.g. 0·95, Mooney 70–90. Medium oil resistance, excellent resistance to low temperatures. Age-Rite Stalite is used as the stabiliser. Easily processed

1022: (formerly Hycar OR 25 NS) approx. 31% acrylonitrile, s.g. 0·98, Mooney 45–70. Non-staining stabiliser (phenol derivative), readily soluble, gel free, very good oil resistance. *Uses:* moulded, calendered and extruded articles which require a low processing viscosity and a relatively low modulus; adhesives, cements

1041: approx. 40% acrylonitrile, s.g. 1·00, Mooney 70–95. Better tack than 1001. Age-Rite Stalite is used as the stabiliser. Easily processed.

Uses: frictioning and coating compounds

1042: 31% acrylonitrile, s.g. 0·98, Mooney 70–95. Cold polymer, good oil resistance. Easily processed

1042 × 69: approx. 30% acrylonitrile, s.g. 0·98, Mooney 70–95. Cold polymer, non-discolouring stabiliser, good oil resistance. Easily processed

1043: approx. 26% acrylonitrile, s.g. 0·98, Mooney 70–95. Cold polymer, non-staining stabiliser, good resistance to oils and low temperatures. Easily processed

1051: approx. 40% acrylonitrile, s.g. 1·00, Mooney 60–85. Improved 1001 type; good processing properties and tack, excellent oil resistance. Age-Rite Stalite is used as the stabiliser. *Uses:* frictioning and coating compounds

1052: approx. 31% acrylonitrile, s.g. 0·98, Mooney 45–70. Similar to the 1042 type; non-staining stabiliser, very good oil resistance. Easily processed. *Uses:* articles with a high elongation

1053: approx. 26% acrylonitrile, s.g. 0·97, Mooney 45–70. Similar to the 1043 type, non-staining stabiliser, good resistance to oils and low temperatures. Easily processed

1072: approx. 31% acrylonitrile, s.g. 0·98, Mooney 35–60. Type 1042, modified with carboxyl groups (5% acrylic and methacrylic acids), non-staining stabiliser, good resistance to oils and low temperatures, very good hot tear resistance, good abrasion resistance. Can be vulcanised with zinc oxide alone, or with TMTM, sulphur and peroxides

1203: mixture of 70% Hycar and 30% p.v.c. resin, s.g. 1·07, Mooney 80–105. Excellent resistance to oils and the effects of weather

1312: approx. 31% acrylonitrile, s.g. 0·98, viscosity 100 000 cp (30°C). Liquid Type 1002. May be vulcanised to solid rubber. *Uses:* softener for NBR compounds to improve the flow properties; in plastisols and liquid phenolic resin compounds

1411: (formerly Hycar OR 15 crumb) approx. 40% acrylonitrile, s.g. 1·00, Mooney 100–130. Crumb form of Type 1001, insoluble. *Uses:* blends with other NBR types to improve extruded and calendering properties, modification of phenolic resins for shock-resistant articles

1432: (formerly Hycar OR 25 TS) approx. 31% acrylonitrile, s.g. 0·98, Mooney 70–95. Cold polymer, crumb form, non-staining stabiliser, very easily dissolved, very good oil resistance. Uses: adhesives, cements, vinyl resins which have fast colours

4021: (formerly PA-21) copolymer of 95% acrylonitrile and 5% 2-chloroethyl vinyl ether, s.g. 1·00. Powder form, vulcanisates are heat resistant up to 200°C, have good ozone, flex cracking and oil resistance. Vulcanises with amines (notably polyethylene polyamine, *e.g.* Vulcacit TR).

Quantity: 2% with 2% sulphur and 1% stearic acid, cure is 60 min at 147°C. *Uses:* gaskets, conveyor belts, V-belts, pressure rollers, hose lines, white and light-coloured articles.

Butadiene-acrylonitrile latices:

1500 × 290: approx. 20% acrylonitrile, 62% total solids, pH 9·5. *Uses:* foamed materials

1512: approx. 31% acrylonitrile, 53% total solids, pH 8·5, viscosity 15 cp, particle size 500 Å

1551: approx. 41% acrylonitrile, 52% total solids, pH 9·5, viscosity 33 cp, particle size 1800 Å. *Uses:* adhesives, oil resistant paints, leather finishes, resin modification

1552: approx. 32% acrylonitrile, 53% total solids, pH 9·0, viscosity 42 cp, particle size 1800 Å. *Uses:* adhesives, bonding agents, paints, impregnations

1561: approx. 41% acrylonitrile, 41% total solids, pH 9·5, viscosity 70 cp, particle size 500 Å. *Uses:* adhesives, abrasion resistant paints, leather finishes, bonding agents, resin modification

1562: approx. 31% acrylonitrile, 41% total solids, pH 9·5, viscosity 38 cp, particle size 500 Å. *Uses:* adhesives, finishes, bonding agents, resin modification

1562 × 103: approx. 31% acrylonitrile, 40% total solids, pH 9·5. *Uses:* paper addition, textile finishes, adhesives

1571: approx. 42% acrylonitrile, 41% total solids, modified

with 5% acrylic/methacrylic acid, pH 8·0, viscosity 12 cp, particle size 1200 Å. *Uses:* oil and fat resistant paints, bonding agents, adhesives, resin modification

1572: approx. 31% acrylonitrile, 50% total solids, modified with approx. 5% acrylic/methacrylic acid, pH 6·5, viscosity 18 cp, particle size 1000 Å. *Uses:* paper saturation, adhesives, finishes, bonding agents, resin modification

1577: approx. 31% acrylonitrile, 40% total solids, contains carbonyl groups, pH 9·5, viscosity 36 cp, particle size 400 Å. *Uses:* water and fat resistant paints, paper impregnation and saturation, leather finishes

1872: approx. 31% acrylonitrile, 40% total solids, pH 8·5. *Uses:* as for 1572

1877: approx. 31% acrylonitrile, 40% total solids, pH 9·5. *Uses:* as for 1577.

Polyacrylate latices:

2600 × 30: 50% total solids, s.g. 1·07, pH 6·5. *Uses:* leather and textile finishes, unwoven fabrics, paper saturation, bonding agents

2600 × 39: 50% total solids, s.g. 1·07, pH 6·5. *Uses:* adhesives, paints, finishes, bonding agents

2601: 50% total solids, s.g. 1·06, pH 6·5, viscosity 47 cp. Particle size 1600 Å. *Uses:* adhesives, finishes, bonding agents, resin modifier.

2671: 50% total solids, s.g. 1·06, pH 5·8. *Uses:* paper saturation, adhesives, bonding agents

4501: 50% total solids, s.g. 1·06, pH 6·5. *Uses:* as for 2601.

Series 2000:
Butadiene-styrene copolymers (formerly Hycar OS).

HYDRALINE Trade name for cyclohexyl acetate.

HYDREX 440 Hydrogenated stearic acid.
Uses: activator (218).

HYDRIN Homo- and copolymers of epichlorhydrin. Fully saturated polymers with excellent resistance to ageing and oxidation, and with resistance to hydrocarbon solvents (14).
— 100: polymer of epichlorhydrin
— 200: copolymer with ethylene oxide.

HYDRITE Japanese produced thioplasts.

HYDROABIETIC ACID Partially saturated abietic acid obtained by hydrogenation of acetic acid. The dihydroabietic acid ($C_{19}H_{31}COOH$) is precipitated, in which only one double bond is saturated, also tetrahydroabietic acid ($C_{19}H_{33}COOH$), which is completely saturated. Salts of hydroabietic acids are used as emulsifiers in the production of synthetic rubber.

HYDROBENZAMIDE
$C_6H_5.CH(N{=}CH{-}C_6H_5)_2$; condensation product of benzaldehyde and ammonia, m.p. 110°C, b.p. 131°C. Soluble in alcohol and ether.
Uses: vulcanisation accelerator, rarely used.
TN: Vulcazol (363).

HYDROFOL Group of hydrogenated fatty acids. Activators for accelerators (44).

HYDROFURFURAMIDE

$C_4H_3O . CH(N=CH . C_4H_3O)_2$, furfuramide. Brown crystals, m.p. 117°C, b.p. 250°C, while decomposing. Soluble in alcohol and ether. Accelerator, rarely used. 5% zinc oxide is necessary.
Quantity: 1·5% as hardening agent for resins.
TN: Furfuramide
Vulcazol A.

HYDROPLAT B Trade name for dibutyl tetrahydrophthalate.

HYDROPOL Hydrogenated polybutadiene with 5–30% of the original unsaturation level. Similar to polyethylene but more flexible at low temperatures.
Uses: cable and wire insulation, films, tubing, extruded articles (130).

HYDROQUINONE

$C_6H_4(OH)_2$, 1:4-dihydroxybenzene, m.p. 170–171°C, b.p. 285–287°C, s.g. 1·332. Soluble in alcohol, ether, and in 14 parts water.
Uses: short-stop in emulsion polymerisation of SBR, basis of various antioxidants, vulcanises natural rubber in combination with lead oxide.
TN: Tecquinol (193).

HYDROQUINONE MONO-BENZYL ETHER

$HO . C_6H_4O . CH_2C_6H_5$, [p-hydroxyphenyl benzyl ether, benzyl hydroquinone, p-(benzyl oxy)phenol]. Yellow powder, m.p. 115–120°C, s.g. 1·26. Soluble in benzene and alkalis, insoluble in petroleum hydrocarbons. Antioxidant for rubber and latex, effective against normal oxidation, flex cracking and weathering, prevents frosting. Non-discolouring, suitable for products which undergo human contact.
Quantity: 0·25–1%. Blooms above 0·5%. Non-staining and nondiscolouring.
TN: Agerite Alba (51)
Monobenzon
Benoquin.

HYDROQUINONE MONO-ETHYL ETHER

White crystalline powder, m.p. 54–56°C, b.p. 243°C. Soluble in benzene, alcohol and acetone.
Quantity: 0·05–1%.
Uses: polymerisation inhibitor in monomers and synthetic latices.

HYDROQUINONE MONO-METHYL ETHER

White flakes, m.p. 53°C, s.g. 1·55. Weak, general purpose antioxidant.

HYDRO RUBBER $(C_5H_{10})_x$. Colourless, viscous or elastic, to tough, non-elastic mass, s.g. 0·8585, mol. wt. 30 000–150 000. Soluble in the usual solvents, insoluble in acetone and alcohol.
Production: catalytic hydrogenation of rubber in solution at 70–80°C, with platinum black as a catalyst. Hydrogenation of solid rubber at 90–100 atm and 250–300°C with nickel or platinum as catalyst gives a highly degraded product. If rubber is hydrogenated as a 2% solution in cyclohexane at 170–175°C and 15–20 atm with 30–40 parts of nickel catalyst on diatomaceous earth, a slightly degraded product with a high mol. wt. is produced. Dry distillation gives methyl ethyl ethylene, a partially hydrogenated isoprene. By hydrogenation of guttapercha and balata the asymmetry of

the molecules is corrected; the result is a similar product, hydro-gutta-percha; s.g. 0·8595.

Uses: include interlayering in safety glass, impregnating agents, bonding agents, and as an addition to high pressure lubricants.

Lit.: J. Le Bras and A. Deland, *Les Derives Chimiques du Caoutchouc Naturel*, Paris, 1950.

2-HYDROXY-p-CYMOL

Isopropyl-o-cresol, carvacrol, b.p. 237°C, s.g. 0·967. Soluble in alcohol and ether, insoluble in water. Obtained from p-cymol by sulphonation and dissolving in water.

4-HYDROXYMETHYL-2:6-DITERT-BUTYL PHENOL

$$(CH_3)_3C \underset{CH_2OH}{\overset{OH}{\bigcirc}} C(CH_3)_3$$

White powder, m.p. 140°C. General purpose antioxidant.
TN: Ionox 100 (2).

p-HYDROXYPHENYL MORPHOLINE

$$O\begin{array}{c} CH_2-H_2C \\ CH_2-H_2C \end{array} N.C_6H_4.OH$$

Pale yellow liquid, m.p. 168°C. Antioxidant. Slightly discolouring.
TN: Solux (6).

12-HYDROXYSTEARIC ACID

Wax-like flakes, m.p. 84–88°C, s.g. 0·99, iodine No. 3, saponification value 180. Processing aid, softener and lubricant, improves extrusion properties, dimensional stability, and dispersion of fillers, decreases the

'nerve' of compounds, gives improved surface finish. May be mixed, *e.g.* with NR, SR, polyethylene, polyacrylates, ethyl cellulose.
TN: Castorwax (9).

HYFAC Group of higher fatty acids and hydrogenated fatty acids. Activators and softeners (196).

HYLENE Group of isocyanates.
Grades:
M: 4:4'-diphenyl methane diisocyanate
DP: 50:50 2:4-tolylene diisocyanate and 2:4:4'-diphenyl ether tri-isocyanate
T: 2:4-tolylene diisocyanate
TM: 2:4-tolylene diisocyanate (80:20 2:4 and 2:6)
TU: 4:4'-dimethyl-3:3'-diisocyanate diphenyl urea
MP: phenol adduct of 4:4'-diphenyl methane diisocyanate (6).

HYSTERESIS, ELASTIC In a stress/strain diagram for an elastic deformation, different curves will be obtained when loading and unloading a test piece. The surface enclosed by the loading and unloading curves is the so-called hysteresis cycle and represents the part of the mechanical energy which has been lost and transferred into heat. The hysteresis increases with increasing time between loading and unloading.
Specification: DIN 53 510, DIN 53 513.

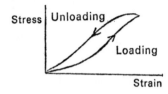

241

HYSTRENE Mixture of higher fatty acids. Trade name for stearic acid (97). Activator and softener.

HYSTRON Reinforcing resin and processing aid. Has a high styrene content and is used for natural rubber, SBR and other elastomers. Increases hardness, stiffness, abrasion resistance and electrical properties.
Uses: soles, floor coverings, mechanical goods (31).

HYTREL Thermoplastic elastomers, similar to polyethylene terephthalate esters, resistant against oils, solvents and hydraulic liquids. Thermoresistant between −45 and 150°C (6).

I

ICCRI Abbreviation for International Committee for the Classification of Rubber Information.

ICI Abbreviation for Imperial Chemical Industries (London, S.W.1).

ICR Abbreviation for initial concentration rubber; natural rubber produced from undiluted latex. Contains more non-rubber constituents than normal and therefore has a higher Mooney viscosity and modulus, and cures faster. Has good ageing properties.

IFC Abbreviation for Institut Français du Caoutchouc.

IGELITE
Grades:
PCU: p.v.c.
MP: copolymer of p.v.c. and the methyl ester of acrylic acid (32).

IGEPALE Group of wetting agents and emulsifiers for synthetic and natural latices; condensation products of alkyl phenols and ethylene oxide. Non-ionic liquids to wax-like solids, s.g. 1·02–1·18 (85, 189, 217).

IGEPONE Group of surface-active materials based on fatty acid condensates. Release agents, stabilisers for synthetic latices (85):
AP-78: oleic acid ester of sodium isothionate
CN-42: sodium-N-cyclohexyl-N-palmityl taurate
T-33: sodium-N-methyl-N-oleyltaurate.

IGETEX Buna latices developed before World War II. Have 35–45% rubber content, corresponding to the equivalent Buna types S, SS, N, NN. *Uses:* include dipped articles, textile impregnations, protective coatings.

IGEVINE Group of polyvinyl ethers.

IIR Isobutylene-isoprene rubber, in the nomenclature of ASTM D 1418-58T. Has excellent resistance to weathering, low air permeability, good physical properties.

IISRP Abbreviation for International Institute of Synthetic Rubber Producers (45 Rockefeller Plaza, New York 20; 40 Square Marie Louise, Brussels 4).

IKELAMBAWII Former trade term for a natural rubber grown in the Congo.

IKIPREEN Group of polyurethanes (220).

IKIRPEN Polyurethane rubber (221).

IMAC C 12 Cationic exchange resin.

IMBUTOL E 50% aqueous emulsion of 2:6-ditert-butyl-4-methyl phenol. Antioxidant for latex (180).

IMELON Polyacrylonitrile fibre (222).

IML-1 Product used to improve the flow properties of highly filled compounds. Free flowing powder. *Quantity:* 5% (6).

INDONEX W-2 Aromatic petroleum derivative. Dark brown liquid, s.g. 0·98, iodine No. 56. Softener and plasticiser for natural rubber, SBR and CR, extender (78).

INDULIN 70 Masterbatch of 70 parts lignin (54% v/v) and 100 parts SBR 1500 (223).

INDUSTRENE AND INDUS-TRENE R Blends of higher fatty acids.
Uses: activators and softeners (97).

INGENHOUSZ, JAN 1730–99, b. Breda, d. London. Member of the Royal Society, practised as a doctor in Holland, Austria and England, personal doctor to Maria Theresa, published a number of scientific articles. In 1775 used rubber flasks and tubes for his gaseometric experi-

ments, as is seen in his letter to the president of the Royal Society.
Lit.: *Phil. Trans.,* 1776, **66**, 257; *ibid,* 1779, **69**, 377.
J. v. Wiesner, *Jan Ingenhousz,* Vienna, 1905.

INGRAL Wax, protective against light ageing (224).

INGRALEN Group of aromatic hydrocarbons. Softeners and extending oils for natural and synthetic rubbers (224).

INGRAPLAST Group of naphthenic softeners for light-coloured compounds of natural rubber, synthetic rubbers and reclaim (224).

INGRAVIS 3 Solid hydrocarbons used to improve extrusion and resistance to light ageing (224).

INHIBITOR Substance which prevents or retards a chemical reaction (negative catalyst).

INHIBITOR
Group of antioxidants:
AT: phenolic antioxidant. White crystalline powder, m.p. 68°C. Non-discolouring, for NR and SR. Suitable for products which come into contact with foodstuffs. Dosage: 1–1·5% (440)
DAHQ: ditert-amyl hydroquinone (193)
OB: diphenyl-p-phenylene diamine (440).

INJECTION MOULDING A process similar to that used for moulding thermoplastics, in which a suitably compounded rubber is transferred from a heated cylinder through a nozzle into a heated mould, where vulcanisation takes place.

INOSITE $C_6H_6(OH)_6$, hexa-hydroxycyclohexane, cyclohexane-hexol, hexaoxyhexahydrobenzene. Although it is an isomer with glucose (termed a cyclic sugar), it is not a true sugar but a cyclic hexa-hydroalcohol. There are nine possible stereoisomers, of which seven are optically inactive. There occur naturally two optically active forms, the racemic form and various cis-trans isomers. The most frequently occurring natural form is cis-1:2:3:5-trans-4:6-cyclohexane hexol.

i-Inosite, m-inosite, mesoinosite, myoinosite, nucit, phaseomannit, Bios I, meat sugar, dambose. White, odourless, non-hygroscopic crystals, m.p. 225–227°C, s.g. 1·752. Optically inactive. Solubility in water at 25°C 14 g/100 ml, at 60°C 28 g/100 ml, sparingly soluble in alcohol, insoluble in ether and organic solvents, crystallises from water or acetic acid at temperatures above 80°C. The dihydrate $C_6H_6(OH)_6$.$2H_2O$ (dambose) crystallises from water at below 50°C in large mono-clinic crystals, m.p. 218°C, s.g. 1·524, becomes anhydrous above 100°C. Does not ferment and does not reduce Fehling's solution. Occurs widely in living tissue and assists the growth of animals and micro-organisms. Frequently occurs in plants as hexaphosphoric acid esters (phytin).

Many rubber and latex types contain mono and di methylinosite. i-Dimethylinosite is equivalent to Dambonit, isolated from gaboon rubber (from *Landolphia* and *Ficus* types).

1-Methylinosite occurs in *Hevea* latex in concentrations of 1–2% and is also called quebrachite or que-brachitol, although quebrachite comes from quebrachor bark and is dextro-rotatory.

d-Methylinosite, Bornesit. Isolated from Borneo rubber, mainly from *Ficus elastica* (Matazit, from Mada-gascar rubber, Pinit, from *Pinus lambertina,* Sennit from sennes leaves).

Lit.: Sebrell-Harris, *The Vitamins*, Vol. 2, New York, 1954, 321–86.

R. Beckmann, '*m-Inosite*, Cantor (Ed.), Aulendorf, 1953.

Moderne Methoden der Pflanzenana-lyse II, 1954.

Davis-Blake, *Chemistry and Tech-nology of Rubber*, New York, 1937.

INSTITUT DES RECHER-CHES SUR LE CAOUTCHOUC AU VIETNAM IRCV, Laikhé, Vietnam. Formerly Institut des Re-cherches sur le Caoutchouc de l'Indochine (IRCI). Founded 1940 as an affiliated institute of the Institut Français du Caoutchouc.

INSTITUT DES RECHERCHES SUR LE CAOUTCHOUC EN AFRIQUE IRCA, Bingerville, Ivory Coast. Founded 1942 for the study of rubber production from *Funtumia* and *Euphorbia* types and from rubber producing lianes. Since 1946, has investigated the possibility of estab-lishing *Hevea* plantations. Works in cooperation with the French 'mother' institute.

INSTITUT FRANCAIS DU CAOUTCHOUC IFC, Paris XIV, Rue Scheffer. Founded 1936, for research into natural rubber; comes under the coordination of the Inter-national Rubber Research and Development Board.

INSTITUTION OF THE RUBBER INDUSTRY
IRI, 4 Kensington Palace Gardens, London, W.8. Founded 1921 by rubber specialists to further the development of the science and technology of rubber. The IRI enjoys great respect because of the high standard of its technical publications, the conferences which it organises, and the specialist diplomas which it awards. Publication: *Transactions and Proceedings of the IRI* (six times per annum). Diplomas: LIRI (Licentiateship) by examination, AIRI (Associateship) by examination, thesis or nomination, FIRI (Fellowship) by nomination, for exceptional services to the industry. The institute awards (yearly) the Hancock and Colwyn medals for research or technological achievement.

INSTITUTO ESPANOL DEL CAUCHO S. A.
Barcelona. Founded 1955 by the Spanish rubber industry (Consorcio de Fabricantes de Articulos de Caucho).

INSTRON MACHINE
Apparatus for determining the tensile strength of rubber.

INTEL M 400
SBR rubber with a high coefficient of friction (225).

INTENE
Trade name for elastomeric polybutadiene.

INTERMEDIARY POLYMER
Transient structural product formed from the monomer before the final polymer is obtained.

INTERNAL MIXER
Machine with profiled rotors rotating in an enclosed chamber in opposite directions. Used to masticate rubber and to mix compounds; includes the Banbury, Werner-Pfeiderer, and Shaw Intermix. The Gordon plasticator which works on the principle of an extremely large extruder belongs in this category; also the reclaimator used to produce reclaim.

INTERNATIONAL RUBBER HARDNESS
Expressed in degrees, IRHD; equivalent to British Standard Hardness. Based on the determination of the penetration of a sphere into an elastomer test piece under standard conditions. The scale, divided into 100 parts, is used as a unit scale whose lower limit (zero degrees) is determined by a material with an elastic modulus nought (no resistance to penetration) and whose upper limit is determined by a material which has an infinite elastic modulus (no penetration). For elastic isotropic materials there is a relationship between the Young's modulus and the hardness expressed in International Standard Units according to the approximate formula:

$$\frac{F}{M} = 0.000\ 17 \cdot R^{0.65} \cdot P^{1.35}$$

where F is the load in kg, M is Young's modulus in kg/cm^2, R is the radius of the sphere of the hardness meter, and P is the penetration in hundredths of a millimetre.

The curve formed gives a relationship between $\log M$ and hardness in International Units. The centre point of the curve is determined by:

$$\log M = 1.372$$

or

$$M = 23.55\ kg/cm^2$$

The International Rubber Hardness Degree corresponds approximately to the Durometer Degree. Measurement determining penetration of a sphere 2·38 or 2·50 mm diameter at an initial load of 30 ± 1 g and a final load of 564 ± 2 g (diameter of sphere 2·38 mm) or 580 ± 2 g (sphere diameter 2·50 mm); penetration of the sphere is measured for each load after 5 and 30 sec respectively. The difference in penetration with the initial and final loads is expressed, using graph or table, in International Hardness Degrees. ASTM D 1415-62T, ISO Recommendation R 48. BS 903: 19: 1950.

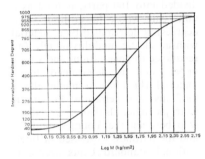

INTERNATIONAL RUBBER REGULATION AGREEMENT Agreement reached by rubber producing countries to regulate the export of rubber, and thus prevent the exceptionally low prices resulting from oversupply. The agreement was nullified in 1944, as the war in the Pacific had already rendered it virtually unworkable.

INTERNATIONAL RUBBER STUDY GROUP A corporation founded by Britain, Holland, and the USA in 1944 to enable producers and consumers to discuss the problems of the rubber industry. In 1945 France joined the corporation, and in 1947 membership became open to all interested countries; there are now twenty-two members. The emphasis of discussion is primarily on economic aspects, and participation has been limited to delegates of the member governments.

INTISY RUBBER Trade term for a mediocre quality natural rubber from *Euphorbia intisy*, formerly grown in southern Madagascar. The trees were exterminated because of the clumsy methods used to obtain the rubber.

INTOL Styrene-butadiene rubber of the International Synthetic Rubber Co. (225).

INTRAL Group of saturated and unsaturated fatty acid esters with long-chain structure. Dispersing agents and emulsifiers.

INTRAL A-E Polar compound with high b.p., pH approx. 10. Stabiliser for latex (226).

INTRAMINES Sodium salt of sulphonised lauryl and myristyl collamides. Wetting agent, softener and release agent (226).

INVINES Group of organic barium/cadmium and barium/cadmium/zinc compounds. Heat and light stabilisers for vinyl polymers and copolymers (191).

IONOMERS Group of transparent, tough thermoplastics. Resistant to oils and solvents, with high tensile strength and elasticity. The polymers contain intermolecular ionic bonds as well as intramolecular covalent bonds. The introduction of ionic bonds in a semi-crystalline polymer suppresses crystallisation

and increases the modulus and resistance to solvents, as well as the stiffness at room temperature. The main component is ethylene. Ionic bonds are introduced by copolymerisation with materials containing carboxyl or metallic groups.
TN: Surlyn A (6).

IONOX Group of antioxidants (2):
100: 4-hydroxymethyl-2:6-ditert-butyl phenol
201: 3,5-ditert-butyl-4-hydroxy-benzyl ether
220: 4:4'-methylenebis(2:6-ditert-butyl phenol)
330: 1:3:5-trimethyl-2-4-6-tri(3:5-ditert-butyl-4-hydroxybenzyl)benzene.

IPO Abbreviation for isopropyl oleate.

IPP Abbreviation for isopentenyl pyrophosphate.

IR Synthetic isoprene rubber, in the nomenclature of **ASTM D** 1418-61T.

IRCA Abbreviation for Institut des Recherches sur le Caoutchouc en Afrique.

IRCC Abbreviation for Institut des Recherches sur le Caoutchouc au Cambodge.

IRCI Abbreviation for Institut des Recherches sur le Caoutchouc en Indochine. Since 1957 Institut des Recherches sur le Caoutchouc en Vietnam (IRCV).

IRCV Abbreviation for Institut des Recherches sur le Caoutchouc en Vietnam (formerly Institut des Recherches sur le Caoutchouc en Indochine).

IRDC Abbreviation for International Rubber Development Committee.

IRGALITE Group of rubber pigments (227).

IRGAPHOR Rubber master batches of Irgalite colours (227).

IRHD Abbreviation for International Rubber Hardness Degree.

IRI Abbreviation for Institution of the Rubber Industry.

IRIA Abbreviation for Indian Rubber Industries Association, Bombay 1, Fort.

IRIUM Trade name for sodium lauryl sulphonate.

IRRA Abbreviation for International Rubber Regulation Agreement.

IRRATHENE Polyethylene irradiated with high energy electrons. Shrinks on heating, but does not melt.
Uses: insulating foil for electronic parts (198).

IRRB Abbreviation for International Rubber Research Board; renamed International Rubber Research and Development Board (IRRDB).

IRS-2000, LATICES Group of butadiene/styrene latices. Monomer ratio 50:50, high degree of conversion, hot polymers, solid content 40–60% viscosity 100 to 28 000 cp. Resin soap used as emulsifier.

Uses: adhesives, textile coatings, and chewing gum (110).

IRSG Abbreviation for International Rubber Study Group.

ISAF Abbreviation for intermediate super abrasion furnace.

ISO Abbreviation for International Organisation for Standardisation. Produces International Standard Regulations. Secretary General, Geneva, 1 Rue de Varembe.

ISOAMYL ETHER
$[(CH_3)_2.CH.CH_2CH_2]_2O$, di-isoamyl ether. Colourless liquid with fruit-like odour, b.p. 172°C, s.g. 0·783. Miscible with alcohol and ether, immiscible with water.
Uses: solvent, reclaiming of rubber.

ISOBUTYL ACETATE
$CH_3COO.CH_2CH(CH_3)_2$. Colourless liquid, b.p. 118°C, s.g. 0·871 (20°C). Miscible with alcohol. Solvent for NR and SR, used for controlling viscosity of cements.

ISOBUTYL ACETYL RICINOLEATE
$C_{17}H_{32}(OCOCH_3)$
                    $.COOCH_2CH(CH_3)_2$.
m.p. −30°C, flash p. 216°C, s.g. 0·928 (25°C). Softener.
TN: Flexiricin 66 (9).

ISOBUTYL ALCOHOL
$(CH_3)_2CHCH_2OH$, 2-methyl-1-propanol. Colourless liquid, b.p. 106–109°C, s.g. 0·802–0·804. Miscible with alcohol and ether. Solvent for synthetic rubber, used for stabilising the viscosity of CR solutions.

ISOBUTYL RICINOLEATE
$C_{17}H_{32}(OH)COOCH_2CH(CH_3)_2$,

m.p. −23°C, flash p. 218°C, s.g. 0·931 (25°C). Softener.
TN: Flexricin 16 (9).

ISOCELL TSH Nitrogen blowing agent based on toluene-sulphonhydrazide. Powder containing 20% mineral oil. For odourless and light-coloured sponge, cellular and microcellular rubbers. Non-staining.

ISOCYANATE Unsaturated compounds $R—(N{=}C{=}O)_n$, mono-, di-, polyisocyanates. Extremely reactive. R is usually an aliphatic, aromatic or heterocyclic group. The aromatic diisocyanates with the NCO group in the p-position are particularly reactive. Polyamines, *e.g.* TETA and TEPA, and other bases and metal salts have a catalytic effect, whereas acids act as inhibitors. Isocyanates react readily with alcohol to form carbamic acid esters (urethanes). Di- and polyisocyanates thus form high mol. wt. polymers.
Uses: include production of polyurethanes, bonding agents with a high bonding strength for metal/rubber and textile/rubber components, fibres.

ISOELECTRIC POINT Electrically neutral point at which flocculation of electrically charged colloidal particles occurs. In natural latex the isoelectric point is pH 4·5–4·8.

ISO HARDNESS (Equivalent to to IHRD). The ISO Hardness is almost identical to Shore A Hardness in the range of 50–100°. In the lower range of hardnesses 0° IRHD characterises a material with an elastic modulus of zero, whereas 0° Shore uses a finite modulus as its basis.

ISOLAC Condensation product of rubber with β-naphthol; obtained catalytically (obsolete). Reddish-brown, thermoplastic solid mass, softening p. 60–90°C, s.g. 1·015. Soluble in benzene. Increases hardness, abrasion resistance and ozone resistance.
Uses: reinforcing agent for lightly loaded compounds, *e.g.* shoe soles and uppers, wire insulation.

ISOLENE Trade term for poly-isobutylene. Type A, mol. wt. 3000–100 000, Type B, mol. wt. above 100 000.

ISOLON Plastic p.v.c. material (228).

ISO-OCTYL ISODECYL PHTHALATE Colourless liquid, b.p. 234–252°C (5 mm), s.g. 0·994–0·980. Softener for NR, SR and vinyl polymers.
TN: Elastex 18-P (107).

ISO-OCTYL PALMITATE
$C_{15}H_{31}COOC_8H_{17}$. Clear, pale liquid with mild odour, m.p. 6–9°C, b.p. 228°C (5 mm), s.g. 0·863. Softener and processing aid for SR and NR, secondary plasticiser for vinyl resins.
TN: RC Plasticiser 0–16 (110).

ISOPAL Copolymer of 75% styrene and 25% isoprene.

ISOPENTENYL PYROPHOS-PHATE IPP. Direct predecessor in the biosynthesis of rubber in plants. When IPP is added to fresh latex most of it is quickly transformed into rubber. A small proportion takes part in a subsidiary reaction to form sterols.

ISOPOL 75:25 copolymer of styrene and isoprene; resin with a high styrene content. Oil soluble softener (229).

ISOPOLYMERISATION Homopolymerisation. Polymerisation of identical molecules, *e.g.* rubber, cellulose, polystyrene.

ISOPOR N,N'-dinitroso-pentamethylene tetramine.

ISOPRENE

$$CH_2{=}C{-}CH{=}CH_2$$
with CH_3 on the carbon

2-methyl-1:3-butadiene. Colourless, oily, unstable, oxidisable liquid, m.p. −120°C, b.p. 34–35°C, s.g. (20°C) 0·681, refractive index 1·4194. Insoluble in water, miscible with alcohol and ether. First obtained in 1860 by C. G. Williams (*Trans. Royal Soc.*, London, 1860, 150, 241; *J. Chem. Soc.*, 1862, 15, 110) from the products of dry distillation of rubber and gutta-percha with a yield of 3–5%. Polymerisation had previously been mentioned by Williams (1860), Bouchardat (1879), Wallach (1887) and Tilden (1892). The structural formula was discovered in 1882 by Tilden, the first synthesis was by Euler (1897) from methylpyrrolidine. Isoprene is the foundation stone of natural rubber. Various productions have been described, such as the pyrolysis of turpentine and terpenes (Tilden 1882, 1884), from dipentene using the Isoprene lamp (Harries, 1910), from p-cresol (F. Hofmann 1909), from isoamyl alcohol (Perkins 1912) and from acetylene and acetone (Merling 1912). A technical pyrolytic process using turpentine was used in 1940 in the USA (Newport Industries process). In the technical synthesis currently used in the USA, the basic substance, 2-methylbutene-2 is subjected to light catalysed oxidation. A mixture of 2-hydroperoxides is obtained and reduced to 2- or 3-methylbut-3-ene-2-ol, which is then dehydrated in a steam phase reaction over magnesium sulphate. Isoprene is obtained in a 99% pure state and free of other C_5 hydrocarbons.

ISOPRENE LAMP

ISOPRENE LAMP Apparatus introduced by C. Harries to produce isoprene by passing dl-limonene (dipentene) over an electrically heated platinum spiral.
Lit.: *Ann.*, 1911, **383**, 228.

p-ISOPROPOXYDIPHENYLAMINE

p-ISOPROPOXYDIPHENYL-AMINE Pale brown to grey flakes, m.p. 80–86°C, s.g. 1·12–1·28. Soluble in acetone, chloroform, carbon disulphide, benzene and alcohol. Antioxidant, particularly for tyres. Gives good protection against heat and flex cracking.
TN: Agerite Iso (51, AO-3161).

ISOPROPYL ACETATE

ISOPROPYL ACETATE
$CH_3COOCH(CH_3)_2$. Colourless liquid, b.p. 89°C, s.g. 0·87. Miscible with alcohol and ether. Solvent for NR and SBR, used for reducing the viscosity of cements.

4-ISOPROPYLAMINO-DIPHENYLAMINE

4-ISOPROPYLAMINO-DIPHENYLAMINE Brownish purple flakes, m.p. approx. 70°C, s.g. 1·17. Soluble in aromatics; solubility in SBR 4%, in NBR 2·5–3%. Antioxidant for NR and SR. Gives good protection against ozone, normal oxidation and flex-cracking under static and dynamic strains. Can be used alone or as a

secondary material with other anti-oxidants. Tends to discolour.
TN: Nonox ZA (60).

ISOPROPYL BENZOATE

ISOPROPYL BENZOATE
$C_6H_5COOCH(CH_3)_2$, b.p. 218·5°C, s.g. 1·016. Miscible with most organic solvents, immiscible with water. Solvent with a high b.p. for plastics, cellulose derivatives, polystyrene.

N-ISOPROPYL-2-BENZ-THIAZOLE SULPHENAMIDE

$$\left[\underset{N}{\overset{S}{\bigcirc\hspace{-0.5em}\bigcirc}} C-S- \right]_2 N-\overset{H}{\underset{}{C}}(CH_3)_2$$

IBS. Accelerator with a strong delayed action, particularly suitable for compounds highly loaded with furnace blacks. Gives good physical properties.
TN: IBS (129).
Lit.: BP 773 178 (1955).

N-ISOPROPYL-N′-PHENYL-p-PHENYLENE DIAMINE

Greyish violet flakes, m.p. 70°C, s.g. 1·14. Antioxidant for NR and SR. Gives good protection against ozone and flex cracking, particularly when used together with a protective wax. Works well against oxygen and heat. Its protective action is at a maximum when used together with a conventional antioxidant.
Uses: particularly in heavy tyres containing nylon cord.
TN: Antioxidant 4010 NA (43).
Cyzone IP (21)
Eastozone 34 (193)
Flexzone 3-C (23)
Nonox ZA (60)

Permanax 115 (20)
Santoflex 77 (5) 50% dispersion in oil
Santoflex IP (5).

ISO RUBBER $(C_5H_8)_x$. α-iso rubber is produced when rubber hydrochloride is reclaimed by heating with pyridine. The β isomer is obtained by reclaiming the hydrochloride of α-iso rubber.
TN: Plastoprene LV (295).

ISOTAXIE (German term). Synthesis of sterically uniform polymers with the aid of stereospecific catalysts.

ISOVYL French produced p.v.c. fibre, with low tensile strength, but relatively good heat resistance (230).

ITAUBA Local term for *Hevea lutea Müll. Arg. var. Cuneata Huber.* Occurs in rainy areas of the Amazon region which are free from flooding. Gives a yellowish latex and rubber of excellent quality. Also known as Seringueira vermelha and, in Peru, Jeve debil because of its reddish bark.

IT PLATES Insulation sheets for high temperature resistance made from 500–700 parts asbestos to 100 parts rubber. The asbestos is mixed with the vulcanising ingredients in a rubber solution and processed on an It plate calenders. These consist of one heated and one cooled roll, which form the sheets, the solvent being evaporated on the hot roll. The rough sheets are plied together and compressed on a doubling calender. Nitrile rubber is preferred where high oil resistance is required.

ITURI Jeserai. Bolivian term for *Hevea Foxii Hub.*, which yields 75% of the country's output.

IZIGA Abbreviation for Industriezweiginstitut Gummi und Asbest (E. Germany).

IZOD APPARATUS Instrumen for determining impact resistance of plastics and electrical insulation material; ASTM D 256–56.

J

JATEX Trade term for various types of concentrated latex:

A: preserved with ammonia. 60% rubber, 61·5% total solids

AA: preserved with ammonia. 64–65% rubber, 65–66% total solids

AS: (Santobrite Jatex) preserved with 0·3% ammonia and 0·1% sodium pentachlorophenate (Santobrite). 60% rubber, 61·5% total solids

K: preserved with approx. 2% potassium hydroxide. 63–64% rubber, 65–66% total solids

RA: highly purified latex with low protein content, produced by repeated dilution and centrifugation. 60·7% rubber. Films have low water absorption (276)

SK: latex with a high stability, containing potassium hydroxide, sodium silicate and a potassium soap of castor oil. 60–61% rubber, 64–65% total solids. Vulcanising ingredients can be mixed in without previous dispersion on a ball mill.

JAZZ Marbled dipped articles, *e.g.* balloons.

JELUTONG Djelotong, djelutung, Melabuai, Pantung, Dead Borneo, Pontianak. Obtained mainly from Borneo, Sumatra and Malaya and occurs as the latex of large forest trees of the genus *Dyera* (*Apocynaceae*). The product is probably produced from several species, of which *Dyera costulata, D. Lowii* and *D. Laxiflora* are the best known. Various members of the genus *Alstonia*, and *Rauwolfia spectabilis*, also give jelutong. Marketed as cubical white blocks, contains approx. 22% dry rubber and 78% resin which may be extracted by acetone. Used as a substitute for chicle in the manufacture of chewing gum.

The local tapping method is to wound the trees by several hacking cuts. First an aqueous liquid emerges, followed by a thick sap. The coagulum is obtained by heating and is trodden out to flat sheets and stored in streams or rivers. In the remilling factories it is washed on heavy rolls and pressed while wet into cubical or cylindrical blocks. Good quality jelutong is often adulterated with products of a lower quality from *Alstonia pneumatophora, A. scholaris* and *A. angustiloba* (yellow latex which is difficult to coagulate).

JEQUIE RUBBER Rubber from *Manihot dichotoma Ule*.

JMH Peptiser for synthetic rubber (40).

JOURNALS Publications on the production and processing of natural and synthetic rubbers are printed in the journals named below. There is also a reference work, *Rubber Abstracts*, which is published regularly and contains references to all the relevant journals.

ASTM Bulletin (American Society for Testing Materials), Philadelphia

Adhesives Age, New York

Adhesives and Resins, London

Agronomie Tropicale, Nogent sur Marne, France

Applied Plastics, London

Archives of Rubber Cultivation, Djakarta

Australian Plastics and Rubber Journal, Sydney

BSI News, London

Boletin del Instituto Espanol del Caucho, Barcelona

British Plastics, London

Bulletin du Laboratoire de Recherches et de Control du Caoutchouc, Madrid

Caucho Boletin de Information del Consorcio de Fabricantes de Articulos de Caucho, Madrid

Chemistry and Industry (Society of Chemical Industry), London

Elastomers Notebook (E. I. Du Pont de Nemours and Co), Wilmington

Gomu, Tokyo

Gomu Geppo, Tokyo

Gummi, Asbest, Kunstoffe (Gummi und Asbest), Stuttgart

Gummibereifung, Bielefeld

Hule Mexicano y Plasticos, Mexico

Indian Plastic Review, Calcutta

Indian Rubber Bulletin (Association of Rubber Manufacturers of India), Calcutta

Indian Rubber Grower (Rubber Growers Association of India), Kottayam

Industrie des Plastiques Modernes, Paris

Informations du Caoutchouc (Syndicat Général des Commerces et des Plastiques), Paris

Journal of Polymer Science, New York

Journal of Research of the National Bureau of Standards, Washington, D.C.

Journal of the Institution of the Rubber Industry, London

Journal of the Rubber Research Institute of Malaya, Kuala Lumpur

Karet (Balai Penjelidikan dan Pemakaian Karet), Bogor

Kauchuk i Rezina, USSR

Kautschuk und Gummi, Kunstoffe, Berlin

Kolloid Zeitschrift, Dresden

Kunstoffe, Munich

Lichtbogen (Chemische Werke Hüls)

Makromolekulare Chemie, Heidelberg

Malayan Agricultural Journal, Kuala Lumpur

Materie Plastiche, Milan

Modern Plastics, New York

Natural Rubber News (Natural Rubber Bureau), Washington, D.C.

Naturkautschuk, Fortschritte und Entwicklungen, Vienna

Nederlandse Rubberindustrie, The Hague

Nippon Gomu Kyokaishi, Tokyo

Notizie per l'industria della Gomma, Milan

The Planter, Kuala Lumpur

Planters' Bulletin of the Rubber Research Institute of Malaya, Kuala Lumpur

Plaste und Kautschuk, Berlin

Plastica (Kunstoffinstitut TNO), The Hague

Plastické hmoty a kaucuk, Prague

Plastics, London

Plastic Institute, *Transactions and Journal*, London

Plastics Technology, New York

Quarterly Journal (Rubber Research Institute of Ceylon), Ceylon

Revista de Plasticos, Madrid

Revue Générale du Caoutchouc, Paris

Rheology Abstracts (British Society of Rheology), London

Rubber Abstracts (Rubber and Plastics Research Association of Great Britain), Shawbury, Shropshire

Rubber Age, New York

Rubber and Plastic Age, London

Rubber Chemistry and Technology (American Chemical Society Division of Rubber Chemistry), Akron, Ohio

Rubber Developments, London

Rubber India (Indian Rubber Industries Association), Bombay

Rubber Journal, London

Rubber News Sheet (Secretariat International Rubber Study Group), London

Rubber Quality Bulletin (RMA), New York

Rubber Statistical Bulletin (Secretariat International Rubber Study Group), London

Rubber Times (Gomu Jiho), Tokyo

Rubber World, New York

Tea and Rubber Mail, London

World's Rubber Position, London.

J RUBBER Original, German produced urethane rubber with a good abrasion resistance, but poor low temperature flexibility.

JSR Japanese produced SBR rubber (232).

JSR BR 21 Cis-polybutadiene extended with 37·5 phr, of a naphthenic oil. Mooney viscosity 25 (232).

JULIAANS, A. Paper 'Dissertatio Chemica Inaugurales de Resina Elastica', read to the University of Utrecht in 1870, is the first publication concerned solely with rubber.

K

KADIMIC Mixture of synthetic fatty acids based on petroleum. Produced by oxidation of a gas oil fraction. Activator, comparable with stearic acid (234).

KALABOND Bonding agents for metal/rubber components. Based on polymers with reactive groups which form bonds to the metal and to rubber during vulcanisation. Consist of a copolymer of a conjugated diolefin (butadiene) with a vinyl compound containing a nitrogen ring (vinyl pyridine), and a nitrogen-free vinyl compound (styrene). The polymers are produced as two parts with varying degrees of hardness (butadiene/styrene A 25:50, B 45:30, vinyl pyridine 25) and are used as a mixture. The harder component has a greater bonding strength to metal whereas the softer component bonds more effectively to rubber. The addition of acids which may be polymerised, *e.g.* maleic or itaconic acids, improves the bonding strength. Prior to use the components are mixed with phosphoric acid or calcium phosphate (10–40%), and with zinc oxide and black (31).

KALAMO Local term for *Landophia trichostigma* (northern Madagascar) and the inferior quality rubber product.

KALITE Precipitated calcium carbonate, surface treated with 1% fatty acid (153).

KALTRON Blowing agent for polyurethane foams. Based on fluorinated chlorohydrocarbons (381).

KAMANO Trade term for a wild rubber formerly grown in Burma.

KAMPTULICON The first floor covering containing textile, and based on rubber. BP 10 054 (1844), 13 713 (1851).

KAOLIN China clay, alumina, porcelain clay. The main constituent is crystalline kaolinite $Al_2O_3.2SiO_2$. $2H_2O$ or $Al_4(OH)_8(Si_4O_{10})$, s.g. 2·1–2·6. Cheap, slightly reinforcing filler, primarily used for the production of hard, high quality compounds with a low rubber content. Kaolins are classified according to preparation and source, as hard types with a definite reinforcing power and which increase hardness and modulus, and soft types with coarser particles which give products with a low modulus.
Types: Dixie clay, McNamee clay, Suprex, Stockalite, Devolite, Rubarite, Spestone.

KAPPA Former trade name for wild rubber from *Clitandra orientalis*. Occurs in the Congo.

KARITE TREE Schi tree. *Butyrospermum Parkii.* A *Sapotaceae* (W. Africa) which gives gutta-percha. The product contains 25% rubber and 65% resins.

KARRER PLASTIMETER Automatic parallel plate plastimeter with a constant load and graphic recorder.

KATALITHIUM n-butyl lithium dissolved in a hydrocarbon. Catalyst in diolefin polymerisation (304).

KATZ, JOHANN RUDOLPH 1880–1938, Dutch doctor and chemist. Discovered, using X-rays, that stretched rubber has a crystalline structure.
Lit.: *Koll. Z.*, 1925, **36**, 300; *ibid.*, 1925, **37**, 19.
Gummi Zeit., 1925, **39**, 1044, 2351.
Rubber J., 1966, **148**, 44.

KAUTEX Rubber/cork compound used as an insulation material; alternatively plastics based on p.v.c. (235).

KAUTSCHOL Dark liquid with tar-like odour, s.g. 1·148. Mixture of hydroxylated sulphur hydrocarbons with approximately 14% sulphur. Obtained by processing petroleum and brown coal-tar by extraction with alcohol or acetone. Plasticiser for highly loaded compounds, swelling agent used in reclaiming natural and synthetic rubbers. DRP 416877/1925.

KAYSAM PROCESS Process used to produce hollow articles from heat sensitive latex. The compound is rotated in two directions in a hollow, non-porous heated mould. A coherent uniform gel forms on the wall of the mould. After cooling, the hollow article is washed, dried and vulcanised.
Uses: primarily shoes and toys.

KEL-F Group of polytrifluoro-chloroethylenes of different mol. wt. Oils, wax-like, fatty substances, moulding powders, dispersions and elastomers:
N-1: non-aqueous dispersion of trifluorochloroethylene with 20% total solids
NW-25: 27% non-aqueous dispersion of trifluorochloroethylene (1).

KEL-F ELASTOMER Copolymer of trifluorochloroethylene with vinylidene fluoride (1).

KEL-FLO Liquid fluoropolymer. Softener (1).

KELTAN Ethylene-propylene copolymers and terpolymers, s.g. 0·87. Resistant to abrasion, ozone and ageing (437).

KELTEX Trade name for sodium alginate (236).

KEMIDOL HYDRATE $Ca(OH_2).xMgO$, slaked dolomite chalk. Filler and activator (8).

KEMPORE Group of powerful blowing agents, which give nitrogen on dissociation; for natural and synthetic rubbers and thermoplastic resins:
150 and R 125: azodicarbonamide
LD: 20% aqueous dispersion of azodicarbonamide (237).

KENFLEX Group of polymerised aromatic petroleum hydrocarbons in liquid and solid form. Soluble in organic solvents and hydro-carbons. Softeners, plasticisers and processing aids. Give good stability at high temperatures and good electrical properties. Improve the heat resistance of neoprene, butyl, Hypalon, SBR and vinyls, and the ozone resistance of butyl rubber; improve extrusion and flow proper-ties. Extender for Hypalon. Types A, L, N; m.p. 80°C, 27°C and 1·7°C, respectively, s.g. 1·01–1·08 (238).

KENPLAST Group of petroleum hydrocarbons. Softeners, plasti-cisers and processing aids for natural and synthetic rubbers and latices. Improve resistance to low tempera-tures, ultra-violet stability and ageing properties (238).

KERBOSCH PROCESS Process used early in the development of the rubber industry to obtain rubber by evaporating water from the latex. The latex was placed in a rotating drum of approx. 2 m in diameter and length. A perforated wing-shaped tube was situated in the axle of the drum and rotated in the opposite direction while hot air was blown through it. The inner surface of the drum picked up a thin layer of latex which dried in half a revolu-tion. The rubber retained all non-rubber constituents, was tacky and hygroscopic with a water content of 4–5%. The disadvantage of the process was its small capacity and the danger of flocculating the latex.

KERILLEX Trade name (Malaya) for air dried sheets with very low dirt content.

KEROBIT TBK 2:6-ditert-butyl-4-methyl phenol (32).

KER S Polish produced SBR.
3012: SKB 30A
3016: SBR 1500
3020:
3090: high styrene content.

KESSCOFLEX Trade name for dibutyl tartrate (132).

KETJEN BLACK Group of blacks.
Types: CF, SAF, ISAF-H, ISAF, LHI, HAF, CR, FEF, FF, HMF, APF, GPF, SRF (313).

KETONE RESINS Conden-sation products of ketones with formaldehyde. Additives for chlori-nated rubber lacquers.

KF RESINS Modified resin esters. Soluble in benzine. Additives for chlorinated rubber lacquers (43).

KG- Russian produced channel black:
100: specific surface area 100 m^2/g. Produced from natural gas
80: specific surface area 80 m^2/g. Produced from oil and natural gas.

K GUTTA Material similar to the para-gutta used for insulating submarine cables. Consists, accord-ing to BP 346 382, of a mixture of purified balata and/or gutta-percha with Vaseline; in certain cases protein-free rubber is added.

KhSPE Russian produced chlorosulphonated polyethylene.

KI Russian produced accelerator; reaction product of 2 molecules of aniline and 3 molecules of acetalde-hyde.

KICKXIA Rubber yielding member of the *Apocynaceae* family. The African and Malayan types were differentiated by Stapf and classified as a new branch of the family: *Funtumia*. Useful natural rubber may be obtained only from *K. elastica*.

KIDROA Local name for *Mascarenhasia kidroa* (Madagascar), and the root rubber obtained from it.

KIETAL Russian produced accelerator, triethanolamine derivate.

KILOGRAM LOAD Quotient of the load capacity Q of a tyre and the weight of the material G used to build it:

$$L = \frac{Q}{G}$$

$$Q = G . L$$

The value has been determined from practical experience and is 30–50 according to the dimensions of the tyre. The higher values are for smaller tyre dimensions. Used to calculate the material weight or load capacity of the tyre.

KIRKSYL Viscose rayon (240).

KNEADED RUBBER An unvulcanised rubber compound with a high softener content; used to alter charcoal drawings, and to clean typewriters. Cannot be used to erase completely.

KO BLEND Series of master batches of SBR and latices with 50% insoluble sulphur (31).

KODEL Polyester fibre identical to Dacron (193).

KOH EXTRACT, ALCOHOLIC Alcoholic potassium hydroxide is capable of extracting oxidation products from rubber and also factice and proteins from vulcanisates preextracted with acetone and chloroform. Some constituents of hardened phenolic resins are also removed. ASTM D 297-59T, DIN 53 559.

KOH NUMBER A term used to denote the amount of ammonium salts in concentrated natural latex. Expressed as the number of potassium hydroxide grams necessary to decompose the ammonium salts contained in 100 g of dry solids in the latex. The results may be determined by potentiometric titration with 0·1 N KOH, and using a glass electrode. The KOH number gives an indication of the processing properties of the latex. In latex which is no longer fresh the KOH number is inversely proportional to the mechanical stability. An abnormally high KOH number indicates either bad production or poor storage. ASTM 1076-59.

KOKOMBA Local term for *Mascarenhasia Geayi* (Madagascar) and the rubber obtained from its roots.

KOK-SAGHYZ *Taraxacum Kok-Saghyz Rodin*. A plant of the dandelion type which yields rubber; discovered by Rodin in 1931 in Kazakhstan (USSR). Similar externally to the common dandelion, and contains up to 25% rubber in the sap tubes of the root. The sap in the leaves contains resin and a little rubber. The plants live for several years and grow in regions with continental climate. In 1942 the seed was sent to the USA but poor

257

yield did not justify commercial cultivation. Tyres were produced and examined by the B. F. Goodrich Co. and the United States Rubber Co. and the results were similar to those for tyres made from natural rubber.

Lit.: *India Rubb. World*, 1946, 113, 517.

India Rubb. J., 1943, 195, 505.

Tr. aus Roy. Hort. Soc., 1943, 68, 305.

Ullmann, *Kautschukpflanzen des Gemassigten Klimas*, Berlin, 1951.

KOMPITRO Local term for *Gonocrypta Grevii* (Madagascar) which contains rubber in the base of the stems and in the ripe fruits.

KONNYAKU FLOUR K flour. Colloidal substance from the bulbous roots of Amorphophallus types. Brown powder, for the most part soluble in water. Has occasionally been used as a thickening and creaming agent for latex (as 1% solution). BP 448 203, 448 244, 448 245, Holl.P 37 288.

KONRAD, DR ERICH b. 1894. Former director of the rubber laboratory of IG Farben, honorary chairman of the German Rubber Society. Pioneered research into polymerisation and development of synthetic rubber (Buna, Perbunan).

KOPPER'S PROCESS Process used for producing butadiene by the pyrolysis of cyclohexane.

KORESIN Reaction product of p-tert-butyl phenol with acetylene in the presence of zinc naphthenate. Pale brown to dark, crumbly solid, m.p. 110–130°C. Soluble in hydrocarbons. Tackifier for SBR and SBR/

NR compounds and cements. Has no effect on physical properties, vulcanisation or ageing. Was used in Germany during World War II as a tackifier for Buna S 3 (32, 43, 85).

KOSMINK Channel blacks which conduct electricity; CC (164).

KOSMOBILES Group of carbon blacks:

Kosmobile:	HPC
77:	EPC. Pelletised
HM:	MPC. Pelletised
S:	HPC. Pelletised
S 66:	MPC. Pelletised (164).

KOSMOS Group of reinforcing carbon blacks.

20 HM:	SRF-HM
20 NS:	SRF (non-staining)
35:	GPF
40:	HMF
45:	GPF
50:	FEF (MAF)
60:	HAF
70:	ISAF
85:	SAF
BB:	CC
CF:	CF
Voltex:	CC (164).

KOSMOTHERM Trade name for thermal blacks.

F:	FT
M:	MT (164).

KP Group of softeners:

23:	butoxyethyl stearate
45:	diethylene glycol dipropionate
61:	mixture of phthalic acid and fatty acid esters
77:	dimethyl thianthrene
90:	epoxy type softener, b.p. 200–250°C, s.g. (25°C) 0·912
93:	chlorinated tricresyl phosphate
120:	methoxyethyl acetyl ricinoleate
140:	tributoxyethyl phosphate

150: mixture of phthalic acid and fatty acid esters

201: dicyclohexyl phthalate (184).

KPNi Nickel isopropyl xanthate (420).

KRALAC Group of styrene-butadiene copolymers with a high styrene content:

A: softener, processing aid
A-EP: general purpose
H: hard product
Latex: impregnation, stiffening, bonding agent for fibres, paints (23).

KRALASTICS Group of plasticised terpolymers of styrene, butadiene and acrylonitrile in latex, powder or granular form (23).

KRALEX Czechoslovakian produced SBR cold rubber. Types 1500 and 1501 (428).

KRYMIX SBR/black master batches (303).

	Carbon Black	Oil
680	50 FEF	—
681	52 HAF	10 HA
683	75 HAF	50 HA
685	75 HAF	37·5 HA
687	75 HAF	50 HA
689	75 ISAF	37·5 HA

KS Hoesch KS. Group of easily dispersed reinforcing silica fillers (411).

KSR-BRO 1 Japanese produced polybutadiene rubber. Constituents: 97·5% cis-1·4, 0·9% trans-1·4, 1·6% 1:2 addition. Mooney 45 (232).

KT-9 Russian produced silicone bonding agent. Used to bond silicone rubber to metals. Heat resistant.

KUALAKEP Trade name for crumb rubber made by the Dynat process:
WF: from latex
CL: from scrap rubber.

KUALATEX Trade name for normal and concentrated latex marketed by Latex Distributors (USA).

KUHN, WERNER b. 1899. Professor of physical chemistry at Basle University. Conducted research into the elasticity of rubber, and the streaming birefringence of high polymers.

KUHN-ROTH DETERMINATION Method for the direct determination of rubber by its methyl groups.

KUM-SAGHYZ Rubber from *Chondrilla ambigua* and *Ch. pauciflora* (southern USSR and the USA). The plant extract is used in the USSR in place of chewing gum.

KURALON Synthetic fibres of polyvinyl alcohol (275).

KUSNETZOWKA Resin rich rubber produced in Russia around 1930. Obtained from wild growing *Chondrilla*.

K VALUE The viscosity/molecular weight constant ($\times 10^3$) which characterises the mol. wt. of a polymer. May be calculated from the equation given by Fikentscher

$$\text{Log } \eta_r = \left(\frac{75\ K^2}{1 + 1\cdot5\ K.c} + K \right).c$$

By differentiation, when $c \to 0$ the relationship with the intrinsic viscosity may be calculated.

$$\left[\frac{\ln \eta_r}{c} \right]_{c\,=\,0} = [\eta] = 2\cdot302\,6(75\ K^2 + K)$$

The K value is directly proportional to the tensile strength, elongation and mol. wt., and inversely proportional to the hardness, plasticity, and elastic deformation.

KYNAR Polyvinylidene fluoride (120).

KYREX A Lubricant.

L

L- Type terminology for Russian produced polychloroprene latex. Types 1, 3, 4, and 7.

L-522 Silicone oil. Pale yellow liquid. Release agent (112).

LACRA Elastic fibres based on 1:4-oxybutylene glycol (6).

LACTOFIL Synthetic fibre made from casein (242).

LACTOPRENE Original term for experimental ethyl acrylate copolymer, developed in 1944 by the Eastern Regional Research Laboratory of the US Department of Agriculture in cooperation with the government laboratories of the University of Akron and the B. F. Goodrich Co. The name Lactoprene was chosen because lactic acid was necessary for the production of acrylic acid esters:

EV: polyacrylic acid ester EV. 95:5 copolymer of ethyl acrylate and 2-chloroethyl vinyl ether. In production since 1948 under the name of Hycar PA 21 (now Hycar 4021) and Acrylon EA-5

BN: 88·5:12·5 copolymer of n-butyl acrylate and acrylonitrile. Produced under the name of Akrylon BA-12.

LACTRON FIBRES Thin round rubber fibres produced by spinning latex compounds in a coagulum bath, then drying and curing.

LA GLU Former trade term for a wild rubber growing in Africa.

LA LATEX Abbreviation for low ammonia latex, concentrated latex with approx. 0·2% ammonia instead of the normal 0·7%. The latices contain other preserving agents:
LA 1, 0.2% sodium pentachlorophenate
LA 2, 0.1% zinc diethyl dithiocarbamate and 0.2% lauric acid
LA 3, 0.2% ammonium borate, 0.4% lauric acid, 0.01% sodium pentachlorophenate or alternative bactericide.

LANDOLPHIA Rubber yielding member (liane) of the *Apocynaceae* family (tropical Africa and Madagascar). Approx. 20 species have been discovered, of which *L. Heudelottii*,

L. Karkii, L. Kleinei, L. madagascariensis, L. mandrianambo, L. owariensis and *L. Thollonii* were the most important producers of wild rubber. The rubber was obtained by making incisions approx. 20 cm apart over one-third of the circumference. The latex was usually allowed to coagulate on the bark. The rubber either kneaded to balls, wound onto spindles or cut into pieces known as thimbles. Good liane rubber contains 80–95 % pure rubber and 5–13 % resins. *L. Tholloni* (30 cm high bush) yields a large part of the root rubber which comes from the Congo and Angola area, and this is obtained by treading out the rubber and boiling.

LANESE Acetate cellular fibre (152).

LANGALORA Vahimainty. Local term for *Secamonopsis Madagascariensis* (Madagascar). A fairly useful wild rubber is obtained from the ripe fruits and stems.

LANITAL (Lana italiana.) Synthetic fibre of casein (244).

LANOGEN 1500 Mixture of polyglycols. Used for impregnating textiles with latex.
Quantity: up to 5 % (217).

LANOLIN Wool fat. Complex mixture of higher alcohol esters. Pale yellow mass of ointment-like consistency, m.p. 36–40°C, s.g. 0·97, iodine No. 12, saponification value approx. 100. Readily soluble in chloroform and ether, sparingly soluble in alcohol, forms stable emulsions with water. Obtained by extraction with volatile solvents or sodium carbonate, ammonium carbonate or soap solutions, then acidification. Softener, improves tackiness, extrusion and the dispersion of fillers. Prevents bloom of stearic acid in natural rubber compounds.
Uses: primarily in compounds used for plasters and adhesive tapes.

LANON Polyester fibre.

LANSIL Acetate yarn (245).

LASTEX Thin spun rubber fibres produced from latex (246).

LATAC Hexamethylene ammonium hexamethylene dithiocarbamate (6).

LATECOLL AS 10 % aqueous solution of the ammonium salt of a polyacrylic acid.
Uses: thickening agent for latex and synthetic dispersions and stabiliser in resin emulsions (32).

LATEX In most cases a colloidal suspension of natural or synthetic polymers, alternatively, the sap of rubber yielding plants or a polydispersed colloidal system of rubber particles in an aqueous phase. *Hevea* latex particles and those of most rubber containing plants have a negative charge, pH approx. 7·0, isoelectric point at pH 4·8. Particle size ranges from 0·1–4 microns, average 0·5 micron; they are spherical and stabilised by an exterior natural protective layer of soaps and proteins. They exhibit strong Brownian movement. With *Hevea* latex the dry rubber content varies between approx. 28–40 %, after a long period of non-tapping it rises to 45–50 %.
Hevea latex: viscosity 4–7 cps, s.g. (rubber) 0·9032–0·9052 (serum)

1·0131–1·0354, surface tension approx. 40 dynes. Proteins 2–2·7%, resins (sterol ester, fatty acids, phytosterol) 1·5–3·5%, sugars 1–2%, ash 0·4–0·7%, sterol glucosides approx. 0·07–0·47%, water 55–65%. Apart from the above, *Hevea* latex contains enzymes such as oxidases and peroxidases; also amino acids (glycine, alanine, isoleucine, phenyl alanine, tyrosin, aspartic acid, glutamic acid, lysine, cystine, tryptophan, valine, proline, threonine). Lutoids, Frey-Wyssling particles and other structural elements, which have not been closely examined, also occur.

The first analysis of a sap of *Hevea Guianensis* was performed in 1791 by A. F. De Fourcroy (*Ann. de Chim.*, 1791, 225). 1826, M. Faraday did research into the latex of *Hevea Brasiliensis* (*Quart. J. Sci., Lit. and Arts, XI*, 1826, 19).

LATEX IVX-5   Natural latex concentrated by centrifuging. Has a low ammonia content and mechanical stability equal to that of standard latex (169).

LATEX LBS   Polish SBR latices:
3030: total solids above 28%, 24·5% bound styrene. *Uses*: impregnation of fabrics, polyamides for conveyor belts, synthetic leather
6041: total solids min. 45%, 60% bound styrene. *Uses*: adhesive for p.v.c. to concrete, anticorrosion paint for steel/concrete construction, paper manufacture.

LATEXOMETER   Hydrometer graduated in rubber dry content; used for the rapid determination of the rubber content of latex. As a result of variation in the constituents of latex and the relatively high viscosity the determinations are not very accurate. Latexometers are primarily used in the collecting and receiving posts to give quick checks on tapping.

LATEX PRESERVATION   As a biological liquid, latex has a tendency to deteriorate rapidly in quality and coagulate within a few hours of tapping because of enzymatic and bacteriological influences. The preserving effect of alkalis was discovered by A. F. De Fourcroy in 1791 [BP 467 (1853), USP 9891 (1853)]. The most widely used preservative has been ammonia, concentration of 1·5–2% ammonia on the aqueous phase preventing the development of bacteria. At a high pH (10–10·5) hydrolysis of the proteins takes place and causes a change in the latex properties on storage. The advantage of ammonia lies in its volatility and the possibility therefore of removing it by aeration for processes in which its presence would be harmful. In some cases 0·1–0·3% formaldehyde is added to the latex before preservation. Recently, certain latices have been in demand with a low ammonia content. Such latices contain 0·1–0·2% ammonia and an additional preservative, sodium pentachlorophenate (Santobrite latex). Other similar combinations include those with zincdialkyl dithiocarbamate, aminophenol and ethylene diamine tetra-acetic acid, ammonium borate, or ammonium pentachlorophenate.

LATEX STABILITY   Affected by the chemical constituents, particularly by water soluble acids and phosphorus and magnesium components. Mechanical stability is determined using a standard stirring apparatus at 14 000 rpm, chemical stability expressed as the resistance to the influence of zinc oxide and determined by a zinc oxide test.

LATEXYL Filler for latex and rubber based on silica.

LAUREX Mixture of zinc salts of fatty acids with lauric acid as the main constituent, m.p. 95–105°C, s.g. 1·10. Activator and softener for natural rubber, synthetic rubbers and latices. Has less tendency to bloom than stearic acid.
Quantity: 0·5–1·5 % (23).

LAURIC ACID
$CH_3(CH_2)_{10}COOH$.   Colourless needles, m.p. 43·5°C (commercial product 20–40°C), b.p. (100 mm) 225°C, s.g. 0·883. Soluble in alcohol and ether. Occurs in coconut oil as a glycerine ester.
Uses: softener and activator for natural and synthetic rubbers, with the exception of butyl rubber.

LAUROX Trade name for lauroyl peroxide (171)

N-LAUROYL-p-AMINO-PHENOL Odourless, white powder, m.p. 123–126°C. Antioxidant for butyl rubber.
TN: Suconox-12 (333).

LAURYL MERCAPTAN Colourless, clear liquid with a faint characteristic odour, s.g. 0·85. Reducing initiator and modifier, controls the mol. wt. in the emulsion polymerisation of **SBR** and **NBR**.

LAVSAN Russian produced polyester fibre.

LAZAN-MAXWELL MACHINE Apparatus used to determine the dynamic properties of rubber, using forced, non-resonant oscillations.

L BLACKS Group of carbon blacks:
NI: semi-reinforcing black. Used for the production of diaphragms, gaskets and rubber footwear
T: reinforcing black. Used in rubber for mechanical goods, coarse-moulded articles, soles, heels
TD: double-sieved black pigment. Used for pigmentation of plastics
TP: pelleted, grit-free, reinforcing black. Used for the production of high quality articles (189).

LCM VULCANISATION (LCM: liquid curing medium.) Continuous vulcanisation process for the wire and cable industry; the extrudate is led directly from the extruder into a liquid heating medium, consisting of molten metal, salts or an oil with a high flash p. The vulcanisation temperature lies between 205–260°C, vulcanisation time is 20–25 sec. When using a metal bath the extrudate must be dusted with talc to avoid adherence to the metal. The process was devised by Du Pont.
Compound example:

Neoprene WHV	100
Neozone A	2
Stearic acid	0·5
SRF black	135
Sundex 85	50
Litharge	5
Zinc oxide	5

Diethylthio-
carbamide 4
S.g. 1·41
Extrusion rate 4·5 m/min
Vulcanisation 20 sec at 232°C
Shore A Hardness 70
Tear resistance 130 kg/cm
Elongation at break 230%

LC SHEET (LC: light coloured). Unsmoked sheet rubber, dried in warm air. Light, yellow brown in colour.

LD-214 Former trade name for N,N'-dicinnamylidene-1:6-hexane diamine. Now sold under the trade name Diak No. 3 (6).

LD POLYETHYLENE Abbreviation for low density polyethylene produced according to the Ziegler process.

LE-46 Silicone emulsion. Release agent (112).

LEAD DIETHYL DITHIO-CARBAMATE

$$\left[\begin{matrix} C_2H_5 \\ C_2H_5 \end{matrix} \!\!\!>\!\! N\!-\!CS\!-\!S\!- \right]_2 Pb$$

Light grey powder, m.p. 206–207°C, s.g. 1·87, mol. wt. 504. Ultra-accelerator for continuous vulcanisation, improves permanent set in silicone rubber.
Loading: 0·1–1%, as stabiliser against discoloration for vinyl resins containing chlorine 0·07–0·4%.
TN: +Ethyl Ledate (51).

LEAD DIMETHYL DITHIO-CARBAMATE
$[(CH_3)_2N.C(S)S]_2Pb$. White powder, m.p. 320–323°C (with decomposition), s.g. 2·38, mol. wt. 447·6.

Insoluble in organic solvents. Ultra-accelerator.
Uses: in black natural rubber, SBR, and butyl rubber compounds, suitable for continuous vulcanisation, usually used with a thiazole.
TN: Ledate, Rodform Ledate (51)
Robac LMO (311)
Van Hasselt MTL (355).

LEAD DIOXIDE PbO_2, lead peroxide. Brown powder with strong oxidising action, s.g. 9·375.
Use: activator for quinone vulcanising agents.

LEAD DITHIOBENZOATE
$(C_6H_5.CS.S)_2Pb$. Accelerator (obsolete).
TN: + Lithex (23).
Lit.: USP 1 522 820 (1925).

LEAD FUMARATE Yellowish white powder, s.g. 6·54. Heat stabiliser for vinyl polymers and copolymers, crosslinking agent and stabiliser for chlorosulphonated polyethylene.
TN: Lectro 78.

LEAD OXIDE PbO, lead monoxide, litharge. Yellow powder, m.p. 870°C, s.g. 9·5. Vulcanisation activator. Except for magnesium compounds, it was the only accelerator available before the discovery of organic accelerators, but now it is of no importance. Has a stronger activating effect than zinc oxide, and is also effective in the presence of aldehyde-amine, guanidine, and thiourea accelerators. Has a stiffening effect on unvulcanised compounds. Lead oxide increases the critical compounding temperature when used with dithiocarbamates and thiurams; can only be added to black or dark

264

compounds, since black lead sulphide is formed during vulcanisation. *Uses:* cable compounds with a low sulphur content, extrusion compounds for open steam vulcanisation, activator for vulcanisation of butyl rubber with quinone dioximes (together with red lead), filler in compounds for protective covering (gloves, aprons), because of the high absorption capacity to X-rays.

LEAD PENTAMETHYLENE DITHIOCARBAMATE

$[(CH_2)_5N.CS.S—]_2Pb$.   White, odourless powder, m.p. 250°C, s.g. 2·29. Insoluble in water or organic solvents. Accelerator with exceptional delayed action and good processing safety, very active after the start of vulcanisation. Light-coloured compounds become somewhat darker. Acts as activator for thiazoles.
Uses: continuous vulcanisation, hot air and press vulcanisation.
TN: Carbamate PB
Kuracap Lead PD (421)
Robac LPD (311)
+ Vulcaid LP (3).

LEAD(PHENYL AMINO-ETHYL)PHENYL DIMETHYL DITHIOCARBAMATE   Grey granules, s.g. 1·51. Accelerator for NR and SBR, thiazole boosters.
TN: SPDX-GH (100)
SPDX-GL (includes 25% mineral oil) (100).

LEAD SALICYLATE   Yellowish white, crystalline powder, s.g. 2·36. 46·5% PbO. Activator and anti-scorch agent for natural and synthetic rubber, ultra-violet stabiliser in vinyl resins.
TN: Normasal (191).

LEAD SILICATE   Basic lead silicate. White powder, s.g. approx. 5·8. Activator for natural rubber, NBR and SBR, vulcanising auxiliary for butyl rubber and neoprene.

LEAD STEARATE

$(C_{17}H_{35}COO)_2Pb$. White powder. Soluble in alcohol.
Uses: activator, softener.

LEBEDEV PROCESS   Process for the production of butadiene by pyrolysis of ethyl alcohol.

$$2CH_3.CH_2OH \xrightarrow[\text{catalyst}]{300-400°}$$
$$CH_2=CH—CH=CH_2$$
$$+ H_2 + 2H_2O$$

Lit.: S. V. Lebedev, *J. Gen. Chem.* (*USSR*), 1953, 3, 698.
Chem. Abstr., 1934, 28, 305.
G. Egloff and G. Hulla, *Chem. Revs.*, 1945, 36, 67.
C. Ellis, *The Chemistry of Petroleum Derivatives*, II, 1936, 173.

LECITHIN   Brownish yellow, hygroscopic, wax-like mass, s.g. 1·03, pH 6·6, isoelectric point 3·5, iodine No. 95, saponification value 196, acid No. 25. Soluble in ether and alcohol; belongs chemically to the phosphatides. Lecithin obtained from soya beans may be used as a softener in NR, SBR and reclaim. Mixes easily; improves the dispersion of fillers. Increases the tack of compounds and is an effective activator for litharge, with which it combines to give an organic lead salt. This is an accelerator similar in effect to that of the lead salt of mercaptobenzthiazole. Also effective as an antioxidant.
Lit.: *Rev. Gen. Caout.*, 1955, **32**, 321.

LECTRO Group of heat stabilisers for vinyl polymers and copolymers, particularly for use in electrical insulation:
60: complex lead chlorosilicate. White powder, s.g. 4·0
77: lead chlorophthalosilicate. White powder, s.g. 4·14.
78: basic lead fumarate. Yellowish powder, s.g 6·54. Also effective as a crosslinking agent for chlorosulphonated polyethylene (191).

LEDA Group of accelerators:
MTZ: zinc dimethyl dithiocarbamate
TMT: tetramethyl thiuram disulphide
ZDC: zinc diethyl dithiocarbamate
Zineb: zinc ethylene bisdithiocarbamate (247).

LEONIL Group of alkyl aryl sulphonates. Anionic active emulsifiers and wetting agents (217).

LEVAFORM Group of release agents:
SiV: based on dimethyl polysiloxane. Viscous emulsion, s.g. 0·96
Si emulsion: aqueous dimethyl polysiloxane emulsion, s.g. 0·97
Si oil: dimethyl polysiloxane. Colourless oil, s.g. 0·97. Vulcanisates possess a high gloss surface and are smooth to the touch
K: alkali salt of a fatty acid derivative. Yellowish powder, s.g. 1·23. Soluble in water. *Quantity*: 0·2–2% solution (43).

LEVAPON Type T, wetting agent used for impregnating textiles with neutral latex, improves the acceptance and ability to adhere to the textile. Type TH, used in combination with other synthetic soaps as a foaming agent in the production of foam rubber. A 2–3% solution prevents the adhesion of unvulcanised sheets (43).

LEVAPRENE 450 Ethylene-vinyl acetate copolymer. May be vulcanised with peroxides. For optimum physical properties fillers are recommended (up to 80 parts reinforcing black or active filler, 200 parts inert black or filler). Vulcanisates are serviceable from −60 to 180°C. Excellent resistance to ozone, oxidation, light ageing, hot air and steam, may be blended with other polymers (43).

LEWA METHOD A tapping method discovered by E. Koehler, former director of Lewa plantation in then German E. Africa. The bark is painted with substances which will coagulate the sap; incisions are made above the painted surface. The sap coagulates quickly in strips on the bark, and these may be removed after a short period. Used frequently in E. Africa for obtaining rubber from *Manihot Glaziovii*.

LIGHT AGEING Light cracking. Degradation of rubber under the influence of direct or indirect sunlight. The effects may be studied in the laboratory using a Fade-O-Meter, Weather-O-Meter, or sunray lamp, according to ASTM D 750-55 T.

LIGHT AGEING INHIBITORS Primarily mixtures of natural or synthetic micro and macrocrystalline waxes, used in concentrations of 0·5–2·5%. Protect rubber articles against the influence of light and prevent atmospheric cracking under static conditions by slowly blooming and forming thin protective layers on the rubber surface.

LIGHT SCATTERING Method for determining mol. wt. of polymers.

LIGNIN $(C_{20}H_{22}O_6)_n$, polydehydro diconiferyl alcohol. High mol. wt. benzene derivative; produced from spent lye, a result of cellulose being obtained from wood. Brown powder, s.g. approx. 1·3. Soluble in acetone, ethylene glycol and triethanolamine, sparingly soluble in benzene, butyl acetate and trichloroethylene, insoluble in water. Used as a low gravity filler (also as dispersing agent for other fillers), gives improved resistance to ageing and a shining, velvety surface, reinforces if it is coprecipitated with latex as a master batch. May be used in light-coloured articles.

LIGRO Mixture of higher fatty acids. Activator and softener (248).

LIGROIN Petroleum distillate in fractions condensing at 20–135°C. Solvent for rubber.

LILION Polyamide fibre (244).

LIMITING VALUE The limiting value of a colloid is the minimum quantity of electrolyte necessary for coagulation, related to the actual final concentration, *i.e.* the concentration of the water phase. The limiting value of latex can be determined using calcium chloride. The determination is carried out by mixing 10 ml calcium chloride solution of varying concentrations with 10 ml latex. After leaving to stand for 1 min, 50 ml water are added to the mixture and some of the compound allowed to flow out onto a watchglass. The appearance of floccul action determines the experimental limiting value. The corrected limiting value is virtually independent of the rubber content. The addition of stabilisers to the latex increases the limiting value.

The corrected limiting value may be calculated from:

$$G_k = \frac{100}{100 + (100 - R)} \cdot G$$

where G is the experimental limiting value, and R is the volume of dry material in 100 ml.

LINDEMANN PROCESS Two-stage vulcanisation process formerly used for the production of sponge rubber. The first stage takes place under pressure, first in water, then in air; this prevents blowing and ensures uniform heat absorption. In the second heating stage uniform cells are formed because the pressure is reduced.

LINEAR POLYMERISATION Polymerisation producing chain-like macromolecules with the monomer units arranged linearly.

LINER Length of textile used as an intermediary layer between, *e.g.* calendered sheets, topped or frictioned fabrics, extruded profiles. Used to prevent adhesion of the rubber articles.

LINOLEIC ACID
$C_{17}H_{29}COOH$. Colourless liquid, s.g. 0·95. Soluble in alcohol and ether. Is an unsaturated fatty acid with three double bonds. Occurs in natural latex in small quantities (acetone extract). Activator; zinc oxide is necessary.

LIPIDS Collective term for a group of inhomogeneous animal and plant substances with similar physical properties, especially in solubility, to fats. Phosphatides and phytosterols fall within this group and occur in small quantities in natural latex, balata, gutta-percha, and the latices of a number of rubber producing trees.

LIPINOL SV Secondary plasticiser, extender (189).

LIRI Abbreviation for Licentiateship of the Institution of the Rubber Industry. Awarded on the basis of examination.

LITCHFIELD, PAUL WEEKS 1875–1959. In 1900 joined the Goodyear Tyre and Rubber Co. (founded in 1898 by Frank A. Seiberling). Director from 1906, appointed chairman in 1930. Played a great part in the development of the rubber industry.

LITEX Emulsion paints based on styrene-butadiene copolymers with a high styrene content.
Types:
F, FM, SB: 40% butadiene
SB 35: 35% butadiene
SB 20: 20% butadiene. The hardness of the film increases with increasing styrene content

MPD 344: emulsion used to increase the hardness of foam rubber (189).

LITHIUM STEARATE
$C_{17}H_{35}COOLi$. Colourless powder, m.p. 221°C. Soluble in water and alcohol. Lubricant for rubber, stabiliser for vinyl resins.

LITHOLITE Lithium stearate (249).

LITHOPONE White pigment with good covering power, coprecipitate of zinc sulphide and barium sulphate. Contains approx. 1% zinc oxide, s.g. 4·2. Produced by the reaction of barium sulphide and zinc sulphate which are soluble in water:

$$BaS + ZnSO_4 \rightarrow BaSO_4 + ZnS$$

Gives good white base for pigments, dyes and colourants. An inert filler, has no effect on vulcanisation and ageing properties. The ageing of cold cured vulcanisates is improved by using lithopone. Stiffens unvulcanised compounds and is suitable therefore for extrusion and calendering compounds; also for dispersion in latex compounds. Oil absorption 12–20%. Lithopone tyres are graded according to their zinc sulphide content:

Yellow seal	15%
Red seal	30%
Lilac seal	35%
Green seal	40%
Bronze seal	50%
Silver seal	60%

The covering power increases with increasing zinc sulphide content.

LITRE LOAD The ratio of the bearing capacity Q of a tyre (in kg) and its internal volume V (in l):

$$L = \frac{Q}{V}$$

$$Q = V \cdot L$$

The value obtained, according to the dimensions of the tyre, is 18–22 kg/l and is used in calculating the load capacity or volume of a tyre.

LLM Abbreviation for low, low modulus.

LM Abbreviation for low modulus (carbon black giving a low modulus).

LMD Abbreviation for lead dimethyl dithiocarbamate.

LOAD CAPACITY The USA Tire and Rim Association has produced the following equation to calculate the load capacity of a tyre.

$$Q = K \cdot p^{0.585} \cdot B^{1.39} \cdot (F + B)$$

where

Q is load in pounds,
p the air pressure in lb/in^2,
B the breadth of the cross-section in inches,
F the diameter of rim in inches, and
K a constant; 0·465 for passenger vehicles, 0·425 for heavy vehicles.

A similar formula for the metric system is:

$$Q = 25 \cdot 9 (p \cdot V)^{0.688}$$

where

Q is the load in kg,
p the air pressure in atm, and
V the volume of the tyre in litres.

LOANDO Natural rubber from various *Landolphia* types (Congo and Angola).

LOANGO Former trade term for natural rubber growing in Africa (Gaboon).

LOMAR D Sodium salt of a high mol. wt. sulphonated naphthalene condensation product. Dispersion aid and viscosity controller.
Uses: dispersion of blacks in latex (261).

LOMAR PW Sodium salt of a condensed mononaphthalene sulphonic acid. Dispersing agent (250).

LOMBIRO Local term for the bush or liane *Cryptostegia Madagascariensis* (Madagascar) and the wild rubber formerly obtained from it. The sap from the base of the tree was supposed to yield a useful product but that from the higher branches was extremely tacky. The latex contains 7–10% raw rubber and is difficult to coagulate. The product consists of approx. 88% rubber and 12% resins.

LONGINOS, JOSE L. MARTINEZ d. 1803, Mexico. Took part in the Spanish botanical expedition into Mexico led by Cervantes; studied *Castilla elastica* and suggested uses for latex and rubber.

LONG TON 2240 lb (1016 kg).

LOPOR Nos. 40, 42 and 80; petroleum fractions. Plasticisers (41).

LOPORI Former trade term for a wild rubber which is obtained from the Congo.

LORIVAL Group of liquid products. Types R 5, R 25, and R 200 have a viscosity of 5000, 25 000 and 200 000 cp respectively. CR types contain sulphur and may be vulcanised to give hard and soft products.
Uses: primarily as softeners, bases for colour pigments, moulded articles, coatings, bonding agents for grinding wheels, CR types are used for impregnations and textile coatings (251).

LORKARIL Terpolymer of styrene, acrylonitrile and butadiene (252).

LOROL Mixture of higher fatty alcohols, chiefly lauryl and myristyl derivatives. The sodium salts of the sulphate monoester are primarily used as emulsifiers, wetting agents and stabilisers for latex.
TN: Aquarex (6).

LOW SECTION TYRES Tyres designed so that the height is only about 88% of the breadth, the carcase taking the form of an elongated ellipse.

LOW TEMPERATURE RESISTANCE Brittle point, brittle temperature. Temperature at which the rubber becomes brittle. The lowest temperature at which a test piece of an elastomer or a plastic can resist sudden mechanical strain. Apart from the type of elastomer and the degree of vulcanisation, the value depends on the test apparatus, the conditions under which the test takes place, the shape of the test piece and the speed and power of the impact.
Determination: ASTM D 736-54T (Thiokol method), investigates the fracture of test pieces bent at −40 to −55°C; gives only an indication of whether the test piece is resistant to fracture at these temperatures.
ASTM D 746-57T defines low temperature resistance as the temperature at which 50% of the test pieces break.

$$T_b = T_h + \Delta T \left(\frac{S}{100} - \frac{1}{2} \right)$$

where

T_b is the temperature in °C at which the test piece breaks,

T_h is the highest temperature at which fracture occurs in all test pieces,

T is the difference between the individual temperatures at which measurements are made, and

S is the sum of the fractures as a percentage of the total tests at each temperature from the temperature at which the first fracture occurred to the final T_h.

Two types of testing machines for breakage (motor and solenoid power) are described in the specifications: ASTM D 797-58, determination of the variation of elastic modulus with temperature, ASTM D 1043-51, (Clash-Berg test), determination from the temperature dependence of the torsion angle of a twisted test piece at constant torque, ASTM D 1053-61 (Gehmann test), determination of the shear modulus from the torsion angle at varying temperatures.

LOXIOL G Group of fatty acid esters, dispersing agents, lubricants, release agents for NR, SR and plastics:

10: fatty acid ester of a polyvalent alcohol, m.p. approx. −4°C. Pigment dispersing agent, lubricant used in the cable industry and for extruded compounds 31: neutral fatty acid esters, m.p. approx. 18°C. Lubricant and release agent for NR and SR, particularly for nitrile rubber (429).

LOXITE 3000 Bonding agent for metal/rubber components, consists of poly-2:3-dichlorobutadiene, or a copolymer of 2:3-dichlorobutadiene and chlorovinyl acetylene with chlorinated polymers, *e.g.* chlorinated rubber, rubber hydrochloride, p.v.c., polyvinylidene chloride. Benzoyl peroxide is used as a curing agent. Xylene or toluene are suitable for Loxite 3000 cements, which must be used warm (approx. 70°C) (129).

LOXITE 4529 Aqueous reclaim dispersion. Black liquid, s.g. 1·08, 43–45% solids. Processing as for natural latex (129).

LP Abbreviation for liquid polymer.

LPD Abbreviation for lead pentamethylene dithiocarbamate.

LP TYRE Abbreviation for low pressure tyre.

LS Abbreviation for low structure black.

LS-53 Flurosilicone rubber. Used in the aircraft industry (7).

LS RUBBER Abbreviation for latex sprayed rubber. Compressed blocks made from rubber obtained by the atomisation of latex, according to Hopkinson.

LTP Abbreviation for low temperature polymerisation (production of cold rubber).

LUBRINE F Slippery rubber used for products requiring a low coefficient of friction (*e.g.* 'O' rings, V-fillings, bearings). Produced by grafting active monomers or oligomers on the elastomer surface. The grafted-polymer is hydrolysed to the equivalent acid and fluorinated. The substitution of hydrogen atoms on the surface by fluorine, lowers the dipole nature of the surface and therefore the force of attraction of surfaces in contact.

LUCEL ADA Azodicarbonamide (65).

LUCIDOL-70 Trade name for benzoyl peroxide (171).

LUCITE Group of acrylic resins (6).

LUDERSDORFF, FRIEDRICH WILHELM German chemist, 1801–86. Known for his research into rubber. In his book *Das Aufloesen und Wiederherstellen des Federharzes, Genannt Gummi Elasticum, zur Darstellung Luft-und Wasserdichter Gegenstaende*, Berlin, 1832, vulcanisation is discussed in a written work for the first time. Apart from various erroneous conclusions, he made three significant discoveries: turpentine oil and sulphur give rubber solutions which are not tacky after drying; the tackiness of fresh rubber coatings may be removed by sulphur powder; and a specially built vessel (a type of plasticising machine) makes the solution process easier. He also vulcanised gutta-percha and produced rubber compounds. Some of his results were used by Berzelius in his chemistry textbook.

LUDOX 30% colloidal solution of polymerised silica.
Uses: to increase the tack of latex and other adhesives (6).

LUEPKE PENDULUM Impact pendulum used to determine the resilience of rubber, according to P. Luepke.

$$\frac{1}{1 - \cos \Phi} = 20 \cdot 0$$

Specification: BS 903: A8: 1963.

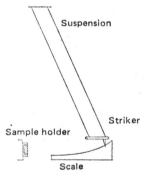

Lit.: *Rubber, Chem. & Techn.*, 1934, 5, 591.

LUMCO Group of furnace blacks produced by M. H. Lummerzheim and Co., Ghent (253). In 1951 the company built the first furnace black plant in Europe.
810P: ISAF
840P: HAF
850P: FEF
VGVP: SRF.

LUMINESCENT COLOURS A luminescent effect can be achieved in rubber compounds and plastics by using luminescent pigments. The maximum effect is only achieved in transparent compounds:
Zinc sulphide, green.
Zinc cadmium sulphide, orange and yellow.

LUMPS Rubber scrap obtained during the processing of latex in factories; results, *e.g.* from spontaneous coagulation, precoagulation, drying. May be processed to brown crêpe.

LUNOMETER Apparatus used to determine the density of fibres in textiles; based on the principle of light interference.

LUPEOL $C_{30}H_{50}O$, triterpene ester. Needle-like crystals, m.p. 215°C. Soluble in warm alcohol, ether, benzene, petroleum ether. Occurs in latex of *Cryptostegia* types (*C. madagascariensis, C. grandiflora*) and various *Ficus* species. Present in balata and jeluton as the acetate.

LUPERCO Group of organic peroxides (415):
101 XL: 2:5-dimethyl-2:5-di(tert-butyl peroxy)hexane on an inert filler
130 XL: 2:5-dimethyl-2:5-di(tert-butyl peroxy)hexene-3 on an inert filler
230 XL: organic peroxide
AST: benzoyl peroxide as a paste in silicone oil
CST: 2:4-dichlorobenzoyl peroxide as a paste in silicone oil
WET: benzoyl peroxide powder with 25% water.

LUPERSOL Group of peroxides
(65):
101: 2:5-dimethyl-2:5-di(tert-butyl peroxy)hexane
130: 2:5-dimethyl-2:5 di(tert-butyl peroxy)hexyne-3
DDM: methyl ethyl ketone peroxide.

LUTOIDS Structureless, voluminous, clear particles which occur in fresh latex and take up approx. 20–25% of the total volume. The particles dissolve when ammonia is added, dilution causes contraction. Lutoids appear to have an important effect on the stability, viscosity creaming and coagulation properties. Discovered by N. L. S. Homans and G. E. van Gils.
Lit.: *Nature*, 1948, 161, 177.

LUTONAL Group of polyvinyl ethers. Sensitisers for natural and synthetic latices (32).

LUV 36 Thermal black (167).

LV RUBBER NR with controlled, low viscosity, stabilised by hydroxylamine hydrochloride. Contains approx. 3% oil, Mooney viscosity approx. 50–60.

LYCRA Elastic fibres based on 1:4-oxybutylene glycol (6).

LYTRONE Group of synthetic polyelectrolytes, polymers and co-polymers which are soluble in water and alkalis.
Uses: thickening agents for latex (5).

M

M-24 Experimentally produced polyacrylonitrile fibre (193).

MAC AND LAI RUBBER Unsmoked sheet rubber with the appearance of smoked sheets. A mixture of oxalic and acetic acids with p-nitrophenol and formaldehyde is used as a coagulating agent (p-nitrophenol 1–3 parts : 1280 parts rubber) and formaldehyde act as fungicides or bactericides and make preservation by smoking unnecessary. As p-nitrophenol causes processing difficulties in rubber by discoloration and blooming, it has limited use. The insulation resistance is lower than that of smoked sheets.
Lit.: BP 826 709.
USP 2 944 990.

MACINTOSH, CHARLES, FRS 1766–1843, b. Glasgow, d. Dunchatton. In 1823 invented the raincoat named after him, in which a rubber layer obtained from a solution in benzene is enclosed between two layers of material. Former methods of impregnation, such as those of Samuel Peal (1791) and Rudolf Ackermann proved to be unsuitable.

MACQUER, PIERRE-JOSEPH 1718–1784. French doctor and chemist. Member of the Academy of Sciences for 27 years, director of the porcelain factory at Sèvres, researched rubber solutions. In 1761 wrote an article on turpentine and discovered ether to be a suitable solvent. He recognised the possibilities for the use of rubber and was the first (in 1768) to suggest the production of rubber probes and medical tubes; was also the first to note that by heating rubber in a Papin's autoclave, a horn-like, hard resin is produced. Macquer was interested in many branches of science and wrote several extensive textbooks on experimental and theoretical chemistry. His writings on rubber appeared in *Histoire et Reports de l'Academie pour l'Annee 1768*.

MACROMOLECULES Molecules which, in solution, have numerous atoms with a minimum mol. wt. of approx. 10^4. The physical properties, particularly the viscoelastic ones, must not be perceptibly changed by increasing the number of basic groups by one. There is no firm boundary between lower molecular and macromolecular materials, as the two overlap to some extent.

MACWARRIEBALLI Worthless, tacky wild rubber from *Forsteronia gracilis*, obtained from British Guiana in 1880.

MADAGASCAR RUBBER Former trade name from Madagascar where numerous rubber yielding plants grew, but most of which yielded products of low quality which were not worth harvesting. *Euphorbia intisy, E. pirahazo*, fourteen *Landolphia* types (including *L. madagascariensis* which gave the so-called Madagascar rouge), ten *Mascarenhasia* types (including *M. anceps, M. longiflora, M. utilis, M. lisanthiflora* which yielded 'Majunga'), *Plectanea elastica, Pl. microphylla, Cryptostegia madagascariensis, Gonocrypta Grevei, Kompitsia elastica, Marsdenia verucosa, Secamonopsis*

madagascariensis, Pentopetia elastica are named as rubber producers. In 1912 the total production of wild rubber from Madagascar was 847 tons.

MAGCARLUV Trade name for magnesium carbonate (167).

MAGLITE D Dense magnesium oxide with a small particle size (255).

MAGNALUV Trade name for magnesium oxide (167).

MAGNESIUM CARBONATE $MgCO_3$, basic magnesium carbonate, white powder, s.g. 3·04. $(MgCO_3)_4$ $.Mg(OH)_2.5H_2O$, magnesia alba, fine, loose white powder. Accelerates vulcanisation, has a reinforcing effect and is used as a reinforcing filler for soles, heels, hose lines, and hot water bottles. Gives stable compounds for open air curing, decreases the tendency to ignite. The refractive index is almost the same as that of natural rubber so that practically no pigmentation occurs. Suitable as a filler for transparent compounds.

MAGNESIUM OXIDE MgO, magnesia usta, light and heavy. Loose, white powder, s.g. 3·2–3·7. Has a slightly accelerating effect because of its basic nature. Prior to the introduction of organic accelerators, it was used with magnesium carbonate and litharge as the sole accelerator. Has a reinforcing and stiffening effect and gives high moduli and hardnesses but poor processability and very low nerve. Vulcanisates have a high permanent set.
Uses: in natural rubber compounds containing factice to overcome the retarding effect of the factice during

vulcanisation, to neutralise traces of acid in cold cured rubberised textiles, to take up moisture, to improve the ageing properties of cold vulcanisates and the electrical properties of cable compounds, activator for hard rubber, stabiliser and vulcanising agent for neoprene, increases the scorch resistance of neoprene.

MAGNESIUM STEARATE

$Mg(C_{18}H_{35}O_2)_2$. White powder, m.p. 125–130°C, s.g. 1·04. Soluble in hot alcohol, insoluble in water. Adheres well, prevents the adhesion of unvulcanised compounds.
Uses: release agent, dusting agent.

MAGNESIUM USTA Magnesium oxide.

MAGNETIC RUBBER Magnetic properties may be introduced into rubber by mixing or pressing, into the surface, barium ferrite powder.

MAJUNGA Trade name for *Mascarenhasia* and *Landolphia* type natural rubbers produced in Madagascar in the early days of the rubber industry. Includes majunga (Madagascar noir) from *Mascarenhasia anceps, M. longifolia, M. mangorensis, M. utilis, M. lisanthiflora,* Madagascar rouge (Piralahi) from *Landolphia Perieri,* majunga noir from *Mascarenhasia arborescens.* Terminology for these was not consistent.

MAKROLON High mol. wt. polycarbonate. Stable at temperatures between −100 to 135°C, transparent, resistant to light ageing, odourless and tasteless, physiologically inert, stable electrical properties up to 140°C. Has good dimensional stability (43).

MAKROMOLEKULARE CHEMIE, DIE Journal for macromolecular research in basic chemistry. Publisher: Verlag Dr A. Hüthig, Heidelberg and Frankfurt.

MALEIC ACID

$$H—C—COOH$$
$$\|$$
$$H—C—COOH$$

cis-1:2-ethylene dicarbonic acid. White crystals, m.p. 130–131°C, s.g. 1·59. Soluble in water, acetone, alcohol, glacial acetic acid.
Use: retarder for chloroprene rubber, has no effect on the physical properties of vulcanisates.

MALEIC ACID ANHYDRIDE

Ortho-rhombic crystals or white mass, m.p. 52·5–52·8°C, may be sublimated easily, b.p. 202°C, s.g. 1·48. Soluble in water, forming maleic acid, in alcohol, forming the ester, and in acetone, ethyl acetate, chloroform, dioxane and benzene.
Uses: retarder for natural and synthetic rubber, does not affect the physical properties. In natural rubber compounds in quantities of 5–15%, together with acrylic acid, acrylonitrile or benzoyl peroxide, it causes an increase in hardness and strength and a reduction in solubility.

L-MALIC ACID

$$HO.CH.COOH$$
$$|$$
$$CH_2.COOH$$

L-malic acid. White crystals, m.p. 99–100°C, s.g. 1·60. Soluble in water.

275

Uses: retarding agent for NR and ST, and for latex at 120°C, anti-gel agent in nitrile rubber solutions, adhesives and cements.

MAMALAWA Local term for the product of *Macaranga Reineckei* (Samoa). Similar to gutta-percha.

MANDRIANAMBO Local term for a natural rubber from *Landolphia mandrianambo* (Madagascar).

M AND S TYRES Mud and snow tyres with a coarse tread pattern for winter driving.

MANGABEIRA Local term for *Hancornia speciosa Gom* and the rubber obtained from it (tropical S. America: Brazil). The trees reach a maximum height of 5 m and were tapped by making a spiral incision around the trunk. The latex was coagulated either by heat or using alum, sodium chloride or the latex of the Casinguba tree which also grows in Brazil and is an excellent coagulant for *Hevea* latex. Only from wild trees because the slow growth rate was unsuitable for plantation work. The trees are ready for tapping after 15 to 20 years. The product came on the market as brownish red bales, pink in the centre. It was also produced as thin sheets ('Santos sheet' or 'Rio') which had a lower content of water and resin. The rubber is of good quality but was often debased by the use of poorer quality latex from *Plumeira drastica*.

MANGANESE In 1921 Bruni and Pelizzola discovered that manganese has a degrading effect on rubber and causes excessive tackiness. The average manganese content of plantation rubber can be taken as approx. 0·0003 % (3 ppm), lower quality types of rubber, particularly the blanket type, often contain much higher amounts. The maximum quantity allowed is fixed by RMA at 10 ppm.
Lit.: *India Rubb. J.*, 62, 13.

MANICOBA DE JEQUIE Local term for *Manihot dichotoma Ule* (Brazil: Bahia) and the wild rubber obtained from it.

MANICOBA DE PIAUHY Local term for *Manihot piauhyensis Ule*, a wild rubber (Brazil).

MANIHOT One of the S. American trees of the *Euphorbiaceae* family, growing mostly in the dry areas of Brazil. Was grown on plantations in the early days of the rubber industry. In 1912, approx. 50 000 ha of these trees were being grown in former German E. Africa. The main species was *M. glaziovii Müll. Arg.* (Ceara rubber tree); also *M. dichotoma, M. heptaphylla* and *M. piauhyensis.*

MANIHOT DICHOTOMA ULE Manicoba de Jequie, Jequie rubber. Wild clumps of trees found in Brazil (SE Bahia). This tree yields a light, tacky rubber. Tapping, as with *Hevea*, is done through spiral or fish-bone incisions.

MANIHOT GLAZIOVII MULL. ARG. Discovered by the French botanist Glaziov. Grows in Brazil (Ceara, Rio Grande do Norte and Para). The rubber brought on the market was known as Ceara rubber or Manicoba Ceara. The trees reach a height of up to 15 m and have knobbly thickenings at the roots.

The sap contains approx. 20% rubber and is rich in protein. In addition to the primitive local tapping methods, in which the rubber was partially coagulated in cuts in the bark or on the ground, or dried ('procede du choro', tear method), it was tapped in a similar manner to *Hevea Brasiliensis*. Processing was done by the smoke method. According to the quality, a distinction was made in Brazil between Borracha fina or Borracha defumado, which was obtained from the latex by using the smoke method. Sernamby came from the rubber already coagulated in the cup; Choro from the rubber coagulated on the bark. E. Africa produced balls which were often processed into crêpe or pressed into blocks. The latex particles are rod-shaped and up to 10 microns long.

MANIHOT PIAUHYENSIS ULE Rubber producing *Euphorbiaceae*, grows in Brazil (mainly SE Piauhy). Annual yield approx. 5 kg of rubber per tree. The most productive method is to tap immediately over the neck of the root, so that the rubber is obtained on the earth at the base of the trunk, the ground having been layered with clay. A few plantations have been established in Piauhy.

MANOSIL White reinforcing filler, VN3: pure, precipitated silica, s.g. 1·95, particle size 15 mμ; AS7: pure, precipitated aluminium silicate, s.g. 1·95, particle size 45 mμ.
Uses: transparent and translucent articles, soles, white walled tyres (256).

MANOSPERSE A Softener and processing aid for highly loaded compounds, particularly with white reinforcing fillers and reinforcing resins (256).

MAPICO Group of iron oxide pigments (159).

MARASPERSE CB AND N Sodium lignin sulphonate. Dispersing agent (257).

MARATEX Basic calcium lignin sulphonate. Filler (257).

MARBON Group of styrene-butadiene copolymers with a high styrene content. V: m.p. 90°C, s.g. 1·27. Reinforcing filler for NR, SBR, NBR, and CR. Processing aid, improves calendering and extruding properties and decreases the shrinkage. Causes a marked increase in hardness, modulus, tensile strength, tear and abrasion resistance, also improves electrical properties. Series 8000: A, AE, E, pure resin. Series QBX: master batch with 67% pure resin and 33% oil extended SBR (172).

MARINE GLUE Solution of rubber, resins, asphalt, tar, or similar solutions, in petroleum, carbon disulphide, turpentine oil and other solvents. Used as a caulking material, in the form of saturated oakum, for filling cracks in joints of ships.

MARK, HERMANN b. 1895. Director of the Institute of Polymer Research, Brooklyn, former professor of physical chemistry at the University of Vienna. Research includes the processes of polymerisation, structure analysis of high mol. wt. materials, crystal structure, X-ray analysis. Co-editor of the *Journal of Polymer Science*, co-author of *Makromolekulare Chemie*, 1950.

MARLEX 50 Highly crystalline polyethylene with a linear molecular configuration, m.p. approx. 127°C. Has high tensile and tear strength, resistant to temperatures of −115 to 120°C, to solvents and chemicals, has good electrical properties and low gas permeability. Was produced using special catalysts (chrome oxide on aluminium oxide and silicon dioxide) at low temperature and pressure, approx. 35 kg/cm^2 (185).

MARMIX Aqueous dispersion of a modified styrene resin. Reinforcing filler for natural and synthetic latices, increases modulus, hardness and tensile strength.
Uses: paper impregnation, protective coatings, adhesives, compounds used for electrical purposes (172).

MARMIX 15720 Aqueous dispersion of a 50:50 butadiene-styrene copolymer. Contains 49% solids, pH 11·0. Processed in the same way as SBR latex.
Uses: impregnation, adhesives, saturation (172).

MAROFIX Group of neoprene adhesives (426).

MAROPLAST Group of butyl and butylene sealing compounds (426).

MARTENS APPARATUS Test apparatus for determining the dimensional stability of hard plastics in heat and under a static bending strain. The temperature at which a test piece may be bent by 6 mm while under load is determined. DIN 53 462, 53 458.

MARTINDALE APPARATUS Apparatus for determining the wear and tear of textiles and frictioned and coated cloth.

MARZETTI PLASTOMETER First simple extrusion plastometer, in which the masticated rubber or compound is compressed through an aperture by air pressure, up to 80 atm. The weight of thread extruded in a given period of time is taken as an indication of plasticity. The thread can be marked by a blade at fixed time intervals. To a certain extent the vulcanisation behaviour can be assessed by the increasing reduction of the distance between each notch.

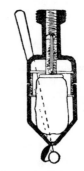

Marzetti plastometer

$$\frac{V}{t} = CP^2$$

where V is the extruded volume in mm^3
t is the time (15 minutes)
P is the pressure in kg/cm^2 and
C is a constant.

MASCARENHASIA Rubber producing member of the *Apocynaceae* family (Kenya, Madagascar, Mozambique). Used to give considerable quantities of wild rubber. Tamarind juice is the usual coagulant.

MASSAI Trade name for an African wild rubber from *Landophia florida* and *L. owariensis* which was marketed in the form of balls (niggers) from Sierra Leone.

MASSICOT Lead monoxide, PbO.

MASTERBATCH Precompound of rubber with a compounding ingredient in a higher concentration than in the final compound.

MASTICATION Plasticisation of rubber by mechanical and thermal means on open rolls or in an internal mixer in which reaction with oxygen causes scission of the molecular chains. Mastication is more effective at low temperatures; at high temperatures a reversible thermal softening occurs which prevents degradation. The mastication process may be accelerated by using peptisers, softeners and plasticisers. The plasticity is taken as a measure of the degree of mastication.

MATEZITE
$C_6H_6(OH)_5.OCH_3$, d-monomethyl inositol, d-inositol-3-methyl ether, m.p. 168°C. Isomer with bornesite. Was isolated from Madagascar rubber (*Mateza roritina*).

MAVOKELI Local term for *Pentopetia elastica* (Madagascar) and for the wild rubber obtained from it.

MAYUMBA Former trade term for wild rubber from the Congo, in particular from *Landolphia* and *Ficus* species.

MAZER WOOD Gutta-percha, term used by the explorer John Tradescant, who brought the first specimen of this to Europe in 1656.

MBI Abbreviation for mercapto-benzimidazole.

MBT Abbreviation for 2-mercaptobenzthiazole.

MBTS Abbreviation for 2:2'-dibenzthiazyl disulphide (6, 23).

MBT-XXX Purified, odourless mercaptobenzthiazole (21).

MCNAMEE CLAY American type of kaolin.

MDI Former trade name for Hylene M.
M-50: 50% solution in o-dichlorobenzene (6).

MEADOL Alkaline lignin. Produced by fractional precipitation of sulphite lye with carbon dioxide. Type MRM: free lignin, acts as a weak acid. Type MWS: sodium salt. Type MLP: purified MRM form.
Uses: reinforcing filler for natural rubber and SBR when coprecipitated from the latex; latex impregnated paper (258).

MEALORUB (From meal of rubber.) Vulcanised rubber powder produced from latex.

MECHANICAL STABILITY Measure of the stability of concentrated latex when subjected to mechanical motion. Determined by stirring the latex with a standard stirrer at 14 000 rpm. The measure taken is the time required for the first appearance of flocculation, which is recognisable by a characteristic wave motion on the surface of the latex at a specified light angle. ASTM D 1076–59, BS 1672:2: 1954.

MEF Abbreviation for medium extruding furnace (carbon black).

MEGUM Groups of bonding agents for metal/rubber components. Used for natural and synthetic rubbers, with the exception of butyl rubber and silicone rubber. Based on solutions of natural and synthetic rubbers with halogenated polymers, thermosetting resins and haemoglobin (180):
D 2064: solution in chlorobenzene
G-Red: solution in aliphatic solvents
SK: aqueous dispersion.

MEKP Abbreviation for methyl ethyl ketone peroxide.

p-MENTHANE HYDROPER-OXIDE Yellowish liquid, s.g. 0·910–0·925. Polymerisation catalyst used in the production of styrene–butadiene copolymers at low temperatures.

$$H_3C-HC \begin{array}{c} CH_2-H_2C \\ CH_2-H_2C \end{array} CH-\underset{\underset{CH_3}{|}}{\overset{\overset{OH}{\underset{|}{O}}}{\underset{|}{C}}}-CH_3$$

Lit.: USP: 2 775 578 (1956), 2 735 870 (1956).
Rubb. Age, February 1957, 860.

MENTOR 28 Refined, aliphatic petroleum hydrocarbon. Softener (41).

MEPASIN $C_{14}-C_{18}$ paraffin hydrocarbons by the Fischer-Tropsch synthesis. Basic material for a wetting agent.

MERAC Aqueous solution of an activated dithiocarbamate and a mercaptobenzthiazole salt. Red-brown liquid, s.g. 1·034. Accelerator for natural and synthetic latices (4).

MERAKLON Trade name for isotactic polypropylene fibres (59, 60).

MERCAPTAN Term derived from 'corpus mercurio aptum' because of the ability to form characteristic mercury salts which are extremely difficult to dissolve. Thioalcohols, thiols and analogues of alcohols which contain sulphur, where the OH group is replaced by the SH group, $C_nH_{2n+1}SH$ react as weak acids and are oxidised by mild oxidising agents and by air, to dialkyl disulphides. The alkali salts are soluble in water.
Uses: vulcanisation accelerators (mercaptobenzthiazole, mercaptothiazole), peptising agents (naphthyl-β-mercaptan, xylyl mercaptan), bleaching agents for latex (xylyl mercaptan).

MERCAPTOBENZIMIDAZOLE

Yellowish powder, m.p. 290°C, s.g. 1·42. Soluble in acetone, alcohol, ethyl acetate, insoluble in benzene, carbon tetrachloride, benzine and water. Antioxidant giving good protection against normal oxidation and overcure. Non-staining. Has a retarding effect on mercaptobenzthiazole, tetramethylene thiuram disulphide and dithiocarbamates; in higher concentrations stiffens unvulcanised compounds. Small additions of phenyl-α-naphthylamine and phenyl-β-naphthylamine improve and enlarge the protection against oxidation. Sulphur-free vulcanisates of natural rubber, nitrile rubber and SBR, containing higher concentrations, have good steam resistance. In combination with aldol-α-naphthylamine and N-phenyl-N′-

cyclo-hexyl-p-phenylene diamine good resistance to hot air ageing is obtained. Unvulcanised compounds without zinc oxide soften and can not be stored. It is unsuitable for sulphur chloride cures. Has a sensitising effect in natural latex, and is unsuitable for articles which come into contact with food.

TN: Antioxidant MB (43)
Antigen MBZ (zinc salt) (86)
Antioxidant MB
Antioxidant ZMB (zinc salt)
Antioxygène MTB (19)
Antioxygène MTBZ (zinc salt) (19)
MBI
Nocrac CGP (with di-β-naphthyl-p-phenylene diamine) (274)
Nocrac MB (274)
Nocrac MBZ (zinc salt) (274)
Nonox CNS [with bis(2-hydroxy-3-α-methyl cyclo-hexyl-5-methyl phenyl)methane]
Permanax 21 (20).

2-MERCAPTOBENZTHI-AZOLE

Pale yellow powder with characteristic odour and bitter taste, m.p. 165–179°C (179°C for chemically pure substance), s.g. 1·4–1·5, mol. wt. 167. Soluble in alkalis, carbon disulphide, aromatic hydrocarbons, methylene chloride and other chlorinated solvents, alcohol, acetone, ethyl acetate insoluble in benzine hydrocarbons, carbon tetrachloride and water. Produced by heating aniline, carbon disulphide and sulphur under the pressure of the hydrogen sulphide formed during the reaction. Introduced into the rubber industry in 1925. A medium-fast accelerator with a long plateau effect, active above 121°C. Has a fairly low critical temperature and can cause scorching during processing. Gives vulcanisates with excellent physical properties. Because of the definite plateau effect variations in cure times have little effect on the rubber properties. Acts as an anti-scorch agent in sulphur-free compounds and as a retarder in neoprene. Compounds with dipentamethylene thiuram tetrasulphide produce a combination with delayed action.

Activation: as a weak acid, by basic accelerators such as aldehyde-amine and guanidine, and thiurams and dithiocarbamates. The critical temperature may be reduced considerably by small additions of thiurams, making a considerable reduction in sulphur and accelerator quantities necessary.

Compounding: 3–5% zinc oxide and 1–4% stearic acid are necessary. May also be activated by other oxides and carbonates. Lead oxide, which has a stronger effect than zinc oxide, reduces processing safety and also the vulcanisation range. Non-staining and non-discolouring on contact; unvulcanised compounds have a tendency to a sulphur bloom. Suitable for press and hot air cures. Vulcanisation temperature 125–160°C. Vulcanisates have good ageing properties, but have a characteristic odour which can be covered by the addition of deodorants.

Quantity:
Natural rubber and SBR
primary 1–2% with 3–2% sulphur, secondary 0·5–1% with 0·3–0·2% tetramethyl thiuram disulphide and 1·5% sulphur, sulphur-free 0·5% with 3–4% tetramethyl thiuram disulphide, for excellent ageing properties 1–1·5% with 0·5–0·25% tetramethyl thiuram disulphide, 0·5%

tellurium and 1·5–0·5% sulphur. Activation 0·1–0·5% dibutyl ammonium oleate, guanidine or aldehyde-amine, or 0·05–0·15% of a dithiocarbamate or thiuram.

Butyl rubber

(only a slight tendency to scorch) 0·5–1% with 0·5–1·5% tetramethyl thiuram disulphide and 1–2% sulphur.

Reclaim

(alkali reclaim shows a high tendency to scorch), 0·8–1·2% with 1–3% sulphur.

Latex

(has only a slight effect on stability and prevulcanisation), 1–2% with 1–3% sulphur and 0·1–0·2% KOH. Activated by thiurams or dithiocarbamates.

Uses: include tyres, conveyor belts, driving belts, mechanical goods, air hose, insulation materials, footwear, and impregnations. Suitable for light-coloured articles. Unsuitable for medicinal articles or for articles which come into contact with food (because of the bitter taste), which is also retained in the vulcanisates.

TN: Accel M (16)
Accelerante MBT (397)
Accélérateur 200
Accélérateur G
Accélérateur MBT (418)
Accélérateur Merkapto
Accélérateur rapide 200 (20)
Accélérateur rapide G (19)
+ Accelerator CX (6)
Accelerator Merkapto (408)
Amizen M (395)
Ancap (22)
Belger M (394)
Captax (51)
Croydax (120)
Diaceller M (392)
Eveite M (59)
Kuracap MBT (241)
MBT

MBT-XXX (purified, odourless)
Mercapto
Mercapto F
Mercapto F/P
Mertax (purified, odourless) (5)
Mitsui M
Nocceler M (274)
+ NUTX (100)
Polyzole MBT (237)
Rapidex KB (390)
Rotax (purified, odourless) (51)
Soxinol M
Thiotax (5)
Vulcadote M
Vulcafor M (60)
Vulcafor MBT (60)
Vulkaid MBT (3)
Vulkacit Merkapto (43)
Vulkacit   Merkapto/C   (surface treated) (43)
Vulkator PX (364)

Blends

with 66·7% tetramethyl thiuram disulphide: Accelerator 108 (23), Accelerator B (5), Butyl accelerator 21 (6), Captax–Tuads Blend (51)

with dipentamethylene thiuram disulphide: Robac Alpha (311)

with diphenyl guanidine tartrate: + Vulcafor DAT

with Acidic diphenyl guanidine salt: Vulcafor DHC (60)

with   dithiocarbamate: Vulcafor DHC (60)

with undisclosed substance: Robac GP (311)

Reaction products:

with monoethanolamine: Accelerator soluble Lat. 2 (19)

with anhydroformaldehyde-o-toluidine and heptaldehyde-aniline: + OVAC (23)

with hexamethylene tetramine and benzyl chloride: + Acrin (6)

with Diisopropyl sulphenamide: Dipac (120)

Derivatives:

Accelerator GRE (19), Takar No.

1–2 (20), + Vulkacit BZ (43), Z-88-P
with aldehyde-amine: BJF (23).
Lit.: USP 1 544 687, 1 631 871 (1927), BP 355 567 (1929).

2-MERCAPTOIMIDAZOLINE

Ethylene thiourea. White, crystalline, odourless powder, m.p. 195°C, s.g. 1·43. Fairly soluble in water and alcohol, insoluble in benzene and benzine. Exists in two structures:

Safe, fast vulcanising accelerator, easily dispersed for the neoprene Types W and GN and Hypalon 20. Effective concentration 0·1–1%. Active above 120°C. Non-staining. Vulcanisates show good ageing and mechanical properties.
TN: Accel 22 (16)
Axeline
MI-12 (180)
NA-22 (6)
Pennac CRA (1207)
Robac 22 (311)
Van Hasselt ETN (355)
Vulcarite 129 (50% aqueous dispersion) (57)
Blends: Kenmix NA-22 (50% with 50% Kenflex N) (238)
PND-70 (70% with 30% polyisobutylene) (282)
PZND-84 (12% with 72% ZnO-XX-78 and 16% polyisobutylene) (282).

MERCAPTOTHIAZOLINE

Yellowish white powder, m.p. 101°C, s.g. 1·50. Fast accelerator, its vulcani-

sates having good ageing properties. Active above 121°C, decomposes at vulcanisation temperatures above 160°C, safe to process with the addition of salicylic acid as a retarder, disperses easily in rubber. In latex compounds has no effect on stability or prevulcanisation. Does not cause discoloration.
Quantity: 1–2%, activated with 0·1–0·5% dibutyl-ammonium oleate or aldehyde-amine, 0·1–0·3% guanidine, 0·05–0·15% dithiocarbamate or thiuram. 1–4% stearic acid is necessary.
Uses: chlorobutadiene copolymers.
TN: Accelerator 2′MT (6).
Lit.: USP 2 269 472, 2 283 334–7.

MERLON Group of aqueous polyvinyl dispersions (5).

MERSOLATE Mepasine sulphonate. Sodium salt of sulphochlorinated C_{14}–C_{18} Fischer–Tropsch hydrocarbons.
Uses: redox polymerisation of Buna, wetting agent.

MESAMOLL Alkyl sulphonic acid ester of phenol and cresol. Yellowish, oily liquid, s.g. 1·05–1·07. Softener and extender for nitrile rubber. Large quantities increase the resilience and the resistance to low temperatures but reduce tensile strength, hardness and modulus.
Quantity: 5–10% for increased plasticity, 20–40% for increased elasticity and resistance to low temperatures (43).

METAL BOND PROCESS

Process for bonding metal parts with two-part adhesives, comprising an elastomer, *e.g.* neoprene, and a resin which hardens on heating, *e.g.* phenolic or melamine resin. The elastomer solution and the resin component are mixed on a glass-fibre or a nylon strip which has previously been dried. The adhesive strip is placed between the metal surfaces which are to be joined and these are then bonded by pressure or heat.

Uses: in the auto and aircraft industries.

METAL/RUBBER BONDING

Soft rubber cannot be bonded to metals or other materials without an intermediary layer, which will bond well to both the metal and the rubber. Suitable intermediary layers are:

Brass. Various rubber types form a strong durable bond by the reaction of sulphur during vulcanisation with the alloy constituents of brass. The strength of the bond depends on the type of rubber, the compound and the brass alloy. The brass must be of the α-form with 70–80% copper and 20–30% zinc. The parts of the metal to be bonded are galvanised with brass, by conventional methods, and painted with a rubber solution; the compound is applied and vulcanised under pressure at 130–175°C. Under certain circumstances vulcanisation can give good bonds without pressure being applied.

Hard rubber. Ebonite gives a strong bond to metal, but is suitable for only limited applications as the flexibility of the rubber is reduced by the hard intermediary layer. Moreover ebonite is thermoplastic and the bond strength is strongly reduced above 60°C and is zero above 100°C. In addition the high sulphur content may cause weaknesses in the soft rubber near the bonding layer. *Uses:* include solid rubber tyres, tank linings, pipes and pumps.

Chlorinated rubber. As a direct bonding agent for butadiene-acrylonitrile rubber and neoprene; for natural rubber and other elastomers an additional intermediary layer of neoprene is necessary. Suitable also for ceramics, glass, wood and other hard materials. The bonds are resistant to acids, alkalis, alcohols, aliphatic hydrocarbons and mineral oils, but not to aromatic hydrocarbons, ketones, esters, vegetable and animal oils and fats. They are heat resistant up to 140°C.

Rubber hydrochloride. Either alone, or, more commonly, in blends with other products, rubber hydrochloride, applied as a thin layer, gives a strong, non-thermoplastic bond between vulcanised compounds of natural and synthetic rubber and metal, ceramics, glass, wood, paper and similar products. The addition of reclaim to the compounds improves the strength of the bonding which is almost as strong as that of brass. The bonding has good resistance to ageing, is resistant to acids, alkalis, alcohols and oils, and shows good properties under dynamic strain.

Cyclised rubber. Bonds unvulcanised compounds and vulcanised rubber to metals (*e.g.* aluminium, lead and steel), ceramics, glass, plastics and similar materials. Cyclised rubber is thermoplastic, loses its bond strength at high temperatures (90–120°C) and will not withstand temperatures above 60°C. It is resistant to oxidation, acids, and water. The bond has excellent elasticity. *Uses:* include linings and impregnation.

Isocyanates. In this group 4:4′-diphenyl methane diisocyanate (50% solution in xylene or o-dichlorobenzene) and 4:4′:4″-triphenyl methane triisocyanate (20% solution in methylene chloride) have been shown to be excellent bonding agents for metals. The cleaned metals, which should have a somewhat roughened surface, are painted with the isocyanate solution, on which the rubber compound is placed and vulcanised. The isocyanate reacts readily but may not be used under damp conditions or where alcohols or amines are present, because the ability to bond will be destroyed.

Miscellaneous. Other products and processes suitable for general or special purposes include mixtures of albumen with latex, chlorinated/hypochlorinated rubber, dichloro-butadiene-chlorovinyl acetylene polymers (Loxite 3000), the Kalobond process based on butadiene-styrene-vinyl pyridine polymers, and various trade products (*e.g.* Chemosil, Megum) for unvulcanised compounds or vulcanised rubber.

Lit.: S. Buchan, *Rubber to Metal Bonding*, 2nd edn, London, 1959.

METAL TYRES Tyres in which the fibre cords are replaced by thin wires to impart greater strength and improve heat conduction. By reducing the number of plies, the vulcanisation period can also be reduced, by approx. 40%. The tyres are advantageous for high speeds and heavy loads but pose difficulties in construction, particularly in producing a good wire/rubber bond. They must be used at high internal air pressure.

METALYNE Tall oil methyl esters, s.g. 0·96. Softener (10).

METHACRYLIC ACID

$$H_2C=C\begin{array}{c}CH_3\\\\COOH\end{array}$$

Clear liquid, m.p. 15°C, b.p. 160°C, s.g. 1·02. Soluble in water, alcohol and ether. Its various esters may be polymerised easily and their uses include plastics, softeners, lacquers.

METHOCEL Group of hydroxypropyl methyl celluloses. Thickening agents for latex, suspension agents in suspension polymerisation of p.v.c. (61).

METHOX Dimethoxyethyl phthalate. Softener (184).

2-(4-METHOXYBENZENE DIAZOMERCAPTO)NAPHTHALENE Initiator in the emulsion copolymerisation of butadiene and styrene at low temperatures (obsolete).

METHOXY BUTANOL
$CH_3OCH(CH_3)CH_2.CH_2OH$, 3-methoxy butanol, m.p. −85°C, b.p. 161°C, s.g. 0·923. Miscible with water, alcohol and many organic solvents. Solvent.

METHOXY BUTYL ACETATE
$CH_3COOCH_2CH_2CH(OCH_3)CH_3$. Colourless liquid, b.p. 170°C, s.g. 0·95. Solvent for chlorinated rubber, resins, cellulose ether and oils.
TN: Butoxyl (217).

2-METHOXYETHANOL
$HO.CH_2.CH_2OCH_3$, ethylene glycol monomethyl ether, b.p. 124°C, s.g. 0·975. Miscible with water, alcohol, ether, acetone.
Uses: solvent for cellulose acetate, natural resins and some synthetic resins.
TN: Methyl cellosolve (112).

METHOXYETHYL ACETYL RICINOLEATE
$C_{17}H_{32}(OCOCH_3)COOCH_2$
$.CH_2OCH_3$,
m.p. −60°C, b.p. (1 mm) 195°C, s.g. 0·96. Softener.

TN: Flexricin P-4C (9)
Flexricin 62 (9)
Staflex IXA (330).

METHOXYETHYL RICIN-OLEATE

$C_{17}H_{32}(OH)COOCH_2CH_2OCH_3$, m.p. $-36°C$, b.p. (1 mm) 186°C, s.g. 0·951. Softener.
TN: Flexricin P-1C (9).
Staflex IX (330).

METHOXYETHYL STEARATE

$C_{17}H_{35}COOCH_2CH_2OCH_3$, b.p. (20 mm) 220–275°C, s.g. (25°C) 0·885. Softener.

METHYL ABIETATE

$C_{19}H_{29}COOCH_3$. Colourless to yellowish, thick liquid, b.p. 360–365°C with decomposition, flash p. 180–218°C, s.g. 1·033, mol. wt. 316. Soluble in most organic solvents, insoluble in water.
Uses: softener and solvent for rubber, plastics, natural resins and ethyl cellulose, also for the production of cements and adhesives.
TN: Abalyn (10).

METHYL ACETATE

CH_3COOCH_3. Colourless liquid, b.p. 57–59°C, s.g. 0·928. Soluble in water, miscible with alcohol and ether. Solvent.

METHYL ACETOXY STEARATE

$C_{17}H_{34}(OCOCH_3)COOCH_3$, m.p. $-7°C$, s.g. 0·926. Softener.
TN: Paricin 4 (9).

METHYL ACETYL RICIN-OLEATE

$C_{17}H_{32}(OCOCH_3)COOCH_3$. Pale yellow liquid with mild odour, m.p. $-30°C$, b.p. (1 mm) 185°C, s.g. 0·938. Softener for natural and synthetic rubbers and latices; improves resistance to low temperatures, increases the elongation at break.
TN: Flexricin P-4 (9).

METHYL BUTINOL

$(CH_3)_2C(OH)—C≡CH$, 2-methyl-3-butinol, 2-hydroxy-2-methyl-3-butin. Colourless liquid, m.p. 2·6°C, b.p. 104–105°C, s.g. 0·867. Miscible with water, acetone, benzene, benzine and other organic solvents, forms an azeotrope with water; b.p. 90·7°C, contains 28·4% water. Intermediary product in the synthesis of isoprene.

METHYL CELLULOSE

Cellulose methyl ether. White kernels. Soluble in cold water (insoluble in hot water), soluble in benzene, acetone, carbon tetrachloride, toluene, xylene, alcohol, butyl acetate, butanol. The solubility is dependent on the degree of substitution. Forms neutral solutions with water which are stable between pH 3–12. The presence of inorganic salts increases the viscosity of the solutions. Commercial methyl cellulose has a methoxyl content of approx. 29% (degree of substitution 1·8).
Uses: primarily as thickening agent for latex, subtitute for water soluble gums, bonding agents, adhesives.
TN: Methocel
Tylose
Hydrolose
Cellothyl
Cethylose
Cethytin
Syncolose
Bagolax
Cellumeth
Cologel.

METHYL CUMATE Copper dimethyl dithiocarbamate (51).

METHYL CYCLOHEXANOL

Colourless liquid, m.p. $-38°C$, b.p. 160–180°C, s.g. 0·93. Solvent.
TN: Methylhexalin
Heptalin
Methylanol.

2-α-METHYL CYCLOHEXYL-4:6-DIMETHYL PHENOL

Colourless liquid, s.g. 1·00. General purpose antioxidant, stabiliser for latex.
TN: Nonox WSL (60).

METHYL CYCLOHEXYL STEARATE

$C_{17}H_{35}OCOC_6H_{10}CH_3$, m.p. 9°C, b.p. 220–240°C (4 mm), s.g. 0·89. Softener.

4-METHYL-2:6-DITERT-BUTYL PHENOL

2:6-ditert-butyl-4-methyl phenol.
TN: Topanol O (60).

METHYLENE CHLORIDE

CH_2Cl_2, dichloromethane, methylene dichloride. Colourless liquid, b.p. 40–41°C, s.g. 1·326 (20°C). Miscible with alcohol and ether. Solvent, primarily for NR, SR and adhesives.

4:4'-METHYLENEBIS(2:6-DITERT-BUTYL PHENOL)

Yellowish white solid, m.p. 154°C, b.p. (40 mm) 289°C, s.g. (20°C) 0·990. Antioxidant for natural and synthetic rubbers and latices, gives good thermal stability. Non-discolouring.
TN: Binox M (2)
Ethyl Antioxidant 702 (88).

METHYLENE BISDIBUTYL THIOGLYCOLATE

Dibutyl methylene bisthioglycolate. Reddish-brown liquid, b.p. 185–190°C (2 mm), s.g. 1·09. Softener for SBR and NBR; increase in quantity increases the elasticity while reducing tensile strength, hardness and modulus, improves the resistance to low temperatures. Suitable as an emulsion in latex compounds.
Uses: in the aircraft industry, refrigerators, highly elastic articles, friction compounds.
Quantity: 5–10% for increased plasticity, 20–60% for high elasticity and low temperature resistance.
TN: Plastikator 88 (43).

METHYLENEBISETHYL BUTYL PHENOL

$CH_2[-(2)-C_6H_2(OH).C_2H_5.C(CH)_3-1:4:6]_2$, 2:2'-methylenebis(4-ethyl-6-tert-butyl phenol). White crystalline powder; m.p. 119–125°C, s.g. 1·10. Antioxidant giving good protection against heat and normal oxidation. Non-discolouring in white compounds and does not bloom on exposure to sun rays. Has no effect on vulcanisation.
Quantity: 0·25–2%.
TN: Antioxidant 425 (21).

2:2'-METHYLENE BISMETHYL BUTYL PHENOL

$CH_2—[(2)—C_6H_2(OH).(CH_3)$
$.C(CH_3)_3—1:4:6]_2$.
White crystals, m.p. 125–133°C, s.g. 1·04. Antioxidant for NR and SR.

Has no effect on vulcanisation, does not bloom, gives good protection against oxidation, improves flex cracking resistance, prevents surface cracking caused by exposure to light. Non-staining and generally non-discolouring but causes slight discoloration in white compounds on exposure to light.
Quantity: 0·24–1·5%.
TN: AC-5 (15)
Antioxidant BKF (43)
Catalin Antioxidant CAO-5
CAO-15
MBP 5 (380).

2:2′-METHYLENE BIS(4-METHYL-6-α-METHYL CYCLO-HEXYL) PHENOL

White powder, m.p. 120°C, s.g. 1·17. Antioxidant for chloroprene rubber.
TN: Nonox WSP (60).

1:1-METHYLENE BIS-2-NAPHTHOL

$CH_2(C_{10}H_6.OH)_2$. Antioxidant.
TN: Antioxygène M2B and WBC
CAD-32 (15).

METHYLENE BISSTEARAMIDE

$(C_{17}H_{35}.CO.NH—)_2.CH_2$. Wax-like solid, m.p. approx. 130°C. Gives moulded and extruded articles a very high glaze.
TN: Armowax (98).

4:4′-METHYLENEBIS(6-TERT-BUTYL-o-CRESOL)

White to yellowish powder, m.p.

102°C, b.p. 294°C (40 mm), s.g. 1·087. Antioxidant for NR, SR and latices, improves flex cracking resistance, resistance to oxidation and prevents surface cracking caused by exposure to light. Non-staining and non-discolouring.
TN: Ethyl Antioxidant 720 (88).

METHYLENE DIPHENYL DIAMINE

Methylene dianilide. Brown, resin-like mass, m.p. 55–60°C, s.g. 1·15. Soluble in benzene, acetone, methanol.
Uses: one of the first organic accelerators.
TN: + Accelerator 6 (6)
FA-mou.

CIS-ENDO METHYLENE TETRAHYDROPHTHALIC ACID ANHYDRIDE

(3:6-endo methylene 1:2:4:5-tetra-hydro-cis-phthalic acid anhydride). White, crystalline powder, m.p. 164–165°C. Soluble in acetone, chloroform, alcohol, ethyl acetate, carbon tetrachloride, benzene, and toluene. Reacts with water to give the corresponding acid. Produced by the reaction between maleic acid anhydride and cyclopentadiene in benzene.
Uses: retarder in SR and NR, the esters act as softeners.
TN: Nadic anhydride (265).

288

METHYLETHYL ACETOXY STEARATE

$C_{17}H_{34}(O.CO.CH_3)COO$
$.CH_2CH_2O.CH_3$,
m.p. $-7°C$, s.g. 0·948. Softener.
TN: Paricin 4C (9).

METHYL ETHYL KETONE

$CH_3CO.CH_2.CH_3$, 2-butanone. Colourless liquid, b.p. 79·6°C, s.g. (20°C) 0·805. Solubility in water 1:4, miscible with benzene, alcohol and ether. Solvent for adhesives and cements of NR and SR.

METHYL ETHYL KETONE PEROXIDE

MEKP. Extremely reactive catalyst for polyesters; appears on the market as a 20–60% blend with dimethyl phthalate.
TN: Lupersol DDM (65).

METHYL GLYCOL ACETATE

Solvent which can be blended; particularly for chlorinated rubber, acetyl cellulose, cellulose ether and resins.

1-METHYLINOSITOL

$C_6H_6(OH)_5OCH_3$, 1-inositol-2-methyl ether, quebrachite, quebrachitol, m.p. 192°C, sublimes in a vacuum at 210°C. Occurs in *Hevea* latex in concentrations of 1–2%. To obtain 1-methylinositol the latex serum is evaporated, the proteins coagulated by heating are removed and the concentrate crystallised out at 0°C; the crystals are separated and recrystallised. Yield is approx. 0·2% from the serum (E. Rhodes and J. L. Wiltshire, RRI, Malaya, 1931, 3, 160). An attempt was made to use 1-methylinositol as a substitute for sugar for diabetics, but it was found unsuitable (R. A. McCance and R. D. Lawrence, *Biochem. J.*, 1933, 27, 986).

METHYL ISOBUTYL KETONE

$CH_3COCH_2CH(CH_3)_2$, isopropyl acetone, 4-methyl-pentanone-2. Colourless liquid with camphor-like odour, b.p. 119°C, s.g. 0·802 (20°C). Solvent for rubber and plastics.

METHYL LEDATE

Lead dimethyl dithiocarbamate (51).

METHYL OLEATE

$C_{17}H_{35}COOCH_3$, b.p. 167–170°C (2 mm), s.g. 0·87. Insoluble in water. Softener.
TN: Emery 2031 (96).

METHYL PENTACHLORO-STEARATE

$Cl_5C_{17}H_{30}COOCH_3$, m.p. $-59°C$, s.g. 1·19. Softener.
TN: MPS-500 (214).

METHYL PHTHALYL ETHYL GLYCOLATE

$CH_3OCO.C_6H_4.COOCH_2$
$.COOC_2H_5$.
Colourless, odourless, non-toxic liquid, s.g. 1·22. Softener and plasticiser for SBR, NBR and cellulose plastics, improves mastication and processing. Has no effect on vulcanisation.
TN: Santiciser M-17 (5).

2-METHYL PROPENE

Isobutylene. Colourless gas, b.p. $-6°C$, s.g. 0·627. Soluble in alcohol, insoluble in water.
Uses: in the production of polyisobutylene.

N-METHYLPYRROLIDINE

Colourless liquid, m.p. $-24°C$, b.p. $205°C$, s.g. $1·03$. Miscible with aromatic hydrocarbons, alcohols, acetone and water. Solvent for polyacrylonitrile, and polyterephthalate, used to extract butadiene from the C_4 fraction of the Houdry cracking process. Butadiene extracted by this method has a high degree of purity and can be used directly for polymerisation without further purification.

$$
\begin{array}{ccc}
CH_2 & \!\!\!—\!\!\! & CH_2 \\
| & & | \\
CH_2 & & CH_2 \\
& \diagdown\;\diagup & \\
& N & \\
& | & \\
& CH_3 &
\end{array}
$$

METHYL RICINOLEATE

$C_{17}H_{32}(OH).COO.CH_3$. Pale yellow liquid with mild odour, m.p. $-30°C$, b.p. $170°C$ (1 mm), s.g. $0·924-0·938$. Softener for natural and synthetic rubber and latices; improves low temperature resistance and elongation.

TN: Flexricin P-1 (9).

METHYL RUBBER The first synthetic rubber; discovered by I. Kondakow (*J. Prakt. Chem.*, 1900, 62, 66) and developed by F. Hofmann in the Farbenfabriken, formerly Friedrich Bayer and Co. [DP 250 690 (1900)]. Production involved polymerisation of dimethylbutadiene obtained by the reduction of acetone to pinacol with the separation of water. The acetone was produced from acetylene *via* acetaldehyde, acetic acid and calcium acetate. In 1915 commercial production was begun when Germany was cut off from raw rubber supplies during World War I; 2350 tons were produced. However, since methyl rubber did not measure up to the properties demanded, production ceased at the end of the war. Two types were produced. For Methyl rubber H (hard vulcanisates) dimethylbutadiene was left in metal containers in the presence of air at approx. $30°C$ for 8 weeks. It formed a solid, white, cauliflower-like mass which could be rolled out to a rubber-like product after the addition of softeners. Small quantities were added as polymerisation nuclei to the dimethylbutadiene. The hard vulcanisates were superior to ebonite in electrical properties. For Methyl rubber S (soft vulcanisates) dimethylbutadiene was heated in autoclaves at $70°C$ for up to 6 months. A tough, rubber-like product was produced which could be processed on a mill. Neither type was resistant to air or light and both were stabilised with piperidine. To increase the elasticity diphenylamine, dimethyl aniline and toluidine were added. The physical properies of both types were much lower than that of natural rubber. Whitby and Katz (*Ind. Eng. Chem.*, 1933, 25, 1204–11) discovered that reinforcing blacks, which were not generally used at that time, considerably improved the properties.

A similar product, called Marke B, was dimethylbutadiene-sodium carbon dioxide rubber produced by the Badische Anilin und Sodafabrik (BASF). In this method sodium wire was brought into contact with carbon dioxide and dimethylbutadiene. After a few weeks a rubber-like product developed which was purified by washing it free of sodium. The addition of organic bases reduced oxidation in the somewhat unstable product. It could be processed on hot rolls, but was only produced in small quantities for cable insulation.

METHYL SALICYLATE Natural or synthetic winter-green oil, etheric oil. Occurs naturally in the leaves of *Gaultheria procumbens* and the bark of *Betula lenta*; produced synthetically by the esterification of salicylic acid with methyl alcohol. Colourless, yellowish or reddish oil, b.p. 220–224°C, s.g. 1·184. Soluble in chloroform and ether, miscible with alcohol.
Use: odorant.

β-METHYLUMBELLIFERON $C_{10}H_8O_3$, 4-methyl-7-oxy-coumarone. White crystals. Soluble in acetone, partially soluble in alcohol. Absorbs ultra-violet light.
Uses: protective against sun-rays for rubber compounds.

METROLAC Hydrometer (marked in lb/gal); used to determine the density, or rubber content of fresh latex.

METSO Cleaning agent for moulds.
Metso 99: sodium sesquisilicate
Metso granular: sodium metasilicate (259).

MEYER, KURT H. 1883–1952. Director of BASF (1921–1932), professor of chemistry at the University of Geneva (1932–1952). Author of *Elasticity of Rubber, High Polymers, Cellulose*, co-author (with H. Mark) of *Makromolekulare Chemie*, co-editor of *Journal of Polymer Science, Biochemische Zeitschrift* and *Zeitschrift für Physikalische Chemie*.

MFPT Abbreviation for mer-captobenzthiazole-formaldehyde-p-toluidine condensate.

MGOA Trade term for a natural rubber from *Mascarenhasia elastica* (E. Africa). Contains 7–8% resin and 85–90% pure rubber.

MG RUBBER Methyl methacrylate/natural rubber graft polymer; group of tough elastomer or plastic products with good physical properties and a high degree of hardness.

MICROBALLOONS Hollow phenolic resin balls filled with an inert gas and having a diameter of 0·05–0·15 mm. Act as a closing valve for liquids, being placed on the surfaces in a layer 1·2–2·5 cm thick, of, for example, solvents or diluents. Evaporation losses can be decreased by up to 90%.

MICROGEL A macromolecule in relation to the sol and gel phases as a structural element of natural and synthetic rubbers. The term was originally used in connection with SBR. Freshly prepared natural rubber consists of two hydrocarbon components, one of which is soluble in benzene while the other forms a colloidal suspension and is called a microgel. It is assumed that the microgel consists of crosslinked hydrocarbons of the latex particles.

MICROHARDNESS A measure of the hardness of rubber articles which are too thin or small for conventional hardness meters. The instrument recommended by ISO is identical in principle, with the ASTM hardness meter but on a smaller and lighter scale, the movement of the indentor being measured microscopically. The penetration of an indentor of 0·396 mm diameter is measured, taking the difference in depth between an initial load of 0·86 g after 5 s and a total load of 15·72 g after 30 s. The reading is given in International Standard degrees (ISO, BS Hardness, IRHD). For a hardness of 60° ISO, a test piece of 1 mm thickness is suitable. The relation between °ISO and °Shore (durometer) is:

°ISO	°Shore
30	25
40	37
50	48
60	59
70	70
80	80
90	90
100	100

MICROMANIPULATOR A sensitive microscopic apparatus, developed by Peterfi and Chambers, used primarily for the isolation of bacteria, chromosomes, and cell nuclei and the examination of microorganisms. Used by E. A. Hauser in his research into the structure of latex particles, in which he discovered a two-phase structure. According to this, latex particles consist of an almost solid elastic shell and a viscous inner material.

MICROMYA Special calcium carbonate for use in latices (422).

MICRONEX Group of channel blacks. MK II: HPC, Standard: MPC, W-6: EPC (3).

MICRONISED SC Sulphur with 5% clay. Used for vulcanisation (260).

MIGRATION Spontaneous migration of molecules and ions, or migration produced by an electric current in a solution or a dispersed system. In rubber, the migration is of compounding ingredients, such as accelerators, antioxidants, pigments, softeners, or sulphur, which are partially soluble in rubber and migrate from higher to lower degrees of concentration (blooming, bleeding).

MIIKI NO. 20 Japanese produced gas black. Type FT.

MIKROFOR N Dinitrosopentamethylene tetramine (428).

MIKROPUR CM N,N'-dinitro pentamethylene tetramine (283).

MIKROPUR CW Blowing agent based on dinitrosopentamethylene tetramine.

MIKROSIL S Very fine particle size silica. Contains approx. 85% SiO_2, s.g. 1·95, litre weight 120–150 g. Highly reinforcing filler for NR and SR (423).

MILL (MIXING ROLLS) A machine having two hollow steel rolls which contain cooling or heating systems and are placed in a frame in a horizontal or diagonal position. The rolls move in opposite directions at different speeds in the ratio 1:1·1 to 1:1·5. The second roll, *i.e.* the rear, moves at the faster speed. The gap between the rolls may be altered. The raw rubber is processed between the rolls by mechanical friction, and plasticised. The degree of mastication is greater on cold rolls than on warm rolls. The rubber band is formed around the front roll, thickness of the band depending on the gap between the rolls, while the superfluous rubber lies between the rolls as a bank. When a sufficient degree of mastication is attained, fillers and vulcanising ingredients are added on the rolls.

MILL OUTPUT The quantity (kg/min) of rubber, of a given plasticity, which a mill can produce; inversely proportional to the temperature.

MIMOTEX Centrifugated concentrated latex from Plantations Réunies de Mimot (Mimot: Cambodia).

MIMUSOPS Member of the *Sapotaceae* family; produces balata and products similar to balata or gutta-percha. The main producer is *M. globosa Gaertn.* (*Mimusops balata Gaertn.*) (N. Brazil, Guiana, Venezuela, Trinidad).

MINERAL RUBBER Solid bituminous substances, such as natural asphalt (Gilsonite) or blown distillation residues of petroleum; which bear no relation to rubber, s.g. 1·0–1·2. Softener (5–10%) and extender (up to 100%) in dark compounds. Gives a low modulus, a slight increase in hardness, high elongation at break, improved flex cracking resistance, and lowers the volume cost. Because of its small amount of unsaturation it absorbs a certain quantity of sulphur. Unsuitable for use in butyl rubber.

MINYADOTANA Local term for *Zschokkea Foxii* (Bolivia) whose sap is often mixed with *Hevea*

and *Castilla* latex during the production of natural rubber.

MIPOLAM Copolymer of vinyl chloride and acrylic acid esters, s.g. 1·34. Has good resistance to alcohols, benzine, acids and alkalis. The addition of softeners gives rubber-like masses.
Uses: seals, gaskets (192).

MISTLETOE RUBBER Rubber obtained from the fruits of *Loranthaceae*, specifically *Phthirusa sp.* and *Struthantus sp.*; discovered by Giordana in 1902. The rubber may be obtained either by crushing the unripe fruits or by grinding dried fruits and boiling the fruit pulp. It can be easily vulcanised and was brought on the market as Tina rubber.

MISTRON VAPOUR Natural magnesium silicate. Reinforcing filler.

MITCHIE-GOLLEDGE PROCESS Process used for the production of rubber from latex in the early days of the rubber industry; uneconomical to use. The latex was centrifuged, after the addition of acetic acid, in a centrifuge containing scoops. The sponge-like coagulum was cut into pieces, rolled flat and ground on a mill into longish crumbs (worms).

MITSUI Group of Japanese produced accelerators:
DPG: diphenyl guanidine
M: 2-mercaptobenzthiazole.

MOBILOMETER Apparatus used to determine the consistency of pastes and highly viscous masses. The time for a heavy perforated plate to fall through the test mass in a cylinder is measured.

MOBILSOL 33 Universal processing oil used for NR and SR (404).

MOBS Abbreviation for N-morpholinyl benzthiazyl sulphenamide.

MOCINO, JOSE MARIANO 1757–1820, b. Mexico, d. Spain. Took part in a Spanish expedition led by Cervantes to collect Mexican flora. Sent latex samples from Vera Cruz to Cervantes in Mexico City and undertook research on rubber production in Orizaba.

MODICOL VD Synthetic polymer stable in alkaline solutions. Thickening agent for latex (261).

MODIFIER Substance which interferes with the chain reaction in polymerisation and controls the average mol. wt. and the viscosity of the polymer.

MODULEX Furnace black. Type HMF (40).

MODULUS Measured during the tensile strength test for soft rubber; the tensile stress is noted at specific elongations (100%, 300%, etc.).
Specification: DIN 53 504.

MODULUS The stress, expressed in kg/cm^2 (or lb/in^2), necessary to attain a specified elongation (normally in multiples of 100%). It is taken from the cross-section of an unstretched test piece, and is a measure of the tightness of cure in the vulcanisate; it is one of the most important measurements which determine the quality of a rubber. The modulus is affected by the various compounding ingredients.
Specification: DIN 53 504.

MOD-U-METER Instrument for determining the degree of pressure deformation in rubber. Described originally by R. W. Brown (*Experimental Stress Analysis*, Cambridge, Mass., 1947, 4 (2), 49–50) and by J. Drogin ('Methods Employed in Compounding Research—1', *India Rubb. World*, 1952, 127, 65).

MODX Dispersion of sodium acetate in tetrahydroglyoxaline. MODX-B: mixture of inorganic and organic acetates with a small amount of oil. Activator, increases the tensile strength and modulus (100).

MOLECULAR SIEVE ACCELERATORS A group of accelerators which have been adsorbed onto molecular sieves and which only regain their activity after separation from the molecular sieve at high temperatures:
CW-1015, piperidine pentamethylene dithiocarbamate. Active above 121°C, 15% active material, s.g. 1·80. Suitable for NR, SBR, NBR. *Quantity:* 0·5–2%.
CW 1115, dibutylamine. Active above 138°C, 15% active material, s.g. 1·64. Used for natural and synthetic rubbers. *Quantity:* 0·5–2%.
CW-2015, ditert-butyl peroxide. Active above 121°C, 15% active material. Crosslinking agent for silicone rubber.
CW-3010, catechol-1:2-dihydroxybenzene. Active above 149°C, 10% active material. Used for neoprene. *Quantity:* 1–4%.
CW-3120, 1:3-diethyl thiourea. Active above 149°C, 20% active material, s.g. 1·82. Used for neoprene. *Quantity:* 1–4%.

CW-3615, water-free hydrocarbon. 15% active material. Used for hard vulcanisation of butyl rubber, according to USP 2 701 895.

CW-9246, ultra-fast accelerator. Active between 27 and 160°C, 46% active material. Used for natural and synthetic rubber. *Quantity:* 0·25–1% (127).

MOLECULAR SIEVES Highly porous crystalline masses with a sieve-like structure and a pore size of 4–5 Å. Produced by removing the water of crystallisation from sodium or calcium aluminium silicates. Strongly absorb water and can take up to 18% of their own weight in water from air with a relative humidity of only 1%. Suitable for the separation of compounds, since molecules which are smaller than the size of the pores may be absorbed in preference to the larger molecules. Molecular sieves have an unusual affinity for polar compounds and unsaturated organic molecules. Chemically loaded molecular sieves form excellent accelerators with a delayed action. The combination of highly active accelerators with molecular sieves gives products which are heat₁ sensitive, give maximum processing safety with good scorch resistance and the shortest possible vulcanisation time.

MOLECULAR WEIGHT Neither natural rubber nor synthetic polymers have a uniform mol. wt. but contain molecules of varying sizes. To determine the average mol. wt. the following methods may be used: osmotic pressure, light rays, intrinsic viscosity, analysis of the end groups, ultra-centrifuging, diffusion, sedimentation. The mol. wt. distribution can be determined by fractional precipitation or analogous process, or by using the ultra-centrifuge.

MOLLAN Group of softeners based on phthalic acid esters (262).

MOLLIT Group of softeners:
I: diethyl diphenyl urea
II: dimethyl diphenyl urea
IV: diethyl-o-toluyl urea
AB: acetyl benzoyl glycerol derivative (217).

MOLTOPRENE Polyurethane foam material, s.g. 0·04–0·5. Produced by the reaction of linear or branched polyester with an isomeric mixture of 2:4 and 2:6 toluylene diisocyanates and water, liberates carbon dioxide, causing the polyurethane to foam. Flexible or rigid foams are produced according to the type of ester and the number of its hydroxyl groups. The foam densities can be controlled by varying the quantity of water (43).

MONDUR Group of isocyanates:
C: isocyanate polyester resin
CA: isocyanate polyester resin
CB: isocyanate polyester resin
HX: hexamethylene diisocyanate
M: 4:4′-diphenyl diisocyanate
N5:   1:5-naphthalene diisocyanate
O: octadecyl isocyanate
P: phenyl isocyanate
S: isocyanate polyester resin
TM: triphenyl methane triisocyanate
TDS: toluylene diisocyanate
TD-80: toluylene diisocyanate (80% 2:4T) (263).

MONGALA Former trade name for wild rubber produced in the Congo.

MONOBUTYL PHTHALATE
$C_6H_4(COOC_4H_9)COOH$. Retarder.
TN: + Retarder B.

MONOETHYLAMINE
$C_2H_5NH_2$, aminoether, m.p.
$-80.5°C$, b.p. $16.5°C$. Miscible with water, ethanol and ether; the aqueous solution acts as an alkali. Stabiliser for natural latex.

MONOETHANOLAMINE
NH_2—CH_2—CH_2—OH, ethanol-amine, 2-aminoethanol; 2-hydroxy-ethylamine. Viscous hygroscopic liquid with ammoniacal odour, m.p. $10.3°C$, b.p. $165–172°C$, s.g. $(25°C)$ $1.012–1.027$. Miscible with water, methanol and acetone. Activator in compounds containing reinforcing fillers, dispersing agent, improves the water-resistance of vulcanisates. Strong base. Used in natural and synthetic rubbers and latices.
TN: Accélérateur soluble F FL (19).

MONOMER X-970 Bifunctional acrylic ester. For production of hard rubber with good compound flow properties (33).

MONOPLEX Group of softeners:

5   dibenzyl sebacate
DBS   dibutyl sebacate
DCP: dicapryl phthalate
DOS: dioctyl sebacate
DIOS: diiso-octyl sebacate (33).

MONOTERT-BUTYL-m-CRESOL
$C_6H_2[(CH_3)_2C—](CH_3)OH-1:3$
Antioxidant for rubber.
TN: MBMC (176).

MONTACLERE Styrene-modified phenol. Clear liquid, s.g.

1.08. Soluble in petroleum ether, benzene, alcohol, acetone, trichloro-ethylene, insoluble in water.
Uses: non-discolouring antioxidant for white and coloured articles, latex dipped articles and latex foam (5).

MONTANE WAX Wax of fossil origin. Black–brown amorphous product, m.p. $70–77°C$, s.g. approx. 1.0. Obtained from bituminous brown coal by extraction with organic solvents or distillation with superheated steam. Consists of up to $50–60\%$ wax (Montane acid esters and free Montane acids), up to $15–25\%$ from fossilised resins and from humus-type, sulphur-containing substances.
Uses: for the production of special effects, *e.g.* wax-like textures, shine in hard rubber.

MOONEY CURE TIME
Measured using the Mooney viscometer, expressed as the time taken for the viscosity to rise from 5 up to 35 Mooney units above the minimum reading, at a given temperature.

$$V_{30} = \frac{(\text{Minimum} + 35) - (\text{Minimum} + 5)}{t_{35} - t_5}$$

$$= \frac{30 \text{ Mooney}}{\Delta t}$$

DIN 53 524; ASTM D 1646-62.

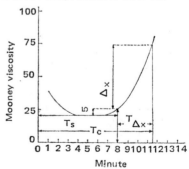

296

where M_v is the minimum viscosity
T_s is the scorch time
$= M_v + 5$, and
T_c is the vulcanisation time
$= T_s + T_{\Delta_x}$.

MOONEY, DR MELVIN

b. 1893, d. 1968. American physicist, specialist in rubber properties. Inventor of various measuring apparatuses, of which the Mooney viscometer for determining the plasticity of rubber has, in particular, come into general use. Was awarded the Goodyear Medal in 1962.

MOONEY SCORCH TIME

Determined by using a Mooney viscometer, the time (in min) from the beginning of measurements to the time at which the viscosity is 5 Mooney units, above the minimum reading at a given temperature (graph Mooney cure time). ASTM D 1646-62, BS 1673:3:1951, DIN 53 524.

MOONEY VISCOMETER A

shearing disc viscometer used to determine the plasticity or viscosity of raw elastomers and compounds. The apparatus is basically composed of a grooved rotor, 38·1 mm (ML) diameter for normal materials, and 30·5 mm (MS) diameter for very hard materials, rotating at 2 rpm in a closed cylindrical chamber. The shearing resistance or torque is transferred *via* a worm shaft to a spring and an indicator passing over a dial calibrated in Mooney units. 1° Mooney equals 8·64 gm. The test usually lasts for 4 min at 100°C (*e.g.* ML-4/100°C). Scorch and cure times can also be determined. A further development of this apparatus is the so-called pendulum elastomer with an oscillating biconical

rotor. This can be used to determine the complete vulcanisation curve. ASTM D 1646-62 (formerly D 927, D 1077), BS 1673: 3: 1951, DIN 53 523, 53 524.

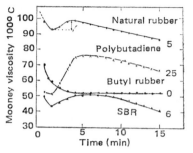

Viscosity time curve

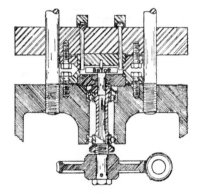

MOPLEFAN Biaxially orientated isotactic polypropylene.

MOPLEN Trade name for polypropylene (59).

MOR 2:4-morpholinemercaptobenzthiazole.

MORFEX Group of accelerator blends:
Standard: mixture of 2-mercaptobenzthiazole with tetramethyl thiuram monosulphide

33: mixture of 75% zinc-2-mercaptobenzthiazole and 25% tetramethyl thiuram monosulphide. Pale yellow powder, s.g. 1·49. Soluble in acetone, partially soluble in benzene and ethylene dichloride. Fatty acids, zinc oxide and sulphur are necessary in the normal quantities. Activated by aldehydeamines, guanidines and other bases. Non-discolouring, fast curing. *Quantity:* natural rubber 1–2% with 0·75–2% sulphur, SBR 2% with 2·5–2·75% sulphur. *Uses:* normal and continuous vulcanisation of cable and wire insulations, mechanical goods

55: mixture of zinc-2-mercaptobenzthiazole with 50% tetramethyl thiuram monosulphide (obsolete) (23).

MORFLEX Group of softeners:
100: diiso-octyl phthalate
110: di(2-ethyl hexyl)phthalate
120: dicapryl phthalate
130: didecyl phthalate
135: butyl decyl phthalate
140: dibutyl phthalate
145: butyl benzyl phthalate
160: dihexyl phthalate
175: octyl decyl phthalate
190: dinonyl phthalate
200: diiso-octyl sebacate
210: dioctyl sebacate
240: dibutyl sebacate
245: butyl benzyl sebacate
250: dibenzyl sebacate
300: di(2-ethyl hexyl)adipate
330: didecyl adipate
350: dibenzyl adipate

375: octyl decyl adipate
410: di(2-ethyl hexyl)phthalate
DNA: dinonyl adipate
IPO: isopropyl oleate
P-(number) group of polyesters, m.p. −37°C to 0°C, s.g. 0·92–0·99 (25°C)
X-926: di-n-octyl phthalate (264).

MORPEL Group of softeners for rubber and plastics:
IPO: isopropyl oleate
X-565: diethylene glycol monooleate
X-570: glycerine monooleate
X-574: glycerine monolaurate
X-710: glycerine monostearate
X-841: dinonyl naphthalene
X-870: diethylene glycol monolaurate
X-928: polyethylene glycol dilaurate
X-929: amyl naphthalene mixture
X-930: butyl oleate
X-931: diethylene glycol dipelargonate
X-932: diethylene glycol distearate (264).

MORPHOLINE C_4H_9NO, tetrahydro-1:4-oxazine, diethylene oximide. Hygroscopic liquid, m.p. −4·9°C, b.p. 128·9°C, s.g. 0·999. Miscible with water, acetone, benzene, ether, alcohol.
Uses: solvent for resins, waxes, basic material for vulcanisation accelerators, antioxidants, softeners.

MORPHOLINE DISULPHIDE

$$\left[O \!\!\begin{array}{c} CH_2\!-\!H_2C \\ CH_2\!-\!H_2C \end{array}\!\! N\!-\!S\!- \right]_2$$

4:4'-dithiomorpholine. Coarse, granular grey powder, m.p. 122°C, s.g. 1·29. Accelerator. Acts as sulphur donor yielding 26–27·5% sulphur.

Stable at processing temperatures, presumably decomposes at vulcanisation temperatures and forms sulphur, morpholine and other compounds. Practically scorch-free in sensitive compounds, *e.g.* tyre treads produced from natural rubber with HAF black. The addition of sulphur is recommended, but the quantity must be low to maintain the modulus and hardness. Acidic compounding ingredients, *e.g.* pine tar, suppress the antiscorch properties. Does not cause discoloration in white compounds and gives good ageing properties. Compound examples:

MOULD RELEASE AGENTS Mould lubricants. Materials which prevent the sticking of a moulding to the mould, facilitate the removal of the article from the mould, and prevent the formation of a crust. Among the materials used are: wetting agents, water soluble polyethylene oxides, emulsions of synthetic waxes and thermoplasts, and more recently, silicones in the form of solutions, aqueous emulsions and pastes. Apart from their effective release properties and the smooth, shiny surfaces they impart, silicones

Product	Morpholine disulphide	CBS	MBTS	DPG	Sulphur
Tyre treads HAF	1·0 –2·0	0·50	—	—	0·25–1·0
Tyre treads EPC	1·0 –2·0	0·75	—	—	0·50–1·50
White articles	0·75–2·0	—	0·40	0·25	0·50–1·50
Mechanical goods	0·75–1·50	1·25	—	0·50	0–0·50

TN: Actor R (16)
Sulfasan R (5)
Vulnoc R (274).

MOTOLINIA, T. DE In his history of New Spain (1536) described various Aztec religious rites in which rubber was used.

MOULD CORE Part of the mould which shapes the inner surface of an article.

MOULDED ARTICLES Rubber articles vulcanised in moulds in a press.

MOULDING STIFFNESS Resistance of a moulded article to plastic deformation immediately after removal from the mould.

are the most effective as they have a higher thermal stability so that longer use is possible.

MOVIL P.V.C. fibre (endless or staple fibre). Produced since 1953 according to the Rhovyl patent under licence to Montecatini (Italy). Types N: Rhovyl, F: Fibravyl, T: Thermovyl.

MOWIOL Group of polyvinyl alcohols with varying degrees of polymerisation and hydrolysis, give highly viscous solutions with water. *Uses:* as protective colloids and gelling agents (217).

MOYAL-FLETCHER APPARATUS A machine to determine

the dynamic properties of rubber subjected to forced resonance oscillations. Lit.: J. E. Moyal and W. P. Fletcher, *J. Sci. Instr.*, 1945, 22, 167.

MOZAMBIQUE Former trade term for wild rubber from *Landophia Karkii* (Mozambique).

MPC Abbreviation for medium processing channel (carbon black).

MPS-500 Stabilised methylene pentachlorostearate (214).

MPT Abbreviation for methylene-p-toluidine.

MPTD Abbreviation for dimethyl diphenyl thiuram disulphide.

MRERB Abbreviation for Malayan Rubber Export Registration Board, Singapore. Founded 1953 to control rubber exports from Malaya.

MRFB Abbreviation for Malayan Rubber Fund Board.

MRMC Tert-butyl-m-cresol. Antioxidant (176).

MT Abbreviation for medium thermal (carbon black).

2-MT Abbreviation for 2-mercaptothiazoline.

MTM Abbreviation for mixed tertiary mercaptans, mixture of tertiary mercaptan with 14 C-atoms. Used as a modifier in emulsion polymerisation.

MTMT Abbreviation (French) for tetramethyl thiuram monosulphide (monosulfure de tétra méthyl thiurame).

MUDAR Mudar rubber. A product, similar to gutta-percha, of *Calotropia procera* and *C. gigantes* (India). Sometimes used in the production of water-resistant textiles.

MUDIE COMMISSION A commission, under Sir Francis Mudie, who, after a visist to Malaya in 1954, decided that the rubber industry was overtaxed. Suggested a new, sliding tax system as a scheme to accelerate the renewal of old plantations.

MUF N-p-tolyl-N′-p-tolyl-sulphonyl-p-phenylene diamine. Antioxidant (23).

MULLINS' EFFECT The reduction in stiffness of loaded vulcanisates after repeated stretching. Caused by breaks in bonds between rubber and reinforcing fillers or agglomerations of filler particles.

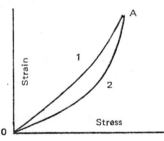

1. Original stress/strain curve.
2. After repeated deformation.

MULTIPORE Rubber sheets with many tiny cells. Produced by painting a latex compound onto a cloth and heating it. The air breaks through the latex layer from the threads of the cloth and produces tiny cells. The cloth is removed after vulcanisation.

MULTOFLEX MM Precipitated calcium carbonate with a particle size of 0·05–0·06 micron. Filler (153).

MULTRATHANE
F-144: polyester. Basic material for polyurethane rubber
M: 4:4′-diphenyl methane diisocyanate
N-5: 1:5-naphthalene diisocyanate (263).

MUTEKE Muteki, Mutecha. Local term for various *Landolphia* types and the wild rubber obtained from them (Rhodesia). The rubber is classed as only slightly tacky and of good quality.

MVP Russian produced methyl vinyl pyridine rubber.

MYRCENE $C_{10}H_{16}$, 7-methyl-3-methylene-1:6-octadiene. Oil with pleasant odour, b.p. 167°C, s.g. 0·8013 (15°C). Soluble in alcohol, chloroform, ether. May be polymerised to dimyrcene and polymyrcene. Dimyrcene reacts with nitrogen trioxide to give a nitrosite identical to rubber nitrosite. Rubber distillates do not contain myrcene.

N

NA Group of accelerators:
11: diaminodiphenyl methane
22: 2-mercaptoimidazoline

33: thioamide. Grey white crystalline powder, s.g. 1·13 above 102°C. Delayed action, particularly suitable for neoprene Type W; good compromise between processing safety and vulcanisation speed. Gives a high modulus and low compression set. *Quantity:* 0·5–1·5%, neoprene latices 1% on the solid polymer (6).

NA-101 Delayed action accelerator for neoprene with increased processing safety, fast cure rate and high state of cure. Non-staining, nondiscolouring, may be used alone or with secondary accelerators.
Du Pont

NACCOMINE Diphenyl guanidine (29).

NACCONATE Group of diisocyanates:
65: toluylene diisocyanate (65:35% 2:4-T and 2:6T)
80: toluylene diisocyanate (80:20% 2:4T and 2:6T)
100: toluylene diisocyanate (100% 2:4T)
200: 3:3′-bistoluylene-4:4′-diisocyanate
300: 4:4′-diphenyl methane diisocyanate (265).

NACCONOL NR Alkyl aryl sulphonate (SF: salt free). Dispersing and wetting agent for latex (265).

NAFTOLENE Mixture of unsaturated, high mol. wt. hydrocarbons. Softener for natural rubber. Gives vulcanisates of high quality (180).

301

NAIRNE, EDWARD Assistant to Priestley, the producer of philosophical instruments. Discovered the erasing property of raw rubber, from which the name of rubber is derived. In 1770 was the first to sell India rubbers (at a price of 3 shillings/cm^3).

1:5-NAPHTHALENE DIISO-CYANATE

m.p. 126·9°C, b.p. (50 mm) 183°C, s.g. 1·43. Strongly toxic.
Uses: production of polyurethane (Vulkollan).
TN: Desmodur 15 (43).
Mondur N 5 (263).

β-NAPHTHOL $C_{10}H_7OH$. Colourless crystals, m.p. 122°C, b.p. 285°C, s.g. 1·22. Soluble in alcohol, ether, benzene, chloroform. Antioxidant.
TN: Antioxygène BN (19).

NAPHTHYL-β-MERCAPTAN β-thionaphthol, 2-naphthalene thiol, 2-mercaptonaphthalene, 2-naphthalene mercaptan. Crystalline powder, m.p. 81°C, b.p. 286°C. Soluble in alcohol, ether, petroleum ether, insoluble in water. Peptiser for NR and SR. Has a slight activating effect in gum stocks with dithiocarbamates, thiazoles and thiurams, has no effect on ageing properties. Non-staining, odourless in vulcanisates. Plasticising and reclaiming agent for scorched compounds and vulcanisates. Causes dermatitis.
TN: RPA-2 (6)

Vulcanel TBN (60)
Lit.: USP 2 064 580, 2 186 714, 2 216 840 (1940), 2 402 641 (1946).

NAPLYWY (Russian) rubber containing excretions from *Chondrilla*. Pale brown, sandy mass, resembling a growth. Its formation is caused by the larvae of the gold beetle (*Sphenoptera foveola Gebl.*) which eat into the bark of the twigs. The excreted latex mixes with earth and coagulates. The mass consists of 80–90% sand, 1·4% rubber and resin. An attempt was made in Russia to use it as a source for rubber.

NAT (Formerly ADR-GR.) Natural rubber from the Ivory coast. Produced in crumb form and pressed into compact bales:
301: (ADR-GR 1) from latex
203: (ADR-GR 2) from cup lumps
105: (ADR-GR 3) from rubber dried on the tapping surface
106: (ADR-GR 4) from scrap collected from the ground (flat bark)
601: (IRC-GR) from undiluted latex treated with sodium bisulphite
604: from undiluted latex.

NATAC Modified resin acid. Prevents the blooming of compounding ingredients from vulcanised articles (40).

NATCOM Natural rubber produced in crumb form and pressed to compact bales of approx. 34 kg; packed in polyethylene and in palettes of $\frac{1}{2}$ or 1 ton.
Type LX: from latex coagulum
CP: from cup lumps.

NATIVE RUBBER Smallholding rubber. A large proportion of the inferior quality rubber on the market does not come from plantations but is produced by native smallholdings in Malaysia, Indonesia and Ceylon, often under primitive conditions. In contrast to the early days of the rubber industry when natural rubber was obtained from a variety of species, natural rubber is now taken exclusively from *Hevea Brasiliensis*. The smallholdings range in size from a few trees to about 25 ha. In Malaysia, Ceylon, and some parts of southern Borneo, the plantations are tended and tapped regularly, and the rubber is processed into smoked sheet of good or medium quality. In Sumatra and in Borneo, however, it is still processed in a primitive manner, the plantations are neglected, and the product is of the poorest quality, since the production of rubber is regarded merely as an additional means of earning to the growing of food. The rubber is delivered to the local traders in four different forms: smoked sheet, unsmoked sheet, slabs and sheety crêpe.

The trees are tapped irregularly, and nonsystematically, cutting being designed to achieve the highest possible yield of latex. The cuts made for tapping are careless and the cambium is often badly damaged. Cans, coconut shells, halved bamboos and other objects are used for collecting the latex. In many areas of Sumatra and Borneo, the latex is 'stolen' and the trunks of trees are damaged by numerous irregular cuts. The latex is coagulated by using formic acid, acetic acid, or with sulphuric acid, alum, salt, acid plant saps or urine.

Sheet rubber. Coagulation is done in square tin or wooden pans, or sometimes in large, subdivided wooden tanks. The coagulum is usually pressed flat on to a board, either by hand or with a bottle, and then rolled out by means of one or two hand-operated rollers. The sheets are smoked or dried in the open. Unsmoked sheets are smoked later at rubber centres by the traders, but the result is of poor quality.

Slabs. Undiluted latex is put into caskets, canisters, tins, or cavities in the ground and coagulated. The rubber is adulterated by addition of stones, sand, earth, and the like. The rubber blocks, often weighing up to 50 kg, are surface dried, then washed on heavy rollers in so-called remilling factories and rolled out as remilled crêpe or blankets of 1–10 mm thickness. This type of rubber comes almost exclusively from Indonesia.

Sheety crêpe. Produced solely in Tapanuli, West Sumatra, in small quantities (50–200 tons per month); is a coarse open crêpe, made by milling. The coagulum is pressed several times through rollers, stretched out and dried in the open air.

In the last few years the respective governments have been attempting to improve the standard of the small holdings and the quality of their products through education and financial support. The percentage of rubber produced on small holdings is from 30–65% in each country.

NATSYN Trade name for synthetic cis-1:4-polyisoprene (115).

NATTA, GIULIO b. 1903. From 1933–1938 was at the Universities of Pavia and Rome and at the Polytechnic Institute of Turin. Since 1938 director of the Institute of Industrial Chemistry, Milan. Has produced over 400 publications and 150 patents on the development of stereospecific polymers.

NATURAL RUBBER (Caoutchouc, from Cahuchu: flowing wood —caa: wood, ochu: flowing, tears.) The word caoutchouc, which comes from the Inca language, was the term adopted by Fresneau for rubber from *Castilla elastica*. The tree was known as Hheve. The inhabitants of Mexico used the term ulli for the rubber, and ulequahuitl for the tree. Other terms are feather resin (from resina elastica) and gummi (from gummi elastica). In German the term Kautschuk is used for the raw product and Gummi for the vulcanised material. Terms in other languages: French: caoutchouc, gomme, Italian: gomma, cauccio, Portuguese: caucho, cauchu, boracha, goma, Japanese: gomu, Malayan: karet, getah, rambong. Natural rubber is a high mol. wt. hydrocarbon which comes from the sap (latex) of many plants, *e.g.* from the *Moraceae, Euphorbiaceae, Apo-*

cynaceae, Asclepiadaceae and Compositae, and is obtained by coagulation with chemicals, by drying, electrical coagulation and other processes. The most important rubber producer is *Hevea Brasiliensis*; small quantities of rubber are also obtained from *Parthenia argentatum* and other Composites. The wild rubber which used to be obtained from other plants is no longer important. The rubber produced from latex contains, besides the hydrocarbon, small quantities of protein, carbohydrates, resin-like substances, mineral salts and fatty acids. These other constituents act in part as natural accelerators and antioxidants and give the product processing properties which do not exist in the pure hydrocarbon. Although primarily occurring in tall plants, rubber has also been identified in smaller plants and in mushrooms and toadstools (*Lactarius* and *Peziza* types) in quantities of up to 1.7%. Natural rubber is cis-1:4-polyisoprene with the empirical formula

	Limits %	Mean values %	
		Smoked sheet	Crêpe
Moisture	0·3–1·2	0·61	0·42
Acetone extract	2·5–3·2	2·90	2·70
Protein	2·5–3·5	2·80	2·80
Ash	0·15–0·9	0·38	0·30
Rubber	92–94	93·8	93·6
Chlorine	0·002–0·01	0·006	0·003
Sulphate	0·02–0·05	0·03	0·04
Sterols		0·5	0·5
Higher fatty acids		1·4	1·1
Copper*	2–10	5	4
Manganese*	0·8–4	1·5	1·0

* Expressed as ppm, not as a percentage.

$(C_5H_8)_n$ and the structural formula

$$\left[\begin{array}{c} CH_3 \\ | \\ -CH_2-C=CH-CH_2- \end{array} \right]_n$$

When vulcanised, rubber becomes highly extensible, shows fast recovery, and is highly elastic. Characteristics: s.g. 0·934 (20°C), refractive index (20°C) 1·5222, entropy 1·875 joule/g/°C, Mooney viscosity 60–90.

NATURAL RUBBER BUREAU NRB. Since 1973 Malaysian Rubber Bureau. Until 1960 was known as the Natural Rubber Development Board (NRDB), part of the Malayan Rubber Fund Board, a non-commercial organisation of natural rubber manufacturers in Malaysia. Founded for the furtherment of research, development and application of natural rubber. Advice and information posts in USA, UK, Germany, Austria, Italy, Spain, Australia, New Zealand, India, Malaysia. Publications: *Natural Rubber News, Rubber Developments. NR–Technology Techn. Bulletins* (Technical information sheets in various languages). European representatives: Frankfurt, Eschersheimer Landstrasse 275, BRD; Vienna II, Praterstrasse 44 (East Europe).

NATURAL RUBBER PRO-DUCERS' RESEARCH ASSOCIA-TION NRPRA. Since 1973 Malaysian Rubber Producers' Research Association. Formerly British Rubber Producers' Research Association (BRPRA), 48–56 Tewin Road, Welwyn Garden City, Herts. Research institute for natural rubber. Founded 1938 on the framework of an international agreement made in 1934 between England, France, Holland, India and Siam.

NAUGAPOL Butadiene-styrene copolymers. The index coding is the same as that of the ASTM code for SBR (23).

NAUGATEX Trade name for a group of compounding ingredients for latex; available as pastes, dispersions and solutions. Also used for butadiene-styrene latices.

1. Compounding ingredients:
112: dimethylammonium dimethyl dithiocarbamate
144: piperidine pentamethylene dithiocarbamate
166: soluble dithiocarbamate
188: zinc-2-mercaptobenzthiazole
209: dispersion of N-p-tolyl-N′-p-tolyl sulphonyl-p-phenylene diamine
215: zinc-2-mercaptobenzthiazole
219: diphenyl-p-phenylene diamine
225: zinc-2-mercaptobenzthiazole
235: tetramethyl thiuram monosulphide
240: aqueous dispersion of Sunproof, a wax-like antioxidant
245: zinc dibenzyl dithiocarbamate
259: acetone-hydroquinone reaction product. Antioxidant
269: phenyl-β-naphthylamine
275: zinc dithiocarbamate
279: diphenyl-p-phenylene diamine
289: diphenylamine-acetone condensation product
500: 40% paste of Sunproof
501: 25% paste of Micronex channel black
503: 40% paste of zinc-2-mercaptobenzthiazole
505: 50% paste of a reaction product of diphenylamine and acetone (Aminex)
509: 40% paste of zinc oxide
510: p(p-tolyl sulphonyl amido)-diphenylamine
511: zinc dibenzyl dithiocarbamate

512: zinc dimethyl dithiocarbamate
513B: zinc diethyl dithiocarbamate
514: zinc dibutyl dithiocarbamate
515: aqueous dispersion of zinc pentamethylene dithiocarbamate
519: 60% paste of a reaction product of diphenylamine and acetone (BLE)
521: sulphur dispersion
535: dimethylammonium dimethyl dithiocarbamate
537: paste of algin and ammonium alginate (Superloid)
539: paste of titanium dioxide
552: tackifier in paste form
571: paste of a coagulant for dipping
578: paste. Yellow pigment
698: paste of an antifoam agent
699: paste. Red pigment
702: paste. Blue pigment
703: paste. Orange pigment
NX-503-C: zinc mercaptobenz-thiazole.

2. Butadiene-styrene latices:

Type	Total solids	Monomer ratio	Viscosity
2000	42%	50/50	17 cps
2001	42%	50/50	28 cps
2002	49%	50/50	300 cps
2006	27%	72/28	10 cps
2105	62%	70/30	800 cps
2107	62%	50/50	500 cps
2108	40%	73/27	60 cps
2113	48%	50/50	800 cps

2711: as 2000 but with a higher film strength
2714: higher modulus, improved resistance to abrasion and tearing
2733: good bonding strength, good chemical and mechanical stability and resistance to light ageing

2734: high styrene content. For paints and colours
2748: medium styrene content. For paper impregnation
J-8174: cold SBR latex. For foam rubber
J-8535: cold SBR latex. for carpet underlays (23).

NAUGATUCK 124 N,N'-dicyclohexyl-2-benzthiazole sulphenamide (23).

NAUGAWHITE Alkylated bisphenol. Brown, clear, viscous liquid, s.g. 0·96. Antioxidant for NR, SBR, NBR, CR and latices. Has no effect on processing safety or on vulcanisation. Does not bloom or stain. Non-discolouring. Protects against normal oxidation, heat, light and combustion gases. Naugawhite powder; mixture of 70% Naugawhite with 30% inert filler (23). 434: alkylated phenol. Yellowish fluid, s.g. 1·0.

NBR Acrylonitrile-butadiene rubber, according to the nomenclature of ASTM D 1418.

NBS Abbreviation for National Bureau of Standards, USA.

NBS ABRASION MACHINE Machine developed by the National Bureau of Standards; used to determine the abrasion resistance. The weight and volume losses for 100 revolutions of the wheel are determined. ASTM D 394, Method B 1.

NCR Nitrile chloroprene rubber, according to nomenclature of ASTM D 1418-58T.

NCRT Abbreviation for National College of Rubber Technology, Northern Polytechnic, Holloway, London, N7. Qualifications in polymer technology awarded by the college include the Licentiateship (LNCRT), Associateship (ANCRT) at first degree level, and Fellowship (FNCRT) for post-graduate research.

NEDAC Delayed action accelerator for chloroprene rubber, m.p. 60–65°C, s.g. 1·40 (21).

NEEDLE TEARING TEST Measurement of the tear resistance of rubber products when subjected to piercing, *e.g.* on sewing. The needle tearing power, under specified conditions of tear testing is the highest force measured on tearing out a needle 1 mm thick. Resistance to tearing with a needle: the needle tearing power in relation to a 10 mm thick tear piece. DIN 53 506.

NEGAMINE 142 A Amino ester of a fatty alcohol with a long-chain structure. Processing aid, mould lubricant (226).

NEGOMEL AL 5 Condensation product of ethylene oxide and fatty acids. Antistatic fluid, s.g. 0·95 (60).

NEGOX Epoxy amine. Antioxidant (39).

NEGOZONE Group of waxes which protect against light ageing.

NEGROHEADS Sernamby. Former trade term for a type of para rubber of inferior quality, produced by pressing together the scraps coagulated on the tree and the lumps from the collecting bowls. Unsmoked.

NEIDHART SPRING Rubber spring which has the same elastic properties unloaded, partially loaded or fully loaded. Consists of a square tube lying in a second square tube set at a 90° angle to the first (principal). In the four spaces which are formed by the inner sides of the outer tube and the outer sides of the inner tube, are loose rubber rollers. These hold both tubes together as a compact element, while allowing them to turn relative to one another. As with torsion filament strained by a turning motion, they resist turning with an increasing force which, in this case, and in contrast to steel springs, follows a progressive and non-linear characteristic curve. A further advantage of this form of spring is the damping: tension and relaxation of tension do not follow the same curve, the tension curve lies somewhat higher. The spring has a high damping capacity with a variable course so that a shock absorber is unnecessary. The use of the spring is widespread, and is included in cars, aeroplanes, the hanging of cable-railway suspensions and tipping-chairs, spring elements for heavy machines, elastic couplings for machines.

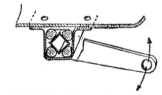

Principle of the spring element

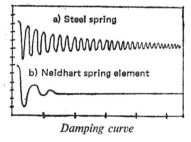

Damping curve

NELLEN TUBING MACHINE PLASTIMETER A machine for determining the tube extrusion properties and extrusion speed of compounds.

NEN Dutch standard.

NEO-FACTICE Reaction product of liquid polybutadiene with sulphur or sulphur chloride; similar to factice.

NEO-FAT Group of higher fatty acids and resin acids. Activators, softeners, emulsifiers, in the reclaiming of rubber:

1-56:　90% palmitic acid
1-60:　technical stearic acid
1-65:　90% stearic acid
11:　　lauric acid
12:　　approx. 95% lauric acid
14:　　approx. 94% myristic acid
18:　　approx. 93% stearic acid
18–61: 79% stearic acid, 21% palmitic acid
94-10:　83% oleic acid, 7% palmitic acid
242:　　stabilised resin acid
D-142: mixture of linoleic and oleic acids
D-242: fatty acid/resin acid mixture
HFO:　hydrogenated fish oil fatty acids (98).

NEOFAX F Factice with good resistance to mineral oils and benzine. Suitable for synthetic rubber types and articles, which are resistant to swelling.

NEOLYNE Alkyl resin from colophony. Additive for adhesives (10).

NEOPAST Suspension of two-thirds finely dispersed zinc oxide and one-third dispersing agent.
Uses: chloroprene rubber (379).

NEOPHAX Special factice for neoprene, s.g. 1·03–1·05. Diluent, softener and plasticiser for natural and synthetic rubbers (74).

NEOPRENE Formerly Duprene. Generic term for synthetic, rubber-like polymers of 2-chlorobutadiene (chloroprene) or copolymers with chlorobutadiene as the main constituent. Developed in 1927 by Carothers, Williams, Collin and Kirby with Du Pont de Nemours, and first came on the market in 1931. Produced in the form of various solid types and latices which differ primarily because of their differing tendencies to crystallise. Each type is characterised by the rate at which the hardness of an unvulcanised compound increases and also by its processing properties and uses. May be distinguished according to use as: general purpose types (including sulphur modified G types and W types modified with mercaptans) and special purpose types. Production is by an emulsion polymerisation process which gives a more uniform mol. wt. distribution than is possible when using block polymerisation which takes place very rapidly (6):

AC, special type with mercaptan as modifier, thiuram disulphide as stabiliser. Has strong tendency to crystallise with rapidly hardening films and a high bond strength. Is more stable and gives better colours than Type CG. Polymerisation at 10°C. Pale yellow to greenish rods, s.g. 1·23. Soluble in aromatic and chlorinated hydrocarbons, some esters and ketones. Available in three different degrees of plasticity. *Uses:* cements, adhesive solutions. Softened by milling for a long period but for less time than CG. Does not become tacky on the mill. Viscosity and bonding strength decrease if milling is carried out too slowly. Hot vulcanisation: 4% magnesium oxide, 5% zinc oxide, thiocarbanilide or butyraldehyde-aniline as accelerator. Cold vulcanisation: 5% zinc oxide, 4% magnesium oxide, 10–20% lead oxide and 2–8% aldehyde-amine accelerator. Non-vulcanising cements also need 5% zinc oxide and 4% magnesium oxide in combination with a resin which considerably improves cohesion of films. 10% calcium silicate may be useful as a stabiliser.

AD, special type with mercaptan as modifier, no stabiliser needed. Polymerised at a low temperature, s.g. 1·23. Soluble in aromatic and chlorinated hydrocarbons and also in some esters and ketones. Has strong tendency to crystallise, good storage stability, good colour stability. Softens quickly when milled, does not become tacky. *Uses:* bonding agent, adhesive solutions. Processing: as for AC.

AF, adhesive type. Has good heat resistance. Does not phase out, limited storage stability.

CG, special type with sulphur as modifier, thiuram disulphide as stabiliser. Pale brown to light green rods, s.g. 1·23. Polymerised at a low temperature. Has strong tendency to crystallise. Good storage stability. May be plasticised faster and to a greater extent than AC. *Uses:* cements and solutions with high solids contents. Hot vulcanisation: 4% magnesium oxide, 5% zinc oxide. Cold vulcanisation: 4% magnesium oxide, 5% zinc oxide, 10–20% lead oxide, 2–8% aldehyde-amine accelerator. Non-curing solutions: 4% zinc oxide in combination with a resin. + E, the oldest neoprene type, stabilised with thiuram disulphide, phenyl-α-naph-thylamine as antioxidant. Brown, plasticised sheets, s.g. 1·23. Vulcanises slowly. Good storage stability. Is fairly stiff, does not soften after long milling.

Vulcanisates have very good heat resistance and flex cracking resistance. Vulcanisation: 10% magnesium oxide, 10% zinc oxide, 1% sulphur, 5% resin.

FB, unmodified type. Viscous liquid, s.g. 1·23. Soluble in aromatic and chlorinated hydrocarbons, some esters and ketones. Has strong tendency to crystallise. *Uses:* bonding agents and non-volatile, vulcanisable softener and processing aid for other types. Has no influence on the hardness of the vulcanisates.

FC, as FB, but is not modified with sulphur and has a lighter colour.

+ FR (FR: freeze resistant), co-polymer of chloroprene and iso-prene, stabilised with thiuram disulphide, phenyl-α-naphthylamine as antioxidant, s.g. 1·15. Has slight tendency to crystallise. Suitable for products used at low temperatures. Vulcanisates have a high elasticity and low compression set. Softens only slightly after long masticating. Vulcanisation: 4–5% magnesium oxide, 5% zinc oxide, 1% sulphur or 1% accelerator (2-mercaptoimidazo-line or di-o-tolyl guanidine dicatechol borate): Permalux. Gum stocks have only low tensile strengths, and therefore most products need a reinforcing filler. Ester softeners give products of good flexibility at low temperatures.

GN: (GR-M) general purpose type, with sulphur as modifier, thiuram disulphide as stabiliser, no anti-oxidant, s.g. 1·23. Soluble in aromatic and chlorinated hydrocarbons, and in some esters and ketones. Has medium tendency to crystallise. Easily plasticised on a mill and in an internal mixer. 0·1–0·25% piperidine penta-methylene dithiocarbamate acts as a plasticiser. Long, cold mastication gives tacky compounds. Used for light and dark compounds. Pro-cessed on normal rubber machines. May be blended with other elasto-mers with the exception of butyl rubber and Hypalon 20. May be obtained in three different grades of plasticity (ML 100°C–2½ min).

M-1: max. 54 Mooney viscosity.
M-2: 55–65 Mooney viscosity.
M-3: 66–75 Mooney viscosity.
Vulcanisation: 4% magnesium oxide, 5% zinc oxide, 0·5–2% mercapto-imidazoline or di-o-tolyl guanidine dicatechol borate (Permalux).

GN-A (GR-M 10), general purpose type, with sulphur as modifier, thiuram disulphide as stabiliser, phenyl-β-naphthylamine as anti-oxidant, s.g. 1·23. Soluble in aromatic and chlorinated hydrocarbons, some esters and ketones and in blends with aliphatic hydrocarbons. Has tendency to crystallise, fast curing. Easily plasticised but not to such a degree as GN, does not become as soft as GN and does not become so tacky. May be blended with other elasto-mers with the exception of butyl rubber and Hypalon. Available in 4 plasticity grades: (ML 100°C–2½ min).

M-1: 54 max. Mooney viscosity.
M-2: 55–65 Mooney viscosity.
M-3: 66–75 Mooney viscosity.
M-4: min. 76 Mooney viscosity.
Vulcanisation: as for GN. Only suitable for dark compounds because of the antioxidant which causes discoloration.

GRT (RT: retains tack), general purpose type copolymer with undis-closed material, sulphur as modifier, thiuram disulphide as stabiliser. s.g. 1·23. Non-staining antioxidant.

Soluble in aromatic and chlorinated hydrocarbons, and some esters and ketones. Only a slight tendency to crystallise. Good storage stability. May be easily plasticised, on being masticated for a considerable period of time becomes very soft and tacky. Vulcanisates have the same proper-ties as GN, but better resistance to hardening as a result of crystallisa-tion. May be blended with elasto-mers in the same way as GN. Vulcanisation: as for GN.

HC, highly crystalline polymer. Similar to trans-polyisoprene rubber in viscosity, plasticity and degree of crystallisation.

ILA, copolymer of chloroprene and acrylonitrile (3·5:1). Has good flame, oil, solvent, heat and ozone resistance.

KNR, special type with sulphur as modifier, thiuram disulphide as stabi-liser, s.g. 1·23. Has strong tendency to crystallise. Soluble in aromatic and chlorinated hydrocarbons, in some esters and ketones. May be plasti-cised more easily and to a greater extent than other types, particu-larly when a peptiser is used (0·5–1% piperidine pentamethylene dithio-carbamate or 1·5–2·5% tetraethyl thiuram disulphide). *Uses:* solution with high solids contents (90%) and low viscosities, protective paints, coatings, dipping solutions for coat-ings. Blends with elastomers as for GN. Vulcanisation: 4% magnesium oxide, 5% zinc oxide, for self-vulcanising compounds 4% mag-nesium oxide, 5% zinc oxide, 10–20% litharge, 2–4% aldehyde-amine accelerator.

+ Q, copolymer with acrylonitrile, sulphur as modifier, thiuram disul-phide as stabiliser. Does not crystal-lise. Non-staining antioxidant. Has good processability and improved oil resistance. Because of its high degree of stiffness, a softener is necessary. Is replaced by WB for extruding because of the good dimensional stability of the latter. Shows no calender grain.

+ RT (RT: retain tack), copolymer with styrene, sulphur as modifier, thiuram disulphide as stabiliser. Has low crystallisation rate, s.g. 1·23. Non-staining antioxidant. Compounds retain good tack, vulcanisates have the same properties as GN, but have a better resistance to hardening due to crystallisation. Replaced by GRT.

S, unmodified special type, s.g. 1·20. Stabilised with thiuram disulphide. High mol. wt. and very tough, cross linked. Non-staining antioxidant. Is not masticated on a mill. Active as a stiffener for unvulcanised compounds of other types. Only absorbs small quantities of fillers because of its high toughness. *Uses:* crêpe soles, stiffeners. Vulcanisation is unnecessary. In crêpe soles 0·5 % magnesium oxide and 0·5 % sodium acetate are added as acid absorbers and protective agents for textiles, 1–2 % nickel dibutyl dithiocarbamate is also added to prevent the discoloration of coloured soles. In blends with other types the normal vulcanisation system is effective.

W, general purpose type with mercaptan as modifier, phenothiazine as short stop, s.g. 1·23. Soluble in aromatic and chlorinated hydrocarbons without premastication, in some esters and ketones and in blends with aliphatic hydrocarbons. Has a strong tendency to crystallise. Good storage stability. Has a more uniform mol. wt. distribution than the G types. Is not easily degraded, compounds do not therefore become either soft or tacky. Good dimen-

sional stability in soft extruded articles. Mooney viscosity 100°C–2½ min. Vulcanises slowly. Vulcanisation is with sulphur and usual accelerators; for compounds with a low compression set, a sulphur-free system should be chosen. Curing systems: 4 % magnesium oxide, 5 % zinc oxide, 0·5–1 % tetramethyl thiuram monosulphide, 0·5–34 % DOTG, 1 % sulphur. Sulphur-free compound 4 % magnesium oxide, 5 % zinc oxide, 0·5–1 % 2-mercaptoimidazoline.

W-M 1, as W, but with a lower viscosity.

WB, 'nerve' free type. Has good processing properties. Particularly suitable for extruded and calendered articles.

WD, type with good low temperature flexibility, to −55°C. *Uses:* cable sheathings for cold areas.

WHM, higher mol. wt. and greater viscosity than WHV.

WHV (HV: high viscosity), general purpose type, s.g. 1·23. Solubility: as for GN. Mercaptan used as modifier, phenothiazine as short-stop. Has medium tendency to crystallise; W type with a high viscosity, Mooney viscosity 115–135. Used in compounds with a high filler and softener content; may be extended with oil. Processed in an internal mixer without previous mastication. Blends with other elastomers: as for GN. Curing systems: 4 % magnesium oxide, 5 % zinc oxide, 0·5–1·5 % DOTG, DPG or tetramethyl thiuram monosulphide, 1 % sulphur. Sulphur-free compound: 4 % magnesium oxide, 5 % zinc oxide, 0·25–1 % 2-mercaptoimidazoline. The quantity of the vulcanising ingredients depends on the filler loading.

WRT, general type with mercaptan as modifier, phenothiazine as a short-stop, s.g. 1·25. Soluble in aromatic and chlorinated hydrocarbonds and in some esters and ketones. Very good storage stability. Properties are a combination of Types W and RT. *Uses:* for products with good heat resistance and low compression set. Processing: as for Type W. Vulcanises somewhat more slowly than W. For compounds with a low degree of water absorption 15% lead oxide is recommended in place of the magnesium oxide/zinc oxide combination, with 0·2–0·4% 2-mercaptoimidazoline.

WX, general purpose type with mercaptan as modifier, phenothioazine as short-stop, s.g. 1·24. Has medium tendency to crystallise (between that of Types W and WRT); very good storage stability. Soluble in aromatic, naphthenic and chlorinated hydrocarbons, some esters and ketones. Processing properties as for W but with a lower stiffening rate due to crystallisation. Is more stable than GN.

NEOPRENE LATEX Alkaline dispersion of poly-2-chlorobutadiene (neoprene). pH 10·5–12·5. The negatively charged particles (except in Type 950) have an average particle diameter of 0·10–1·13 μ with a range of 0·05–0·19 μ and show strong Brownian movement. Produced by the emulsion polymerisation of 2-chlorobutadiene. Has good mechanical and chemical stability which is reduced after long storage because of the development of acids and a reduction of the pH. Below 10°C the latices thicken, but if carefully warmed to room temperature regain their original viscosity; below 0°C the latices freeze and gel. Quick coagulation may be achieved by acids and salts, also *e.g.* fluoro-

silicates, nitroparaffins, alcohol, acetone, polyvinyl methyl ether, positively charged dispersions, drying, freezing, filtration through porous surfaces, electric currents. The same compounding practice is used for neoprene as for natural latex, in general zinc oxide, sulphur and an accelerator are necessary for each compound although vulcanisation is possible without an accelerator or sulphur. Accelerators used are dithiocarbamates, thiocarbanilide and poly-p-dinitrosophenol. To achieve good ageing properties at least 5% zinc oxide and 2% antioxidant are necessary. For light-coloured articles up to 15% zinc oxide and nonstaining antioxidants, 15% zinc oxide is necessary with 2% antioxidant in cellulose materials. The individual types vary in use, being either general or special types; uses include adhesives, dipped articles, paper impregnation, paints, foam rubber, bonding agents for fibres, leather, cork.

Types:

60: special type, s.g. 1·12, pH 10·2, viscosity 33 cp. 59% solids, s.g. (solids) 1·23. Good stability. Sodium resin soap as emulsifier. *Uses:* foam rubber with good strength of the hot, wet vulcanisate. Processing: as for 571 but vulcanises more slowly

400: special type, s.g. 1·15, pH 12·6. 50% solids, s.g. (solids) 1·42. Very little evolution of hydrochloric acid. Excellent resistance to heat and ozone. *Uses:* impregnations, paints

450: copolymer of 2-chlorobutadiene and acrylonitrile, approx. 40% solids

571: general purpose type, s.g. 1·10, pH 12·4, viscosity 8 cp. 50% solids, s.g. (solids) 1·28. Sodium resin soap as emulsifier. Films have a higher strength than other types. Compounding: 5% zinc oxide, 1% sulphur, 1–2% accelerator, 2% antioxidant. For better ageing properties zinc oxide may be increased to 15%. Vulcanisation: approx. 30 min at 140°C

572: special type for adhesives, s.g. 1·10, pH 12·4, viscosity 9 cp. 50% solids, s.g. (solids) 1·23. Sodium resin soap as emulsifier. Films solidify at room temperature, but have a high bonding strength and are tougher and stronger than those from other neoprene latices. For adhesives vulcanisation is unnecessary, but zinc oxide and antioxidants must always be used. For exceptional resistance to water and oil this type is vulcanised with 1% sulphur and 1–2% accelerator

601-A: (improved Type 601) general purpose type, s.g. 1·12, pH 12·2, viscosity 35 cp, 59% solids, s.g. (solids) 1·23. Sodium resin soap as emulsifier. Produced by creaming. Type 842-A: products of this latex possess special properties at low temperatures, in particular better resistance to stiffening. Compounding: dipped articles 5% zinc oxide, 1% sulphur, 0·5–2% accelerator; vulcanisation: 14–30 min at 100–120°C. Foam rubber: 5% zinc oxide, 2% sulphur, 1% catechol, 2% poly-p-dinitrosobenzene 25% (Polyac), 1% sodium dibutyl dithiocarbamate (47% aqueous solution); vulcanisation: 15 min at 126°C. Low density foam: 5% zinc oxide, 2% thiocarbanilide, 1% sulphur; vulcanisation: 30–45 min at 126°C

635: (creamed Type 735) special type, s.g. 1·12, pH 12·2, vis-

cosity 32 cp. 50% solids, s.g. (solids) 1·230. *Uses:* similar to 735. Increases the elongation and film strength of other neoprene latices; up to 25% added. In adhesives with a long tack life, prevents the cracking of drying films. Processing: as for 571

650: (creamed Type 750) general purpose type, s.g. 1·11, pH 12·9, viscosity 35 cp. 60% solids, s.g. (solids) 1·23. Sodium resin soap as an emulsifier. Excellent storage stability. Forms strong, highly extensible films. Only slight tendency to crystallise. According to the compounding technique products can be obtained with either a high or a low modulus. Only a very slight shrinkage. *Uses:* for very soft, elastic foam rubber, dipped articles, paints, moulded articles, extrusion compounds. Processing: as for 571

673: special type, s.g. 1·12, pH 12·2, 58% solids, s.g. (solids) 1·23. Has very strong tendency to crystallise. *Uses:* adhesives

+ 700: special type, s.g. 1·09, pH 12·4, viscosity 14 cp. 50% solids, s.g. (solids) 1·19. Saturation material for paper and cellulose; gives very good physical properties. Films remain flexible at low temperatures, but are too soft for general use. Recommended as a softener for other neoprene latices. Compounding: 5% zinc oxide, 1% sulphur, 1% thiocarbanilide

735: special type for the treatment of paper; is added to the pulp, s.g. 1·06, pH 12·2, viscosity 6 cp. 34% solids, s.g. (solids) 1·23. Impregnated papers have very good physical properties.

Gelled films are stronger than those of other latices and have plastic properties and solubility in neoprene solvents. *Uses:* include special papers, sand papers, shoe inserts, technical filter papers, cartons, sealing, wallpapers. Processing: as an additive to pulp, it is not normally vulcanised, but the addition of zinc oxide and antioxidant is necessary. Compounding ingredients can be added to the pulp separately. For good resistance to water and oil it is vulcanised with 1–2% sulphur and 1–2% accelerator

736: (modification of Type 735, by the addition of a stabiliser which facilitates the mixing of the latex with the paper pulp): special type for paper. Has high strength in a wet and dry condition, s.g. 1·06, pH 12·4, viscosity 5 cp. 34·5% solids, s.g. (solids) 1·23. Processing: as for 735

750: general purpose type s.g. 1·10, pH 12·5, viscosity 8 cp. 50% solids, s.g. (solids) 1·23. Sodium resin soap as an emulsifier. Has only slight tendency to crystallise. Excellent storage stability. Wet films have a high strength, low modulus. *Uses:* dipped articles. Processing: As for 571

842-A (improved Type 842): general purpose type. s.g. 1·10, pH 12·4, viscosity 9 cp, 50% solids, s.g. (solids) 1·23. Films have good resistance to stiffening at low temperatures, only very slight tendency to crystallise. Compounding: 5% zinc oxide, 1% sulphur, 0·5–2% accelerator, vulcanisation 15–30 min at 100–120°C. *Uses:* dipped articles,

impregnation of textiles and paper, paints, fibre bonding agent

950: cationic latex for treatment of fibres, textiles, and substances which cannot be treated with negative latices, s.g. 1·11, pH 9·3, stable between pH 2–12. 50% solids, s.g. (solids) 1·23. Behaves when compounded as other neoprene latices. Because of its positive charge some materials, particularly surface active agents, cannot be used. For dispersions of compounding ingredients cationic or nonionic emulsifiers must be used.

NEOPRENE PEPTISER P-12 Odourless, white powder, s.g. 1·04. Peptising agent with excellent heat stability for *neo*prene (266).

NEO-SPANGOL Salve-like, paste-like colloidal softener, improves the elasticity and strength of vulcanisates because it is partially vulcanisable (267).

NEOTEX Group of unclassified blacks with a low carbon structure.

Type	Particle size mμ	Surface area m²/g	Furnace Black of equivalent particle size	pH
100	25	107	HAF	8·0
130	20	130	ISAF	8·0
150	17	150	SAF	8·0

The blacks give a normal modulus as opposed to the high modulus of the equivalent particle size furnace blacks. The low structure makes its dispersion in synthetic polymers more difficult (159).

NEOTEX Adhesive compound of neoprene latex.

NEOTHANE Polyurethane rubber (115).

NEOZONE Group of antioxidants (6):

Neozone: mixture of 50% phenyl-α-naphthylamine, 25% diaminotoluene and 25% stearic acid. Dark grey granulated material, s.g. 1·15. Gives good protection against normal ageing and heat. Stiffens unvulcanised compounds, activates aldehyde-amines, guanidines, thiazoles and thiazolines. Discolours in sunlight. Blooms above 3%. Quantity: 1–2%

A: phenyl-α-naphthylamine

B: 2:4-diaminotoluene

C: mixture of 92·5% phenyl-α-naphthylamine and 7·5% diaminotoluene. Grey-brown, resin-like lumps, m.p. above 45°C, s.g. 1·19, solubility in rubber approx. 4½%. Gives good protection against heat and normal ageing, has a slightly stiffening effect in unvulcanised compounds. Slightly activates thiazoles and thiazolines. Discolours on contact with light rays

D: phenyl-β-naphthylamine

E: phenyl-β-naphthylamine (mixture with 2:4 diaminotoluene oxalate)

HF: undisclosed composition (obsolete)

L: ethyl-β-naphthylamine

S: mixture of 50% phenyl-α-naphthylamine, 25% m-toluylene diamine and 25% stearic acid

T: o-tolyl-β-naphthylamine.

'NERVE' Term widespread in the rubber industry, and which refers to a condition of toughness and resistance to deformation in raw rubber and compounds. J. Ball (1947) defined 'nerve' as synonymous with elasticity in the unvulcanised condition.

NEUBURGER KIESELERDE A non-black filler consisting of 25–30% kaolinite and 70–75% quartz; occurring at Neuburg/Donau (W. Germany).

NEUTRONYX 300 Condensation product of a polyalkyl ether with fatty acids. *Uses:* foaming agent for the production of foam rubber (268).

NEUVILLE, A. J. DE LA Jesuit, French Guiana. In 1723, published a description of an elastic rubber from which the natives produced containers, bracelets, necklaces, and strings for bows.
Lit.: 'Sur une poire faite de gomme et qui sert aux Indicus de Seringue', *Mémoires de Trévoux* 1723, 527–28.

NEVASTAIN Non-staining antioxidant for synthetic and natural rubbers; based on an alkylated phenol:

A: liquid, s.g. 1·09. Soluble in most organic solvents, insoluble in water

B: pale brown powder or flakes, m.p. 55°C, s.g. 1·137 (269).

NEVILLAC Group of softeners, plasticisers and tackifiers:

Nevillac: alkylated phenolic resin, s.g. 1·11. Tackifier and softener. Has no effect on vulcanisation. Used for NR, SR and latices, particularly for adhesives and cements

10°: phenol derivative. Pale brown viscous liquid, b.p. 300–370°C, s.g. 1·07–1·09. Softener for NR, SR and vinyl resins

TS: phenol derivative. Pale brown viscous liquid, s.g. 1·08. Used in NR and SR. Diluent for rubber and latex, softener for rubber and vinyl resins (269).

NEVILLE Group of softeners, plasticisers, tackifiers and reclaiming agents:

LX-685, 125: solid hydrocarbons. m.p. 98–109°C. s.g. 1·113 (25°C), mol. wt. 600–1000, iodine No. 130–155. Processing aid and softener for NR and SR, improves tensile strength, elongation at break and colour stability

LX-685, 135: solid hydrocarbon, m.p. 110–120°C, s.g. 1·113, iodine No. 130–155. Processing aid and softener for NR and SR, improves tensile strength, elongation at break and ageing properties

LX-685, 180: solid hydrocarbons, m.p. 145–155°C, s.g. 1·113, iodine No. 130–155. Softener and plasticiser for NR

and SR, diluent; improves dispersion of pigments and fillers and also resistance to ozone

LX-777: hydrocarbon, s.g. 1·03–1·04. Reclaiming oil, softener for NR and SBR reclaims

LX-782: solid hydrocarbons, m.p. 102–109°C, iodine No. 160. Plasticiser for NR and SR

LX-809: reclaiming oil and softener for NR and SBR reclaims. Liquid, s.g. 1·03

R-16-A: coumarone-indene resin. m.p. 94–107°C. Processing aid and softener for NR and SR; improves dispersion of pigments

Resins (R-12, R-16, R-17, R-29): group of coumarone-indene resins. Pale brown to dark brown, viscous liquids, m.p. between 5–117°C, s.g. 1·08–1·15. Softeners, plasticisers, tackifiers, reinforcing materials (269).

NEVINOL Coumarone-indene oil. Pale brown, viscous liquid with characteristic odour, b.p. 300–370°C, s.g. 1·03–1·08. Softener and plasticiser for natural, and synthetic rubbers and, in emulsion form, for latices. Stains and is unsuitable for light-coloured goods (269).

NEVTEX Modified coumarone resin. Softener for rubber (269).

NG-200 Pelleting machine for rubber chemicals; used as an aid for mixing (270).

NIBREN WAXES Group of synthetic, chlorinated waxes, m.p. 90–125°C, s.g. 1·5–1·7. Soluble in many organic solvents. May be blended with polymers, synthetic resins and waxes. Resistant to alkalis and acids (43).

NICKEL DIBUTYL DITHIO-CARBAMATE

$$Ni\begin{cases} S-C-N(C_4H_9)_2 \\ \| \\ S \\ S-C-N(C_4H_9)_2 \\ \| \\ S \end{cases}$$

Green powder, m.p. 87–90°C, s.g. 1·29. Soluble in chlorinated hydrocarbons, benzene, acetone. Protects SBR, nitrile rubber and neoprene against ozone attack and light-ageing under static and dynamic conditions. Improves heat resistance of neoprene compounds and retards discoloration of neoprene vulcanisates by sunlight, improves the ageing properties of neoprene S compounds. Used alone, is unsuitable for natural rubber or elastomers in contact with NR vulcanisates, because it causes a serious deterioration in the ageing properties of NR. However, it improves the protection given by other antioxidants in natural rubber. Has little effect as an antioxidant in SBR and must always be used together with an antioxidant. Slightly boosts acidic accelerators in SBR and Hypalon compounds, but has a slightly retarding effect in neoprene. Blooms when used in excess. Is retarded by large quantities of fillers.
TN: Antage NBC (16)
BTN-Henley (123)
DBDTKNi
NBC (6)
Nocrac NBC (274)
Robac Ni-BUD (311)
Van Hasselt BIN (355).

NICKEL DIMETHYL DITHIO-CARBAMATE
Antioxidant for neoprene and Hypalon. Green flakes.
TN: Niclate (51).

NICKEL PENTAMETHYLENE DITHIOCARBAMATE
$(H_{10}C_5=N.CS.S-)_2Ni$, NPD. Pale green, odourless powder, s.g. 1·42. Contains approx. 14·8% nickel. Improves the heat resistance of neoprene and Hypalon and retards discoloration caused by light rays.
Quantity: 2–4% on the polymer.
TN: Robac NiPD (311).

NICLATE Nickel dimethyl dithiocarbamate (51).

NIGROMETER INDEX Arbitrary unit used to characterise the colour depth of blacks. With the nigrometer developed by G. L. Cabot Inc. (USA), the reflection from the rearside of a glass plate painted with a black linseed oil paste is determined. Increasing values on a logarithmic scale show an increase in the amount of reflection corresponding to a decrease in the depth of colour. Calibration of the instrument is by means of a ceramic standard.

NIIAT Russian Motor Transport Research Institute.

NIIRP Scientific Research Institute of Mechanical Goods Industry, u. Trubezkaja, Moscow.

NIIShP Research Institute of the Tyre Industry, 1st Sokolinoy gora No. 25, Moscow E-275.

NIPOL Japanese produced synthetic rubber:
N: butadiene-acrylonitrile copolymer

HS: butadiene-styrene copolymer with a high styrene content

P-70: 70:30 blend of nitrile rubber with p.v.c. (271).

NIPPON Japanese produced chlorinated rubber (272).

NIPPON GOMU KYOKAI Association of the Japanese Rubber Industry. Founded 1932.

NISSIM SD Japanese produced di-o-tolyl-guanidine.

NITREX Group of butadiene-acrylonitrile copolymers in latex form with high mechanical stability and good oil resistance. Used primarily as coatings for papers, textiles, leather, softeners for resin latices, *e.g.* Bralec 2713-C, and p.v.c. dispersions, binding agents for pigments:

2612: butadiene-styrene-acrylonitrile; terpolymer with 10% styrene and 28% acrylonitrile. 50% solids, pH 10·0, viscosity 35 cp, Mooney viscosity 90. Has low water absorption. *Uses:* impregnations, saturating agent for paper and cellulose, softener for resin latices

2614: butadiene-acrylonitrile-methacrylic acid. Can be vulcanised at low temperature with zinc oxide

2615: butadiene-acrylonitrile-methacrylic acid. Vulcanised with zinc oxide at low temperatures. *Uses:* coatings for paper and textiles, adhesives, self-adhesive tapes

2616: butadiene-acrylonitrile, high acrylonitrile content. 42% solids, pH 10·5, Mooney viscosity 70, very small particle size. Has low water absorption. *Uses:* binding agent for unwoven textiles, leather impregnations, paper industry

2617: butadiene-styrene-acrylonitrile. Good heat and ageing properties

2619: styrene-acrylonitrile. Compatible with other nitrile latices

2620: butadiene-acrylonitrile. 42% solids, pH 10·5, Mooney viscosity 70, very small particle size. Good stability and binding properties. *Uses:* saturation of paper

2625: butadiene-styrene-acrylonitrile, 29% styrene and 33% acrylonitrile. 48% solids, pH 10·0, Mooney viscosity 135. Self-vulcanising, has high tensile strength

6849: 32% acrylonitrile. High solids content. Used for oil resistant foam materials (23).

NITRILE RUBBER General term for butadiene-acrylonitrile copolymers. ASTM code NBR (formerly BR-A).

NITRILE SILICONES By grafting nitrile groups onto dimethyl silicone, compounds with a higher polarity are obtained; these have increased resistance to solvents and improved electrical properties. Stable up to 150°C. Nitrile content approx. 23%.
Uses: primarily as additives for solvent resistant lubricants, antifoam agents, softeners.
TN: NS-fluids.

NITRILON Russian produced polyacrylonitrile fibre, according to the process of the Textile Institute Kirov, Leningrad.

NITROBENZENE VULCANISATION I.

Ostromislensky discovered in 1915 that NR may be vulcanised by heating with approx. 5% 1:3:5-trinitrobenzene, 1:3-dinitrobenzene and other polynitrobenzenes in the presence of lead monoxide.

2-NITRO BIPHENYL

o-nitrobiphenyl, ONB, o-nitrodiphenyl. Ortho-rhombic crystals with a sweet odour, m.p. 36·7°C, b.p. 325°C, s.g. 1·44 (25°C). Soluble in alcohol, methyl alcohol, acetone, tetrahydrofurfuryl alcohol, dimethyl formamide, carbon tetrachloride, turpentine, acetic acid, insoluble in water. Softener for resins, cellulose, acetate and nitrate, polystyrene.

NITROBUTANOL
$CH_2 . CH_2 . CH(NO_2) . CH_2OH$. Liquid, m.p. $-47°C$, b.p. $105°C$. Soluble in water. Stabiliser for latex compounds.

NITROCYCLISED RUBBER
$(C_5H_7NO_2)_4$. Yellow substance obtained by the reaction of concentrated nitric acid with rubber in carbon tetrachloride. The structural formula is unknown.

2-NITRO-2-METHYL PROPANOL
$(CH_3)_2C(NO_2) . CH_2OH$. White powder, m.p. 90–91°C. Retarder for rubber and latex.

NITRON Trade name for 1:4-diphenyl-3:5-enol-aniline dihydrotriazol.

p-NITROPHENOL
$1:4-C_6H_4(OH)NO_2$. Colourless, odourless crystals, m.p. 114°C, s.g. 1·28. Sparingly soluble in cold water, readily soluble in hot water, alcohol, chloroform, ether, alkali hydroxides and carbonates. Fungicide for smoked sheet, preventing mould growth caused by too long a period of storage or during the monsoons. The wet sheets are dipped in 1% solution before the smoking process. Should not be added to the latex. p-Nitrophenol is toxic and there is a danger of discoloration due to the yellow alkali salt, *e.g.* with rubber coated materials, therefore its use as an additive has become obsolete (forbidden in Ceylon since 1955). It is still used on the plantations as a disinfectant, *e.g.* factory rooms and smoking chambers.

1-NITROPROPANE
$CH_3CH_2CH_2NO_2$. Liquid, m.p. $-108°C$, b.p. 131·6°C, s.g. 0·9934 (25°C). Sparingly soluble in water, miscible with most organic solvents. *Uses:* as for 2-nitropropane.

2-NITROPROPANE
$CH_3 . CHNO_2 . CH_3$. Liquid, m.p. $-93°C$, b.p. 120·3°C, s.g. 0·9821 (25°C). Sparingly soluble in water, miscible with most organic solvents. *Uses:* solvent for synthetic rubbers, cellulose acetate, vinyl resins, fats, oils. Vulcanisation retarder.

NITROPRUSSIDE-HYDROPEROXIDE PROCESS
Redox emulsion polymerisation of SBR cold rubber with the nitroprussic

ion $[Fe(CN)_5NO]$ as a reducing agent and hydroperoxide as the oxidising agent.

NITRO RUBBER Produced by the violent reaction of concentrated nitric acid with rubber when a brownish red vapour develops. Dittmar obtained a yellow reaction product which was given the formula $C_{10}H_{12}N_2O_6$. Harries introduced nitrogen trioxide into a rubber solution and obtained a green reaction product, Nitrosite A, of the probable formula $(C_{10}H_{16}N_2O_3)_x$. After standing for some time, this became Nitrosite C $(C_{10}H_{16}N_3O_7)_x$. According to some recent research by Bruni and Geiger, the reaction of nitrosobenzene with rubber as a solution in benzene, gives a yellow product $(C_{11}H_{11}ON)_x$ with one nitro group per isoprene molecule and with the following structure.

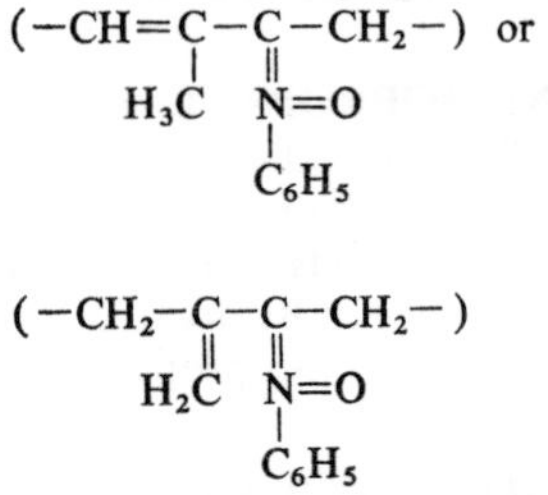

By heating rubber with 5% of a nitroso derivative a tacky product can be obtained which is a suitable base for adhesives.

NITROSAN Mixture of 70% N,N′dimethyl-N-N′-dinitrosoterephthalimide and 30% inert filler. Decomposition temperature 80–100°C, s.g. 1·2. Blowing agent for natural and synthetic rubbers (6).

p-NITROSODIMETHYL ANILINE

Green crystals, m.p. 85–93°C, decomposes in air. Soluble in alcohol and ether, insoluble in water. Accelerator. Discovered and introduced by Peachey in England, 1914. Vulcanisation properties are similar to those of diphenyl guanidine. One of the most important accelerators used during World War I. Vulcanisation temperature 142°C.
Quantity: 1·5–2·5% with 3–4% sulphur.
Uses: particularly for hard rubber.
TN: + Accelerator 1 (6)
Accelerene (19)
Accelerine
Accinelson
+ B-Naphtol
+ Vulcafor I (60)
+ Vulcaniline (60).

N-NITROSODIPHENYLAMINE Diphenyl nitrosamine. Brown, crystalline powder, m.p. 64–66°C, s.g. 1·24–1·27. Soluble in acetone, ethyl acetate, benzene, methylene chloride, carbon tetrachloride, alcohol, sparingly soluble in benzine, insoluble in water. Retarder for rubber and latex; increases the critical temperature and improves the processing safety of 'scorchy' compounds. Activates slightly at high vulcanisation temperatures. Causes slight discoloration of light-coloured vulcanisates. Particularly effective in blends with mercapto, thiuram, dithiocarbamate and guanidine accelerators and their mixtures, but is unsuitable for preventing scorching of sulphur-free compounds containing tetramethyl thiuram disulphide; is not

very effective in compounds containing aldehyde-amines. Also used as a reclaiming agent for the processing of slightly scorched compounds on cold rolls.
Quantity: 0·2–1 % as retarder, 1 % as reclaiming agent.
TN: Antifix D
Delac J (23)
Diophene SD (19)
Good-Rite Vultrol (14)
Redax (51)
Retarder J (246)
TJB (23)
Vulkalent A (43)
Vulcatard (60)
Vultrol (14).

NITROSOHYDROFURFUR-AMIDE $(C_5H_3ONO)_3N_2$, nitroso furfurine. Accelerator of the aldehyde-ammonia group. Used particularly for hard rubber.
TN: Vulcazol N (363).

1-NITROSO-2-NAPHTHOL
α-nitroso-β-naphthol. Yellowish brown, crystalline powder, m.p. 109–110°C. Soluble in hot alcohol, benzene, ether, carbon disulphide, alkalies and acetic acid. Softener in rubber.
TN: JMH (40).

NOBS Accelerator with a delayed action, for natural rubber and SBR.
NOBS No. 1: mixture of 90% N-oxydiethylene benzthiazole-2-sulphenamide and 10% 2:2′-dibenzthiazyl disulphide. Has a less pronounced delayed action and offers less processing safety than **NOBS** Special. Gives a high modulus in compounds containing furnace blacks; 3–5% zinc oxide and stearic acid are necessary

NOBS   Special: N-oxydiethylene benzthiazole-2-sulphenamide (21).

NOCCELER Group of accelerators:

8:	butyraldehyde-aniline
22:	2-mercaptoimidazoline
AC:	butyraldehyde-aniline condensation product
BG:	o-tolyldiguanidine
BZ:	zinc dibutyl dithiocarbamate
C:	thiocarbanilide
CZ:	N-cyclohexyl-2-benzthiazole sulphenamide
D:	diphenyl guanidine
DM:	dibenzthiazyl disulphide
EZ:	zinc diethyl dithiocarbamate
F:	mixture of dibenzthiazyl disulphide, 2-mercaptobenzthiazole and diphenyl guanidine
H:	hexamethylene tetramine
K:	acetaldehyde-aniline
M:	2-mercaptobenzthiazole
MS:	dibenzthiazyl disulphide
MSA:	N-oxydiethylene-2-benzthiazole sulphenamide
MZ:	zinc-2-mercaptobenzthiazole
PX:	zinc ethyl phenyl dithiocarbamate
PZ:	zinc dimethyl dithiocarbamate
SDC:	sodium diethyl dithiocarbamate
TS:	tetramethyl thiuram monosulphide
TT:	tetramethyl thiuram disulphide (274).

NOCRAC Group of antioxidants:

100:	diphenyl ethylene diamine
200:	alkyl phenol
224:	polytrimethyl dihydroquinoline
500:	mixture of N,N′-diphenyl-p-phenylene diamine and N-phenyl-α-naphthylamine

810: N-phenyl-N'-cyclohexyl-p-phenylene diamine
A: Aldol-α-naphthylamine
AW: 6-ethoxy-2:2:4-trimethyl-1:2-dihydroquinoline
B: acetone-diphenylamine reaction product
C: Aldol α-naphthylamine. Powder
D: phenyl-β-naphthylamine
DP: N,N'-diphenyl-p-phenylene diamine
HP: mixture of diphenyl-p-phenylene diamine and phenyl-β-naphthyl amine
MB: 2-mercaptobenzimidazole
MBZ: zinc mercaptobenzimidazole
NBC: nickel dibutyl dithiocarbamate
NP: mixture of N-phenyl-β-naphthylamine and N,N'-diphenyl-p-phenylene diamine
PA: phenyl-α-naphthylamine
SP: phenol modified with styrene
White: di-β-naphthyl-p-phenylene diamine (274).

NOCTISER Group of peptising and reclaiming agents (274):
SM: mixture of Types SS and SZ
SS: 2:2'-dibenzamidodiphenyl disulphide
SZ: zinc-2-benzamidodithiophenate.

NOIR DU CONGO Trade term for rubber from *Clitandra orientalis (Cl. Arnoldiana)* (Congo). High quality rubber known locally as Kappa.

NOMENCLATURE: ELASTO-MERS With the development of numerous synthetic rubber types and the emergence of a variety of trade names, a standard nomenclature became essential. During World War II, types produced in the government factories in the USA were designated according to a GR (government rubber) code, as follows:

GR-I: GR isobutylene (butyl rubber)
GR-M: GR monovinyl acetylene (neoprene)
GR-P: GR polysulphide (thiokol rubber)
GR-S: GR styrene (butadiene-styrene rubber)
GR-S + nos: account for differences in catalyst emulsifier, stabiliser, polymerisation temperature, monomer ratio
GR-S + X nos: experimental products produced in limited quantities.

Of the above, only GR-S has found widespread use. When the production of synthetic rubber was given over to private enterprise (1954–55) many trade terms and codes were introduced. The ASTM therefore created an intermediary specification D 1418 (1956), a general classification system for elastomers which was produced as an easily comprehensible short terminology for literature references and a complement for the various trade products. The elastomers and plastics used in the rubber industry are divided into:

Class I, elastomers:
A: vulcanisable elastomers
1: diene-rubber types
2: non-diene rubber types
B: non-vulcanisable plastics and other elastomers
Class II: hard plastics
Class III: reinforcing resins
Class IV: basic materials used for paints.

Elastomers and rubber types in solid form and as latices are classified on the basis of the chemical structure of the polymer chain:

M: elastomers with a saturated polymethylene chain

N: elastomers with nitrogen in the polymer chain

O: elastomers with oxygen in the polymer chain

P: elastomers with phosphorus in the polymer chain

R: rubber or elastomers with an unsaturated carbon chain, *e.g.* natural and synthetic rubber are derived at least partially from diolefins

Si: elastomers with silicon in the polymer chain

T: elastomers with sulphur in the polymer chain.

The R family is characterised more precisely by putting the monomer term before the word 'rubber' (with the exception of natural rubber). The first letter preceding the 'R' shows the diolefin used in the production of the rubber (with the exception of natural rubber). Additional letters before the diolefin term show the comonomers:

BR: butadiene rubber

IR: synthetic isoprene rubber

CR: chloroprene rubber

NR: natural isoprene rubber (natural rubber)

ABR: acrylate-butadiene rubber

IIR: isobutylene-isoprene rubber

NBR: nitrile-butadiene rubber

NCR: nitrile-chloroprene rubber

PBR: pyridine-butadiene rubber

SBR: styrene-butadiene rubber

SCR: styrene-chloroprene rubber

SIR: styrene-isoprene rubber.

A similar abbreviation is used under the code for latices, *e.g.* **SBR** latex.

In the Si family the group name is placed first:

Si: silicone with methyl groups alone in the polymer chain, *e.g.* dimethyl polysiloxane

PSi: silicone with methyl as well as phenyl groups in the polymer chain

VSi: silicone with methyl and vinyl groups in the polymer chain

FSi: silicone with methyl and fluoro groups in the polymer chain

PVSi: silicone with methyl, phenyl and vinyl groups in the polymer chain.

The 'M' family includes elastomers with a saturated chain of polymethylene groups.

IM: poly*iso*butylene

EPM: ethylene propylene copolymers

CSM: chlorosulphonated polyethylene

CFM: polychlorotrifluoroethylene

FPM: copolymers of vinylidine fluoride and hexafluoropropylene

ACM: ethyl or other acrylates and 2-chloroethyl vinyl ether copolymers

ANM: ethyl or other acrylates and acrylonitrile copolymers.

NOMENCLATURE: PLASTICS

In the literature, the various plastics have been given abbreviated terms and names which were collected together in the ASTM code (D 1600) under the following code terminology:

Plastics and resins

Acrylonitrile-butadiene-styrene	ABS
Carboxymethyl cellulose	CMC
Cellulose acetate	CA
Cellulose acetate butyrate	CAB
Cellulose nitrate	CN
Diallyl phthalate plastics or resins	DAP
Ethyl cellulose	EC
Melamine-formaldehyde	MF
Phenol-formaldehyde	PF
Polyacrylic acids	PAA
Polyacrylonitriles	PAN
Polyamides	Nylon

Polybutadiene-acrylonitriles	PBAN
Polybutadiene-styrene	PBS
Polychloroprene	PC
Polyethylene	PE
Polyhexamethylene	Nylon 66
Polyisobutylene-isoprene	PIBI
Polyisobutylene	PIB
Polymethylchloroacrylate	PMCA
Polymethylmethacrylate	PMMA
Polymonochlorotri-fluoroethylene	PCTFE
Polystyrene	PS
Polytetrafluoroethylene	PTFE
Polyvinyl acetate	PVAc
Polyvinyl alcohol	PVA
Polyvinyl butyral	PVB
Polyvinylchloride	PVC
Polyvinylchloroacetate	PVCAc
Polyvinyl formal	PVF
Urea-formaldehyde	UF
Softeners, additives	
Dibutyl phthalate	DBP
Dicapryl phthalate	DCP
Diisodecyl adipate	DIDA
Diisodecyl phthalate	DIDP
Diiso-octyl adipate	DIOA
Diiso-octyl phthalate	DIOP
Dinonyl phthalate	DNP
Di-n-octyl-n-decylphthalate	DNODP
Dioctyl adipate	DOA
Dioctyl azelate	DOZ
Dioctyl phthalate	DOP
Dioctyl sebacate	DOS
Tricresyl phosphate	TCP
Trioctyl phosphate	TOF

NON-FER-AL Precipitated iron- and aluminium-free calcium carbonate; particle size 5–10 micron. Filler (153).

NONFLEX Group of Japanese produced antioxidants:
BA: reaction product of acetone and diphenylamine
WS: phenol modified with styrene

NONISOLES Group of non-ionic polyglycol fatty acid esters. Emulsifiers and stabilisers for synthetic latices (69).

NONOX Group of antioxidants produced by ICI:
A: phenyl-α-naphthylamine
AN: phenyl-α-naphthylamine
B: diphenylamine-acetone condensation product. Dark brown powder, s.g. 1·14. For general use: protects against natural ageing, flex cracking and heat, particularly in black compounds. *Quantity:* 0·5–1% in solid rubber and latex. *Uses:* cables, hose lines, air bags, conveyor belts, mechanical goods
BL: diphenylamine-acetone condensation product. Dark brown liquid, s.g. 1·10. Soluble in benzene, carbon tetrachloride and alcohol. Protects against heat ageing and flex-cracking. *Quantity:* 1–2%. *Uses:* tyre treads and side walls. Has no effect on vulcanisation. Up to 4% may be added without blooming. Causes a brown discoloration and stains due to migration
CC: phenolic sulphide. Pale yellow powder, m.p. approx. 118°C, s.g. 1·30. Soluble in alcohol, insoluble in benzene and water, solubility in rubber approx. 2%. Causes discoloration. Unsuitable for light-coloured articles. *Quantity:* 1–2%. *Uses:* for liquid or gaseous sulphur chloride vulcanisates

CGP: mixture of mercaptobenz-imidazole and di-β-naphthyl-p-phenylene diamine. General purpose antioxidant and copper and manganese inhibitor for NR, SR and uncured NR. Suitable for cable compounds

CI: di-β-naphthyl-p-phenylene diamine

CNS: mixture of mercaptobenzimidazole and bis(2-hydroxy-3-α-methyl cyclohexyl-5-methyl methane). Yellowish, crystalline powder, s.g. 1·25. Copper and manganese inhibitor. Does not cause staining. Small quantities reduce the modulus, but the effect is reduced by a higher dosage. Increases the scorching tendencies of thiurams and sulphenamides, whereas the 'scorchiness' of MBT and ZDC is decreased. *Quantity:* 0·5–1% as inhibitor, 0·5–2% (in extreme cases up to 4%) as inhibitor and antioxidant, can be combined with other antioxidants. *Uses:* include impregnations, cable insulations, foam rubber, adhesives, unvulcanised compounds

D: phenyl-β-naphthylamine

DED: diphenyl ethylene diamine

DPPD: diphenyl-p-phenylene diamine

EX: phenol condensation product. Resinous material, m.p. 45–50°C, s.g. 1·07. Soluble in benzene and alcohol, solubility in rubber approx. 4%. Non-staining. Suitable for light-coloured, white and transparent ar-

ticles, for natural rubber in contact with p.v.c., cellular rubber articles, neoprene, sulphur chloride vulcanisates. Does not bloom. Protects against heat, flex cracking, copper and manganese. *Quantity:* 0·5–1%

EXN: phenol condensation product. Pale yellow powder, m.p. 70–80°C, s.g. 1·17. Soluble in benzene, alcohol, chloroform, solubility in rubber approx. 1·5%. Non-staining. Suitable for light-coloured and white articles, foam rubber, neoprene, cellular rubber articles and sulphur chloride vulcanisates. Does not bloom. Protects against heat, flex cracking, copper and manganese. *Quantity:* 0·5–1%

EXP: phenol condensate. Dark brown, granular material, m.p. approx. 60°C, s.g. 1·11. Non-staining. Suitable for white, light-coloured and transparent articles, cellular rubber articles, neoprene articles in contact with p.v.c. and sulphur chloride vulcanisates. *Quantity:* 0·5–1%

H: undisclosed compound (obsolete)

HF: mixture of amines

HFN: mixture of aryl amines. Brown grey flakes, m.p. approx. 100°C, s.g. 1·22. Sparingly soluble in benzene and alcohol, insoluble in water, solubility in rubber approx. 2%. Protects against normal ageing and flex cracking. *Quantity:* 1–2%, blooms above 1·25%; used in tyres, conveyor belts, hose lines, footwear, impregnations

HO: mixture of phenols. Yellow liquid, s.g. 0·94. Soluble in benzene, chloroform, alcohol, insoluble in water. Effective general purpose, non-discolouring antioxidant. *Dosage:* 1–2%

NS: phenol-aldehyde amine. Yellowish, resinous powder, m.p. 60–65°C, s.g. 1·09. Soluble in alcohol, sparingly soluble in benzene and benzine, insoluble in water, solubility in rubber approx. 2%. Non-staining. Protects against normal ageing and heat ageing, activates thiazoles and dithiocarbamates. Unsuitable for cold vulcanisates but may be used for hot vulcanised rubber and latex compounds. *Quantity:* 1–2%

NSN: phenol-aldehyde amine. Resinous, brown mass, m.p. 55°C, s.g. 1·08. Soluble in alcohol, sparingly soluble in benzene and benzine, solubility in rubber approx. 2%. Non-staining. Protects against normal ageing and heat ageing, activates thiazoles and dithiocarbamates. Suitable for hot vulcanisates of rubber and latex, unsuitable for cold vulcanisates. *Quantity:* 1·5%

OD: octyl diphenylamine. Pale brown, waxy lumps, m.p. approx. 90°C. Soluble in benzene, ethylene dichloride, acetone and petroleum ether. Disperses easily. Causes slight staining. Protects against heat, normal oxidation and flex cracking in compounds of NR, SBR, NBR and neoprene with and without blacks. *Quantity:* 0·5–2%

S: aldol-α-naphthylamine

SP: phenol modified with styrene. Pale brown, viscous liquid, s.g. 1·08. Soluble in aliphatic and aromatic hydrocarbons, ethanol, acetone, and trichloroethylene. Cheap antioxidant and gel inhibitor used in the production of SR. Non-staining. *Quantity:* 0·5–2%

T mixed phenols. Brown fluid, s.g. 0·92. Easily soluble in benzene, chloroform and alcohol. Non-discolouring, general purpose antioxidant with good protection against heat, dynamic conditions and normal ageing. *Dosage:* 1–2%

TBC: 2:6-ditert-butyl-4-methyl phenol

W: mixture of 90% β-naphthol and 10% stearic acid

WSL: alkylated phenol. Colourless liquid, s.g. 1·00. Soluble in benzene, alcohol, acetone and trichloroethylene. Crystallises during storage without affecting its properties. Non-staining. Protects against normal oxidation and flex cracking. *Quantity:* 0·5–1%. Does not bloom in quantities under 2%. *Uses:* for white and light-coloured articles made from natural rubber and neoprene; white side walled tyres; floor coverings, cables, footwear, impregnations, latex compounds

WSO: phenol condensation product. White crystalline powder, m.p. approx. 168°C, s.g. 1·00. Non-discolouring, non-staining, of some effect against light ageing but weak in black compounds, good heat resistance. *Dosage:* 0·5–1%

WSP: phenol condensation product. White, crystalline powder, m.p. approx. 120°C, s.g. 1·17. Soluble in alcohol, benzene, benzine and trichloroethylene, insoluble in water. Non-staining. Protects against normal oxidation and copper, does not prevent flex cracking. Used in natural rubber and neoprene. *Quantity:* 0·25–2%

ZA: 4-isopropyl aminodiphenylamine (60).

NONOXOL CM 50% solution of catechol in methanol. Brownish violet liquid. Soluble in benzene and water. Accelerator in neoprene compounds, particularly for films. Will cure at room temperature in 7 days (60).

NOPCO Group of rubber processing aids.

1807-L: soluble stearate. Antifoam agent for latex

1285: sulphonated ester. Dispersing and wetting agent for latex

GMO: tackifier for rubber (261).

NOPCOVIS Sodium polycarboxylate. Thickening agent for latex (261).

NORDEL Ethylene-propylene terpolymer with a non-conjugated diene. S.g. 0·85, saturated elastomer, vulcanisable with sulphur, with excellent resistance to ozone, weather and heat (6).

NORDOPREN Polyurethane (275).

NOREPOL Abbreviation for northern regional polymer. A rubber-like polyester produced by the reaction of polymeric fatty acids or their esters (from soya or linseed oils) with ethylene glycol. The process was used from 1939 to 1945 in the USA for the production of some rubber goods. The product has a low tensile strength and has achieved no further importance.

NORMASAL Lead salicylate. Activator (191).

NOR RUBBER Term for the first of F. Hofmann's synthetic rubbers. Produced in 1909.

NOVAC A-13 Mixture of 3 parts dibenzthiazyl disulphide and 1 part selenium dibutyl dithiocarbamate (89).

NOVAPERCHA Emulsion of fatty acid esters with hydrophilic groups; converted into granular form with 50% light-coloured reinforcing filler, s.g. 1·5. Dispersing agent boosts acceleration, improves ageing.
Quantity: up to 10% on the filler.

NOVODUR Terpolymer of styrene, butadiene and acrylonitrile with high impact resistance.

NOVOLACQUERS Phenol-aldehyde resins. Remain permanently

thermoplastic in the absence of methyl groups.

NOVOPLAS Group of thioplasts (276).

NPD Abbreviation for nickel pentamethylene dithiocarbamate.

NR Abbreviation for natural rubber, according to the ASTM D 1418 nomenclature.

NRB Abbreviation for the Natural Rubber Bureau. Formerly Natural Rubber Development Board (NRDB).

NRPRA Abbreviation for Natural Rubber Producers' Research Association. Formerly British Rubber Producers' Research Association (BRPRA).

NRPRA STRAIN TESTER Simple apparatus used to determine the tensile stress/strain properties of natural rubber for technical classification, according to the method given by the US National Bureau of Standards. The elongation is determined using a standard test piece under a load of 5 kg.
Lit.: W. L. Holt, *India Rubb. World*, 1948, 118, 614.
W. P. Fletcher, *Rubber, Chem. & Techn.*, 1950, 23, 107.

NS Abbreviation for nonstaining. Used in the terminology of non-staining gas black types and antioxidants.

NS ESSEX A non-staining SRF carbon black (40).

NS FLUID Abbreviation for nitrile silicone fluids.

NSR-X Nitrile-silicone rubber:
4803: Shore A Hardness 80°
5602: Shore A Hardness 60°
8701: Shore A Hardness 70° (198).

NUBUN Modified Buna S latex. *Uses:* electrical insulation of wires (129).

NURAC Trade name for diphenyl guanidine.

NUTROSE Water soluble sodium caseinate.

NWEDO Former trade name for natural rubber which grows in Burma.

NX PASTES Groups of accelerators and antioxidants in the form of pastes (23):
500: Sunproof wax (40%)
503-C: zinc-2-mercaptobenz-thiazole (40%)
503-E: aminox (50%)
511: zinc dibenzyl dithio-carbamate (50%)
513-B: zinc diethyl dithiocarbamate (50%)
514-A: zinc dibutyl dithiocarbamate (45%)
535: 38% solution of dimethyl-ammonium dimethyl dithio-carbamate.

NYASSA Former trade term for a natural rubber from *Landolphia Karkii* (Africa).

NYGENE A prestretched nylon fibre with increased strength. Used for nylon cord.

NYLON Group of polyamides used in a wide variety of fields; discovered in 1932 by W. H. Carothers and developed by Du Pont. The term was originally used as a trade name but since 1954 has been used as a generic term for linear polyamides:

Nylon 66, polycondensation product of hexamethylene diamine and adipic acid, m.p. 240°C, s.g. 1·14. Soluble in phenols, cresols, concentrated formic acid; attacked by oxidising acids, lactic acid, thioglycolic acid, chlorhydrin, hydrogen peroxide. The starting materials give low mol. wt. polyamides (α and β) upon heating while giving off water. At high temperatures polycondensation results in long molecules, known as super-polyamide (ω form). The mol. wt. can be controlled by monofunctional acids or amines which act as chain stoppers and also by restricting the amount of water removed during polymerisation. May be spun to fibres of up to 0·2 denier. The term 66 comes from the 6 C atoms in the diamine chain and the 6 C atoms in the acid chain

$$NH_2—(CH_2)_6—NH_2$$
Hexamethylene diamine
$$+ \; HOOC—(CH_2)_4—COOH \;\rightarrow$$
adipic acid

$$NH_2—(CH_2)_6—NH—$$
$$—CO—(CH_2)_4—COOH \;\rightarrow$$
Monoadipyl hexamethylene diamine (AH salt)

$$[-HN(CH_2)_6—NH—CO$$
$$—(CH_2)_4—CO—]_n \;\rightarrow$$
Polyhexamethylene adipamide (Nylon 66)

Fibres are produced using a melt spinning process at approx. 300° C with subsequent stretching to 3–5 times the original length. The fibres have good stabilities, very low water absorption, high resistance to moisture and good resistance to alkalis and chemicals. For use as tyre cord the fibres are further treated thermally to reduce the extensibility.

Nylon 610, polycondensation product of hexamethylene diamine with sebacic acid. Better resistance to moisture, more flexible and more easily processed than Nylon 66 but does not have such good heat resistance.

Nylon 6, polymerisation product of amino caprolactam. Has a controllable crystalline structure. May be processed to give films, and has a high melt viscosity.

Nylon 11, polycondensation product of hexamethylene diamine and 1:12-aminoundecanoic acid. High resistance to moisture. Easily plasticised.

BCI Nylon, alkoxy-substituted nylon. Used for injection moulding, binding agents for fibres, paints in solution. Cross linking occurs on heating with an acid and the product may be thermally hardened Nylon 12, polydodecanolactam 12.

TN: Nylon 66: Zytel 101, Zytel 105 (black compound, resistant to weather), Perlon T
610: Zytel 31, Zytel 33 (heat stabilised)
6: Zytel 211, Perlon (L)
11: Rilsan
12: Polyamide 12.

NYLON TYRES Tempered tyres. Tyres in which the carcase is built up on nylon cords. When treated by a tempering process, the nylon fibres which are sensitive to strain and heating, are made dimensionally stable. The advantages of these tyres are their very low weight, improved fatigue resistance and increased resistance to over-loading under pressure and at high speeds.

O

O-124 Xanthate complex. Accelerator (22).

O-164 Xanthate complex. Accelerator (22).

OCTADECENE NITRILE ODN. Plasticiser for nitrile rubber and vinyl plastics.

OCTAL Trade name for dioctyl phthalate.

OCTAMINE Condensation product of diphenylamine and diisobutylene. Wax-like material, m.p. 75–85°C, s.g. 0·99. Soluble in benzene, benzine, ethylene dichloride and acetone, insoluble in water. Antioxidant which protects against heat and normal oxidation in natural rubber, neoprene, butadiene-styrene and butadiene-acrylonitrile copolymers. Causes a slight discoloration in vulcanisates on exposure to light. Stains by migration.
Uses: tyres, inner tubes, footwear, linings, insulation, sponge rubber, moulded articles (obsolete) (23).

N-OCTYL-2-BENZTHIAZYL SULPHENAMIDE

White pellets, m.p. 100°C, s.g. 1·14. Delayed action accelerator for NR and SBR.
TN: Vulcafor BSO (60).

OCTYL DECYL ADIPATE
$C_8H_{17}OCO(CH_2)_4COOC_{10}H_{21}$,

m.p. −60°C, b.p. 210–232°C (4 mm), flash p. 200–235°C, s.g. 0·914–0·924. Insoluble in water. Softener.
TN: Adipol ODY and 180 (184)
Cabflex ODA (139)
Hercoflex 190 (10)
Morflex 375 (264)
PX-218 (305)
Staflex DIODA (330).

OCTYL DECYL PHTHALATE
$C_8H_{17}OCOC_6H_4COOC_{10}H_{21}$, m.p. −40°C, b.p. 235–248°C (4 mm), flash p. 228–232°C, s.g. 0·967–977. Softener.
TN: Cabflex ODP (139)
Darex IOIDP (174)
Dinopol IDO and 235 (184)
DNODP (5)
Hercoflex 150 (10)
Morflex 175 (264)
PX-118 (305)
Staflex DIODP (330).

p-OCTYL PHENYL SALICYLATE Ultra violet stabiliser for polyethylene and polypropylene.
TN: OPS.

ODA Abbreviation for iso-octyl decyl adipate.

ODN Abbreviation for octadecene nitrile.

ODORANTS Aromatic substances used to disguise or destroy (deodorants) the characteristic odour of vulcanisates or to give a specific odour (odorants), *e.g.* in odourless medical articles and products which come into contact with food, a leather odour for shoes, perfumes in household articles, and as a sales stimulus. The characteristic odour of rubber articles was described as unpleasant by American housewives, participating in a survey;

apart from this odorants are specified for many articles which require total lack of odour. The most commonly used products are complex organic compounds, frequently methyl salicylate (winter-green oil), hydroxy-methoxybenzaldehyde (vanillin), anthranilic acid methyl ester, o-oxybenzaldehyde, eugenol, cedar oil, etheric oils.

Quantity: 0·05–0·25%.

Lit.: B. J. Wilson, *British Compounding Ingredients for Rubber*, W. Heffen, Cambridge, 1963.

F. Jacobs, *Rev. Gen. Caout.*, 1935, **12**, 27.

R. A. Engels, *India Rubb. World*, 1936, **94**, 37.

ODP Abbreviation for iso-octyl decyl phthalate.

OEI Abbreviation for One Essential Ingredient, the first chain modifier used in the production of butadiene-styrene copolymers to control chain length during polymerisation.

OENR Abbreviation for oil extended natural rubber.

OENSLAGER, GEORGE 1873–1956. Chief chemist with B. F. Goodrich, Akron. Developed the first organic accelerator containing nitrogen and initiated the use of blacks as reinforcing fillers. Undertook research into car tyres and vulcanisates. Holder of the Perkin Medal, 1933, and the Goodyear Medal, 1948.

OEP Abbreviation for oil extended polymer (usually SBR).

OER Abbreviation for oil extended rubber.

OGOWAY Former trade term for natural rubber from the Congo, mainly from *Landolphia* and *Ficus* species.

OHOPEX Group of softeners:
Q-10: fatty acid phthalate, m.p. −47°C, b.p. 215–135°C (4 mm), s.g. 0·952
R-9: octyl fatty acid ester, s.g. 0·864 (184).

OIL EXTENDED RUBBER SBR cold rubber with a high Mooney viscosity, extended with up to 62·5% oil added as an emulsion to the latex before coagulation. Easily processable rubber/softener blends are produced by this method. The high viscosity of the rubber enables a high amount of oil to be added without a deterioration of the physical properties, the oil replacing the low mol. wt. constituents of the normal cold rubber. Hot rubber with the high Mooney value necessary for oil extension is not easily processed because of the high content of insoluble gel (up to 50%). The processing properties of the rubber depend on the composition of the oil. Naphthenic or highly saturated oils give relatively poor processing but have far better resistance to discoloration than highly aromatic oils. Suitable oils are: Naphthenic (Circosol 2 XH), highly aromatic (Califlux TT, Dutrex 20), aromatic (Sundex 53, Shell SPX 97). Oil extended polymers were brought on the market in 1951 in the USA.

OIL RESISTANCE The resistance of rubber to swelling caused by oils under given conditions of time and temperature; expressed as the percentage increase in volume. If rubber is placed in liquid hydrocarbons both swelling and a decrease of the physical properties occur. The significance of these effects is dependent on the use of the specific rubber. In all cases where dimension stability plays a role, *e.g.* O-rings, gaskets, seals, the slightest swelling is undesirable even when the physical properties remain unchanged. In other cases slight changes in volume can be allowed so long as such properties as elasticity and abrasion resistance remain within the necessary limits. For practical purposes therefore oil resistance may be defined as the property which enables a rubber article to be used for its intended purpose even in contact with oil.

Specification: ASTM D 471
BS 903: A16: 1956
DIN 53 521.

OKC RESIN Soft resin of viscous liquid consistency, s.g. approx. 1·0. Soluble in most organic solvents. Stabiliser and antioxidant, particularly for Buna S and similar synthetic types, also for SBR latices. Improves processability of SR and prevents stiffening during milling, increases tack and flow properties. *Quantity:* 3% (217, 278).

OKSINON Russian produced 2:4-diaminodiphenylamine.

OLEFINIC POLYMERISATION Addition polymerisation of ethylene and its derivatives, in which the double bond in each monomer is replaced by two single bonds in the polymer chain.

OLEIC ACID
$CH_3(CH_2)_7CH=CH(CH_2)_7COOH$, 9-octadecenoic acid, m.p. 4–10°C,

b.p. 286°C (100 mm), s.g. 0·895. Soluble in alcohol, benzene, chloroform, ether. Activator and softener.

OLEYL NITRILE Pale yellow liquid with faint odour, b.p. 303–360°C, s.g. 0·847, flash p. 180°C, saponification value 205. Softener for NBR, reduces the 'nerve' and modulus, helps mixing and gives good properties at low temperatures. Unsuitable for articles coming into contact with food.
TN: Plasticiser OLN.

OMYA BSH Natural, surface-treated champagne calcium carbonate of fine particle size and uniformity. Easily dispersed. Diluent filler for NR and SR (279, 422).

OMYALITE R Extender for carbon blacks and other rubber reinforcing fillers.

OPEX Blowing agent for natural and synthetic rubbers:
40: mixture of 40% dinitrosopentamethylene tetramine and 60% inert filler, s.g. 1·91
42: 42% active material
80: 80% active material
93: 93% active material
100: 100% active material
PL-80: mixture of 80% dinitrosopentamethylene tetramine and 20% ester softener. Moist crumbs with a characteristic odour, s.g. 1·38 (237).

OPPANOL Group of polyisobutylenes:
B: polyisobutylene. Types: B 8, B 15, B 30, B 50, B 100, B 150, B 200 (the indices correspond to approx. 1/1000 of the mol. wt.)
C: polyvinyl isobutylene ether. Thermoplastic, for adhesives.

Oppanol is produced also in other forms, *e.g.* as foils (O, OL, ORG, BA), for linings, cable insulations, damp resistant seals (32).

OPPASIN A Group of colours which are fast to vulcanisation; granulated master batches, approx. 60% of pigment being dispersed in a polyolefin copolymer. May be mixed with NR and SR (217).

OP RUBBER (Oil extended polymer.) Oil extended butadiene-styrene copolymers.

OPS Abbreviation for p-octyl phenyl salicylate.

OPTIMUM CURE The period of vulcanisation necessary to achieve maximum physical properties. The term has not been exactly defined, but max. tensile strength or max. modulus is usually used as an unknown, although the max. product of tensile strength and elongation at break is sometimes used. The following formula is suggested:

$$\text{Optimum} = \frac{4T + 2S + M + H}{8}$$

where T is the time for maximum tensile strength, S the time for minimum tension set, M the time for highest modulus, and H the time for highest hardness.

ORGANOSOL Suspension of a finely divided resin in a volatile organic solvent, sometimes being blended with a softener. The resin is only soluble at increased temperatures. After the solvent has been evaporated, the residue, which has been allowed to cool, forms a homogeneous plastic mass.

ORIENTE Local term for the product of *Sapium stylare Müll. Arg.*; named after the province Oriente (E. Ecuador). A rubber of little value.

ORLON Formerly Fibre A, ANP; polyacrylonitrile fibre. Made by a dry spinning process. Produced experimentally since 1942 and commercially since 1950.
Types: 81, continuous fibre, 42 (improved Type 41), staple fibre.

ORONITE Group of emulsifiers, wetting agents and softeners (179).

ORR Abbreviation for Office of Rubber Reserve, a post established in 1945 by the Reconstruction Finance Corporation (RFC) as the successor to the Rubber Reserve Company (RRC) to administrate the state synthetic rubber factories in the USA. During the course of passing the synthetic rubber industry to private concerns, the ORR was terminated. Its activities were taken over by the Synthetic Rubber Division (SRD) of the RFC, and later by the Federal Facilities Corporation.

ORTHEX Blend of anhydroformaldehyde-p-toluidine and zinc-2-mercaptobenzthiazole (23).

ORZAN Group of surface active lignin sulphonates.
Uses: binding agents for fibres; emulsifiers, stabilisers, dispersing agents (280).

OSCILLANT ELASTOMETER Oscillating disc rheometer. Instrument with an oscillating biconical rotor, and based on the principle of the Mooney viscometer; can be used to determine the total vulcanisation curve.

OSOLAN SB Polish butadiene-methylmethacrylate copolymers, aqueous dispersions with anionic emulsifiers. Total solids 40%, pH 5–6·5.
Uses: finish for artificial leather, giving good elasticity up to −15°C.

OSTROMISLENSKY, IVAN I. Russian chemist, 1880–1939. Did pioneer work on polymerisation and synthetic rubbers; developed the synthesis of butadiene from alcohol and discovered the sulphurless vulcanisation accelerators.

OSTROMISLENSKY REACTION Production of butadiene from ethanol and acetaldehyde at 300–450°C with alumina as a catalyst:

$$CH_3CH_2OH \rightarrow CH_3CHO + H_2$$
$$CH_3CH_2OH$$
$$+ CH_3CHO \rightarrow H_2C=CH$$
$$-CH=CH_2$$
$$+ 2H_2O$$

Lit.: I. I. Ostromislensky, *J. Russ. Phys. Chem. Soc.*, 1915, **47**, 1472–1506.
G. S. Whitby, *Synthetic Rubber*, New York and London, 1954, 86
G. Egloff and Hulla, *Chem. Rev.*, 1945, **36**, 73.

OTID Mixture of 70% dibenzthiazyl disulphide and 30% diethyl benzthiazole sulphenamide (23).

OVAC Reaction product of 2-mercaptobenzthiazole with hept-aldehyde and anhydroformaldehyde-o-toluidine (23).

OVIEDO Y VALDEZ, GONZALO FERNANDEZ DE In his *General History of the Indes* ('Historia General y Natural de las Indes', Seville, 1535; Madrid, 1851) he relates how Columbus on his second journey to Haiti (Hispaniola), saw the natives playing with elastic balls.

OXIDE WAX A Activator for light-coloured reinforcing fillers in quantities of 5–10% of the filler, release agent as 0·5–1% aqueous solution (32).

OXIRANE HS 35/40 Polyglycol. Gelling agent for latex used in the production of foam rubber. Substitute for sodium silicofluoride. Gelling takes place only above 60°C, below this, the material is stable for several days (2).

OXOZONIDE Addition product according to C. Harries, of oxygen, ozone and unsaturated (olefinic) hydrocarbons. On passing ozonised oxygen through a rubber solution a slightly explosive product of rubber ozonide is produced. The rubber oxozonide $(C_5H_8O_4)_x$ gives the same decomposition products as the ozonide when hydrolysed, but in a different ratio.

p-p'-OXYBISBENZENE SULPHONYL HYDRAZIDE

White, crystalline powder, m.p. 135–140°C, decomposition temperature 130–160°C, s.g. 1·56. Blowing agent which liberates nitrogen. Non-staining. Should be added to compounds as the last constituent and should not be used in an internal mixer. Suitable for NR, SR, polysulphides, polyethylene, p.v.c. Gives a fine, uniform cell structure in all polymers. Has a retarding effect in NR, SBR, IIR and NBR and an activating effect in chloroprene rubber. Can be used with sodium bicarbonate to even out the irregularities of the latter.

Quantity: 2–10% depending on the type of the vulcanisate, suitable for press vulcanisation and open air curing.

Uses: include cellular and microcellular articles, sponge rubber, mats, rolls, extruded cords made from NR and SR.

TN: Celogen OT (23)
Genitron OB (209)
Porofor BSH (43)
Porofor DO-44 (43)
Treibmittel OB (5).
Lit.: USP 2 741 624.

N-OXYDIETHYLENE-2-BENZTHIAZOLE SULPHENAMIDE

2-(4-morpholinyl-mercapto) benzthiazole). Gold powder or brown flakes, m.p. 70–90°C, s.g. 1·34–1·36. Soluble in benzene, chloroform and methanol. A structurally modified thiazole accelerator with a delayed action and little tendency to scorch at processing temperatures. Gives excellent physical properties and flex cracking resistance on ageing. Suitable for applications where other thiazoles would cause scorching during processing or storage. Used generally in combination with ultra-fast accelerators. Is strongly boosted by diphenyl guanidine and other basic ingredients; boosting is necessary in the absence of furnace blacks. Processing safety is increased by 0·5–1% nitrosodiphenylamine. Large vulcanisation range. Does not bloom, discolours slightly, 3–5% zinc oxide is necessary, and normal quantities of stearic acid.

Uses: in natural rubber and SBR compounds with furnace blacks (HAF or SAF) with or without DPG as a booster.

Quantity: 0·5–1% with 2–2·5% sulphur.

TN: Accelerator MOZ (408)
Amax (51)
NOBS Special (21)
Santocur MOR (5)
Vulcafor BSM (60)
Blends with 10% dibenzthiazyl disulphide:
Amax No. 1 (51)
NOBS No. 1 (21).

OXYETHYL NYLON Stable, flexible superpolyamide derivative in which the usual properties of nylon are combined with a greater flexibility, rubber-like properties and a higher water absorption. Graft copolymers of polyethylene oxide onto nylon do not have the average characteristics of a statistical copolymer but retain to a certain degree, the individual properties of both components. An oxyethyl-6:6 nylon with 50% ethylene oxide has a softening p. of 221°C and an apparent second-order transition below −40°C.

OXYGARD Canadian produced ditert-butyl-p-cresol.

OXYSTOP Trade name for dinaphthyl-p-phenylene diamine.

OY Abbreviation for 'Ong Yem' clones.

OZOKERITE Earth wax, mineral wax. Product of the petroleum industry; consists primarily of paraffin hydrocarbons with a branched chain, s.g. 0·920–0·940.
Uses: protection against light crazing (similar to paraffin wax).

OZONIDE Explosive oils with suffocating odour. Formed, according to C. Harries, by the addition of ozone to unsaturated (olefinic) hydrocarbons, when the double bond is split. In the presence of water, ozonides decompose forming hydrogen peroxide and aldehyde or ketone decomposition products of the original compound. The chemical structure of the compound can thus be determined. Rubber ozonide with the empirical formula $C_5H_8O_3$, is a glass-like explosive mass, m.p. approx. 50°C. It is hydrolysed to laevulinic acid and laevulinic aldehyde showing the derivation of rubber hydrocarbons from isoprene.
Lit.: C. Harries, *Ann.*, 1905, **43**, 311.
L. Long, *Chem. Rev.*, 1940, **27**, 437.

OZONIDE REACTION The reaction of ozone with organic substances; the addition of ozone to unsaturated groups and decomposition of the ozonides formed to give carbonyl derivatives

$$R\!\!\diagdown C = C \diagup R'' + O_3 \longrightarrow$$
$$R'\diagup \qquad \diagdown R'''$$

$$R\!\!\diagdown C - C \diagup R'' \xrightarrow{+H_2O}$$
$$R'\diagup \underset{O}{\diagdown \diagup} \diagdown R'''$$

$$\underset{\|}{\overset{O}{R-C-R'}} + \underset{\|}{\overset{O}{R''-C-R'''}} + H_2O_2$$

Lit.: C. Harries, *Ann*, 1905, 343, 311.

OZONOLYSIS Analytical process for organic substances with olefinic bonds; achieved by the addition of ozone at the double bonds, followed by hydrolysis of the ozonides and determination of the decomposition products. Rubber hydrocarbons yield laevulinic acid and laevulinic aldehyde.

OZONOMETER Apparatus used to determine the ozone concentration in ozonised air when assessing the ozone resistance of elastomers. The result is based on the absorption of 2537 Å radiation by ozone.

OZOTEST Apparatus used to determine the resistance of rubber to ozone (401).

P

P- Group of softeners. P-1 methyl ricinoleate, P-4 methyl acetyl ricinoleate (9).

P-2 Piperidine pentamethylene dithiocarbamate (22).

P-21 Russian produced disecbutyl hydroquinone. Antioxidant.

P-23 Russian produced tritertbutyl phenol. Antioxidant.

P-33 FT thermal black (51).

PAD-60 Masterbatch of 60% dibenzthiazyl disulphide and polyisobutylene (282).

PADANG RUBBER Trade term for a former wild rubber obtained from *Ficus elastica* and *Parameria glandulifera* (west coast of central Sumatra).

PAINTER MACHINE Apparatus used to determine the dynamic properties of rubber using forced, non-resonant oscillation of 2–60 c/s. Lit.: G. W. Painter, *Rubb. Age*, 1954, **74**, 701.

PALAMOLL Copolymers of butadiene and diethyl fumarate. Vulcanise faster than SBR and have a higher modulus but lower tensile strength, tear resistance and elongation at break:
I: monomer ratio 25:75
II: monomer ratio 35:65.

PALAQUIUM *Sapotaceae* family. Various types are producers of gutta-percha and occur wild throughout Borneo, Malaya, Sumatra, Burma and in a cultivated form in Java (Tjipetir). They grow as tall, thin trees over 30 m in height and approx. 2 m in circumference and have leather-like leaves. A damp, warm climate is necessary, with annual rainfall preferably evenly distributed, in excess of 2 m, and less than 600 m above sea-level.

PALATAL Group of unsaturated polyester resins (32).

PALATINOL Group of softeners:
A: diethyl phthalate
AH: di(2-ethyl hexyl)phthalate
BB: butyl benzyl phthalate
BF: 1:4-butanediol phthalic acid ester

BH: dihexyl phthalate
C: dibutyl phthalate
CN: butyl propoxyethyl phthalate
CV: butyl C_4–C_6-phthalate
DP: di-'Lorol'-phthalate (from sperm oil and coconut oil)
E: diethyl ethyl phthalate
F: C_7–C_9 phthalate, alcohols from the Fischer–Tropsch synthesis
FK: C_7–C_8 phthalate
FN: C_4–C_6 phthalate
FO: oxo-alcohol phthalate
G: butyl-C_4–C_6 phthalate
HC: phthalate of 'Intrasolvan HS' and 2-ethyl butanol (1:1)
HS: 'Intrasolvan HS'-phthalate
JC: dibutyl phthalate
K: dibutoxy phthalate
L: 'Intrasolvan E'-phthalate
M: dimethyl phthalate
O: dimethoxyethyl phthalate
P: diphenyl phthalate
T: dipropyl phthalate
UB: phthalic acid ester of higher alcohols derived from coal
UV: phthalic acid ester of a C_6 alcohol
V: phthalic acid ester of C_4–C_6 alcohols
W: dichloroethyl phthalate
Z: oxo-alcohol (decyl alcohol) phthalate.
Uses: plastics, adhesives and lacquer industries (32, 43).

PALAY RUBBER Former trade term for a wild rubber which used to be obtained from *Willoughbeia martabanica* and *Cryptostegia grandiflora* (E. India). Contains approx. 90% pure rubber and 8% resins. *Cr. grandiflora* contains approx. 5% pure rubber in the leaves and 1% in the twigs.

PALMALENE Blend of higher fatty acids. Activator and softener (113).

PALO AMARILLO Former trade term for a wild rubber which used to be obtained from *Euphorbia fulva* (*E. elastica*) (Mexico). Contains approx. 70% resin. The latex contains 7% rubber and 19% resin.

PAMANUKAN TJIASEM LANDEN The oldest rubber plantation in the world (Java). Originally a coffee plantation of the Anglo-Java Plantation Co.; in 1872 approx. 30 ha of the *Ficus elastica* type were planted.

PAN Polyacrylonitrile fibre, s.g. 1·16. Produced by the dry spinning process since 1943 (IG Farben). Stable up to 180°C, resistant to water and acids (283).

PANA Abbreviation for phenyl-α-naphthylamine.

PAPAIN Proteolytic enzyme obtained from the fruits of *Carica Papaya* (melon tree). Splits proteins, *via* polypeptides, into amino acids. Most favourable pH 4–7.
Uses: reduction of the protein content in rubber containing serum (skim) of centrifuged latex.

PAPI Group of polymethylene polyphenyl isocyanates. Dark brown liquids, s.g. 1·2, viscosity 1500–7500 cp, isocyanate equivalent 135. *Uses:* crosslinking agent for urethane rubbers and butyl rubber, bonding agent for synthetic fibres to NR and SR (148).

PARACRIL Group of butadiene-acrylonitrile copolymers with a varying acrylonitrile content and varying Mooney viscosity (ML-2, 100°C). The resistance to aliphatic hydrocarbons and mineral oils increases with increasing acrylonitrile content, but the low temperature properties decrease. Have good resistance to plant and animal fats and oils. Processing and compounding is similar to that of natural rubber and SBR. Reinforcing fillers, (blacks, aluminium silicate) are necessary to achieve good physical properties. Organic esters are used as softeners to improve the processing and low temperature properties. The products have good ageing properties at normal and higher temperatures (up to 150°C), and very low compression set in sulphurless compounds or compounds with a low sulphur content; they also have good to excellent abrasion resistance:

18–80: approx. 23% acrylonitrile. Mooney viscosity 70–80. Slightly discolouring stabiliser. Less easily processed than the other types. Excellent low temperature properties. Oil absorption 62–70 vol. %. *Uses:* mechanical goods for low temperatures

AJ: approx. 23% acrylonitrile, s.g. 0·96. Mooney viscosity 50–60. Hot polymer. Non-staining stabiliser. Easily processed. Excellent low temperature properties, good oil resistance. Oil absorption 60–75 vol. %. *Uses:* mechanical goods, packaging for low temperatures

ALT: approx. 25% acrylonitrile content, s.g. 0·96. Mooney viscosity 80–90. Non-staining stabiliser. Cold polymer. Easily processed. Scorch resistant, very good low temperature properties, good to medium oil resistance, low water absorption and good resistance to inorganic chemicals. *Uses:* mechanical goods, highly loaded oil and fuel hose; an additive to polymers with high acrylonitrile content to improve the low temperature flexibility

B (formerly 26-NS-90): approx. 28% acrylonitrile, s.g. 0·97. Mooney viscosity 80–90. Non-staining stabiliser. Easy processing. Good low temperature properties. Oil absorption 33–60 vol. %. *Uses:* articles with a good compromise between properties at low temperatures and oil resistance

BJ (formerly 26-NS-60): approx. 28% acrylonitrile, s.g. 0·97. Mooney viscosity 50–60. Non-staining stabiliser. Easy processing. Good low temperature properties, medium oil resistance, good solubility. Oil absorption 33–60 vol. %. *Uses:* adhesives, cements, coatings, polymeric softener for vinyl resins, moulded articles

BJLT: approx. 32% acrylonitrile, s.g. 0·99. Mooney viscosity 50–60. Cold polymer. Easy processing, no plasticisation is necessary. Scorch resistant, fair low temperature properties, good resistance to fuels and oils, good abrasion resistance. *Uses:* moulded articles, the low Mooney viscosity improves mould flows and allows the use of high filler loadings, adhesives, cements, highly loaded hose

BLT: approx. 32% acrylonitrile, s.g. 0·99. Mooney viscosity 80–90. Cold polymer. Easily processed. Scorch resistant, fair low temperature properties, good resistance to oils and fuels, good abrasion resistance, good solubility. *Uses:* adhesives, cements, mechanical goods

C (formerly 35-NS-90): approx. 28% acrylonitrile, s.g. 0·99, Mooney viscosity 80–90. Non-staining stabiliser. Easy processing. Medium low temperature resist-

ance, good resistance to aliphatic hydrocarbons and mineral oils, can be blended with some synthetic resins, low water absorption, very good resistance to abrasion. Oil absorption 12–30 vol. %. *Uses:* articles requiring good oil resistance at normal temperatures

CLT: approx. 40% acrylonitrile, s.g. 1·01, Mooney viscosity 80–90. Cold polymer. Easily processed. Scorch resistant, very good resistance to oils and fuels, low water absorption, excellent abrasion resistance, very poor low temperature resistance, good film strength. *Uses:* mechanical goods, structure adhesives

CV: approx. 36% acrylonitrile, s.g. 1·00, Mooney viscosity 75–85. Non-staining stabiliser impregnated with vinyl resin in crumb form. Readily soluble in ketones and aromatic and chlorinated hydrocarbons without previous plasticisation. Gel free, has good tack. *Uses:* adhesives (together with phenolic resins); the crumb form can be dry blended with phenol, vinyl and styrene resins to give products which are shock resistant

D: approx. 45% acrylonitrile, s.g. 1·02, Mooney viscosity 50–60. Non-staining stabiliser. Hot polymer. Easily processed. Excellent resistance to aliphatic hydrocarbons, mineral oils, plant and animal oils and fats, good resistance to aromatic oils and fuels, diester-lubricants and some chlorinated hydrocarbons, intermediate water absorption, good resistance to inorganic chemicals, very good abrasion resistance, very poor low temperature properties

OHT: butyl acrylate ester-acrylonitrile copolymer with excellent resistance to high temperatures, s.g. 1·05, Mooney viscosity 35–45. Processability and compounding as with natural rubber and SBR. It requires, however, a special technique to prevent tackiness and self-adhesion, amines act as curing agents; HAF blacks give the best possible properties. For low compression set, a post-cure is necessary at 150°C. *Uses:* articles requiring good high temperature resistance and good resistance to hot oils, high pressure lubricants, ozone, oxidation and flex-cracking, also low gas permeability

OZO: blend of a Paracril and vinyl resin. s.g, 1·065, Mooney viscosity 70–90. Non-staining antioxidant, crumb form. Scorch resistant. Processing as for natural rubber and SBR. Can be blended with synthetic rubber and various plastics, the use of ester softeners and fillers is recommended. Vulcanisates have rubbery characteristics, good physical properties, good abrasion resistance, oil resistance, excellent resistance to ozone and weathering, and very low inflammability. Suitable for bright colours (32).

PARA ENTREFIN Borracha entrefina. Trade term for a second class para rubber which grew in Brazil and which often had a spongy nature because of careless preparation, moisture and the decomposition of the non-rubber constituents.

PARAFFIN P wax. Mixture of solid aliphatic petroleum hydrocarbons of the general formula C_nH_{2n+2}. Solid, white, somewhat translucent, fatty, odourless mass, m.p. 45–65°C, s.g. 0·90–0·91. Soluble in benzene, chloroform, ether, carbon disulphide, insoluble in water and alcohol. Blooms in rubber compounds at percentages above 1 % and under static conditions protects against ozone. Prevents tackiness in latex films, improves extrusion properties and mould release, transparency of chlorinated rubber films and the dispersion of fillers. Used as a protective paint in iron containers and ship holds for the transportation of latex, prevents frosting of highly loaded compounds.

PARA FIN Borracha fina. Trade term for a first class para rubber. Bales of 20–50 kg in weight. The cut cross-section shows a layered structure, with layers approx. 1 mm thick formed as a result of the production method. Contains all the non-rubber constituents of the latex.

PARAFLUX Saturated polymerised petroleum hydrocarbons. Black liquids, s.g. 0·95–1·02. Softeners for natural and synthetic rubbers which have no effect on vulcanisation. Improves the tack of compounds (100).

PARAFORMALDEHYDE $(CH_2O)_n$. White powder, with formaldehyde odour. Soluble in alcohol and hot water.
Uses: disinfection of rubber drying and storage rooms on plantations, production of plastics.

PARAGUTTA According to BP 353 518, a blend of equal parts of deresinified gutta-percha or balata with protein-free rubber and up to

40% montane wax. The rubber was freed from protein by heating the latex with sodium hydroxide under pressure (BP 307 966). Paragutta was used for a long time as an insulating material for underwater cables (dielectric constant 2·6) but has been replaced by polyethylene.

PARAOXYNEOZONE Russian produced p-hydroxy phenyl-β-naphthylamine.

PARAPLEX Group of softeners based on modified alkyd resins and polyesters, for chlorinated rubber, chloroprene, nitrile rubber, polyvinyl resins and cellulose resins. The polyesters have a stabilising effect on vinyl polymers and improve their resistance to ultra-violet rays, heat and oils (33).

PARAPLEX RUBBER Formerly Paracon. A polyester rubber produced in the USA during World War II:
X-100: polyester of ethylene glycol, 1:2-propylene glycol, adipic acid and sebacic acid with 3% maleic acid
S-200: type which may be vulcanised with sulphur.

PARAPOL Styrene-isobutylene copolymers (131).

PARA RUBBER Former trade term for *Hevea* wild rubber (Brazil) which was traded in the form of smoked bales; also used occasionally for plantation rubber but there is doubt whether the term should apply to *Hevea Brasiliensis* or to a closely related type (*Hevea benthamiana*), as the trees were primarily scattered in small clumps, or singly among numerous species in the jungle.

PARA-STONETEX Trade name for chlorinated rubber (284).

PARATEX Trade name for chlorinated rubber (284).

PARAVAR Trade name for chlorinated rubber (23).

PARICIN Group of softeners:
4: methyl acetoxy stearate
4-C: methoxyethyl acetoxy stearate
6: butyl acetoxy stearate (9).

PARKESINE The first synthetic plastic; predecessor of the celluloids. Produced from a mixture of nitrocellulose, camphor and alcohol. Patented by Henry Parkes, 1861.

PARKES' PROCESS Cold vulcanisation. Vulcanisation of thin walled articles by sulphur chloride; discovered, 1846, by A. Parkes (BP 11 147). The articles for vulcanisation can be dipped in a 2–5% solution of sulphur chloride in carbon tetrachloride, benzine or benzene, or exposed to sulphur chloride vapour in lead chambers. After vulcanisation, the articles must be treated with ammonia to neutralise the hydrochloric acid formed during the process and to decompose excess sulphur chloride.

PARLON X Chlorinated synthetic polyisoprene which has similar properties to chlorinated natural rubber. Produced during World War II in the USA as a substitute for the natural product (10).

PARMR Mineral rubber. Softener and extender (3).

PARTHENIUM ARGENTATUM A. GRAY Guayule,

Huayule. A very strong bush of the *Compositeae* family (Mexico, southern USA, Soviet Union); contains rubber in the living cells of the bark, wood, and roots. Apart from rubber, the bark contains balsam-like aromatic substances and resins. The species was discovered in 1852 by Bigelow and described by Asa Gray. In Russia eight varieties are known (*angustifolium, brevifolium, deltoideum, dissectum, gracile, latifolium, longifolium, marioloides*). The phylogenetic younger forms (*latifolium* and *angustifolium*) gave richer quantities of rubber than the phylogenetic older forms (*brevifolium* and *marioloides*), which possess strongly lignified cells.

A plant of moderate and subtropical climates; needs specific climatic conditions and soil conditions for good development and high product yield. In Mexico the natural occurrence of this plant is confined solely to the limestone areas in semi-desert regions. The tree can withstand long periods of dryness and maintain a fairly good rate of growth with a low water content. Sandy soil decreases the yield of rubber. On the best soil where there is sufficient water, the plant grows prolifically, but the formation of rubber is strongly retarded. A certain seasonal cycle, between growth and secretion, yields higher harvests, and a medium growth yields a higher amount of rubber. The plants are harvested normally after 4–5 years. The rubber content in the dry material can be up to 20% in the best stems, but on average yields only 15–17%. Even

as far back as 1876 there was a Guayule product known as Caoutchouc Durango in Philadelphia. In 1888 a small amount of industrial production was started with the extraction of rubber from dried and ground plants. In 1904 the first large factory was built in Torreon in Mexico. The extraction method was replaced by mechanical recovery, the material being ground and finally washed with hot water on ribbed rolls. The pulp was then suspended in water, and the greater part of the cell material was precipitated, while the rubber collected on the surface. It was then purified by steam and processed to crêpe on smooth rolls. Guayule rubber contains approx. 70% pure rubber, 20% resin and 10% insoluble materials (lignin and cellulose). The rubber hydrocarbon is chemically identical with *Hevea* rubber (1:4-cis-polyisoprene) but has a lower mol. wt., higher plasticity and may be oxidised more easily. It is more tacky. Guayule rubber is used to a certain extent as an additive to *Hevea* rubber in friction compounds and compounds requiring high tack. It came on the market in Mexico at an early date under the names of Hule crudo and Hule refinado. Current production in the USA is approx. 7000 tons per annum, and that of the Soviet Union estimated as still higher.

PATRAMOLD PROCESS
Process used for the production of electro forms; ebonite facsimiles. An ebonite sheet at 70°C is pressed against the original form. After production of the electroform the ebonite is replasticised by heating for a short time in water, and re-used.

PAUCHOUTEC Local term for gutta-percha-like product from *Isonandra acuminata* (India).

PAUS Abbreviation for pale amber unsmoked sheet.

PB Abbreviation for 'Prang Besar' clones; produced on a plantation of the same name in Malaya (Kajang, Selangor) from 1922–23 and tapped for the first time in 1927–1928.

PBD Abbreviation for cis-polybutadiene.

PBD-75 Masterbatch of 75% bismuth dimethyl dithiocarbamate with polyisobutylene (282).

PBN Abbreviation for phenyl-β-naphthylamine.

PBNA Abbreviation for phenyl-β-naphthylamine.

PBR Pyridine-butadiene rubber, according to the nomenclature of ASTM D 1418.

PCP NO. 5 Antioxidant of undisclosed composition (176).

PDA Abbreviation for polydiarylamine.

PDA-10 Polydiarylamine. Reddish brown solid, m.p. approx. 50°C, s.g. 1·18. Antioxidant for NR, SR and latex; roughly equivalent to phenyl-β-naphthylamine in its protective power (286).

PDD-70 Masterbatch of 70% p-dibenzoyl quinone dioxime and polyisobutylene (282).

PDOTGD-65 Masterbatch of 65% di-o-tolyl guanidine and polyisobutylene (282).

P(DPG)D-65 Masterbatch of 65% diphenyl guanidine and polyisobutylene (282).

PDSD-715 Masterbatch of 65% p-dibenzoyl quinone dioxime and 6·5% sulphur with polyisobutylene (282).

PE Abbreviation for polyethylene.

PEACHY PROCESS Vulcanisation process developed in 1918 by S. J. Peachy and A. Skipsey (BP 129 826), in which an active form of sulphur is formed by reaction of hydrogen sulphide and sulphur dioxide but is active only at the moment of formation. It is assumed that vulcanisation is caused by atomic sulphur or an intermediary, trithiozone, $(S_3)_3$. When this process is used commercially, the articles are first exposed to gaseous sulphur dioxide, for 10 minutes, and treated with hydrogen sulphide. The process is only suitable for thin walled articles (1 mm max.), but gives vulcanisates equal in properties to hot cured articles, and can be used with NR and other polymers.
Lit.: *J. Soc. Chem. Ind.*, 1921, **40**, 5.

PEAL, SAMUEL Obtained the first known rubber patent for a method of impregnating textiles with a solution of rubber in turpentine. BP 1801 (1791).

PEELCOTE Removable vinyl resin covering used for corrosion protection, *e.g.* for metal parts and machines. Applied by spraying or dipping (287).

PEGASOL Group of petroleum distillates with different boiling ranges: 1425 b.p. 60–120°C, 2025 93–120°C. Solvents for rubber (178).

N-PELARGONOYL-p-AMINOPHENOL

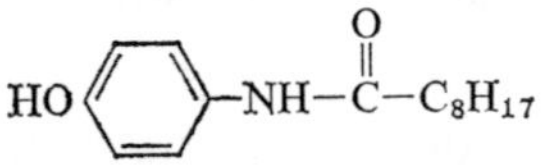

$$HO-\langle\text{benzene ring}\rangle-NH-\overset{\overset{\text{O}}{\|}}{C}-C_8H_{17}$$

White powder, m.p. 121–125°C. General purpose antioxidant. *TN:* Suconox 9 (333).

PELLETEX Pelletised SRF gas black. Pelletex NS: non-staining (39).

PELLETISER Granulating machine for producing materials in pellet form.

PENANG RUBBER Trade term for a natural rubber, formerly shipped from Penang, which grows in Malaya and Sumatra. A plant of the *Willoughbeia firma, Parameria glandulifera, Leuconotis eugenifolia*, and *Ficus elastica* types.

PEND Russian produced low pressure polyethylene.

PENETROMETER Apparatus used to determine the hardness of plastic or elastic materials, asphalt, bitumen, tar, rubber and similar materials by pressing in loaded standard needles under precisely controlled conditions of time, temperature and load.

PENDULUM IMPACT MACHINE Machines by which a bent test piece, normally suspended on two points, is subjected to impact by a falling pendulum. DIN 51 222.

PENNAC Group of accelerators (120):

AM: dithiocarbamate of indeterminate structure. Fluid, s.g. 0·98. Ultra-accelerator for room temperature vulcanisation of latex

CTA: 2-mercaptoimidazoline

MS: tetramethyl thiuram monosulphide

SDB: sodium dibutyl dithiocarbamate. 47% solution

ZT: zinc-2-mercaptobenzthiazole

ZT-W: zinc-2-mercaptobenzthiazole with 10% inert material.

PENNOX Group of antioxidants based on alkylated diphenylamine and bisphenolene (120).

PENNSALT 901 Chlorine-containing solvent with organic sulphur compounds; used to dissolve vulcanised Chemosil and Chemlock rubber/metal bonding agents (120).

PENNZONE Group of antiozonants (120):

B: 1:3-dibutyl thiocarbamide

C: 1:3-diethyl thiocarbamide

E: 1:3-diethyl thiocarbamide

L: mixture of alkyl thiourea. Brown liquid.

PENTACHLOROETHANE

$CCl_3.CHCl_2$, Pentaline. Colourless liquid, m.p. −29°C, b.p. 161–162°C, s.g. 1·67. Miscible with alcohol and ether, insoluble in water. Solvent for resins.

PENTACHLOROPHENOL

White, crystalline powder, m.p. 190°C, b.p. 310°C, with decomposition, s.g. 1·978 (22°C). Soluble in alcohol, ether, benzene. Preservative for NR and SR; the sodium salt is used as a preservative for latex.

PENTACHLOROPHENOL LAURYL ESTER

Fungicide, insecticide and bactericide for vulcanised rubber. Odourless liquid. Soluble in non-polar solvents and insoluble in polar solvents, *e.g.* alcohol. Withstands vulcanisation temperatures.

Quantity: 1%, for tropical products 2%.

TN: Mystox LPL.

PENTACHLOROTHIO-PHENOL

C_6Cl_5SH. Grey powder, s.g. 1·15–1·79. Soluble in benzine, benzene, chloroform and carbon tetrachloride, insoluble in water. Chemically active plasticising agent (peptiser), particularly for synthetic rubbers. Processing temperature 100–160°C on a mill or in an internal mixer. Has no effect on the properties or ageing of vulcanisates. Produces negligible yellow discoloration of white articles. Cannot be used in sulphurless compounds with tetramethyl thiuram disulphide.

Quantity: natural rubber 0·04–0·2%, SBR 1–2·5% at 120–130°C, nitrile rubber 2·5–3·5%.

TN: Peptazin Tiol (428)
Renacit V, VR (43)
RPA-6 (6).

PENTACISER

Group of softeners for SBR and NBR, based on pentaerythritol (288).

PENTAERYTHRITOL

$C.(CH_2OH)_4$, tetra(hydroxymethyl) methane. White, crystalline powder, m.p. 254 °C. Quarternary alcohol. *Uses:* production of maleic and alkyd resins, as a softener in the form of higher fatty acid esters.

PENTAMETHYL DISILOXANE METHYL METHACRYLATE

Monomer developed by the Dow Corning Corporation (USA), Syl-Kem 21. Clear liquid, b.p. 86·5°C, s.g. 0·903, refractive index 1·4202 (25°C). Recommended for use for polymers or copolymers (7).

PENTEX

Tetrabutyl thiuram monosulphide.

Pentex flour: 12·5% Pentex with 87·5% clay. Pale brown powder, s.g. 2·16.

Pentex O, Thiuram (23).

PENTON

Chlorinated polyether. Good physical properties and chemical resistance, stable up to 100°C.

Uses: surface protection of metals, corrosion resistant linings (10).

PEPTAZIN

Group of peptisers (428):

BAFD: o,o′-dibenzamidodiphenyl disulphide

BFT: N,S-substituted derivative of o-aminothiophenol

NA: sodium pentachlorthiophenate

TIOL: pentachlorthiophenol

ZN: zinc pentachlorthiophenate.

PEPTISANT 10

Dithiobisbenzanilide (20).

PEPTISED RUBBER

A rubber which is chemically softened during processing by the addition of 0·1–0·5% peptiser, thus needing far less power for mastication. Suitable peptisers include xylyl mercaptan (RPA-3) and Renacit IV which are effective at relatively low temperatures and give a very soft raw rubber with a Mooney viscosity of approx. 40–50, which is similar in consistency to masticated rubber. Di-o-benzamino phenyl disulphide is active only at 120°C and the rubber produced with the aid of this peptiser has normal hardness.

TN: Plastorub (400)
Peptorub (400).

PEPTISER 620 40% solution of thiocresol in hydrocarbons. Peptiser for natural rubber and butadiene-styrene copolymers. Has no effect on scorch, vulcanisation or ageing properties.
Quantity: 0·1–0·4% (289).

PEPTISER 640 40% solution of thioxylenol in hydrocarbons. Peptiser for natural rubber and butadiene-styrene copolymers. Has no effect on vulcanisation or ageing properties.
Quantity: 0·1–0·4% (289).

PEPTISING AGENT Mastication aid, reclaiming agent, chemically effective plasticising agent with a powerful softening effect on rubber, enabling the mixing time to be reduced, plasticity increased, and the viscosity of solutions reduced. While normal softeners have primarily a mechanical lubricating effect, peptisers act as oxygen carriers and thus increase the oxidative decomposition of the gel structure of rubber. By using a peptiser the same degree of mastication may be achieved with a much lower degree of oxidation, and in a much shorter time, than by purely mechanical mastication. While the conventional softener decreases the hardness of the vulcanisates a peptiser has practically no effect on hardness. Examples of effective pep-

tisers are: xylyl mercaptan, naphthyl-β-mercaptan, pentachlorothiophenol, dibenzamidodiphenyl disulphide and their zinc salts, phenyl hydrazine, phenyl hydrazones, thiosalicyclic amide, dinitrosodiphenyl guanidine, α-nitroso-β-naphthol, 2-mercaptobenzthiazole. A power saving of 40–50% is achieved.

PEPTORUB Rubber preplasticised with Renacite IV on the plantations; the starting material is brown crêpe. Mooney viscosity max. 65, impurities max. 0·05%, ash max. 0·6%. Good uniformity of properties. Enables a marked reduction in mastication time.
Uses: as for brown crêpe, particularly suitable for cellular rubber (400).

PERBUNAN Formerly Buna N. A low modified butadiene-acrylonitrile copolymer, produced in Germany before and during World War II; developed from 1930–1932:
Perbunan (Buna N): 26% acrylonitrile, 75% conversion, continuous polymerisation, polymerisation temperature 29°C, time 28 h, rate 10 kg/h/m^3, Defo plasticity 2400–2800
Perbunan Extra (Buna NN): 40% acrylonitrile (36% bound), 75% conversion, batch polymerisation, temperature 24°C, time 60 h, rate 10 kg/h/m^3, Defo plasticity 2600–3000.
The equivalent latices were Igetex N (formerly Perbunan Special): 30% solids, 26% acrylonitrile.
Igetex NN, 45% solids: 35–40% acrylonitrile and was also produced with a higher monomer concentration to give a 45% latex. The product has now been replaced by Perbunan N.

PERBUNAN C Former designation for chloroprene rubber produced by Bayer Leverkusen.

PERBUNAN N Highly modified German butadiene-acrylonitrile copolymers with varying acrylonitrile contents and varying plasticities. Unvulcanised, they are soluble in aromatic and chlorinated hydrocarbons and ketones. The resistance to aliphatic hydrocarbons and oils increases with increasing acrylonitrile content, while resilience and low temperature resistance decrease. Have good resistance to heat and normal ageing, good abrasion resistance and dynamic properties. In the numerical indices of the various types, the first two numbers indicate the acrylonitrile content and the following two equal one-hundredth of the Defo value. All types are stabilised with Antioxidant DDA.

2807 NS: 28% acrylonitrile, s.g. 0·98, Mooney viscosity 45. Non-staining stabiliser. Good processability. Suitable for articles which come into contact with food

2810: 28% acrylonitrile, s.g. 0·98, Mooney viscosity 65. Better processing than 2818. Pale vulcanisates tend to discolour on exposure to light; good resistance to heat and normal ageing. Very good resistance to low temperatures and good resilience, medium oil resistance

2818: 28% acrylonitrile, s.g. 0·98, Mooney viscosity 95. Slightly discolouring stabiliser. Properties as for 2810, but is tougher, therefore having greater capacity for absorbing softeners and fillers. Compounds have better dimensional stability.

2818 NS: as 2818 but non-staining. Suitable for articles which come into contact with food

3302 NS: 34% acrylonitrile, s.g. 0·99, Mooney viscosity approx. 30. Non-discolouring stabiliser. Very good processing, resistant to cyclisation. Good resistance to oil and benzine, good low temperature properties, suitable for contact with food

3307 NS: 34% acrylonitrile, s.g. 0·99, Mooney viscosity 45. Good balance of oil resistance and other properties. Suitable for articles which come into contact with food

3310: 34% acrylonitrile, s.g. 0·99, Mooney viscosity 65. Stabiliser causing a slight discoloration. Properties of the vulcanisates as for 3307

3312 NS: 34% acrylonitrile, s.g. 0·99, Mooney viscosity 80. Non-staining stabiliser, has better stability and surface finish when open-air cured. Less easy to process than other types. Suitable for articles which come into contact with food

3805: 39% acrylonitrile, s.g. 1·0, Mooney viscosity 45. Non-staining stabiliser. Properties as for 3805. Suitable because of its high polarity, for blending with polar plastics. Suitable for articles which come into contact with food

3810: 39% acrylonitrile, s.g. 1·0, Mooney viscosity 65. Slightly discolouring stabiliser

32: 38% acrylonitrile. Liquid. Can be vulcanised to solid rubber. Softener for NBR compounds; improves the extruding and calendering properties

4M 3620 (Ultramoll 362D): 36% acrylonitrile, modified with 4% methacrylic acid. *Uses:* for blending with plastics

N/VC: blend with 70 parts (parts by weight) N 2807 NS and 30 parts (parts by weight) p.v.c., stabilised and gelled (43).

PERBUNAN N LATICES In the latices, the first two index numbers of the various types denote the acrylonitrile content while the last two numbers equal one-hundredth of the Defo plasticity of the films. All types are stabilised with Antioxidant ZKF and are non-staining. The properties of the films are dependent on the acrylonitrile content.

Uses: include bonding agents, protective coatings, impregnations, rubberised hair. Types 4M and 15M contain additional carboxyl groups (methacrylic acid), which improve their ability to blend with polar plastics (43).

Uses: carpet backing, fibre binding, impregnation for paper and textiles.

PERDEUTERIO POLYISO-PRENE Synthetic polyisoprene, the hydrogen of which is replaced by deuterium. The properties differ from those of normal polyisoprene, s.g. 1·007. Resilience and tensile strength are considerably higher. In the B. F. Goodrich Research Center, approx. 250 g were produced but production is both complicated and expensive.

PERDUREN (Formerly Ethanite.) Polysulphide rubber; condensation products of sodium polysulphide with:

G: dichlorodiethyl ether

H: dichloroethyl formaldehyde acetal (43).

PERFLECTOL Group of antioxidants:

Type	Solids %	Acrylonitrile %	Methacrylic acid %	Styrene %	pH	Particle size	Film Defo hardness
N latex 4M	35	34	4	—	6–7	600–800	6 000
N latex 4M	45	34	4	—	6–7	600–800	1 500
SN latex 15M	35	5	15	20	6–7	600–800	8 000
N latex 2818	45	28	—	—	8–9	600–800	1 600–2 000
N latex 3310	45	33	—	—	8–9	600–800	800–1 200
N latex 3310-HD	40	33	—	—	8–9	50–100	800–1 200
N latex 3810	45	38	—	—	8–9	600–800	800–1 200

PERBUNAN SM LATEX Polymer of approx. 50% (60% butadiene, 40% styrene) with methacrylic acid. Has good mechanical and chemical stability. Requires curing by sulphur and accelerators. Vulcanisates age well.

Standard: wax-like product; mixture of 65% acetone-aniline condensation product (Flectol H) and 35% diphenyl-p-phenylene diamine

H: polytrimethyl dihydro-
quinoline (5).

PERGUT Group of bonding
agents based on chlorinated rubber.
Have good resistance to chemicals,
acids and water.
Uses: protective paints (43).

PERGUT S Chlorinated rubber
used for bonding elastomers to
metals. S 40, s.g. 1·54, S 90, s.g. 1·60
(43).

PERKADOX Group of organic
peroxides (171):

BC 40: 40% dicumyl peroxide on
calcium carbonate. For vulcanising NR and SR and
crosslinking polyolefins

PDS 40: 40% 2:4-dichlorobenzoyl
peroxide in silicone oil.
Used to crosslink silicone
rubber

SB: 95% dicumyl peroxide. For
vulcanising NR and SR and
crosslinking polyolefin

SC: p-chlorobenzoyl peroxide

SD: 2:4-dichlorobenzoyl
peroxide

SE 8: octanoyl peroxide. Polymerisation catalyst for ethylene

Y 12: polyfunctional peroxide,
40% on calcium carbonate.
May be used in place of
dicumyl peroxide for vul-
canising NR and SR and
crosslinking polyolefin

Y 440: bifunctional peroxide. May
be used as Y 12.

PERLON (NYLON 6)
[—NH(CH$_2$)$_5$CO—]$_x$, polyamide
fibre based on caprolactam. Produced
by the melt spinning process; discovered in 1937 (DRP 748 253),
m.p. 215°C, s.g. 1·14.

Production: caprolactam, produced
from phenol *via* cyclohexanonoxime,
is split catalytically at 250–260°C
under vacuum, and polymerised into
long chains. The monomeric lactam,
remaining in a quantity of approx.
10% is removed by washing with hot
water. The dried polyamide is spun
at approx. 270°C stretched and produced as a continuous fibre or as a
staple fibre after crimping.

PERMANAX Group of antioxidants (20):

18: N,N'-diphenyl-p-phenylene
diamine

21: mercaptobenzimidazole

Z 21: zinc mercaptobenzimidazole

24: mixture of alkyl aryl phenols.
Yellow, viscous liquid, s.g.
1·08. Miscible with benzene,
acetone and chloroform. For
NR and SR

45: polytrimethyl dihydroquinoline

47: diphenylamine-acetone condensation product. Dark viscous liquid, s.g. 1·09. Used
particularly against heat ageing

115: N-isopropyl-N'-phenyl-p-
phenylene diamine.

PERMANENT SET The residual
deformation due to the plastic part
of a vulcanisate under pressure (compression set) or under tensile stress
(tension set). May be determined
either at constant stress or constant
strain. ASTM D 395-61, D 1229-62,
BS 903: A5: 1958, 903: A6: 1957,
DIN 53 510, 53 511, 53 517.

PERMATEX Lacquer based on
chlorinated rubber (291).

PERMEABILITY The process
of permeation depends on the

solubility of gases in natural and synthetic rubber. The gases diffuse through the rubber sheet and evaporate on the far side. For hydrogen the diffusion constant in natural rubber at 20°C is 11·4 × 10 cm/sec and the absorption coefficient at normal pressure is as follows (hydrogen = 1):

Air	0·21
Oxygen	0·46
Carbon dioxide	2·46
Carbon monoxide	0·20
Helium	0·31
Ammonia	11·3

Specifications:

Liquids	ASTM D 814
Gases	ASTM D 815.

The quantity of gas passing in a given time through a membrane with a pressure difference between the two sides of the membrane is:

$$\frac{\mathrm{d}Q}{\mathrm{d}t} = \frac{A \times P(p_2 - p_1)}{d}$$

where

Q is the quantity of gas,
t the time,
A the surface area of the membrane,
P the permeability,
d the thickness of the membrane, and
p_1, p_2 the pressure on the two sides.

For pure gases (oxygen, hydrogen, nitrogen) the permeability depends on the membrane, the gas and the temperature. For water and many organic vapours the permeability is also dependent on p_1 and p_2.

PERMEAMETER Apparatus used to determine the permeability of a gas through, for example, impregnated textiles, films.

PERMYL B 100 Ultra-violet absorber, *e.g.* for foam materials, plastics and cellulose (203, 292).

PERNAX Gutta-Gentzsch. Mixture of hardened bitumen with rubber, waxes and a small quantity of gutta-percha and other ingredients, according to DP 24 590. Substitute for gutta-percha and balata. Has poor resistance to water.

PEROXAMINE PROCESS Redox polymerisation process for SBR cold rubber, using a hydroperoxide as the oxidising agent and a polyethylene polyamine as the reducing agent in the presence of traces of iron.

PERSPEX Plastic made from polymethylmethacrylate (60).

PERU RUBBER Peruvian balls. Trade term for a natural rubber from *Hevea Peruviana, Castilla elastica* and *Hancornia speciosa* types, formerly grown in Peru.

PeTD-70 Masterbatch of 70% tetraethyl thiuram disulphide with 30% polyisobutylene (282).

PETROTHENE Series of polyethylenes (99).

PEVEAUTEX Types G and T. Aqueous p.v.c. dispersions containing softeners:
G: may be used as a coating for textiles
T: low viscosity liquid used for the impregnation of textiles (189).

PG Russian produced furnace blacks made from natural gas.
40: 40 m²/g, specific surface area
33: 33 m²/g, specific surface area.

PGAD-72 Masterbatch of 48% dibenzthiazyl disulphide and 24% p-quinone dioxime with 28% polyisobutylene (282).

PGD-50 Masterbatch of 50% p-quinone dioxime and polyisobutylene (282).

PHENALDINE Diphenyl guanidine (20).

PHENEX Condensation product of aniline and ethyl propyl acrolein. Non-toxic, clear, brown liquid, s.g. 1·02. Accelerator for natural rubber, synthetic rubbers and latices, similar in effect to aldehyde-amine accelerators. Gives good ageing properties and does not cause scorching. Activation is by zinc oxide and MODX (mixture of organic and inorganic acetates); it is retarded by clays and blacks. Minimum vulcanisation temperature 135°C (100).

PHENODUR 809 U Reinforcing phenolic resin for NR, NBR and SBR, m.p. 75–88°C. Has a plasticising effect in vulcanised compounds and assists the absorption of fillers (58).

N-PHENYL-N'-CYCLOHEXYL-p-PHENYLENE DIAMINE
$C_6H_5 . NH . C_6H_4 . NH . C_6H_{11}$.
Grey white powder which discolours in light and air, m.p. above 115°C, s.g. 1·29. Soluble in methylene chloride, benzene, acetone, ethyl acetate, carbon tetrachloride and alcohol, insoluble in water. One of the most effective antioxidants, giving very good protection against heat, normal oxidation, weather crazing and flex cracking; also effective against metallic poisoning. Suitable only for dark compounds because of the strong discoloration of light-coloured articles in daylight. Because it is very soluble in rubber, it can migrate and stain by contact. Contact of vulcanisates with oxidising agents or solutions containing iron causes a blue stain. Strengthens unvulcanised compounds, particularly synthetic rubber. Blooms less than other p-phenylene diamine compounds. Prevents the hardening (by cyclisation) of SBR compounds during hot processing.

Uses: tyres, hose lines, cables, mechanical goods which are subject to considerable strain.

TN: Antioxidant 4010 (43)
Antigen 4010
Flexzone 6-H (23)
Nocrac 810 (274)
Santoflex CP (5).

m-PHENYLENE DIAMINE

m-diaminobenzene. White crystals which turn red on contact with air, m.p. 62–63°C, b.p. 284–287°C, s.g. 1·139. Soluble in water, alcohol and ether.

Uses: increases the adhesion of tyre cords to rubber.

PHENYL-α-NAPHTHYL-AMINE

Lemon, crystalline flakes or pellets, m.p. 50°C, s.g. 1·17–1·22. Soluble in acetone, alcohol, ethyl acetate, benzene, carbon tetrachloride, rubber, insoluble in water. Antioxidant for natural rubber, SBR and CR giving very good protection against normal ageing; improves the heat resistance and also protects against copper and manganese. Discolours to dark brown or purple in air. Has no effect on vulcanisation, very slightly softens unvulcanised compounds. Slightly retards neoprene compounds. Causes a dark brown discoloration in vulcanisates on exposure to light and stains light-coloured articles by migration. Does not bloom.

Solubility: up to 10% in all elastomers.

Quantity: 1–2%, in neoprene up to 5%.

Uses: mechanical goods, conveyor belts, tyres, inner tubes, soles, footwear.

TN: Aceto PAN (25)
Antioxidant PAN (43)
Antigen A (86)
Antigen PA
Antioxygéne A (19)
Neozone A (6)
Nonox A (60)
Nonox AN (60)
Nocrac PA (274)
PANA
Blends with 2:4-diaminotoluene:
Antigen C (86)
Antigen PC
Antioxygéne CAS (19)
Neozone C (6)
Blends with 35% N,N′-diphenyl-p-phenylene diamine:
Akroflex C (6)
Akroflex CD (6).

PHENYL-β-NAPHTHYLAMINE

Pale grey powder which discolours on contact with air, m.p. 105–106°C,

s.g. 1·18–1·24. Soluble in acetone, ethyl acetate, methylene chloride, benzene, alcohol, carbon tetrachloride; sparingly soluble in benzine, insoluble in water. Antioxidant for natural and synthetic rubbers and latices, giving excellent protection against normal ageing; improves heat and flex cracking resistance. Has no effect on plasticity and vulcanisation. Causes a dark brown stain on exposure to light. Vulcanisates stain light-coloured articles on contact. Has little effect in preventing oxidation in the presence of copper or manganese. Unsuitable for light-coloured articles.

Quantity: 1–2%, tends to bloom at quantities above 2%.

Uses: mechanical tools, conveyor belts, tyres, air tubes.

TN: 66 (3)
Aceto PBN (25)
Agerite Powder (51)
Antioxidant PBN (43)
Antage D (16)
Antigen D (86)
Antioxidant 116 (22)
Antigen D (86)
Antioxygéne MC (19)
Antioxygéne PBN (418)
Neozone D (6)
Nocrac D (274)
Nonox D (60)
PBN (5)
PBNA
Rionox (418)
Stabilisator AR, STD (364)
STD (286)
Van Hasselt PBN (355)
Blends with 2:4-diamino toluene oxalate:
Neozone E (obsolete) (6)
Blends with N,N′-diphenyl-o-phenylene diamine:
Age-Rite HP (51)
Akroflex CD (6)
Akroflex F (6)
Antigen HP (86)

Antioxidant 108 (22)
Antioxidant HP
Nocrac HP (274)
Nocrac NP (274)
Santoflex 9010 (5)
Santoflex HP (5)
Blends with 25% N,N'-diphenyl-p-phenylene diamine and 25% di-p-methoxy diphenylamine:
Thermoflex A (6)
Blends with 25% N,N'-diphenyl-p-phenylene diamine and 25% p-isopropoxy diphenylamine:
Agerite Hipar (51)
Reaction product with acetone:
Betanox, Betanox Special, Naugatex 506 (23)
Blend with dinaphthyl-p-phenylene diamine:
STD-X (286).

PHENYL-β-NAPHTHYL-AMINE-ACETONE REACTION PRODUCT

Yellowish powder, m.p. approx. 120°C, s.g. 1·16–1·18. Soluble in acetone, benzene, ethylene dichloride, insoluble in water and benzine. Antioxidant for natural rubber, SBR, CR and their latices, giving excellent protection against normal oxidation, heat ageing and flex cracking. Disperses well, slightly retards vulcanisation and causes a dark brown stain in light. Vulcanised products do not stain on contact.

Quantity: 0·25–1%, 2–3% for wire insulation with excellent ageing properties.

Uses: particularly for wire insulation, tyre treads, carcases, inner tubes, dark footwear and impregnations, mechanical goods.

TN: Betanox
Betanox Special
Naugatex 506 (50% paste) (23).

p-PHENYL PHENOL White powder, m.p. above 163°C, s.g. 1·20.

Antioxidant for light-coloured articles, giving relatively good protection against normal ageing. Has no effect on plasticity or vulcanisation. Does not disperse easily but dispersion can be improved by adding stearic acid or a softener at the same time. Causes a slight, pale discoloration on exposure to light. Blooms in natural rubber in quantities over 0·75% and in neoprene over 1%.

Quantity: 0·5–1%.

TN: + Parazone (6).

PHENYL-o-TOLYL GUANIDINE

White powder, s.g. 1·10. Accelerator. Vulcanisation properties lie between those of diphenyl guanidine and di-o-tolyl guanidine.
TN: Portsmouth Accelerator No. 3
POTG (19).

PHENYL TOLYL XYLYL GUANIDINE

Brown, resinous mass, s.g. 1·08. Accelerator. Properties as for di-o-tolyl guanidine. Soluble in benzene.
TN: Accélérateur PTX (19).

PHILBLACK Series of furnace blacks (185):

A:	FEF (MAF)
E:	SAF

I: ISAF
O: HAF
55:   EPC black which retards
        vulcanisation
N 110: SAF
N 220: ISAF
N 242: ISAF-HS
N 285: I-ISAF-HS
S 315: HAF-LS-SC
N 330: HAF
N 347: HAF-HS
N 550: FEF
N 660: GPF.

PHILCURE 113 Tert-butyl sulphonyl dimethyl dithiocarbamate. Accelerator.

PHILPRENE Group of butadiene-styrene copolymers. The index numbers conform with the ASTM code for SBR; apart from these the following types are produced:

1010: hot rubber. For sponge rubber
1104: hot rubber. Pigmented with Philblack
1605: cold rubber. Masterbatch with Philblack A. Contains a non-staining antioxidant
1803: oil extended cold rubber containing 25% Philrich 5; identical to SBR 1801
1805: cold rubber extended with 37·5% naphthenic oil; masterbatch with 75% HAF black
1806: cold rubber/oil/black masterbatch with 25% processing oil and 50% ISAF black
6604: masterbatch with SAF black
6661: cold rubber masterbatch based on SBR 1500 latex; free from dispersing agent
6662: cold rubber masterbatch based on SBR 1500 latex; free from dispersing agent
VP-15: 85:15 copolymer of butadiene and 2-methyl-5-vinyl pyridine. For oil resistant articles at low and high temperatures
VP-25: 72:25 copolymer of butadiene and 2-methyl-5-vinyl pyridine. *Uses:* as for VP-15
VP-A: terpolymer of butadiene-vinyl pyridine-acrylonitrile (130).

PHILRICH 5 Highly aromatic processing oil and extender (130).

PHOSPHORUS NITRILE CHLORIDE POLYMERS Condensation products of phosphorus nitrochloride ($PNCl_2$) with polyhydroxy aromatic compounds. Hydrolytically stable resin with good high temperature resistance.
TN: AP Resin XHU
Dynalak HU (5).

pHR Glycine, alkalinity regulator for neoprene latex (6).

PHTHALIC ANHYDRIDE

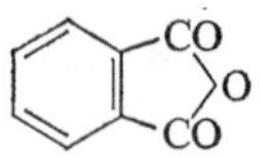

White, crystalline powder, m.p. 124°C, s.g. 1·52. Soluble in acetone and ethanol, sparingly soluble in benzene, carbon tetrachloride, methylene chloride and ethyl acetate, insoluble in water. Effective retarder in compounds with neutral, acidic and basic accelerators and their blends, has no effect as a scorch retarder in sulphurless compounds. Practically no influence on the vulcanisation optimum and the ageing properties. Does not stain on exposure to light. Is more effective in natural rubber than in synthetic rubber. Increases the blowing power of sodium bicarbonate in highly accelerated sponge rubber compounds.

Quantity: 0·1–1·0%.
TN: AFB
Esen (23)
Retarder PA (5)
Retarder PD (22)
Vulcalent B (43)
Witrol P (surface treated) (237).

PHYTOSTEROLS Group of plant sterols; complicated polycyclic alcohols with four hydrogenated carbon rings, of which three are cyclohexane rings. The latter are probably generically connected to simple terpenes. White, usually optically active products. Soluble in most organic solvents, insoluble in water. Occur in small quantities in the acetone extract of *Hevea* rubber and in large quantities in the resins of guttapercha, balata and jelutong. Act as powerful natural antioxidants and are usually responsible for the good storage stability of raw rubber. Sheet rubber, because of its method of production, contains more phytosterin and has a better resistance to oxidation than crêpe.

PIB Abbreviation for polyisobutylene.

PICCO Group of solvents and coumarone resins:
Picco: coumarone-indene resin
Picco 21:   technical xylene
Picco 402 resin: polymerised p-
             coumarone-indene
Picco 517:   solvent oil with high b.p.
Picco 534:   naphtha solvent
Picco 535:   isomers of tetramethyl-
             benzene
Picco SOS: aromatic oil (126).

PICCOCISER Identical with Panaflex. Group of polymerised aromatic hydrocarbons, softeners and plasticisers for NR and SR; increase flexibility at low temperatures, retard crystallisation in neoprene compounds and can be used to replasticise scorched compounds. Have little effect on physical properties (126, 127).

PICCODIENE 14215-SRG Low mol. wt. thermoplastic polymer based on coumarone-indene resins with polyaromatic alkylated diene structures, s.g. 1·11. Increases viscosity and tack and improves processing of SBR (126).

PICCOLYTE Group of synthetic resins, diluents, plasticisers and tackifiers (126).

PICCOPALE EMULSION Group of anionic, cationic and non-ionic emulsions of petroleum resins. Improve the adhesive properties, film elasticity and ageing properties. Types: A, anionic; C, cationic; N, non-ionic.
Uses: include diluents, modifiers and tackifiers in natural latex and synthetic latices for emulsified colours, bonding agents, adhesives (126).

PICCOPALE RESINS Polymeric aliphatic olefins; petroleum hydrocarbon resins, softeners, plasticisers and tackifiers for natural and synthetic rubbers, diluents for butyl rubber. Brown resinous masses, m.p. 70, 85, 100 and 110°C, s.g. 0·970–0·975. Improve the ageing properties of NR and SBR (127).

PICCOPLASTIC Polymer of styrene and substituted styrene. Group of products which can be used as diluents, plasticisers, tacki-fiers and processing aids (126).

PICCOUMARON Group of coumarone-indene resins. Plasticisers, diluents and tackifiers (126).

PICCOVAR Group of polyin-dene resins. Plasticisers and tackifiers (126).

PICCOZIZER 30 Polymerised aromatic oil. Dispersing agent and softener (126).

PICO ABRASION TEST Test developed by B. F. Goodrich to determine the abrasion resistance, *e.g.* of tyres and conveyor belts. Suitable for comparative tests of different types of rubber or black. Instead of the conventional abrasive medium a special blade is used under which the test pieces are rotated. The test can be performed with varying loads, speeds of rotation and angles of the abrader, and gives results which correlate with actual performance.
Specification: ASTM D 2228.

PIGMENTS The dye-stuffs used in the rubber industry may be separated into two groups: inorganic pigments and organic dyes and lacquers. The first group includes iron oxide, cadmium sulphide, titanium dioxide, chrome oxide, antimony sulphide, ochre, lithopone and zinc oxide. Many fillers, among them carbon blacks, act as powerful pigments. The organic pigments include a vast range of shades. The organic pigments frequently come in the form of rubber master batches, which facilitate their dispersion.

PINACOL

$$(CH_3)_2C—OH$$
$$|$$
$$(CH_3)_2C—OH$$

Tetramethyl ethylene glycol. Crystalline solid, m.p. 38°C. Produced by reduction of acetone; by splitting off a molecule of water, it is converted into dimethylbutadiene, an unsaturated diolefin which can be polymerised to methyl rubber. During World War I this was an intermediate product in the synthesis of methyl rubber in Germany.

PINANHYDROPEROXIDE Oxidising agent in redox polymerisation systems for the production of cold rubber.

PINE RUBBER Former trade term for wild rubber from *Landolphia Karkii* (Mozambique).

PINE TAR Softener containing turpentine, a complex mixture of acid and neutral substances, residue of the dry distillation of pine wood. Effective softener for rubber; improves the tackiness of compounds and the dispersion of carbon blacks. Ageing properties improve with increasing phenol content. May activate compounds containing accelerators of the mercapto type but may also cause scorching. The constituents of pine tar vary according to its origin.
Uses: primarily in tyre tread compounds.

PINGUAY Local term for the composite *Hymenolophus floribundus* (USA). Contains approx. $3·6\%$ rubber in the base of its trunk and roots and $0·8\%$ in the leaves.

PIPERIDINE PENTAMETHY-LENE DITHIOCARBAMATE

Piperidine-1-piperidine carbodithionate, N-pentamethylene ammonium pentamethylene dithiocarbamate. Yellowish white, crystalline powder with slight amine-like odour, m.p. (technical) 161–170°C, (pure substance) 179°C, s.g. 1·13–1·15. Soluble in water, benzene, carbon tetrachloride, methylene chloride, sparingly soluble in alcohol, acetone and ethyl acetate. Ultrafast accelerator for the vulcanisation at low temperatures for natural rubber, SBR, NBR and latices, gives odourless, non-staining vulcanisates. Decomposes on contact with iron and moisture. Without metal oxide it is a relatively slow accelerator but with zinc oxide becomes one of the most active. In neoprene it is effective as a plasticiser. The degree of processing safety in dry rubber is very low; compounds show a high tendency to scorch. For this reason it must be processed using a split-batch technique. In latex it has a strong tendency to cause prevulcanisation, but only a slight influence on stability. Damp latex films may discolour when handled. Used as a secondary accelerator in mixtures with tetramethyl thiuram monosulphide, mercaptobenzthiazole and other thiurams.

Uses: self vulcanising cements and compounds, dipped articles, textile coatings, latex compounds.

Accelerator %	Sulphur %	Zinc Oxide %
Solid rubber 0·25–0·75	2·5–1·5	5
Latex, primary 0·5–2·00	1·5–1·0	1–3
secondary 0·25		
Neoprene, plasticiser 0·1–1		

TN: Accélérateur rapide P (19)
Accelerator 552 (6)
Accelerator P (408).
Accelerator 2 P (22)
Anchoracel 2 P (22)
+ Carbamat P
CW 1015 (molecular sieve) (112)
Dules 557
Grasselerator 552
Kuracep PPD (241)
Naugatex 144 (23)
P 2 (22)
Pipidi
+ Pip-pip (5)
PPD (Robac PPD)
Robac PPD (311)
+ Suparac (25% and 75% kaolin)
+ Suparac Z (25% and 75% zinc oxide)
Superaccélérateur 5010 (20)
Velosan
Vulcaform P (362)
Vulcaid P (3)
Vulcaid ZP (Zinc salt) (3)
+ Vulcafor P (60)
Vulcarite 103 (57)
Vulkacit P (43).

PIPERYLENE

$CH_3CH=CH.CH=CH_2$, 1:3-pentadiene, α-methyl butadiene, α-methyl divinyl. Isomers of isoprene with the methyl group in the α-position, b.p. (cis-form) 43·8°C (750 mm), (trans-form) 41·7°C (745 mm), s.g. (cis-form) 0·6911 (20°C), (trans-form) 0·6771 (20°C). Polymerises to a rubber-like product (piperylene rubber). The decomposition products of the ozonides differ from those of *iso*prene.

PIPIDI Trade term for piperidine pentamethylene dithiocarbamate.

PIPPD Trade term for N-phenyl-N′-isopropyl-p-phenylene diamine.

PIPSOLENE Reaction product of methylene piperidine and carbon disulphide in emulsion. Ultra-fast accelerator for latex (obsolete) (5).

PIPSOL X Reaction product of methylene piperidine and carbon disulphide. Pale brown crystals, m.p. 59°C. Ultra-fast accelerator for latex (obsolete) (5).

PIRALAHI Local term for *Landolphia Perieri* (Madagascar) and the natural rubber obtained from it. The sap contains only 6–12% rubber, which is obtained through spontaneous coagulation by boiling or by addition of acid or salts. The rubber appeared on the market under the name Majunge rouge.

PIRIPIRI Local term for *Holarrhenia microteranthera* (tropical Africa) and the natural rubber which used to be obtained from it. The trade product contained approx. 80% pure rubber.

PISTOL Group of sulphonamide softeners for synthetic rubbers and plastics.

PITT-CONSOL Group of peptisers and reclaiming agents (289). 620: 40% thiocresol 640: 40% thioxylenol.

PIX (Latin) tar or pitch.

PLACIDOL Group of softeners. A diamyl phthalate, B dibutyl phthalate, E diethyl phthalate.

PLANHAS DE CAUCHO Local term for wild rubber from *Castilla elastica* (S. America); black sheets.

PLANTING The maintenance of the rubber plantations and the raising of new plants is a vital part of the rubber industry. In earlier times new plantations were raised solely from arbitrarily collected seeds. The value of the product and the secondary qualities of the trees naturally varied considerably. Up to 400 trees per hectare would be planted and the least yielding ones would be thinned out when ready for tapping, so that about 200–250 trees remained. The productivity of these plantations was relatively small, barely reaching 300 kg/ha per year. Today there is an extensive process of selection, based primarily on grafts from parent trees with good productivity, grown especially for this purpose, the so-called clones, or using appropriately selected clone-seeds. The seeds are planted in shaded seedbeds to allow for germination and after about 14 days the seedlings are transplanted for more space. Grafting takes place as soon as the stems are about finger thick, and the plant is then put out into the field. Grafting can also be performed in the field, by which three stems per plant hole are planted out and grafted. The most successful graft is retained and the other two are removed. For some time double grafting, *i.e.* trunk and crown grafting, has also been used.

PLASKON Group of polyester resins (107).

PLASOLEUM Linoleum-like product made from latex and fine sand.
Uses: floor coverings, fillers for dykes.

PLASTICELL P.V.C. foam materials (293).

PLASTICITY A deformation is classed as plastic if the stress is greater than the flow limit so that a plastic movement of the structural elements results; after the stress has been removed recovery is incomplete and a permanent deformation remains. For liquids, without a flow limit, the velocity gradient is given by:

$$D = \frac{dy}{dv} = \Phi\tau = \frac{1}{\eta}\tau$$

where dv is the increase in velocity in one plane in respect to another at a distance dy, Φ is the fluidity and τ the shear strain. In practice, the reciprocal value of fluidity

$$\frac{1}{\Phi} = \eta$$

is used and termed the coefficient of viscosity. Liquids to which this equation applies are called Newtonian liquids or pure viscous liquids. There is a linear relationship between D and τ. For materials with a flow limit, when flowing occurs only at a certain shear level and not at the beginning of stress, the Bingham equation is used:

$$D = \frac{1}{\eta}(\tau - f)$$

If D does not increase linearly with τ, then flow at an exponent <1 is known as structure viscosity and at >1 as rheopexy. In cases of an elastic after-effect where flow occurs after the stress has been removed, the term used is flow elasticity.

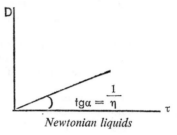

Newtonian liquids

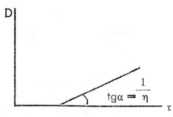

Plastic solids

PLASTICISER A material which may be added to a polymer or polymers to improve flexibility, particularly at low temperatures. Trade name for a large group of softeners produced by various manufacturers:

1:        diphenyl-p-tert-butyl phenyl phosphate
2:        monophenyl-di(p-tert-butyl phenyl) phosphate
3:        diphenyl-o-chlorophenyl phosphate
4:        monophenyl-di-o-chlorophenyl phosphate
5:        diphenyl-o-xenyl phosphate
6:        phenyl-di-o-xenyl phosphate
7:        tri-p-tert-butyl phenyl phosphate

11:	di-p-tert-butyl phenyl-ditert-butyl-2-xenyl phosphate
12:	mono-p-tert-butyl phenyl-ditert-butyl-2-xenyl phosphate
35:	fatty acid dimethyl amide
36:	fatty acid amide
3452-A:	butyl pelargonate (Cellosolve)
4141:	fatty acid ester. Yellowish liquid, s.g. 0·97. For natural and synthetic rubbers, improves flexibility at low temperatures
4300:	diisobutyl azelate
C-311:	ester. Clear, oily liquid, s.g. 0·89. For synthetic rubber, chlorinated rubber, ethyl cellulose, nitrocellulose, vinyl copolymers. Primary and secondary softener; gives good properties at low temperatures
C-325:	fatty acid ester. Brown, oily liquid, s.g. 0·89. For synthetic rubbers; chlorinated rubber, ethyl and nitrocelluloses, vinyl copolymers. Primary and secondary softener; gives good properties at low temperatures
LP:	naphthenic oil. Clear, yellow liquid, s.g. 0·95–0·97. Softener and diluent for NBR, CR, IIR, and NR, vinyl resins. Does not bloom
MP:	naphthenic oil. Pale brown liquid, s.g. 1·00. For CR, SBR, IIR and NR. May be mixed easily. *Quantity:* in NBR up to 25%, in p.v.c. up to 30%
MT-511:	condensation product of a polyalcohol and an α-dicarboxylic acid. Pale

	brown liquid, s.g. 1·17–1·19. For NBR; gives good properties at high and low temperatures, good dimensional stability and cannot be extracted by most solvents
ODN:	octadecene nitrile
OLN:	oleyl nitrile
P-1:	methyl ricinoleate
P-1C:	methoxy ricinoleate
P-2:	ethyl ricinoleate
P-2C:	ethoxyethyl ricinoleate
P-3:	butyl ricinoleate
P-3E:	ethyl butyl ricinoleate
P-4:	methyl acetyl ricinoleate
P-4C:	methoxyethyl acetyl ricinoleate
P-5:	ethyl acetyl ricinoleate
P-6:	butyl acetyl ricinoleate
P-6C:	butoxyethyl acetyl ricinoleate
P-6E:	ethyl butyl acetyl ricinoleate
P-7:	methyl undecenoate
P-8:	acetylated castor oil
P-9:	acetylated polymerised castor oil
P-11:	methyl ester of polymerised castor oil
P-12:	ethyl ester of polymerised castor oil
P-13:	butyl ester of polymerised castor oil
P-14:	methyl ester of acetylated polymerised castor oil
P-15:	ethyl ester of acetylated polymerised castor oil
P-16:	butyl ester of acetylated polymerised castor oil
PG-16:	butyl ester of acetylated polyricinoleic acid
P-Y3E:	ethyl butyl ester of soya bean fatty acids
SC:	triglycol ester of vegetable fatty acid. Colourless

liquid, s.g. 0·97. For natural and synthetic rubbers, latices; improves processability and flexibility at low temperatures

X.66R: triethylene glycol dipelargonate

X.282: polyglycol fatty acid ester.

PLASTICS High mol. wt. organic compounds which may be plastically formed or deformed under certain conditions.

PLASTIFIANT Group of older accelerators:

A: anhydroformaldehyde aniline

B: anhydroformaldehyde-p-toluidine (19).

PLASTIGEL A plastisol with gel-like flow properties.

PLASTIKATOR Group of softeners:

85: ether, thioether. Softener for Perbunan N and C (NBR, CR) (43)

88: methylene bisbutyl thioglycolate (43)

FH: aromatic polyether, s.g. 1·08. Resin-like softener and tackifier for synthetic rubbers, particularly for NBR, SBR, neoprene. Suitable for articles requiring silica. Has no effect on vulcanisation speed. Improves building tack. *Quantity:* 2–30% (43).

OT: thioether. Yellow liquid, s.g. 0·965. Miscible with most organic solvents. Good compatibility with NR and SBR. Dosage: up to about 20%, low concentrations produce

considerable softening. *Uses:* hot-air resistant products with good low temperature flexibility; hoses, gaskets, for vehicles and mechanical devices (43)

RA: softener for thermally plasticised Buna (189)

PLASTIKOL TAH Di-2-ethyl hexyl thiodibutyrate. Low temperature, ester type plasticiser for p.v.c. and p.v.c./nitrile rubber blends (42).

PLASTIMETER Plastometer. Apparatus used to determine the viscoelastic properties (plasticity and elastic recovery) of unvulcanised rubber and their compounds. Plasticity cannot be defined exactly and the measured values are empirical numbers, dependent on the method of measurement. The values obtained using different machines can be compared only to a limited extent. The following test methods are used:

1. Determination of the change in thickness of a test piece between two parallel plates under constant load: Williams, Hoekstra, Goodrich, Scott, Wallace, Karrer plastimeters.

2. Determination of the load necessary to give a constant change in thickness of a test piece between two parallel plates: Defo apparatus. DIN 53 514.

3. Determination of the time necessary for a test piece to flow through an aperture under constant pressure: Marzetti, Behre quick plastimeters, Nellen, Firestone extrusion plastimeters.

4. Determination of the shearing force: Mooney viscometer. DIN 53 523.

5. Determination of the power required to masticate a small test piece: Brabender plastograph.

PLASTIMETER PMG i Apparatus used to determine the Mooney plasticity (398).

PLASTISOLS Dispersions of vinyl resins in softeners with the addition of stabilisers and pigments. The resin dissolves at increased temperatures in the softener and, after cooling, forms a homogeneous plastic mass.
Uses: include paints for rubbers; plastics, metals.

PLASTOFLEX Group of softeners (45).
520: dimethyl isobutyl carbinyl phthalate
DGB: diethylene glycol dibenzoate
MGB: dipropylene glycol dibenzoate.

PLASTOGEN Blend of a high mol. wt. sulphonic acid, which is soluble in oil, with a paraffin oil, s.g. approx. 0·85, acid value 1·1. Plasticiser dispersing agent for highly loaded compounds. Recommended for sponge rubber compounds and very soft vulcanisates.
Quantities of up to 40% can be used without bloom (51).

PLASTOLEIN Group of softeners and plasticisers:
9050: di-2-ethyl butyl azelate
9055: diethylene glycol dipelargonate
9057: diiso-octyl azelate
9058: tetrahydrofurfuryl oleate
9404: triethylene glycol dipelargonate
9715: polyester, s.g. 1·075, mol. wt. 850

9720: polyester, s.g. 1·031, mol. wt. 850 (196).

PLASTOMERS Collective term for high mol. wt. materials which are normally plastic during processing between certain temperature limits. They cannot be changed into a highly elastic condition by crosslinking. A distinction is made between thermoplastics, *i.e.* thermally plastic products, and thermosets, *i.e.* thermally hardened substances which may no longer be returned to a plastic state.

PLASTOMOLL Group of softeners for plastics and lacquers (32).

PLASTONE Activator, softener and plasticiser for rubber (127).

PLASTOPRENE LV Isomerised natural rubber. Resinous mass, m.p. 120–130°C.
Uses: for corrosion resistant colours and paints, printing colours, reinforcing fillers in aldehyde resins (295).

PLASTOR EG 33 33% dimethylimidazole.

PLASTORUB A plantation rubber, preplasticised by addition of a peptiser to the latex before coagulation (440).

PLASTORYL Filler for rubbers and thermoplastics (296).

PLASTYLENE Group of polyethylenes (48).

PLATEAU EFFECT The plot of tensile strength *v.* cure time shows a level portion near the optimum over a longer time interval. The plateau effect depends on the accelerator system and prevents a deterioration of the physical properties of vulcanisates on overcure.

PLD-90 Masterbatch of 90% lead monoxide with polyisobutylene (282).

PLEOGEN Group of polyester resins (297).

P L E X I G U M G r o u p o f polyacrylic acid esters (33).

PLI-ASTIC Rubber/bitumen compound. Filler mass for horizontal expansion joints in concrete constructions such as floors, streets, bridges, water containers (298).

PLICHLOR Trade name for chlorinated polyisoprene (115).

PLIOBOND Group of bonding agents for rubber/metal components (115).

PLIOFILM Rubber hydrochloride; in foils of 0·019–0·064 mm:
N: normal
P: plasticised
FF: for permeable products and products cooled at low temperatures
NF: clear films (115).

PLIOFLEX Group of butadiene-styrene copolymers. Cold and hot polymers and types extended with oil; 23·5% bound styrene, 60% conversion. The type numbers are the same as those of the ASTM code for SBR (115).

PLIOLASTIC Elastic fibres of rubber hydrochloride.

PLIOLITE Group of butadiene-styrene latices (115):
100 series: the latices of this group have good film forming properties and are primarily used unvulcanised as stated below:
101A: 45:55. Adhesives, dipped articles, binding agent for cements, impregnation
140: 40:60. Impregnations, linings, inks
150: 14:86. Reinforcement of other latices
151: 15:85. Reinforcement of other latices
160: 33:67. Textile and paper impregnation, leather finishes
165: 33:67. Pigments
170: 30:70. Paper processing.
2000 series: the index numbers correspond to the ASTM code for SBR latices.

PLIOLITE LATEX VP 100 Vinyl pyridine-butadiene-styrene terpolymer. 40% solids, s.g. (polymer) 0·97, pH 11·0. Fairly good mechanical stability.
Uses: include adhesion of rubber to textiles; tyre cords, conveyor belts, hose lines (115).

PLIOLITE NR Natural rubber derivative. Processing aid and reinforcing agent for rubber compounds (115).

PLIOLITE S Cyclised rubber made from synthetic polyisoprene rubber and used as a substitute for the natural rubber product during World War II.
S-1: cyclised, using stannic chloride. Brittle, hard resin with the same solution properties as those of natural rubber; good solubility in waxes, good chemical resistance and low viscosity. Film strength and adhesion properties are variable. *Uses:* reinforcing agent and extrusion aid in insulation compounds

S-2: cyclised using boron trifluoride. Same properties as natural rubber. *Uses:* carrier for fast drying printing aids (115).

PLIOLITE S Group of styrene-butadiene resins with a high styrene content. White granules, s.g. 1·04. Reinforcing filler for NR and SR; assists extrusion and calendering properties, improves abrasion resistance, tear resistance, flex cracking resistance and dimensional stability, and increases hardness.
Quantity: 10–30% (115).
Uses: include soles, artificial leathers, floor coverings, insulations.

Type	Styrene-butadiene	Softening p. (°C)
S-3	85:15	
S-5	86:14	
S-6	90:10	65
S-6B	82·5:17·5	49
S-6E	87:13	48

PLIOPRENE Modified SBR rubber (115).

PLIO-TUF Group of modified resins with a high styrene content, s.g. 1·01, Shore D Hardness 70 (115).

PLIOVIC Group of p.v.c. resins, copolymers and aqueous dispersions (115).

PLURONIC Group of polyoxy-propylene glycol ethylene oxides with 20–90% C_2H_4O in the general compounds:

$$HO(C_2H_4O)_x.(C_3H_6O)_y.(C_2H_4O)_z.H$$

s.g. 1·02–1·05, mol. wt. 2000–8000. Softener (299).

PLY RATING NUMBER PR No., code for the strength of equal size tyres. The number does not give the number of textile layers.

PM Russian produced furnace blacks obtained from a liquid basic material. The index number for each type indicates the specific area in m^2/g.

PMM Abbreviation for poly-methyl methacrylate.

PMTD-70 Masterbatch of 70% tetramethyl thiuram disulphide and 30% polyisobutylene (282).

PMZD-75 Masterbatch of 75% zinc dimethyl dithiocarbamate and 25% polyisobutylene (282).

PND-70 Masterbatch of 70% 2-mercaptoimidazoline and polyiso-butylene (282).

PNDA Abbreviation for p-nitro-sodimethyl aniline.

PNEUMAX Group of accelerators (428):
DM: 2:2'-dibenzthiazyl sulphide
MBT: technical 2-mercaptobenz-thiazole
U: mixture of diphenyl guanidine and 2:4-dinitrophenyl benz-thiazole sulphide. Yellow powder, m.p. approx. 120°C, for NR and SR. *Uses:* mechanical goods, footwear.

PNPD-72 Masterbatch of 48% 2-mercaptoimidazoline and 24% di-o-tolyl guanidine dicatechol borate with polyisobutylene (282).

POISSON'S RATIO The ratio of the change in the lateral dimension to the change in the length of a material on elastic deformation. Determined from the elasticity modulus and from the shear modulus

$$v = \frac{E - 2G}{G}$$

or by measuring the extent to which a test piece is stretched and its corresponding contraction in width. For rubber $v = 0.5$.

POK-TA-POK Ancient Mexican ball game played by the Mayas and Aztecs in which a rubber ball had to be hit with the shoulder to pass a stone ring fastened to a wall.

POLURANE Polyurethane rubber (300).

POLYACETALDEHYDE RUBBER Brown, tacky, elastic mass formed by cooling acetaldehyde to $-123.5°C$.

POLYACRYLIC ACID ESTER An elastomer based on an acrylic acid ethyl ester which is easily polymerised with peroxide catalysts (hydrogen and benzoyl peroxide, cumyl hydroperoxide, ammonium persulphate, sodium perborate). It was developed in 1944 by the Eastern Regional Research Laboratory of the US Department of Agriculture together with the University of Akron Government Laboratories and B. F. Goodrich. Polymerisation is much simpler with this product than with dienes and may take place either as emulsion polymerisation (3 h, 90% conversion, without modifier or short stop) or as granulation polymerisation in which 8 parts monomers with 1 part water, with potassium persulphate as catalyst, are

polymerised in a mixer. The polymer is obtained directly in granular form. Although polyacrylates are saturated they contain esters and α-hydrogen as reactive groups and may be crosslinked by a Claisen condensation between an ethoxy group of one molecule and an α-hydrogen atom of another, ethyl alcohol being removed.

Crosslinking can take place, using basic materials such as polyamines, alkali hydroxides, lead oxide, sodium metasilicate containing water of crystallisation, aldehyde-amine condensation products, also *e.g.* litharge + mercaptobenzthiazole, p-quinone dioxime + lead chromate and 2:6-dichloroquinone chlorimide. To achieve better vulcanisation reactive groups are introduced by copolymerisation with small quantities of a suitable monomer, *e.g.*

butadiene, isoprene, 2-chloroethyl vinyl ether, methyl vinyl ketone, acrylonitrile. Polymers made in this way are vulcanisable with sulphur and certain accelerators, *e.g.* triethylene tetramine, tetramethyl thiuram monosulphide, trimene base= reaction product of ethyl chloride, formaldehyde and ammonia; butyraldehyde-aniline-condensation products. Vulcanisation, however, takes much longer (30–60 min at 150–155°C in a press or an autoclave). Suitable copolymers are:

1. 95:5 ethyl acrylate and 2-chloroethyl vinyl ether (Lactoprene EV, Hycar PA 21, Hycar 4021, Acrylon EA-5).
2. 87·5:12·5 n-butyl acrylate and acrylonitrile (Lactoprene BN, Acrylon BA-12, Paracril OHT).
3. Copolymer of 1:1-dihydroperfluorobutyl alcohol and acrylic chloride (poly-1:1-dihydroperfluorobutyl acrylate, Poly FBA).
4. 72:25 ethyl acrylate and butadiene.

The acrylic elastomers have excellent heat resistance (up to 200°C) in air and solvents and are surpassed only by silicone rubber. They are resistant to ozone, oxygen, oils and have good electrical properties. Their disadvantages are mediocre physical properties, affinity to water, slow vulcanisation and fairly high brittle p. (−27°C for the butyl acrylate-acrylonitrile polymer, −18°C for the ethyl acrylate-chloroethyl vinyl ether polymer, −60°C for the ethyl acrylate-butadiene polymer).

Poly-1:1-dihydrofluorobutyl acrylate has the highest resistance to oxidation, heat and solvents and is soluble only in fluorocarboxylates and fluoroalcohol fatty acid esters.

Uses: include O-rings, oil and transformer gaskets, conveyor belts, electrical insulation, hose lines, hot air hose, rollers, heat resistant impregnations and coating for textiles, white and coloured articles.

Polyacrylic acid esters are also produced in latex form. Have excellent stability.

Uses: include leather and textile substitutes, paints, floor coverings.

TN: Acronal
Acrylin
Hycar 4021
Paracril OHT
Plexigum
Poly-FBA
Lit.: DRP 654 989, 615 219, 642 574. USP 2 123 599.

POLYACRYLONITRILE FIBRES

Synthetic fibres, produced from polyacrylonitrile or copolymers with at least 85% acrylonitrile; m.p. 250–325°C, s.g. 1·125–1·18, mol. wt. 60 000–200 000. General formula: $(CH_2=CH—CN)_x$.

The first attempts to spin polyacrylonitrile were made in 1931 at IG Farben in Wolfen by H. Rein. α-pyrolidine and dimethylformamide were found to be suitable solvents (DRP 915 034). In 1943 experimental production was started (using the name PAN fibre). Independent of this, the fibres were produced in the USA by Du Pont and in 1942 Fibre ANP (later Fibre A) was used for military purposes. As polyacrylonitrile is not thermoplastic and decomposes below its m.p. the melt spinning process normally used in the production of other synthetic fibres could not be used; a solution of 20–25% in dimethylformamide was used to facilitate spinning. It was spun either dry at 130–140°C in dry air or in an inert gas (PAN, Orlon), in a coagulating bath of glycerol (130–140°C), or in a calcium chloride solution (BP 583 939, 584 548: ICI).

The fibres were stretched (to ten times their original length) at 145–180°C to orientate the crystals and increase the strength.

TN: Acrilan (formerly Chemstrand acrylic fibre) (36)
Courtelle (England)
Creslan (USA)
Crylor (formerly Fibre D) (165)
Dolyn-Acryl (formerly KSF Kel·heim) (186)
Dralon (formerly Bayer acrylic fibre)
Dynel (USA)
Imelon (222)
M-24 (193)
Nitrilon (Russia)
No. 53 (Holland)
Nymcrylon (Holland)
Orlon (formerly fibre A)
PAN (283)
Redon (formerly PK 11) (Germany)
X-51 (formerly Creslan) (21)
X-54 (21)
Wolcrylon (367).

POLYADDITION Process for the production of high mol. wt. products by the reaction of di- and polyisocyanates with polyhydric alcohols and other molecules containing reactive hydrogen atoms. The reaction is a border case between polymerisation and polycondensation in which addition at the double bond of the *iso*-cyanate group takes place

$$O=C=N-R-N=C=O + HOROH$$

$$O=C=N-R-N-C-OROH$$
$$\qquad\qquad\ \ |\ \ \ \ ||$$
$$\qquad\qquad\ \ H\ \ \ O$$

Polyurethanes, polymers of ethylene imine, ethylene oxide and tetrahydrofuran can be formed in this reaction.

POLYALDENE RESINS Polymerisation products of croton-aldehyde, m.p. approx. 55–100°C. Dispersing agents and hardeners. Assist mixing and processing and the production of smooth sheets from highly loaded compounds, *e.g.* floor tiles. Because of a strong polar character, they improve the dispersion of light-coloured reinforcing fillers. Improve the tack of adhesives (400).

POLYAMIDES Polycondensation products with —CO.NH— groups which follow one another, *e.g.* Nylon, Perlon.

POLYAMIDE 12 Polydodecanolactam 12, Nylon 12. When butadiene is polymerised using a Ziegler catalyst a trimer cyclododecatriene-(1:5:9), may be obtained under certain conditions. This may be further transformed into dodecanolactam.

$$HN-(CH_2)_{11}-C=O$$
$$\qquad |\underline{\qquad\qquad\qquad}|$$

Dodecanolactam may be polymerised to a polyamide having similar properties to Perlon (Nylon 6) or Rilsan. The basic molecule contains twelve C atoms, thus Polyamide 12, whereas Perlon and Rilsan contain six and eleven C atoms respectively.

POLYAZINE Polymers made from dialdehydes and hydrazines. Mol. wt. 500–1000.

CIS-1:4-POLYBUTADIENE Butadiene may be polymerised to a cis-1:4 structure by means of Ziegler catalysts. Polybutadiene is superior to natural rubber because of its high abrasion resistance, excellent heat resistance and low temperature properties and good resistance to light crazing. Early polymerisation processes (for Buna) yielded only a mixture of 1:4-cis, 1:4-trans and isotactic 1:2 polybutadienes, whose properties were inferior to those of natural rubber. Polybutadiene has a considerably higher abrasion resistance and a better flex cracking resistance than those of natural rubber but a lower tear strength and a higher dynamic modulus.

Uses: include tyre tread compounds, conveyor belts, V-belts, in blends with NR and SBR.

cis-1:4-Polybutadiene

trans-1:4-Polybutadiene

TN: Ameripol CB
Budene
Buna CB
Cariflex BR
Cis-4
Diene
Europrene CIS
x Intene
Plioflex 5000S
Synpol EBR
Taktene
Unafor.

POLYBUTYLENE ADIPATE
Light coloured viscous fluid, s.g. 1·12. Plasticiser for nitrile and chloroprene rubbers; also for vinyl chlor-

ide polymers and copolymers. Resists oils and fats, low extractability. Does not migrate.
TN: Hexaplas BUT (60).

POLY-2-CHLOROBUTADIENE
Polychloroprene. Modified and unmodified polymers of 2-chlorobutadiene (chloroprene) or copolymers with a major proportion of 2-chlorobutadiene. Produced by emulsion polymerisation

$$CH \equiv CH \rightarrow$$
Acetylene
$$CH \equiv C{-}CH{=}CH_2 + HCl$$
Monovinyl acetylene
$$CH_2{=}C{-}CH\ CH_2$$
$$|$$
$$Cl$$
2-Chlorobutadiene
$$\left[-CH_2{-}C{=}CH{-}CH_2{-} \atop \qquad\quad | \atop \qquad\quad Cl \right]_n$$

The linking of molecules is mainly by 1:4 addition. Following the work of Pater J. A. Nieuwland on condensation reactions of acetylene, poly-2-chlorobutadiene was developed technically from 1927 onwards, by W. H. Carothers and his colleagues at Du Pont and has been produced commercially since 1931. The polymerisation temperature has a marked influence on the crystallisation rate of the polymer and the physical properties of the vulcanisates. General purpose types are normally polymerised at 40°C. Polymers polymerised at 10°C are more easily dissolved and give harder vulcanisates, while those polymerised at 100°C are insoluble and yield soft vulcanisates. Copolymers with styrene, isoprene, and acrylonitrile are, or were produced. Polymerisation is carried out in two stages. The raw polymer is a plastic, soluble, linear α-polymer. In the presence of modifying agents, at room temperature, it becomes an insoluble branched polymer. Vulcanisation by metallic oxides is a further polymerisation to the μ-polymer, which is identical to vulcanised rubber.

Vulcanisation. The compounding principle for poly-2-chlorobutadiene is somewhat different to that for natural rubber. Metallic oxides are used as catalysts for further polymerisation, the activity increasing in the order magnesium, lead, zinc. The fast rate of polymerisation (vulcanisation) by zinc oxide is retarded by magnesium oxide and colophony. PBN and sulphur also have a retarding effect. Sulphur is unnecessary but gives the product better properties and increases the hardness. Aldehydeamines in the presence of lead oxide act as accelerators for fast vulcanisation. In addition to the metal oxides, certain types need other accelerators, *e.g.* 2-mercaptoimidazoline or a complex accelerator (di-o-tolyl guanidine dicatechol borate). Stearic acid can be added as a processing aid. Softeners and fillers are used in the same manner as for natural rubber. The compounds need antioxidants. Vulcanisates have good oil resistance, which is only slightly inferior to that of nitrile rubber, and excellent resistance to alcohols, ketones, and other solvents and organic chemicals. Because of the high chlorine content (37%) the vulcanisates have excellent flame and heat resistance. Despite the different vulcanisation systems poly-2-chlorobutadiene may be blended with other elastomers. The addition of 10% to other elastomers can increase their heat resistance. Vulcanisates are resistant to ozone and only slightly permeable to gases and steam. Poly-2-chlorobutadiene is marketed as solid and liquid polymers and as latices.

Uses: include oil resistant and heat resistant articles, gaskets, hose lines, conveyor belts, cable sheathings, cements, painting, spraying and dipping solutions, sponge rubber, foam rubber, bonding agents, adhesives.

TN: Butaclor
Denka Chloroprene
Neoprene (formerly Duprene)
Perbunan C
Sowpren
Svedopren
GR-M (Neoprene GN).

POLYCIZER Group of softeners:
162: dioctyl phthalate
332: dioctyl adipate
532: octyl decyl adipate
562: octyl decyl phthalate
632: didecyl adipate
662: didecyl phthalate
DBP: dibutyl phthalate
DBS: dibutyl sebacate
DCP: dicapryl phosphate (127).

POLYCO Group of synthetic latices, plastic emulsions and latex thickening agents based, for example, on vinyl-vinylidene chloride, styrene-butadiene-acrylonitrile terpolymers, soluble sodium and potassium polyacrylates (34).

POLYCONDENSATION The building of high mol. wt. compounds by a condensation reaction of monomers with the separation of water or a simple molecule, *e.g.* alcohol, hydrochloric acid.

POLY-CONE Group of silicone oil emulsions. Release agents (237).

POLY-1-CYANOBUTADIENE
Tough, insoluble, resinous mass. A tough, leathery polymer when co-polymerised with styrene. With buta-diene it gives a polymer similar to SBR but with improved oil resistance and inferior elongation and hyster-esis.

POLY-1 :1-DIHYDROPER-FLUOROBUTYL ACRYLATE
Rubber-like product of 1:1-dihydro-perfluorobutyl alcohol with acrylic acid chloride. Has a low volume swell in aliphatic and aromatic solvents, good ageing properties and good resistance to oxidising agents, in-cluding ozone. Polyfunctional amines are used as vulcanisation agents.
TN: Poly-FBA (1).

POLY-p-DINITROSOBENZENE

$$\left[O\!=\!N\!-\!\langle\bigcirc\rangle\!-\!N\!=\!O \right]_x$$

Brown, odourless powder. Activator for thiuram–thiazole compounds, ac-celerator in butyl rubber compounds, stiffening agent for butyl rubber, extremely active accelerator in neo-prene latices at low vulcanisation temperatures. Sold as a 25% blend with an inert substance to which the following statements refer. The active constituent can cause cyanosis or dermatitis, but if the blend is used carefully it is not dangerous. An add-ition of 0·1% prevents the softening of butyl rubber during processing; higher concentrations result in a hard-ening of the compounds. Gives good processing safety in most butyl rubber types. Has good ageing properties. Causes slight discoloration in white and light-coloured articles. Has no effect on the stability of neoprene latices.

Activity: in butyl rubber above 121°C, in neoprene latex above 100°C.
Quantity: butyl rubber 0·15–1% as activator in thiuram–thiazole com-pounds, stearic acid acts as a retarder and should be replaced by paraffin. Neoprene 1–2%, activated by 1% mercaptothiazole or zinc mercaptobenzthiazole or 1·5% sodium dibutyl dithiocarbamate (50%).
TN: Polyac (6)
Butalene (19).

POLY-2:3-DIPHENYLBUTA-DIENE
A synthetic resin, similar to polystyrene but thermostable. Has good resistance to the effects of irradiation and is an electrical insu-lator up to 130°C.

POLYDISPERSIONS
Trade name for dispersions (master batches) of rubber chemicals and polyiso-butylene (282).

POLY-EM
Polyethylene latex with 40% solids.
Uses: include softeners, impregna-tions (301).

POLYESTER FIBRES
Synthetic fibres made from the condensation products of di-carboxylic acids, *e.g.* terephthalic acid, naphthalene di-carboxylic acids with glycols, m.p. 256–260°C, s.g. 1·36–1·42, mol. wt. 8000–10 000. Resistant to alkalis, acids, oxidising agents, heat and light. Developed from 1939 by the Calico Printers' Association (Lanca-shire) and since 1944 in conjunction with ICI.
TN: Diolen (Germany)
Dracon (Fibre V) (USA)
Tergal (France)
Terital (Italy)
Terlenka (Holland)

Terylene (England)
Trevira (Germany).

POLYESTER RUBBERS
Polycondensation products of high
mol. wt. (10 000–20 000) linear polyesters with hydroxycarboxylic acid or
mixtures of dicarboxylic acids and
glycols which may be crosslinked with
peroxides (1–6%) to solid, rubber-
like masses. The production of the
polyesters takes place at approx.
250°C in a vacuum. The addition of
small quantities of maleic acid anhydride gives polymers which are
capable of reacting and may be
vulcanised with sulphur and accelerators. Because of the poor strengths of
the pure polymers reinforcement with
mineral fillers is necessary; blacks
are unsuitable because of their
retarding effect on peroxide crosslinking. Dimeric acids and esters of
plant oils, *e.g.* soya oil, which contain unsaturated groups may be condensed with glycols to give polymers
vulcanisable with sulphur and accelerators. Polyester rubbers were
produced in small quantities in the
USA during World War II. However, they were not produced commercially because of the high cost.
They are resistant to high temperatures, oxidation, weathering and
hydrocarbon solvents, have excellent
low temperature properties without
the use of softeners and have good
electrical properties. They are not
water-resistant above 60°C nor steam
resistant.
TN: Paraplex Rubber (formerly
  Paracon)
Horepol
ACME.

POLYETHYLENE
$(CH_2—CH_2)_n$. Plastic material with
a milky translucence; m.p. 110–
125°C, s.g. 0·91–0·96, mol. wt.
10 000–140 000 (approx. 25 000 for
normal use), hardness 40–70, shows
crystalline fracture at $-50°C$. Is
tough and flexible at room temperature. Inert to alcohols, esters, ketones, ether, non-oxidising acids and
alkalis, attacked by perchloric acid,
nitric acid, free halogens, aromatic
and chlorinated hydrocarbons. Has
good electrical insulating properties.
Produced by polymerisation of liquid
ethylene at high pressure (1000 atm)
and temperatures (200–300°C) (high
pressure polyethylene) or at low
pressure (low pressure polyethylene)
using Ziegler catalysts. Low pressure
polyethylene is superior in many
properties to the high pressure product. By irradiating black compounds
with high energy electrons, products
may be obtained with 5–8 times the
normal tear resistance, excellent corrosion resistance, low temperature
flexibility, and which do not soften.
By modification or polymerisation of
chlorinated ethylenes vulcanisable
products are obtained, *e.g.* Hypalon.
Chlorofluorinated products can be
produced by polymerisation of trifluorochloroethylene   or   copolymerisation with a diolefin or vinylidene fluoride. Such products may be
vulcanised by organic peroxides,
polyamines or diisocyanates.
TN: Alkathene
Polythene
Alathon
Lupolen
Petrothene
Poly-Eth
Hostalen
Marlex-50.

**POLYETHYLENE/BUTYL
BLENDS** Butyl rubber blended
with 20–40% polyethylene has good
resistance   to   oil   and   lubricants,

particularly aromatic substances, good electrical properties, high ozone resistance, remains elastic at low temperatures and has a low degree of water absorption.

POLYETHYLENE FIBRES

Fibres made from polyethylene with a mol. wt. of approx. 20 000–30 000. Produced using the dry spinning or wet spinning process.
Uses: include cables, chemically resistant materials (*e.g.* filters), buoyant ropes.
TN: Northylen
Reevon
Wynene
Polythene
Courlene
Avisco PE

POLYETHYLENE GLYCOL

$HO(CH_2CH_2O)_nH$. Depending on the mol. wt. (ranges from 200–20 000) it may be a viscous liquid (up to approx. 600) or a soft to hard, odourless and colourless wax-like product, m.p. -15 to $65°C$, s.g. 1–1.15. Soluble in water, aromatic hydrocarbons, esters, ketones, alcohols, insoluble in benzine hydrocarbons. Softener, lubricant, release agent.
TN: Carbowax
Polywachse
Polyglykol.

POLYETHYLENE GLYCOL DIBENZOATE

$C_6H_5CO(OCH_2CH_2)_xOCOC_6H_5$. Group of softeners, s.g. 1.14–1.16 ($25°C$), mol. wt. approx. 400–800.
TN: Benzoflex P- (index No. + 208 = mol. wt.).

POLYETHYLENE GLYCOL DILAURATE

$C_{11}H_{23}CO(OC_2H_4)_xCOOC_{11}H_{23}$,

m.p. 4–$14°C$, s.g. 0.97, mol. wt. approx. $750°$. Softener.
TN: Morpel X–928 (264).

POLYETHYLENE TEREPH-THALATE

Polycondensation product of terephthalic acid and ethylene glycol, polyester artificial silk and staple fibre, m.p. 250–$260°C$, s.g. 1.38–1.42, mol. wt. 8000–10 000. Heat resistant. Discovered in 1941 by I. R. Whinfield and I. T. Dickson, produced industrially since 1946.
TN: Terylene (6, 60)
Diolen
Tergal (165)
Enkelene.

POLY-F

Group of poly-1:1-dihydroperfluoroalkyl acrylates.
FAA: amyl
FBA: butyl
FDA: decyl
FEA: ethyl
FHA: hexyl
FOA: octyl
FPA: propyl (1).

POLYFLEX

Canadian produced 6-ethoxy-2:2:4-trimethyl-2-dihydroquinoline.

POLY-2-FLUOROBUTADIENE

Non-crystallising polymer, similar in its low temperature properties to SBR and better than Neoprene GN and FR. By copolymerisation with small quantities of acrylonitrile, good oil resistance is attained.

POLYFOAM

Rigid foam made from high mol. wt. branched polyoxy compounds and toluylene diisocyanate.

POLYGARD

Mixture of alkylated aryl phosphites. Brown liquid, s.g. 0.99. Soluble in acetone, alcohol,

benzene, carbon tetrachloride, ligroin, insoluble in water, but may be hydrolysed. Non-staining stabiliser for butadiene-styrene copolymers, natural rubber, NBR, latices and plastics. Prevents gel formation and improves heat resistance. Can be added to the latex before coagulation in the production of SBR.
Quantity: 1·5% (23).

POLYGEN Group of oil extended SBR types, produced by the coagulation of latices with a high Mooney value, together with processing oils (115).

POLYGLYCOL Group of polypropylene glycols with the code no. giving the average mol. wt. (approx. 400–1200). Softeners (61).

POLYGLYCOL ETHER
Brown, viscous liquid, s.g. 1·14. Antistatic softener which prevents the build-up of static electricity; particularly suitable for Perbunan.

POLYISOBUTYLENE

$$CH_3-\left[\begin{array}{c} CH_3 \\ | \\ -C-CH_2- \\ | \\ CH_3 \end{array}\right]_n \begin{array}{c} CH_3 \\ | \\ -C \\ \| \\ CH_2 \end{array}$$

Obtained by solution polymerisation at low temperatures with boron trifluoride or aluminium trichloride (Friedel-Crafts catalyst) as activator. Polymerisation occurs cationically *via* ion chains. The reaction is exothermic (approx. 10 000 cal/mol) and the heat liberated is removed by evaporating the excess liquid solvent (ethylene). The mol. wt. is approx. 15 000–500 000, the lower the polymerisation temperature, the higher the mol. wt. The lower mol. wt. products are liquid to syrup-like and

become rubber-like at a mol. wt. above 60 000. Completely pure isobutylene cannot be polymerised and n-butylene is necessary as a cocatalyst to initiate the reaction. The quantity of n-butylene used affects the mol. wt. (1%: approx. 185 000, 10%: approx. 80 000). Polymerisation is carried out either as a continuous process in endless steel-band channels with liquid ethylene as a solvent (Oppanol B) or batch wise in kettles with methylene chloride as solvent and liquid ethylene as cooling agent (Vistanex). Polyisobutylene forms a thread-like molecule with two methyl groups on every second C atom and has the property of a saturated hydrocarbon; it is relatively inert and cannot be vulcanised. The final double bond of the long chains is of no significance. Polyisobutylene is insoluble in strongly polar solvents, but swells, or dissolved in non-polar solvents (aliphatic, aromatic and chlorinated hydrocarbons, fats and oils). Resistant to acids and alkalis but may be slightly oxidised. Stabilisers and blacks may be added to give excellent resistance to oxidation. Products are odourless and tasteless, have good electrical insulating properties, good low temperature resistance and may be used over a temperature range of $-60°C$ to 160°C. The low mol. wt. products are oily and have a relatively constant viscosity over a wide temperature range; uses include softeners, additives for adhesives, viscosity stabilising diluents for lubricants and impregnating oils. The solid products have a mol. wt. of 60–200 000. They can be blended with natural and synthetic rubbers, thermoplastics, and particularly with polyethylene and waxes. In rubbers they improve the ageing properties, flex cracking and ozone resistance and the electrical properties. Uses include corrosion resistant coatings, sealants and gaskets in the building industry, adhesives, adhesive tapes, beer and wine hoses, textile proofings, bonding agents, diluents.

TN: Decelith O
Isolene
Oppanol B
Wistanex.

CIS-1:4-POLYISOPRENE

Synthetic natural rubber. The first observations on the polymerisation of isoprene were made by G. Bouchardat (1879), G. Williams (1860), Wallach (1887) and Tilden (1892). F. Hofmann and C. Coutelle (1909) achieved the hot polymerisation of isoprene. Matthews and Strange in England and C. Harries in Germany discovered simultaneously (1910) the catalytic effect of alkali metals. Because no commercial process was available for the production of isoprene, isoprene rubber aroused little interest. Apart from this, polymers produced using the normal processes were not in demand because of their poor mechanical properties. The early polymers were mixtures of 1:4, 1:2 and 3:4 addition with branched chains. It was necessary, therefore, to find specific catalysts which would lead to a predominantly cis-1:4 structure. By polymerisation of isoprene with litium catalysts or with aluminium alkyl/titanium chloride (Ziegler catalysts) a polyisoprene can be produced identical to natural rubber. The following methods may be used to produce this isoprene: dehydration of isopentane or isopentene; dehydration and cracking of dipropylene; the addition of formaldehyde to isobutylene; *via* acetone from acety-

lene; and as by-products of a cracking process.

Lithium catalysts. By the polymerisation of highly purified isoprene at 30–40°C in a dry, oxygen-free atmosphere, using lithium (0·1 %) as a catalyst a product is obtained with a cis-1:4 structure and a small amount (6–8 %) of 3:4 structure, but without the trans isomer and the 1:2 combination. The other alkali metals (Na, K, Rb, Cs) give irregular polymers with 1:4-cis, 1:4-trans, 1:2 and 3:4 structures in the molecule. The rubber is a white to yellowish mass and is stabilised with 3 % polytrimethyl dihydroquinoline.

Complex catalysts. By using Ziegler catalysts of aluminium alkyl/titanium chloride cis-1:4 polyisoprene is similarly obtained by anionic polymerisation with only a small amount of the trans isomer. By changing the ratio of the catalysts a cis-trans ratio can be produced in any required quantity. If the aluminium alkyl constituent is increased the cis content also increases while increased titanium or vanadium contents increase the trans constituent. Complex catalysts give a regular structure, improved processing properties and heat resistance but a lower mol. wt. than do the lithium catalysts. Because of the lower mol. wt. the elastic and dynamic properties are slightly lower. The synthetic polyisoprenes differ from natural rubber in their amounts of crystalline components and in their speeds of crystallisation. While the formation of the crystalline phase at −25°C in natural rubber is completed after 6 h, crystallisation is prolonged for more than 50 h in synthetic types. Vulcanisates of cis-1:4-polyisoprene have almost identical physical properties to those of natural rubber. They have low hysteresis,

Properties:	Natural rubber	Lithium catalyst	Complex catalyst
cis-1:4 %	97·8	94	92–99
trans-1:4 %	0	0	0–8
1:2 %	0	0	0–3
3:4%	2·2	6	1–3
Glass transition temperatures	−70°C	−68·5°C	−70°C
Solubility in benzene %	100	90·4	97–99
Unsaturation %	96	98	94–98
Mooney viscosity	60–85	60–80	50–75

good heat resistance, high elasticity and stability of physical properties at high temperatures. The heat build-up is the same as that of natural rubber but less flex cracking occurs, and the rate of wear is slightly less. Oxygen is absorbed at a slower rate than in natural rubber but the synthetic product is more sensitive to light. The X-ray diffraction pattern is similar to that of natural rubber. The vulcanisation period is relatively longer because of the absence of non-rubber constituents but can normally be reduced to that of NR by adding 1·5% soya lecithin and 0·075% triethanolamine. May be blended with NR and SBR.
Uses: primarily in tyres, mechanical goods, extruded profiles.
TN: Ameripol SN (complex catalyst)
Cariflex IR
Coral rubber (lithium catalyst)
Natsyn (complex catalyst)
SKI-3 (complex catalyst)
Soviet polyisoprene (lithium catalyst).

POLYLITE Alkylated diphenyl-amine. Dark brown liquid, s.g. 0·95. Soluble in petroleum, benzene, ethylene dichloride and acetone. Antioxidant giving protection against heat, oxygen and flex-cracking in natural and synthetic rubbers. Non-blooming.
Quantity: 1–2%.

Uses: tyres, inner tubes, footwear, insulation, latices (23).

POLYMEL DX Modified styrene resin. May be blended with natural rubber, SBR, nitrile rubber and neoprene; decreases both tack and shrinkage of highly loaded compounds (136).

POLYMERS Long-chain, branched or linear, crosslinked or agglomerated molecules, built up from one or more molecular units (monomers) by addition or condensation polymerisation. The final product has the same chemical composition as that of the component monomers but a much higher mol. wt. and different properties. If more than one monomer is contained in the polymer, it is termed a co-polymer (with similar ease of polymerisation) or heteropolymer (with varying ease of polymerisation). The inclusion of monomers takes place regularly. Polymers with a large number of units are termed high polymers. Methods of production are known as suspension, pearl, emulsion, bulk and solution polymerisation. On the basis of their composition high polymers can be classified as:
Linear polymers: in which the basic monomer units are combined as a

chain without branching or cross-linking.

Block polymers: high polymers in which the structurally different chains are linked together linearly.

Branched high polymers: branched chain molecules in which the parts between the branching positions are linear.

Graft polymers: high polymers in which the subsidiary chains are different from the main chains.

Spatially crosslinked polymers: rings connected so as to form a three-dimensional system. As they can exist only in a solid state they are known as single aggregate substances. Intramolecularly linked polymers: single molecules crosslinked to each other.

Layer high polymers: molecules with a two-dimensional structure which produce a network of condensed rings with or without subsidiary chains.

Lit.: 'Richtlinien für die Nomenklatur auf dem Gebiet der makromolekularen Stoffe', *Makromol. Chem.*, 1960, **38**.

POLYMER CPP 1469

Development coding for a cis-1:4 polyisoprene of the Firestone Tire and Rubber Co.

POLYMER HOMOLOGUES

Blends of macromolecules of the same structure but of different mol. wt.

POLYMER ISOMERS Mixtures of macromolecules of different mol. wt. and structures.

POLYMERISATION Reaction in which the molecules of a monomer are linked to give a macromolecule. If two or more dissimilar monomers are involved the reaction is termed copolymerisation or heteropolymerisation.

POLYMERISATION INHIBITOR A substance which, in small quantities, retards or prevents the polymerisation of monomers, and stabilises the monomers.

POLYMERISATION NUMBER The average number of structural units (monomers) in a macromolecule. In most elastomers and plastics the polymerisation number must be several thousand to give suitable physical properties.

POLYMETHACRYLIC ACID ESTERS (Polymethacrylates). Solid, glass-like thermoplastic polymers of the methyl, ethyl, propyl and butyl esters of methacrylic acid. Production is according to the following equation:

$$\begin{array}{c} H_3C \\ \diagdown \\ H_3C \end{array} C=O + HCN \longrightarrow \begin{array}{c} H_3C \diagup OH \\ C \\ H_3C \diagdown CN \end{array}$$

$$\text{acetone} \qquad\qquad \begin{array}{c}\text{acetone}\cdot\\ \text{cyanohydrin}\end{array}$$

$$\xrightarrow[H_2SO_4]{R.OH} CH_2=C\diagup{CH_3}\diagdown{COOR}$$

methacrylic acid ester

Polymerisation proceeds by the chain growth of the vinyl units, *i.e.* by means of the free radicals which link the non-activated molecules to the practically non-branched chains.

$$\begin{array}{c} CH_3 \\ | \\ CH_2=C \\ | \\ COOR \end{array} \rightarrow \left[\begin{array}{c} CH_3 \\ | \\ -CH_2-C- \\ | \\ COOR \end{array}\right]_n$$

The most important product is polymethyl methacrylate. The polymers are processable mechanically, may be poured, are resistant to water, dilute acids and alkalis, mineral oils and benzine, soluble in or swollen by chlorinated hydrocarbons, benzene, esters and ketones.

Polymethyl methacrylate can be used in dispersions as a thickening agent for natural latex. Has no effect on the vulcanisation and stability of the latex, but causes thickening in compounds in the presence of zinc oxide. The pH of the compound should be approx. 10·5.

Quantity: 30–70% on the rubber.
TN: Acronal
Bedacryl
Plexiglas
Perspex.

POLYMETHYLENE POLY-PHENYL ISOCYANATE

$(-C_6H_4.NCO.CH_2-)_n$. Bonding agent for synethic fibres or polymers.
TN: PAPI.

POLY-α-METHYL STYRENE

Odourless, colourless, viscous liquids, b.p. 150–300°C, s.g. 1·01–1·04. Softeners and plasticisers for NR and SR. Have no effect on vulcanisation.
TN: Dow 276-V-2
276-V-9 (61).

POLYNITROSO-2:2:4-TRI-METHYL-1:2-DIHYDRO-QUINOLINE

Pale brown powder, m.p. approx. 110°C, s.g. 1·20. Soluble in benzene, toluene, acetone and chloroform. Vulcanisation retarder.
Quantity: 0·5–1% depending on the accelerator used and the processing safety required.
TN: Curetard (5).

POLYOLEFIN RUBBER

Synthetic rubber, currently at the developmental stage, with good prospects as a constituent of mechanical goods.

POLYOXYPROPYLENE GLYCOL ETHYLENE OXIDE

$HO(C_2H_4O)_x(C_3H_6O)_y$
$.(C_2H_4O)_x.H.$
S.g. 1·02–1·05. Mol. wt. 2000–8000. Softener with a varying C_2H_4O content.
TN: Pluronic (299).

POLY-PHEN S-244 Condensation product of p-tert-butyl phenol with formaldehyde. Brown resin, m.p. 70–80°C, s.g. 1·2. Soluble in aromatic hydrocarbons, alcohols and ketones.
Uses: vulcanising agent for butyl rubber (237).

POLYPHOSPHONITRILE

CHLORIDE Inorganic rubbers produced by heating phosphonitrile chlorides. Externally it is scarcely distinguishable from slightly pre-vulcanised rubber.

POLYPLASTS Term suggested for plastic, synthetic resins. DIN 7731.

POLYPOL S-70 Co-precipitate of lignin and a styrene-butadiene copolymer. A combination of 1·5% benzthiazyl disulphide (on the rubber) and 0·6% copper dimethyl dithiocarbamate (on the lignin) is recommended as an accelerator system.

POLYPREN Polyurethane foam material based on a polyether (302).

POLYPROPYLENE Plastic similar to polyethylene. Produced from oil refinery gases: developed by G. Natta in combination with Montecatini. Has higher heat resistance, higher tensile strength and higher resistance to organic solvents than polyethylene.
Uses: moulded and extruded articles, fibres.

POLYPROPYLENE GLYCOL $HO(C_3H_6O)_xH$. Mol. wt. 250–4000. Soluble in water at a mol. wt. of 250, products with higher mol. wt. are readily to sparingly soluble in organic solvents.
Uses: include softeners, production of plastics, solvents, antifoam agents, lubricants.
TN: Polyglycol (61).

POLYRECOMBINATION Process for the synthesis of polymers from saturated compounds. The reaction is based on a radical mechanism but proceeds in individual units in which the radicals and the saturated units occur alternately. An example is p-diisopropyl benzene: tert-butyl peroxide is added and gives methyl radicals on decomposition, these react with diisopropyl benzene giving a radical which forms a dimer. The dimer reacts analogously to a polymer with the structure:

Reaction temperature: 170–200°C. The product is a white powder. The yield depends on the amount of peroxide used, with excess (3:1) the yield is 100%; lower ratio also gives a low mol. wt. Other products which may be polymerised similarly include diphenyl, p-xylene and p-dichlorobenzene.

POLYRUBBER Rubber-like substance produced by mixing a polyester resin solution with an isocyanate catalyst. The product is moulded into the required shape by pouring into a mould, and is obtained in various hardnesses and cell structures. Stable from $-40°C$ to 150°C, resistant to oxidation, dilute acids, solvents, oils, aromatics and brine.
Uses: include tyres and aircraft industries, soles, flashless moulded articles.

POLYSAR Trade term for synthetic elastomers of the Polymer Corp., Sarnia, Ontario, Canada (303).

POLYSAR ABS 970 Acrylonitrile butadiene-styrene polymer. May be modified with fillers, plasticisers and elastomers to assist sheeting (303).

POLYSAR BUTYL Group of isobutylene-isoprene copolymers (butyl rubber):
100 (GR-I-35): S.g. 0·92, Mooney viscosity 45, molecular % of unsaturation 0·65. Contains a discolouring stabiliser. Slow vulcanising, excellent resistance to ozone and good heat ageing properties. *Uses:* cables, insulating materials
101: S.g. 0·92, Mooney viscosity 60, molecular % of unsaturation 0·65. No stabiliser, slow vulcanising. *Uses:* food packaging, pharmaceutical goods. Blended with polyethylene and waxes

200 (GR-I-50): S.g. 0·92, Mooney viscosity 45, molecular % of unsaturation 1·4. Contains a discolouring stabiliser. Faster vulcanising than the 100 series, better heat ageing properties but less resistance to ozone than the 100 type. *Uses:* hose lines, moulded and extruded articles, electrical goods

201: as 200. Mooney viscosity 70–80. No stabiliser

202: as 200. Mooney viscosity 70–80. Contains a stabiliser

300 (GR-I-15): S.g. 0·92. Mooney viscosity 45. Molecular % of unsaturation 1·8. Contains a staining stabiliser. Faster vulcanising than the 200 types, medium resistance to ozone and heat. *Uses:* hoses, extruded articles, insulation

301 (GR-I-18): S.g. 0·92, higher mol. wt., molecular % of unsaturation 1·6, Mooney viscosity 75. Non-discolouring. Medium rate of vulcanisation, medium resistance to heat and ozone. *Uses:* general purpose, air bags, inner tubes and linings for tyres, extruded and calendered articles, cables, gaskets, proofings, corrosion resistant linings

400 (GR-I-25): S.g. 0·92, Mooney viscosity 45, molecular % of unsaturation 2·2. Contains a discolouring stabiliser. Fast vulcanising, excellent resistance to heat and some resistance to ozone. *Uses:* heater hose, conveyor belts for hot articles, moulded articles

401: S.g. 0·92, Mooney viscosity 45, molecular % of unsaturation 2·2. Contains a non-discolouring stabiliser. *Uses:* as 400, but for light-coloured articles.

POLYSAR KRYFLEX Non-pigmented, non-discolouring styrene-butadiene copolymers produced by copolymerisation:

200: S.g. 0·95, Mooney viscosity 46–65. 28% styrene. Low water absorption and low ash content. *Uses:* wire and cable insulation

252: S.g. 0·96, 44·5% styrene. Easily processable, does not need preheating. *Uses:* soles, floor coverings, hard articles with a low specific gravity

202: S.g. 0·96, high mol. wt., Mooney viscosity 61, 23·5% styrene (303).

POLYSAR KRYLENE Non-pigmented, cold polymerised styrene-butadiene copolymers:

Krylene: S.g. 0·94, Mooney viscosity 49–61. Contains discolouring antioxidant. Equivalent to SBR 1500. *Uses:* tyres, conveyor belts, articles requiring good abrasion resistance, in blends with NBR for articles which come into contact with low aromatic hydrocarbons

602: S.g. 0·94, Mooney viscosity 35, styrene content 28%. Contains non-discolouring stabilisers. Equivalent to Type NS, but with a lower Mooney viscosity. *Uses:* 12NS

NS: S.g. 0·94, Mooney viscosity 49–51, 28% styrene content. Contains discolouring antioxidant, equivalent to SBR 1503 but with a normal ash content. *Uses:* light-coloured articles requiring good abrasion resistance, soles, floor coverings, sponge rubber, extruded articles and mechanical goods (303).

POLYSAR KRYNAC Butadiene-acrylonitrile copolymers polymerised at a low temperature:

800:     S.g. 0·97, Mooney viscosity 83, medium acrylonitrile content. Non-staining antioxidant, medium oil resistance. *Uses:* general purpose, printing rolls, hose, gaskets, textile coatings, moulded articles, and in blends with p.v.c. and styrene-butadiene copolymers

801:     S.g. 0·97, Mooney viscosity 83, high acrylonitrile content. Non-staining stabiliser, maximum resistance to oil, lubricants and heat. *Uses:* hose lines, oil seals, O-rings and other oil resistant articles, and in blends with p.v.c. and SBR

802:     S.g. 0·96, Mooney viscosity 83, low acrylonitrile content. Non-staining antioxidant, medium to low oil resistance, good low temperature properties. Gives easy processing. *Uses:* moulded, extruded and calendered articles, and in blends with p.v.c. and SBR

803:     S.g. 0·97, Mooney viscosity 47, medium acrylonitrile content. Non-staining antioxidant, medium oil resistance. As type 800 but a lower Mooney viscosity. *Uses:* general purpose, articles without a softener and with a high filler content; also with p.v.c. and SBR

XPDR: terpolymer of acrylonitrile, butadiene and divinyl benzene. Mooney viscosity 60. *Uses:* added to NBR compound to decrease their 'nerve', improve calendering and extrusion properties and dimension stability by reducing the shrinkage. Easily processable (303).

POLYSAR KRYNOL Cold polymerised, oil extended styrene-butadiene copolymers:

651:     S.g. 0·95, Mooney viscosity 52, extended with 37·5% oil. Discolouring antioxidant, equivalent to SBR 1710. *Uses:* general purpose, tyres, camelback, conveyor belts, mechanical goods, footwear, articles requiring good abrasion resistance, and in blends with other elastomers

652:     S.g. 0·95, Mooney viscosity 52, extended with 37·5% oil. Non-staining antioxidant, equivalent to SBR 1778. *Uses:* light-coloured articles requiring good abrasion resistance, sponges, extruded articles, and blends with SBR with a high styrene content for cellular rubber soles (303).

POLYSAR S Group of butadiene-styrene copolymers:

S: S.g. 0·95, Mooney viscosity 46–54. 'Hot' polymerised. General purpose elastomer. *Uses:* tyres, camelback, in blends with natural rubber and other elastomers, soles, hose lines, conveyor belts, mechanical goods. Unsuitable for light-coloured articles

S-20:   as S, but Mooney viscosity 41–49

S-50: S.g. 0·95, Mooney viscosity 46–54. Hot polymer. Non-staining antioxidant. General purpose elastomer for light-coloured articles. *Uses:* rolls, floor coverings, soles, gaskets, proofed cloths, mechanical goods and in blends with natural rubber and other elastomers

S-65: for wire and cable insulation. Has a very low ash content (*see* Polysar Kryflex) (200).

SS-250: S.g. 0·99, high styrene content. Cold polymer. Gives a high modulus and high hardness with a low specific gravity, very good resistance to flex cracking and abrasion. *Uses:* floor coverings, soles, sponge rubber, cellular rubber, coloured ebonite

SX-370: similar to Type 1009, with 26% styrene

SX-371: terpolymer of butadiene, styrene and divinyl benzene, s.g. 0·95. Hot polymer. Non-staining antioxidant. *Uses:* in blends with SBR or natural rubber in quantities of 10–50% improves the processing properties, extrusion and calendering, and also dimensional stability by decreasing the shrinkage capacity. Less sulphur is necessary than for standard SBR (26% styrene, cross-linked). *Uses:* extruded and calendered articles, gaskets, footwear, textile proofings.

SX-630: similar to Type 1006 with 28% styrene. Mooney viscosity 47 (303).

POLYSAR S LATEX Butadiene/styrene latices:

II: pH 9·2–10, 26–28% solids, 25% styrene. No antioxidant. *Uses:* adhesives and general purpose, and in blends with natural latex to improve ageing properties

IV: pH 10–11, 39–45% solids, 50% styrene. Equivalent to SBR 2000. Resinous soap as stabiliser. *Uses:* impregnation of tyre cords, adhesives, textile impregnation

722: pH 10·3, 64% solids, 25% styrene. Cold polymer. Fatty acid soap as stabiliser. *Uses:* sponge rubber

XPRD-833: pH 10·5, 42% solids, 25% styrene. Cold polymer. Fatty acid soap as stabiliser. *Uses:* impregnation of cords (303).

POLYSAR TAKTENE High grade cis-1:4-polybutadiene. Types 1200, 1220 and 1250. *Uses:* production of tyres, footwear, rolls, conveyor belts, V-belts, hose, mats (303).

POLYSAR TRANS-PIP Synthetic trans-polyisoprene (balata), s.g. 0·95, tensile strength 352 kg/cm^2, elongation at break 475%, 300% modulus 163 kg/cm^2, ML-4/100°C 70, Shore C Hardness 70. Because of its good abrasion and tear resistance, even on shock strain, it is particularly suitable for golf ball covers and adhesives (303).

POLYSOLVAN Group of solvents; esters of aliphatic alcohols and glycolic acid esters (217).

POLYSTAL Adhesives based on hexamethylene diisocyanate (Desmodur O) or triphenyl methane triisocyanate (Desmodur R).

Uses: butadiene-styrene bonding to synthetic cords, metals (43).

POLYTERPENES Terpene hydrocarbons of the general formula $(C_5H_8)_x$ with x greater than 4. Rubber, balata and gutta-percha belong to this group.

POLYTETRAFLUOROETHYL-ENE

$$\left[\begin{array}{c} F \quad F \\ | \quad | \\ -C-C- \\ | \quad | \\ F \quad F \end{array} \right]_n \quad n = ca\ 1000$$

PTFE. Greyish white, tough, thermoplastic polymer of tetrafluoroethylene; transparent in thin layers, s.g. 2·25. Shore Hardness 55°. Insoluble in all solvents. Chemically inert, stable from −80°C to 250°C, gels at 325°C, and decomposes at 400°C into a gaseous monomer. Resistant to oxidation, chemicals, acids and alkalis, has zero water absorption. The crystalline units have a helical structure containing thirteen C atoms below 20°C and fifteen C atoms at higher temperatures. Has extremely low friction coefficient, excellent dielectric properties, is a good electrical insulator, non-flammable. Processed by extrusion and moulding at approx. 205°C, has no tack. Bonding, *e.g.* to plastics, metals, wood, using conventional adhesives is possible only when the surface to which the adhesive is applied is treated with sodium metal dissolved in liquid ammonia. First produced in 1943 by Du Pont, and in 1947 by ICI.

Uses: include linings, gaskets, hose lines, high frequency insulation materials, cast films, filled mouldings, impregnations of glass fibres and asbestos.

*TN:*Algoflon (59)
Ekafluin (195)
Fluon (60)
Fluon GP 1 (Dispersion) (60)
Fluoroflex (204)
Soreflon (326)
Teflon (6).

POLYTRIFLUOROBROMO-ETHYLENE

$(Br—FC—CF_2)_n$, BFE, bromotrifluoroethylene polymers. Group of polymerised products of different mol. wts. Production and use as for polytrifluorochloroethylene.

POLYTRIFLUOROCHLORO-ETHYLENE

$(F_2C—CFCl)_n$. Group of polymerised products of different mol. wt. Oils, waxes and fat-like substances, compressed powders, dispersions and elastomers. Resistant to oil, low and high temperatures. Chemically inert products with good electrical properties and zero water absorption.

Elastomer. Polymers or copolymers of trifluorochloroethylene and vinylidene fluoride with over 50% fluorine, s.g. 1·85. Soluble in ketones, esters and ethers. Elasticity of the fairly stiff fluoroelastomer chains is obtained by means of methyl groups. Stable from −50°C to 200°C and for a short time at 240°C. Vulcanisates are resistant to oxidising mineral acids, peroxides, alkalis, alcohols, aliphatic solvents and silicone oils. Highly halogenated hydrocarbons cause swelling. Vulcanisates containing no softener behave like plastics; softeners give the products a rubber-like character. Between 77 and 80°C the pure elastomers assume a non-elastic powder form, and temperatures must be maintained below 77°C while processing. Organic peroxides, polyamines and polyisocyanates may be used as vulcanising agents.

Peroxide vulcanisation: 1·5–3% benzoyl peroxide gives the best results. Metal oxides, *e.g.* zinc oxide, act as accelerators. Vulcanisation takes approx. 30 min at 110°C followed by post-cure at 150°C for 1–16 h according to the thickness of the article.

Polyamine vulcanisation: with *e.g.* 1·5–6% triethylene tetramine, tetraethylene pentamine, hexamethylene diamine. Vulcanisation takes 1 h at 130°C with 1 h at 150°C postcure. In contrast to peroxide vulcanisation, plasticisation in this case must be done with the normal softeners.

Isocyanate vulcanisation: with 5–10% diisocyanate, and metal oxides as accelerators. Vulcanisation is for 1 h at 130°C, the post cure is 32–72 h at 100°C.

Reinforcing silicate fillers, particularly those surface treated with silicones, are suitable as cross-linking occurs between the silicone polymer and the polytrifluorochloroethylene. As a result the strength properties are improved. Blends with natural rubber and synthetic rubbers are possible.

Properties of the vulcanisates: tensile strength 140–250 kg/cm^2, elongation at break 400–600%, tear strength 10–30 kg/cm^2.

Processed by extruding, calendering or as solutions or dispersions.

Uses: include gaskets, thermal and electrical insulations, packing materials, protection against corrosion, adhesives, dipping solutions, coatings.

TN: CFE
Daiflon (173)
F-3 (Russia)
Fluorothene (137)

Hostaflon (217)
Kel-F elastomer (1).

POLYTRIMETHYL DI-HYDROQUINOLINE Poly-2:2:4-trimethyl-1:2-dihydro-quinoline.

Dark brown, viscous liquid, s.g. approx. 1·05. Soluble in benzene and acetone, also exists as a light-brown powder, m.p. 110–120°C, s.g. 1·08. Partially soluble in aromatics, insoluble in water. Antioxidant. Gives good protection against oxidation and heat, and to a certain extent improves flex cracking resistance. Unsuitable for articles which require very high flex-cracking resistance. Has no effect on the vulcanisation of natural rubber, SBR and NBR, but is a fast accelerator in neoprene. Also suitable as an antioxidant for latex. The liquid products cause a strong discoloration and are unsuitable for light-coloured articles. The powdered products cause only slight discoloration in concentrations below 0·5%; they can be used in relatively high quantities without blooming.

Quantity: 0·5–2%.

TN: Aceto POD (25), Flectol A (obsolete) (5)
Flectol B (5)
Solid products:
Agerite AK (22)
Agerite Resin D (51)
Antioxidant 184 (22)
Flectol H (5)
Nocrac 224 (274)
Perflectol H (mixture with diphenyl-p-phenylene diamine) (5)

Permanax 45
Santoflex R (5)
Vulcarite 117, 134 (dispersion).

POLYURETHANES Poly-addition products of di- and polyiso-cyanates and materials with reactive hydrogen atoms according to the equation

$$R.NCO + R.OH \rightarrow$$
$$R—NH—CO—O—R$$

The reaction of isocyanates with poly-esters, polyethers or polyester amides (mol. wt. approx. 2000–7000) gives isocyanate polycompounds (pre-polymers). These may be further re-acted with polyhydric alcohols, water or diamines to cause a chain lengthen-ing and crosslinking. In the produc-tion of polyesters from dicarboxylic acids and di- or trihydric alcohols, deficiency of dicarboxylic acid (Vulk-ollan) can be used so that hydroxyl end groups are available for chain length-ening and crosslinking; alternatively, molar ratios are used so that the end groups are hydroxyl and organic acid groups. In the case of polyester amides, amino end groups will also be available for further reactions (Vulcaprene). The reaction of diiso-cyanates with compounds having two functional groups gives linear poly-mers (fibres); if there are more than two functional groups crosslinked polyurethanes are obtained. The reaction with linear polyesters results in elastic products. Depending on the method of production, either a cast or millable polyurethane is obtained. *Cast polyurethanes.* The reaction of a polyester with an excess of diiso-cyanate results in a prepolymer, with isocyanate end groups (isocyanate polyester), which is liquid at approx. 100–130°C. By reacting this pre-polymer with a compound containing reactive hydrogen atoms, such as a di-

or polyhydric alcohol (*e.g.* 1:4-butane-diol, cyclohexane-1:4-diol, trimeth-ylol-propane), an aromatic diamine (*e.g.* methylene bis-o-chloroaniline, o-dichlorobenzidine), or water, chain lengthening *via* urethane or urea groups occurs, with crosslinking by polymerisation of the remaining free isocyanate groups to give an elastic end-product. These prepolymers, which have limited stability above m.p. are mixed with polyalcohols or diamines and shaped to give elastic products by pouring or centrifugal casting, then heating without pres-sure. After the substance has gelled, it is taken out of the mould and post-cured for 10–15 h at approx. 100°C. If water is used, carbon dioxide is evolved and a foamy, crumbly material is produced which may be rolled into sheets and moulded in a press at 150°C. Because of the isocyanate groups, the stability of the sheets is limited (approx. 10 h). The quantity of isocyanate used for the production of the prepolymer and the type of crosslinking agent affects the hardness of the final product. A hardness range of approx. 65–95° Shore is possible. The pro-cessing of liquid polyurethanes takes place without the use of fillers. Suit-able isocyanates include: 1:6-hexa-methylene diisocyanate, 1:5-naph-thylene diisocyanate, 2:4-toluylene diisocyanate, 4:4′-diphenyl methane diisocyanate.
Millable polyurethanes. When there is a deficiency of diisocyanate (max. 0·99:1) in the prepolymer tough, rubber-like, thermoplastic masses are produced from the reaction with dihydric alcohols or diamines which are stable because of the absence of free isocyanate groups and the pre-sence of terminal hydroxyl or amino groups. The mass can be processed

on normal rubber machines and is crosslinked by addition of, for example diisocyanates, formaldehyde, peroxides, and heating under pressure. Preferred cross-linking agents include: 2:4-toluylene diisocyanate and 4:4'-diphenyl methane diisocyanate. The polyurethane obtained can be processed to give soft to hard products with a wide field of use.

The inclusion of double bonds in the propolymer, *e.g.* maleic acid anhydride, fumaric acid, gives an unsaturated polyurethane which may be vulcanised with sulphur and normal accelerator systems, as with rubber.

Polyurethane foams. Because of the formation of carbon dioxide by reaction of an isocyanate with water, foamed materials may be produced. The polyester is reacted, in the presence of a catalyst (basic compound) with a mixture of diisocyanate and water. Carbon dioxide is liberated and a foamed, high mol. wt. polyurethane obtained. Depending on the type of polyester (linear or branched) and the number of hydroxyl groups in the polyester, products may be produced which are slightly crosslinked and flexible or highly crosslinked and rigid. The foam volume may be controlled by the quantity of water. There are two types based on polyesters and polyethers. Polyurethanes have good resistance to ozone, oxygen, benzine and mineral oils, good mechanical properties, tear and abrasion resistance, good high modulus and high damping ability, but less resistance to hot water, steam, acids, alkalis and amines. The resistance to low temperatures is limited to approx. $-20°C$, heat resistant depending on the type, to 80–150°C.

Uses: include substitutes for rubber, leather, plastics; mechanical goods, O-rings, gaskets, conveyor belts, damping mounts, buffers, rollers, foam materials, fibres, bonding agents, *e.g.* for metals, textiles.

TN: Adiprene
Aeropreen (foam material)
ALPCO polyrubber
Alporex (foam material)
Apcothane (prepolymer)
Chemigum SL (115)
Cyanaprene 4590
Daltocel (polyester and prepolymers)
Daltoflex I
Disoprene
Estane
Genthane-S (31)
Ikipren (221)
Ikipreen (foam material) (220)
Moltopren (foam material)
Neothane
Nordopren
Polypren (foam material)
Semperpren (foam material)
Solithane 113
Sorboprene (foam material)
Spenkel (prepolymer)
Stafoam (foam material)
Texin
Vibrathane
Vulcaprene
Vulkollan (43).

POLYURETHANE FIBRES

Fibres produced using the melt spinning process from a polyaddition product of 1:6-hexamethylene diisocyanate and 1:4-butylene glycol (Polyurethane U).

TN: Lycra (6)
Perlon U
Spandex.

POLYVINYL ACETATE

$$\left[\begin{array}{c} -CH-CH_2- \\ | \\ O-CO-CH_3 \end{array} \right]_n$$

Thermoplastic material, m.p. (depending on mol. wt.) 40–100°C. Soluble in aromatic and chlorinated hydrocarbons, alcohols, ketones, esters. Degree of polymerisation up to approx. 6·000.

Uses: include adhesives, lacquers, synthetic fibres, floor coverings.

POLYVINYL ALCOHOL

$$\left[\begin{array}{c} -CH_2-CH- \\ | \\ OH \end{array}\right]_n$$

White to yellowish powder, m.p. approx. 200°C. Normally soluble in hot or cold water or in alcohol and water, insoluble in petroleum hydrocarbons. Aqueous solutions are colloidal, neutral to weakly acidic. Products vary in solution viscosity with the amount of acetyl end groups.

Uses: protective colloids, adhesives, impregnations, oil-resistant seals.

TN: PVS
Vinol
Vinarol
Polyviol
Mowiol
Alvyl
Elvanol

POLYVINYL CHLORIDE

$$\left[\begin{array}{c} -CH_2-CH- \\ | \\ Cl \end{array}\right]_n$$

P.V.C.; emulsion or suspension polymer of vinyl chloride obtained using a peroxide as catalyst. Depending on the polymerisation process the vinyl units may occur in a head-to-head, tail-to-tail or block structure. Copolymers with, *e.g.* vinylidene chloride, vinyl acetate, acrylic esters, maleic ester, fumaric esters, may be produced. Colourless, odourless,

thermoplastic mass, m.p. above 70–85°C, s.g. 1·35–1·38, 53–55% chlorine. May be plastically deformed at 130–150°C, polymerisation degree 300–3000. Soluble in, or swollen by aromatic hydrocarbons, chlorinated hydrocarbons, tetrahydrofuran, esters and ketones. Resistant to mineral oils, benzine hydrocarbons, alcohols, dilute acids and alkalis, ozone and ageing. Cannot be vulcanised. Used as rigid p.v.c. in a pure form, as flexible p.v.c. after the addition of softeners or as a dispersion. Rigid p.v.c. can be processed below its softening p. by machining. Is also processed by thermal deformation, painting, casting, dipping.

Uses: include coatings, hose lines, in a variety of apparatus, mechanical goods, foils, fibres, brushes, insulating materials, simulated leather materials, floor coverings, cellular articles, paints.

POLYVINYL CHLORIDE FIBRES

FIBRES Fibres made from p.v.c. which is soluble in acetone and post-chlorinated or non-chlorinated, also from copolymers with vinyl acetate (85:15–90:10) and acrylonitrile (60:40). Produced using either the wet or the dry spinning process. The fibres have high chemical resistance but poor heat resistance, and are non-flammable.

TN: Fibravyl (formerly Phofibre, France)
Isovyl (France)
Pe–Ce– (formerly WK fibres, E. Germany)
Thermovyl, France
Phovyl (formerly Phofil, France)
Vinyon HH (USA) (all post-chlorinated).
Copolymers with acrylonitrile:
Vinyon N (USA)
Chlorin (Russia)

Rhovyl-Iso (Germany)
PCU-FASER (Germany)
Movil (Italy).

**POLYVINYLIDENE CHLOR-
IDE FIBRES** Fibres produced by
the high pressure process from a
polymer of 85–90% vinylidene chlor-
ide, 10–15% vinyl chloride and
0–2% acrylonitrile.
Uses: industrial textiles and brushes.
TN: Saran
Diorid
Bexan
PeCe-120
Permalon
Velon
Vestan.

**POLYVINYL METHYL
ETHER** PVME. Heat sensitive coagulant for latex; also increases the
mechanical stability of latex preserved with formaldehyde (BP
677 876).

POLYVIOL Polyvinyl alcohol
(359).

POLY ZOLE Group of accelerators:
MBT: 2-mercaptobenzthiazole
MBTS: dibenzthiazyl disulphide
(237).

POPCORN Slang term for hard,
tough, insoluble polymer particles
which are similar in appearance to
popcorn. Formed by the continued
polymerisation of dienes in fraction-
ating columns, storage containers or
other parts of the plant where
polymerisation should not occur. The
pressure developed on formation of
'Popcorn' can cause steel containers
to burst. This undesirable side
reaction can be prevented by adding
nitrogen monoxide.

POPG Abbreviation for poly-
oxypropylene glycol.

POROFOR Group of blowing
agents used for the production of
foam, sponge, moss and cellular
rubbers from natural and synthetic
rubbers; also porous plastic masses:

254: azohexahydrobenzonitrile
476: azodicarboxylic acid diethyl ester
505A: azodicarboxylic acid diamide
ADC: azodicarbonamide
B 13: mixture of 60% benzene disulphonhydrazide with 28% liquid factice in paraffin oil and 12% inert filler. Yellowish crumbs, m.p. 163°C, decomposition temperature 145°C, s.g. 1·24. Decomposes into nitrogen in rubber compounds at temperatures between 110–140°C. Has no effect on ageing, no taste and does not discolour. *Quantity:* 1–5%. With increasing quantity the size of the pores decreases and the number of pores increases; suitable for articles requiring a closed cell structure
BSH: benzene sulphonhydrazide
D 33: diphenyl sulphon-3:3′-disulphonhydrazide
DB: diazoaminobenzene
DNO: dinitrosopentamethylene tetramine
DO-44: p,p′-oxybis benzene sulphonyl hydrazide
D-1074: azodicarbonamide
N: azoisobutyronitrile
S 44: sulphonhydrazide basis. Decomposes at approx. 175°C

TR: substituted thiotriazole. Decomposes at approx. 115°C, liberates 130 cm^3 nitrogen/g.

Experimental products:

LK 1067: diphenyl oxide-4:4′-disulphonhydrazide

LK 1074: azodicarbonamide

LK 1107: sulphonhydrazide derivative (43).

PORRIT-ALLAN APPARATUS

Apparatus used to determine the hydrogen permeability of coated fabrics, *e.g.* balloons, textiles.

POSITEX Positively charged natural latex produced by treating dilute latex with cationic surface active materials, *e.g.* cetyl pyridine bromide. Used to impregnate textile fibres, the latex particles being deposited on the negatively charged fibres; impregnation with normal latex is difficult because the charges are the same. With Positex the treated fibres have a better tensile strength and better resistance to abrasion and cracking during use. The poor ageing properties of the rubber layer do not favour commercial use of this product.

POTASSIUM COBALT NITRITE $CoK_3(NO_2)_6 \cdot \frac{3}{2}H_2O$, Fischer's yellow, Indian yellow. Yellow pigment.

POTASSIUM DIBUTYL DITHIOCARBAMATE

$$\begin{array}{c} C_4H_9 \\ \diagdown \\ C_4H_9 \diagup \end{array} N-C-S-K \quad \overset{\|}{S}$$

Straw yellow liquid, s.g. 1·08–1·12. Accelerator, used as 50% solution. *TN:* Butyl kamate (51).

POTASSIUM DIMETHYL DITHIOCARBAMATE

$$\begin{array}{c} H_3C \\ \diagdown \\ H_3C \diagup \end{array} N-\overset{\overset{\displaystyle S}{\|}}{C}-S-K$$

Short-stop for emulsion polymerisation.

TN: Sharstop 268 (50% aqueous solution)

Thiostop K (50% aqueous solution).

POTASSIUM IODIDE KI.

White crystalline powder, m.p. 680°C, s.g. 3·12. Readily soluble in water.

Uses: catalyst for various polyacrylic rubbers.

POTASSIUM ISOPROPYL XANTHATE

$(CH_3)_2CH-O-SC-S.K.$ Yellow powder. Accelerator (obsolete). *TN:* + Enax (23).

POTASSIUM 2-MERCAPTO-BENZTHIAZOLE

$$\text{C}-\text{S}-\text{K}$$

Soluble in water. General purpose ultra-fast accelerator for latex compounds. Gives good ageing properties.

TN: Accelerator 85 (6)

Accelerator 122 (6).

POTASSIUM PENTAMETHYLENE DITHIOCARBAMATE

$$\begin{array}{c} H_2 \; H_2 \\ C-C \\ H_2C \diagup \qquad \diagdown \\ C-C \\ H_2 \; H_2 \end{array} N-CS-S-K$$

Pale brown liquid, s.g. 1·22. Soluble in water. General purpose ultra-fast

accelerator. Gives good ageing properties in latex compounds; usually
added as a 45% aqueous solution.
Gives fast cures. Active above room
temperatures; may be boosted by
aldehyde-amines, thiazoles and thiurams. Has a strong effect on prevulcanisation of latex without affecting the stability. Wet latex films
discolour when handled.
Uses: 0·5–2% (in solution 1:1 with
water) on the rubber content, with
1–2% sulphur.
TN: Accelerator 87 (obsolete) (6)
Accelerator 89 (6).

POTASSIUM PERSULPHATE
$K_2S_2O_8$. White, odourless, crystalline powder, decomposes at 120°C.
Solubility in cold water 1:50, in warm
water 1:25.
Uses: catalyst in emulsion polymerisation of dienes with styrene and
acrylonitrile. Strong oxidising agent.

PPA Abbreviation for polypropylene adipate.

PP CREPE Abbreviation for
partially purified crêpe. Produced by
diluting (10%) and coagulating centrifuged latex. Low protein, ash, and
acetone extract, nitrogen approx.
0·22%, ash approx. 0·06%. Has low
water absorption.

PPD Abbreviation for piperidine
pentamethylene dithiocarbamate.

PRD-90 Masterbatch of 90% lead
tetraoxide and polyisobutylene (282).

PRDSD-7875 Masterbatch of
27% p-dibenzoyl quinone dioxime,
45% lead tetraoxide and 6·75%
sulphur in polyisobutylene (282).

PRECOAGULATION Spon-
taneous or controlled coagulation of
the yellow fraction of natural latex.
To produce pale crêpe, the yellow
fraction, consisting of lutoids, Frey-
Wyssling particles and carotinoids,
must be removed. As this fraction
is less stable than the rubber fraction,
it can be removed fairly easily by
adding to the latex approx. $\frac{1}{6}$–$\frac{1}{3}$ of
the acid quantity necessary for com-
plete coagulation, stirring slowly.
The yellow fraction then collects in
lumps on the surface. Spontaneous
coagulation may occur because of
bacterial influence, the high magnesium content of the latex, in
relation to the phosphorus content,
or impurities, *e.g.* from rain. As
spontaneous coagulation is undesirable, unstable latex is stabilised by
the addition of so-called anticoagulants (ammonia, soda, sodium
sulphide, formaldehyde).

PREPOLYMER Stabilised reaction product of a polyester, polyether,
or polyester amide with excess diisocyanate. Is converted to a pourable
polyurethane by a dihydric alcohol or
diamine.

PRGAD-78 Masterbatch of 39%
lead tetraoxide, 26% dibenzthiazyl
disulphide and 13% p-quinone dioxime with polyisobutylene (282).

PRIESTLY, JOSEPH
English chemist, 1733–1804. In his
book 'Familiar Introduction to the
Theory and Practice of Perspective',
1770, he pointed out the ability of
rubber to erase lines drawn with a
lead pencil. Because of this ability the
product was given the name India
Rubber (to rub).

Lit.: *New Royal Cyclopaedia and Encyclopaedia or Complete Modern and Universal Dictionary of Arts and Science*, Vol. 1, London, 1788, p. 441.

PROCESSABILITY A general term which embraces processes and associated factors such as mastication and mixing time, energy consumption, scorching tendencies, dispersion of fillers, extrusion, and dimensional stability. Testing methods exist for the various processes and are carried out under specified conditions. Several properties may be tested simultaneously on some machines, *e.g.* the CEPAR apparatus and the Brabender plastograph.

PROCESSING OILS Mineral oils used to increase the plasticity of compounds, facilitate the dispersion of fillers, improve processability and as diluents. The physical and chemical properties of the oils affect the processability of the compounds and can alter the properties of the vulcanisate. The oils are classified according to their chemical composition into three main groups: paraffinic, naphthenic and aromatic, with others intermediate between these groups. Naphthenic and aromatic oils give the best processing properties, whereas paraffinic and naphthenic oils give the best low temperature properties. Aromatic oils have a strong oxidising effect and result in poor ageing properties and high hysteresis, but the compounds have high strength and tack. Paraffinic oils result in a decrease in strength and tack, but give good low temperature properties, good ageing and low hysteresis. The part which is extractable by 85% sulphuric acid indicates the constituents which may be reactive and affect the stability of the

oil; the tendency to scorch increases as the amount which is extractable increases. In general it should not exceed 15%. For the evaluation of processing oils, the viscosity–gravity constant, mol. wt. and refractive index are determined. The viscosity–gravity constant, unlike the aniline p., is independent of mol. wt., and thus a simple classification is possible: highly aromatic 5–15% saturation, aromatic 15–32%, naphthenic 48–56%.

PROCESS OIL GB 2095 Synthetic aromatic process oil for use in chloroprene rubbers and as a secondary plasticiser for nitrile rubbers (438).

PRODUCERS' ORGANISATIONS The first organisations concerned with improving the control of raw rubber productions were the Internationale Vereeniging voor de Rubbercultuur in Nederlandsch Indie (founded 1922) and the British 'Rubber Growers' Association' (1923). The organisations were limited in their activity because their membership did not include all the producers. However, the depression of 1929 and the 1930's prompted an agreement (1934) to be made in London between Britain, France, Holland, India and Thailand, controlling rubber production. The active organ used to set the limits for production and export was the International Rubber Regulation Committee. Besides these limits, regulations were also necessary to increase sales (Article 19 of the agreement) through scientific experiments and propaganda; such projects were financed by a specific tax on production areas, imposed by the individual governments. An International Rubber Research Board and an International Propaganda Committee (later, the International Rubber Development Committee) were set up to coordinate the research work of the institutes. The following institutes were established within the framework of this action:

Natural Rubber Producers' Research Association (NRPRA), formerly British Rubber Producers' Research Association (BRPRA).

Institut Francais du Caoutchouc (IFC), France.

Rubber Stichting, Holland (discontinued 1956).

Rubber Research Institute of Malaya.

Institut des Recherches sur le Caoutchouc au Vietnam (IRVC).

Institut des Recherches sur le Caoutchouc au Cambodge (IRCC) formerly Institut des Recherches sur le Caoutchouc en Indochine (IRCI).

Institut des Recherches sur le Caoutchouc en Afrique (IRCA).

Balai Penjelidikan dan Pemakaian Karet (BPPK), formerly Indonesisch Instituut voor Rubber Onderzoek (INIRO).

The main object in the work of the European institutes lies in developing uses and in the field of fundamental research, while the institutes in the tropics are concerned with production stages and growing of suitable plant materials. Apart from these institutes there were also regional institutes which, to a certain extent, worked in close contact with the other institutes:

Rubber Research Institute of Ceylon (RRIC), formerly Rubber Research Scheme (RRS).

Research Institute of the Sumatra Planters' Association (RISPA), formerly Algemeen Proefstation d. AVROS (APA).

Balai Penjelidikan Perkebunan Besar (BPPB), Java, formerly Proefstation der Centrale Proefstations Vereeniging (Proefstation CVP).

Several institutes also existed in S. America which were primarily concerned with plant development.

PRODUCT GLG Extremely glossy polyurethane lacquer for rubber articles which is applied to the rubber prior to vulcanisation and forms a protective layer (43).

PRODUCT GLI Solution of Desmophene in a non-flammable solvent.

Uses: in a blend with Desmodur R (100:18) or if the colour is important, Desmodur TH (in a slightly higher concentration) to give extremely glossy lacquer paint for rubber articles. The lacquer can be applied to the rubber either prior to, or after vulcanisation (43).

PROFAX Polypropylene, m.p. 166°C, s.g. 0·9. Tensile strength 360 kg/cm^2. Suitable for steam sterilisation (10).

PROKATALITH n-butyl lithium in a hydrocarbon solvent. Catalyst for diolefin polymerisation (304).

PROMOTER A substance which, in small quantities, either starts or increases the activity of a catalyst.

PROPATHENE Trade term for polypropylene (60).

PROPYLENE GLYCOL DIRICINOLEATE
$[C_{17}H_{32}(OH)COO]_2(CH_2CHCH_3)$, m.p. $-51°C$, s.g. 0·938 (25°C). Softener.

1 : -2 PROPYLENE GLYCOL MONOLAURATE

$C_{11}H_{23}COOCH_2CHOHCH_3$, m.p. 5–11°C, s.g. 0·91 (25°C). Softener. Insoluble in water.

1 : 2-PROPYLENE GLYCOL MONO-OLEATE

$C_{17}H_{33}COOCH_2CHOH.CH_3$, m.p. −20°C, s.g. 0·92 (25°C). Softener.

1 : 2-PROPYLENE GLYCOL MONOSTEARATE

$C_{17}H_{35}COOCH_2CHOH.CH_3$, m.p. 39°C, b.p. 182–210°C (11 mm), s.g. 0·93 (25°C). Softener.

PROPYLENE GLYCOL RICINOLEATE

$C_{17}H_{32}(OH)COOCH_2CHOH.CH_3$, m.p. −15°C, s.g. 0·96 (25°C). Softener.
TN: Flexricin 9 (9).

PROPYL GALLATE

White crystalline powder, m.p. 146–148°C, decomposes above 148°C. Antioxidant for natural and synthetic rubber. Used for articles which come into contact with foodstuffs.
TN: Tenox PG (191).

n-PROPYL OLEATE

$C_{17}H_{33}COOC_3H_7$, b.p. 181–189°C (2 mm), s.g. 0·861 (25°C). Insoluble in water. Softener.
TN: Emery 2303 (196).

PROREX OIL

Group of highly refined white mineral oils with low powers of dissolution, s.g. 0·871–0·882. Softeners (178).

PROSTHESE SYNESE

A vulcanisation process discovered in 1940 by the Institut Français du Caoutchouc. Consists in principle of the primary addition (prosthese) of a phenol with a group capable of reaction, *e.g.* resorcinol, to rubber, followed by condensation with formaldehyde (synese) during which the rubber molecules resinify while crosslinking takes place. Analogous to this process is the one developed by Rubber Stichting in which a phenol-formaldehyde resin, capable of direct reaction, is used as a crosslinking agent.

PROTOX

Group of surface treated zinc oxides (56).

PSANNSD-70

Masterbatch of 70% N-cyclohexyl-2-benzthiazole sulphenamide with polyisobutylene (282).

PSD-75

Masterbatch of 75% sulphur in polyisobutylene (282).

PSEUDOPLASTICITY

A liquid is termed pseudoplastic when its viscosity or consistency suddenly decreases when the shearing power (stirring) is increased.

PSI

Abbreviation for pounds per square inch. 1 psi is equivalent to 0·0703 kg/cm^2.

PSID-65

Masterbatch of 65% polymerised, insoluble sulphur in polyisobutylene (282).

PT

Trade term for a group of softeners, plasticisers and solvents based on dipentene, pine oil, pine tar and pine tar oil (139).

PTBP

4-tert-butyl phenol (176)

PTD

Abbreviation for dipentamethylene thiuram disulphide.

PTD-75 Masterbatch of 75% tellurium diethyl dithiocarbamate with polyisobutylene (282).

PTetD-70 Masterbatch of 70% dipentamethylene thiuram tetrasulphide with polyisobutylene (282).

PTFE Abbreviation for polytetrafluoroethylene.

PTM Abbreviation for dipentaethylene thiuram monosulphide.

PUCA SIRINGA Local term for *Hevea viridis* which occurs in the flood areas of the Amazon. Rubber produced from this plant is of inferior quality.

PULLMAN Butyraldehydeaniline (14).

PUREA Modified urea. 50% active content, s.g. 1·47 (136).

PURIFIED RUBBER Rubber with a low degree of water absorption. Obtained by removal of most of the non-rubber constituents (protein, resin, quebrachitol and salts) during production.
Hydrolysis by boiling with a solution of caustic soda. Fresh latex is boiled with approx. 1% caustic soda solution for a specified time, after which the excess caustic soda, the hydrolysis products and the salt are removed by dialysis. The dialysed latex is then diluted with water to a rubber content of 16%, coagulated by adding formic acid and processed to crêpe rubber. This process was used on the Tjipetir (Java) plantation up to 1940.
H rubber, by bacteria decomposition. (H after de Haan–Homans.) Fresh, diluted latex is stirred gently at pH 7·0 until the yellow fraction precipitates. The white fraction, which has excellent stability, is then stirred for 48 h, during which the proteins are hydrolysed by the action of bacteria. The latex is then stabilised with ammonia, centrifuged to remove hydrolysed products, diluted to a rubber content of approx. 16%, coagulated with formic acid and processed into crêpe rubber.
Lit.: Holl. P 63 023, 67 927 (1951).
G-rubber, by removal by soaps. (G after van Gils.) Even after constant centrifugation and dilution some protein remains in latex, since a certain percentage is adsorbed on the rubber particles. However, the protein can be removed by adding a surface active substance which is adsorbed more strongly than protein, *e.g.* by adding soap or an anionic wetting agent. In this process, fresh or concentrated latex which has been preserved with ammonia, is diluted to a rubber content of 10–12%, and 7·5–10% soap solution is added. After 12 h the latex is centrifuged, diluted again to 5%, coagulated and processed to crêpe rubber.
Holl. P. 53 090, 62 277.
Rubber which has little protein content takes almost twice as long to dry as normal crêpe. The ageing

	% N	% Ash	Water absorption mg/100 cm^2
Normal crêpe	0·35	0·30	500
With hydrolysed NaOH	0·10	0·10	100
H-rubber	0·18	0·07	100
G-rubber	0·02	0·05	60

properties deteriorate with the removal of the natural antioxidants, but may be improved, according to USP 2 560 744, by adding 0·002–0·2% sequesterol before coagulation. Compounds of these rubber types require somewhat higher than usual amounts of antioxidants.

Uses: electrical insulating materials, articles with a low water absorption.

PURUB PROCESS Historic coagulation process for latex using hydro-fluoric acid, according to Sandmann (*ca.* 1908).

PUSSEY-JONES PLASTO-METER Apparatus used to determine the hardness of rubber under spherical pressure, according to ASTM D 531–56. The penetration of a sphere of ⅛ in diameter with a load of 1 kg is measured to a hundredth of a mm. The term plastometer is incorrect but is used in the specification.

P.V.C. Abbreviation for polyvinyl chloride.

P.V.C. FIBRE Polyvinyl chloride staple fibre produced by a wet spinning process from a tetrahydrofuran solution (32), according to DRP 737 954 and 802 263.

PVME Abbreviation for polyvinyl methyl ether.

PX- Group of softeners:
104: dibutyl phthalate
108: diiso-octyl phthalate
114: decyl butyl phthalate
118: iso-octyl decyl phthalate
120: diiso-decyl phthalate
138: dioctyl phthalate
208: diiso-octyl adipate
209: dinonyl adipate
220: diiso-decyl adipate

238: dioctyl adipate
404: dibutyl sebacate
438: dioctyl sebacate
800: epoxy compound
912: tricresyl phosphate (305).

PYCAL Group of softeners:
29: phenyl ether of polypropylene glycol
40: bis(phenyl polyethylene glycol) diglycolate
60: diphenoxyethyl diglycolate
70: phenoxyethyl oleate
71: phenoxyethyl laurate
94: phenyl ether of polyethylene glycol
140: modified Type 40
170: modified phenoxyethyl fatty acid ester
194: modified Type 94 (97).

PYD-80 Masterbatch of 80% tellurium with polyisobutylene (282).

PYLEEN Aqueous polyethylene dispersion.
Uses: include the impregnation of paper, textiles, leather (306).

PYRANTON A Diacetonyl alcohol (217).

PYRATEX Vinyl pyridine latex used for the impregnation of tyre cord; improves adhesion between cord and the carcase, decreases the tendency for ply separation (23).

PZD-85 Masterbatch of 85% zinc oxide with polyisobutylene (282).

PZND Group of masterbatches with polyisobutylene:
75: 60% zinc oxide and 15% 2-mercaptoimidazoline
78: 65% zinc oxide and 13% 2-mercaptoimidazoline
84: 72% zinc oxide and 12% 2-mercaptoimidazoline (282).

Q

QUALITEX Trade term for a latex concentrated by centrifuging.

QUATRAX Tetramethyl-ammonium formate (6).

QUINOIDINE Mixture of alkaloids (81).

p-QUINONE DIOXIME

HO—N=⟨ ⟩=N—OH

Dark-brown powder, decomposes at 215°C and above, s.g. 1·21. Partially soluble in acetone, insoluble in water, benzine, benzene, and chlorinated hydrocarbons. Very fast vulcanising agent in the absence of sulphur and in combination with an inorganic oxide, *e.g.* lead peroxide; used for natural rubber, SBR, butyl and thiokol ST. Gives a high modulus and good heat resistance to vulcanisates, but stains and discolours on contact. Use of carbon black is advisable to obtain good physical properties. Optimum physical properties are achieved by combination with sulphur. Compounds have a fairly low critical temperature and tend to be 'scorchy'. Sulphur has a slight retarding effect. Phthalic anhydride, p,p′-diaminodiphenylmethane, N-nitrosodiphenylamine, some thiazoles, thiurams, and dithiocarbamates also increase processing safety. Acidic materials, including stearic acid and channel blacks, should be avoided as they increase the tendency to scorch.
Quantity: 2% with 10% red lead or 6% lead peroxide; Thiokol ST compounds 1·5% with 0·5% zinc oxide.
Uses: particularly butyl rubber, inner tubes, heating bags and other products in which fast vulcanisation, high modulus and good heat resistance are desirable.
TN: Actor Q (16)
GMF (23)
GMF 117 (33% stabilised with 67% alumina) (23)
Vulcafor BQ (60)
Vulcafor BQN (50% stabilised with an inorganic base) (60)
PQD
PGD-50 (50% master batch with polyisobutylene) (282)
Vulcanising agent CDO (43)
Vulnoc GM (274).

R

R-2 Group of accelerators:
+ Standard: reaction product of methylene dipiperidine and carbon disulphide. Ultra-fast accelerator for latex (obsolete)
+ —C: undisclosed composition (obsolete)
— Crystals: reaction product of methylene dipiperidine and carbon disulphide. Pale brown, coarse crystals with characteristic odour, m.p. 59°C, s.g. 1·22. Soluble in alcohol, acetone, benzene and benzine. Ultra-fast accelerator of a complex structure for latex and cements. Gives vulcanisates with excellent properties. *Quantity:* latex 0·75% with 1% zinc and 1·5% sulphur; cements 1·25% with 5% zinc oxide and 3% sulphur (5).

R-23 Sodium salt of mercapto-benzthiazole (5).

RA Abbreviation for 'Raja Allang' clones.

RABBIT BRUSHES Term used in the USA for *Chrysothamnus* types (*Compositae*) which yield rubber.

RABRM Abbreviation for Research Association of British Rubber Manufacturers (later known as Rubber and Plastic Research Association of Great Britain).

RADAX Thiazole-thiurame blend. Pale yellow powder, s.g. 1·37. Soluble in benzene, chloroform, acetone and carbon disulphide, sparingly soluble in alcohol, water and dilute alkalis. Accelerator for natural rubber and isobutylene-isoprene co-polymers, particularly for compounds with a low sulphur content and articles in which a matt surface appearance should be avoided.
Uses: inner tubes, cable compounds (5).

RADIAL TYRES Tyres in which the plies of the textile are placed at right angles to the direction of movement; in a normal carcase the plies are arranged diagonally to the direction of movement, and alternate plies cross each other at an acute angle.

RADIUS Static—the distance of the centre of a wheel from the ground at normal, maximum load and corresponding air pressure, while in a static position. Dynamic—the distance between ground and centre in moving tyres, in practice for speedometer calculation, for

commercial vehicles (50 km/h) and private cars (60 km/h), determined from the circumference, divided by 2. The dynamic radius is greater than the static, as the central point of the wheel is raised by the diametral growth caused by the rolling action. It is dependent on the construction of the tyre and the speed.

RAMFLEX V-17 A reclaim soluble in asphalt. Used in road building (117).

RAM MIXER Rubber mixing machine, with a closed mixing chamber, in which the ingredients to be mixed are pressed down in the mixing chamber by a ram, *e.g.* Banbury internal mixer, Werner and Pfleiderer ram mixer.

RAOLIN Chlorinated rubber (308).

RAPID H-2 Accelerator of undisclosed composition (309).

RAPIDEX Group of Japanese produced accelerators:
GR: tetramethyl thiuram disulphide
H: hexamethylene tetramine
KA: diphenyl guanidine
KB: mercaptobenzthiazole
KY: dibenzthiazyl disulphide (391).

RAPRA Abbreviation for Rubber and Plastic Research Association of Great Britain (formerly RABRM).

RAYON Term for artificial wool or artificial silk produced using the viscose process. The term is also widely used for copper and acetate silk.

RC PLASTICISERS Group of softeners for natural and synthetic rubbers and vinyl resins:

B-17: technical butyl stearate, b.p. 220°C (5 mm), s.g. 0·866
D-31: blend of diisodecyl phthalate and adipate
DBP: dibutyl phthalate
DOA: di-2-ethyl hexyl adipate
DOP: di-2-ethyl hexyl phthalate
DOS: di-2-ethyl hexyl sebacate
ES: softener on epoxy basis
O-16: iso-octyl palmitate
TG-B triethylene glycol dicaprylate
TG-9: triethylene glycol diperlargonate
TG-85: triethylene glycol caprylate caprate (110).

READY RUB Stimulant used to increase latex production; based on 2:4:5-trichlorophenoxyacetic acid.

RECLAIM Reclaimed rubber. Obtained from ground vulcanised scrap from the production of natural and synthetic rubber articles, old rubber tyres, tubes and other such articles. Reclamation is achieved by replasticisation (depolymerisation), using heat and/or pressure and/or chemical agents; 'devulcanisation' does not occur because the bound sulphur is not removed. A compounding ingredient of considerable value whose advantageous properties are, according to some specialists, partially due to the presence of bound sulphur. Reclaim assists the mixing, extruding and calendering processes, by reducing the mixing time, decreases the power consumption, reduces heat development, decreases the thermoplasticity, reduces die swell on extrusion and calender shrinkage, results in excellent dispersion of fillers.

Reclaim is a 'premixed' rubber, and gives compounds improved shape stability (important for extruded tubes and profiles). It decreases the volume cost and is particularly important because of its stable price and uniform properties. As well as reducing the cost of raw materials, the processing costs of rubber articles are lessened, and the tendency to produce poor quality or second rate goods is decreased. Vulcanised reclaimed rubber shows a decrease in tensile strength, elongation at break and abrasion resistance, but when blended with new rubber, a marked fall in properties occurs only when a high proportion of reclaim is used in the compound; but even these values are acceptable for many different groups of articles, however. Hardness and resilience are similar to those of fresh rubber. Reclaim has good ageing properties which result from the intensive treatment to which it is subjected during the reclaiming process (oxidation, heating, washing, mastication in air), which stabilises the hydrocarbon in the rubber against further change. First quality whole tyre reclaim contains approx. 50% w/w rubber hydrocarbon. Non-rubber constituents (the remaining 50%) consist of zinc oxide, antioxidants, blacks and softeners which remain virtually unchanged by the reclaiming process and therefore retain their functions in further processing as in fresh materials. Compounds which contain reclaim generally cure faster than compounds which are free from reclaim; applies to natural and synthetic rubbers, and again is due to the accelerators which remain in the reclaim. The effect is frequently used to reduce the quantity of accelerator needed. Since reclaimed rubber mainly retains the properties of the original elastomer, pure NR reclaim always vulcanises more quickly than pure SBR reclaim. On reclaiming NR vulcanisates the plasticity of the reclaim increases as reclaiming proceeds, whereas synthetic rubbers show a decrease in plasticity after an initial marked increase because of cyclisation.

The first patent for a reclaiming process was taken out by A. Parkes [BP 11 147 (1846)]; the scrap rubber was heated under pressure with calcium hypochlorite until the mass could be kneaded and was then washed with alkali and hot water. According to W. Christopher and G. Gidley [BP 1461 (1853)] the scrap rubber was boiled with soda or lime to remove the free sulphur, and copper oxide was used to take up the hydrogen sulphide. The rubber was finally dissolved in naphtha or turpentine. A. V. Newton treated the scrap for 2–14 days with camphor oil [BP 1687 (1854)] or caused finely crumbed rubber to swell in wood tar, raw turpentine, resinous oil or pine oil for 4–5 days [BP 158 (1860)]. M. Henry [2634 (1862)] treated the scrap with petroleum residue and tar under pressure. According to C. Heinzerling and H. Liepmann [BP 2495 (1875)] finely ground scrap was boiled for a few hours in a 10% soda solution. The resulting dry product was dissolved by heating at 80–100°C, e.g. in benzene, turpentine, then freed by sedimentation of mineral fillers. BP 525 (1878) and BP 2340 (1878) suggest the use of hot hydrochloric acid to destroy textiles and remove metallic oxides. The pretreated rubber is dissolved by heating in test benzine, carbon disulphide or benzene and the solvent removed by evaporation.

Technical processes Before process-ing, the different scraps are sorted according to their original production, freed from metals and ground to a fine powder or to crumb. The process used depends on the possible presence of textiles. After reclamation and resting, the disintegrated rubber, mixed with softeners, is homogenised in an internal mixer or on rolls. In special cases, fillers are added and mixed until the mixture will form a sheet; treatment is concluded with refiners and strainers. Various oils, softeners and chemically effective peptisers are used as reclaiming agents, depending on the processes used.

1. *Pan (Heater) process.* Depends in principle on the process of H. L. Hall [USP 22 217 (1858)] and is used to reclaim textile-free scrap. Finely ground rubber is mixed with a small quantity of softener and with caustic soda and heated in pans in horizontal autoclaves for several hours in steam at 175–210°C.

2. *Acid process.* According to N. C. Mitchell [USP 249 970 (1881)] this process is based on the treatment of the rubber with a 10–25% sulphuric acid without pressure, or under slight pressure, at approx. 95°C for 4–10 h. Apart from destroying any textile present the mineral fillers are also partially dissolved out. The reclaim is carefully washed, neutralised with alkali and plasticised in autoclaves with saturated steam at 175–210°C.

3. *Alkali process.* According to A. H. Marks and R. B. Price [USP 635 141 (1899)]; for a long time this was the most important commercial process, but since the increase in the use of synthetic rubber, which is cyclised by alkalis, it has lost its significance. The coarsely ground scrap is treated in autoclaves for 8–24 h at 150–220°C with a 4–10% caustic soda solution

and reclaiming oils to destroy the fabric present in the scrap. The reclaim is washed well and dried.

4. *Neutral process.* Based on the patent of D. A. Cutler [USP 673 057 (1913)] but, apart from the use of 1–2% zinc chloride solution, similar to the alkali process. Because synthetic rubber hardens to a lesser than usual degree in this process, it is also suitable for reclaiming blends. The reclaim is more tacky than that produced by the alkali process.

5. *Bemelman's process.* The unground rubber is cut into fairly large pieces and heated in autoclaves with 2% naphthalene in a dry atmosphere of carbon dioxide and ammonia (produced from ammonium carbonate) at 200–250°C for approx. 2 h [BP 435 890 (1934)]. The fibres are charred by the naphthalene.

6. *Superheated steam process.* Because most synthetic rubbers harden when exposed to high temperatures, this is suitable only for natural and butyl rubbers. The scrap, which is coarsely ground, is heated at 3–15 atm in superheated steam at 230–250°C for 5–6 h.

7. *Palmer (high pressure steam) process.* The scrap is ground to coarse particles and treated in small, vertical retorts with steam at 40–50 atm, so that the textile is carbonised. The reclaim is normally more plastic than that produced by other processes. This process can only be used for NR (SR would become cyclised).

8. *Banbury–Lancaster process.* The scrap is plasticised in a specially designed internal mixer having a considerably increased rotor speed and pressure, reclaiming agents being added [BP 577 829 (1942) and USP 2 221 490 (1937)]. Friction results in a temperature rise to 240–290°C.

This process is used almost exclusively for the reclaiming of scorched compounds and highly loaded scrap. Kneading time per load is approx. 30 min.

9. *Reclaimator (dip process).* Depends on a plasticity phenomenon. If vulcanised natural or synthetic rubber is heated under specified conditions considerable increase in plasticity quickly occurs and is followed by a sharp decrease in plasticity, after which the plasticity slowly increases as treatment proceeds (USP 2 415 449). The process depends on the sudden termination of the reclaiming process at the plasticity optimum. The finely ground, textile-free scrap is plasticised in a plasticator (reclaimator), which is similar to an extruder with increasing screw pressure for a few minutes at 155–205°C. It is then quickly cooled before discharge.

10. *Solution processes.* According to various processes the scrap is treated with suitable solvents at 130–160°C so that a large proportion dissolves and is plasticised. This process is uneconomic because of the recovery of solvents and is rarely used.

RECLAIMATOR Extruding machine fixed with special screws and used for the reclamation of ground scrap rubber according to the dip process, has a capacity of 150–600 kg/h.

RECLAIM DISPERSION Aqueous dispersion of reclaim with 30–70% dry content. Has a greater tack than latex.

Uses: include impregnation, adhesives, dipped articles, carpet binding, addition to bitumen paints.

RECLAIMING Transformation of vulcanised rubber into a plastic mass which may then be revulcanised. According to current theories the breakdown of a three-dimensional network may be involved.

RECONSTRUCTION FINANCE CORPORATION Appointed by the US Government during World War II to obtain strategic materials; included the Bureau for Synthetic Rubber.

RED LEAD Pb_3O_4, probably a lead-orthoplumbate with the formula $Pb_2(PbO_4)$. Loose bright red powder, m.p. 830°C, s.g. 8·8–9·2. Insoluble in water. The trade products normally contain small quantities of carbonate and lead oxide.
Uses: vulcanising agent, activator and pigment.

REDOX POLYMERISATION Reduction-oxidation polymerisation; the initiation of a polymerisation reaction by an oxidation-reduction reaction giving free radicals, *e.g.* in the polymerisation of SBR cold rubber. A peroxide, such as cumyl hydroperoxide, benzoyl peroxide and other hydroperoxides, is used as the oxidising agent, reducing agent consisting of a heavy metal complex in combination, for example, with an aliphatic acid (levulinic acid, succinic acid) reducing sugar, a sulphur compound (cystine, β-mercapto-ethanol), cholesterol, hydroxyamino acids, dicyanodiamidine, hydantoin.

REDUX Blend of phenol-formaldehyde resin and vinyl polymers. Hardens in 4–20 min at 140–185°C. Resistant to water, oils and normal solvents.

Uses: bonding agent for metals, plastics and wood (94).

REFERENCE FLUID Immersion liquids used for testing synthetic rubbers. No. SR-6, blend of 60% diisobutylene with 40% aromatic material, No. SR-10, diisobutylene (131).

REFINER A machine having two rolls with a space (nip) between each other, a high speed of revolution and a high friction ratio (1:2–1:3). Used to crush and flatten impurities in rubber compounds, *e.g.* grit and nodules of rubber. Coarser impurities are forced to the sides of the rolls and removed. Apart from this, the refining gives good homogenisation of the compounds. The sheets which are taken from rolls are 0·5–1 mm thick. Impurities and nodules which have been flattened by the rolls are removed from the compound by straining after refining.

REGAL Non-classified oil blacks with a low carbon structure:
300: gives similar physical properties to channel blacks but a faster cure
600: similar in particle size to ISAF but gives a lower modulus, lower hardness and higher tensile strength (139).

REGITEX Group of natural latices which have been prevulcanised by peroxides. Type OR (Japanese produced) is a blend with an acrylonitrile-chloroprene copolymer.

REIABO Local term for *Landolphia spaeocarpa* (W. Madagascar) and the natural rubber obtained from it.

REINFORCEMENT Addition of fillers with a small particle size or resins to compounds increases abrasion resistance and tear strength and stiffens the vulcanisate. Reinforcing fillers are carbon blacks (colloidal carbon), calcium silicate, aluminium compounds (alumina) zinc and iron oxides, magnesium and calcium carbonates, lignin and various synthetic resins.

REITHOFFER, JOHANN NEPOMUK 1781–1872. Founder of the Austrian rubber industry. Originally a tailor, became interested in chemistry while in Paris for a year, and attended lectures at the Sorbonne. Settled in Nikolsburg, and made attempts at proofing materials. In 1824 received an Imperial Warrant authorising him to make woolcloth into waterproof clothing. Further patents resulted following the production of threads for elasticated fabrics (1828), rubber shoes, etc. He established a small factory and shop in Vienna; later a factory was built at Wimpassing, Lower Austria, and is now the focal point of the Austrian rubber industry.

RELAXATION If a rubber strip is held stretched at a constant length, the stress decreases as a function of time. The relaxation of the stress is explained by internal flowing, a rearrangement of the groups of molecules. With vulcanised rubber the relaxation is slight because mobility is reduced by the linear molecules being bound by crosslinks. In unvulcanised rubber, however, relaxation of the molecule occurs rapidly. Relaxation is dependent on temperature and time.

RELAXOMETER RelGi Apparatus used to determine the deforma-tion properties, stress relaxation at constant deformation, and Defo plasticity of elastomers (398).

REMILLING The processing of rubber scrap or the rubber slabs produced on the small holdings, into crêpe rubber, normally of a lower quality. The rubber is washed on ribbed rolls running at friction speed for at least 8–15 revolutions, and rolled out to thick or thin sheets which may be pale brown to black after drying (remilled thick brown crêpe, remilled thin brown crêpe, ambers, blankets). Scrap of sheet rubber or poor quality material is processed into remilled smoked blanket. Although there are remilling factories in Sumatra, Borneo and Singapore to process the slabs which normally contain a great deal of impurities, scrap is frequently processed on the plantations.

RENACIT Group of chemically active plasticising and reclaiming agents (peptisers) (43):

I: thio-β-naphthol

II: mixture of 67 parts trichlorothiophenol isomers with paraffin

III: mercaptoanthracene with paraffin and Plasticator RA

IV: zinc pentachlorothiophenate

IV/R: blend with 67 parts zinc pentachlorothiophenate with stearic acid and paraffin

V: pentachlorothiophenol with a stabiliser

VI: mixture of dixylyl disulphide. Brown, odourless liquid. Reclaiming agent for NR, SBR, CR and butyl; also an extender. *Quantity:* NR 0·15–0·5% and 5% reclaiming oils,

SBR 1·5–2% and 20–25% reclaiming oils, butyl 0·5%, NR/SBR blends 0·35–1% and 10–15% reclaiming oils

VI/N: dixylene disulphide mixture. Yellow fluid, s.g. 1·12. Reclaiming agent for NR and SR. *Dosage:* NR, IR 0·15–0·5, NR and SBR 0·5–1·0, SBR 1·5–2·0 BR 2·0, butyl 0·5–1·0, NBR 2·0–4·0, CR 2·0–4·0. Also used for scorched compounds. Has no influence on ageing

VII: pentachlorothiophenol with activating and emulsifying additives, s.g. 2·33 for NR and SR. *Quantity:* NR 0·05–0·5%, SBR 1·5–2%.

REOGEN Blend of a high mol. wt., oil soluble sulphonic acid with paraffin oil. Red brown liquid, s.g. 0·82–0·85, acid No. 8·0–8·4. Non-blooming and non-staining softener used for all rubber types. Retards scorching in natural rubber and reclaim.
Quantity: natural rubber 1–3%, reclaim up to 5%, in concentrations of 5% and above it is used as a degrading and plasticising agent for synthetic rubbers (51).

REOMOL Group of softeners:
DBP: dibutyl phthalate
DBS: dibutyl sebacate
DCP: dicapryl phthalate
DEP: diethyl phthalate
DIOP: diiso-octyl phthalate
DIOS: diiso-octyl sebacate
DMP: dimethyl phthalate
DOS: di(2-ethyl hexyl) sebacate
DOP: di(2-ethyl hexyl) phthalate
E: alkyl glycol phosphate
G: phenol ether
J: alkyl glycol phosphate
NF-1: trichlorophenyl phosphate
NF-2: trichloroethyl phosphate
P: di-2-methoxyethyl phthalate (227).

REPEATING UNIT Mer. The smallest molecule or molecular group of a macromolecule.

REPPE PROCESS Process for the production of butadiene from acetylene. Acetylene and formaldehyde react in the presence of copper acetylide to form 2-butyn-1:4-diol which is hydrogenated catalytically to 1:4 butanediol. This is then dehydrated to butadiene; tetrahydrofuran being formed as an intermediate product.

$$HC \equiv CH + 2HCHO \xrightarrow{Cu_2C_2}$$
$$HOCH_2C \equiv CCH_2OH \xrightarrow{+H_2}$$
2-Butyn-1:4-diol

$$HO(CH_2)_4OH \xrightarrow{-H_2O}$$
1:4-Butanediol

$$CH_2CH_2CH_2CH_2 \xrightarrow{-H_2O}$$
$$\lfloor\!\!\underline{\qquad O \qquad}\!\!\rfloor$$
Tetrahydrofuran

$$CH_2 = CH - CH = CH_2$$
Butadiene

Lit.: USP 2 232 867, 2 319 707, 2 300 969, 2 251 835.

RESILIENCE Rebound elasticity. To determine the elastic behaviour of soft rubber; under impact the ratio of recovered work to expended work is measured, *i.e.* the ratio of the height of rebound H_r of a pendulum to the height of fall H_f:

$$R = \frac{H_r}{H_f}$$

In the above formula the damping effect is not taken into account and the result must therefore be corrected. Various instruments may be

used to measure resilience, *e.g.* the Dunlop pendulum, Lüpke pendulum, Dunlop tripsometer, Zwick's pendulum, Goodyear pendulum, Schob's pendulum. Highly elastic rubber test pieces have 60–75% resilience, moderately elastic rubber test pieces, 40–60% resilience, low elasticity rubber test pieces, 40% resilience. The elastic behaviour of rubber depends on temperature; in the most important temperature range of $-20°$ to 20°C elasticity increases linearly.
Specification: BS 903: 22: 1950, DIN 53 512.

RESINEX Reinforcing, polymerised resin in liquid and solid forms, m.p. up to 115°C.
Uses: reinforcing filler and processing aid for natural rubber, SBR, reclaim, neoprene, nitrile rubber, polyvinyls and other thermoplasts (127).

RESINITE Russian produced thioplasts.

RESISTANCE TO CHAFING Resistance to disintegration of the insulation wound round a wire.

RESISTOX Aldol aniline. Antioxidant (3).

RESORCINOL m-dihydroxybenzene. White, crystalline powder, m.p. 109–110°C, b.p. 280°C, s.g. 1·272. Soluble in alcohol, water and glycerol. Preservative for NR, SR and latices. Has no effect on vulcanisation and physical properties of the vulcanisates.

RETARDAC Canadian produced phthalic anhydride derivative. Vulcanisation retarder.

RETARDER In general a substance used to retard the speed of chemical reactions; in the rubber and polymer industry, a substance used to retard vulcanisation and polymerisation speeds and to prevent scorching during the mixing process.

RETARDER Group of retarders produced by various manufacturers:
A: acetyl salicylic acid
AFB: phthalic anhydride (6)
AN: N-nitroso diphenylamine (440)
ASA: acetyl salicylic acid (5)
B: monobutyl phthalate (6)
BA: 90% benzoic acid and 10% stearic acid, m.p. 120°C, s.g. 1·16. *Quantity:* 1–2%
BAX: benzoic acid (54)
D: octadecylamine (98)
J: N-nitrosodiphenylamine (246)
PA: phthalic anhydride (5)
PD: blend of phthalic anhydride with a modifier and anti-dust agent, m.p. 125–130°C, s.g. 1·48. Used for NR, SBR and NBR (22)
RM: undisclosed composition
TCM–25: blend of 25% trichloromelamine with 75% barium sulphate. Used for NR and SR (3)
TSA: 90% technical salicylic acid and 10% stearic acid (5)
UTB: phthalic acid
W: salicylic acid with a dispersing agent (6).

RETARDEX Dispersion of benzoic acid in an oil. Retarder (100).

RETILEX F 40 Peroxide crosslinking agent. Contains 40% bistertbutyl peroxide of diisopropyl benzene and 60% calcium carbonate as carrier.

RETREADING Recapping. The renewal of the treads and cushion of worn tyres.

REVERTEX A latex concentrated by evaporation and having high mechanical and chemical stabilities, contains all non-rubber constituents. Particularly suitable if a high amount of friction may be expected on processing, as in coating compounds and compounds with a high filler loading, *e.g.* cements. Is excellent for carpet underlays and as a basis for adhesives in which the non-rubber constituents prove useful. Standard type: total solids 75%, rubber 66%, preserved with 1·5% potassium hydroxide and a stabiliser. Paste-like consistency

Revertex T: total solids 60–62%, rubber 56–58%, preserved with ammonia.

REVULTEX Prevulcanised Revertex latex.

REX HARDNESS TESTER Instrument used to determine the hardness of rubber on the Shore A Hardness (durometer) scale; conforms to ASTM D 676. The instrument is in the form of a small cylinder similar to a fountain pen. The hardness indicator springs out of the upper end when a measurement is made and remains in that position so that the maximum value may be read.

REYON Endless filament synthetic cellulose fibre produced by the viscosity process.

RFC Abbreviation for Reconstruction Finance Corporation.

RFS SYSTEM HRH system. Bonding procedure for rubbers; based on a synergistic action of resorcinol, formaldehyde and silica. The system, directly mixed into polymer batch, can eliminate the need for textile pretreatment.

RH 40 AND 50 D Acetaldehyde-aniline condensation products. Accelerators.

RHC Abbreviation for rubber hydrocarbon or rubber hydrocarbon content.

RHENOCURE Group of rubber chemicals (42):

BQ: dibenzoyl-p-quinone dioxime
S: N,N′-dithiobis(hexahydro-2H-azepinon). Non-blooming, non-discolouring sulphur donor
TP: non-discolouring special accelerator for EPDM
ZMC: zinc dimethyl dithiocarbamate
PA: semi-ultra accelerator of the alkyl amine type.

RHEOGONIOMETER (Weissenberg), instrument for precise measurement of viscosity, elasticity, and flow properties of materials. An advanced model is known under the code DDV-II-B (Sangamo Controls Ltd).

RHEOMETER Instrument used to determine the viscoelastic properties of elastomers.

RHODESTER Group of polyester resins (20).

RHODIFAX Group of accelerators:

2: accelerator with delayed action
3: accelerator with delayed action
6: MBTS, DPG and an acidic adipate
7: MBTS, DPG and a neutral adipate
10: MBTS, DPG and HMT
11: blend of three accelerators
12: diethyl benzthiazyl sulphenamide and diethylamine
14: 2 parts MBTS and 1 part tetramethyl thiuram disulphide
15a: accelerator with delayed action
16: N-cyclohexyl-2-benzthiazole sulphenamide (20).

RHODOPAS Group of vinyl polymers and copolymers, including dispersions (20).

RHODORSIL ELASTOMERS Silicone rubber.
RP-60: Shore A Hardness 60°C
RP-70: Shore A Hardness 70°
RP-80: Shore A Hardness 80° (20).

RHODOVIOL Group of polyvinyl alcohols (20).

RHOVYL Formerly Rhofil. P.V.C. artificial silk (230).

RIA 66 66% dispersion of urea in oil. Homogeneous paste, s.g. 1·20. Activator for blowing agents, lowers the decomposition temperature (237).

RIA NC Urea (237).

RIDACTO Liquid amino compound, s.g. 1·045. Used in natural and synthetic rubbers as a booster for thiazoles, thiurams and dithiocarbamates. Balances the cure of SBR/ natural rubber blends and improves ageing properties. Vulcanisates have a flat modulus curve (310).

RIDLEY, HENRY NICHOLAS 1855–1956. From 1888, director of the Botanical Gardens, Singapore; introduced *Hevea Brasiliensis* into plantation economy, through a result of his tapping experiments on the plants brought into the country by Wickham, and through his continuous propaganda. On his hundredth birthday was awarded the Colwyn Medal in recognition of his services. As a botanist he is known for his standard works on the Malayan flora, 'Flora of Malaya', 1922–25, 5 vols., 'The Dispersal of Plants throughout the World', 1930, 'Spices'; editor of the *Agricultural Bull.* (from 1891). His first scientific work *Mammals and Coleoptera of Haileybury* was written at the age of 16.

RIEMANN'S GREEN Cobalt green. Bluish green pigment obtained by heating zinc oxide with cobalt salts. Contains approx. 5 parts zinc oxide to 1 part cobalt oxide.

RIKEN 200 Zinc ethyl phenyl dithiocarbamate.

RILSAN Polycondensation product of 1:12-aminoundecanoic acid. Transparent plastic. Has good resistance to water, fats, oils and benzine and is stable from −50°C to 120°C. *Uses:* include surgical instruments, gaskets, insulating materials, cables, filters, textiles, corrosion resistant linings (Rilsenit) and in the car industry (48, 244).

RIM STRIP Flap. Smooth or corrugated strip, either V- or U-shaped, with a valve opening, used for protection of the inner tube against chafing by the rim.

RIONOX N-phenyl-β-naphthylamine (418).

RIO RESIN Mixture of resinous substances and protective agents. Orange-red to dark red resin, m.p. approx. 70°C, softening p. 54°C, s.g. 1·13. Protective agent against heat, light and ozone, for butadiene-styrene copolymers, nitrile rubber and neoprene. Increases the corona resistance of neoprene and butyl rubber. *Quantity:* neoprene 5–10%, other polymers 10–20% (51).

RMA Abbreviation for Rubber Manufacturers Association, New York.

RMB Abbreviation for resorcinol monobenzoate.

RMD-4511 Styrene-acrylonitrile copolymer with high strength and high chemical resistance (112).

RN-2 AND RN-2 CRYSTALS Reaction products of carbon disulphide and methylene di(N-methyl-cyclohexylamine). Ultra-fast accelerator (5).

RN-7 Reaction product of carbon disulphide and N-methylcyclohexylamine. Ultra-fast accelerator (5).

ROBAC Trade term for a group of accelerators:
22:    mercaptoimidazoline
44:    diaminodiphenyl methane

70:    neoprene accelerator of undisclosed composition. Insoluble in water, aliphatic and aromatic hydrocarbons. Non-discolouring. Physical properties as those of Robac 22. *Dosage:* 1%.

alpha: mixture of 2-mercapto-benzthiazole with dipentamethylene thiuram disulphide. Yellow powder with characteristic odour, m.p. 135–145°C, s.g. 1·05. Fast to ultra-fast, has a delayed action and gives good processing safety, zinc oxide is necessary. *Quantity:* 0·5–1% in compounds which contain no absorbing fillers

CAMD: calcium dimethyl dithiocarbamate

CPD: cadmium pentamethylene dithiocarbamate

CS: undisclosed composition, s.g. 1·32. Soluble in alcohol, acetone and benzene, insoluble in water. Has a good delayed action and gives good processing safety. Non-staining. Suitable for press cures and hot air vulcanisation of proofed textiles. *Quantity:* 0·5–1% with 2·5–1% sulphur

Cu DD: copper dimethyl dithiocarbamate

DAMD: dimethylaminomethylene dimethyl dithiocarbamate

DBMD: di-N-butylammonium di-N-butyl dithiocarbamate

DBUD: dibutylammonium dibutyl dithiocarbamate

DETU: N,N'-diethyl thiourea

DFTU: substituted thiourea, m.p. 67–72°C, s.g. 1·23

gamma: mixture of zinc salts with 27% piperidine. White powder with a pungent odour, m.p. 200–205°C, softening p. 90–95°C, s.g. 1·49. Soluble in benzene, insoluble in water. Extremely active, suitable for low temperature vulcanisation, zinc oxide is necessary in solid rubber. May be retarded by Revertex. Suitable for centrifuged latex. *Quantity:* solid rubber 1–2%, latex 0·25–0·5%. *Uses:* self-vulcanising adhesives, textile coatings

GDMA: glycol dimercaptoacetate

GP: mixture of 20 parts 2-mercaptobenzthiazole with 80 parts of a formaldehyde-aniline condensation product

J: lead dimethyl diphenyl thiuram disulphide

LMD: dimethyl dithiocarbamate

LPD: lead pentamethylene dithiocarbamate

MZ 1: zinc-2-mercaptobenzthiazole

MZ 2: blend of zinc-2-mercaptobenzthiazole with 9% di-o-tolyl guanidine. Pale yellow powder, s.g. 1·61. Sparingly soluble in alcohol. Extremely active above 120°C, boosted by thiuram and guanidine; stearic acid is necessary. Gives good processing safety on mills. *Quantity:* 0·5–1% with 2·5–1·5% sulphur

NiBUD: nickel dibutyl dithiocarbamate

NiDD: nickel dimethyl dithiocarbamate

NiPD: nickel pentamethylene dithiocarbamate

NPD: nickel pentamethylene dithiocarbamate

PPD: dipentamethylene thiuram disulphide

PTM: dipentamethylene thiuram monosulphide

SBUD: sodium dibutyl dithiocarbamate

SPD: sodium pentamethylene dithiocarbamate

TBUT: tetrabutyl thiuram disulphide

TBZ: zinc thiobenzoate. Peptiser

TET: tetraethyl thiuram disulphide

Thiuram P 25: dipentamethylene thiuram tetrasulphide

TMS: tetramethyl thiuram monosulphide

TMT: tetramethyl thiuram disulphide

ZBED: zinc dibenzyl dithiocarbamate

ZBUD: zinc dibutyl dithiocarbamate

ZBUD extra: zinc dibutyl dithiocarbamate complex with dibutylamine in an equimolecular ratio. Golden brown liquid, s.g. 1·09. Immiscible with water, miscible with benzene. Cures at low temperatures (18–100°C)

ZDC: zinc diethyl dithiocarbamate

ZIX: zinc isopropyl xanthate

ZL: zinc lupetidine dithiocarbamate

ZMD: zinc dimethyl dithiocarbamate

ZPD: zinc pentamethylene dithiocarbamate

ZPD extra: zinc pentamethylene dithiocarbamate. Contains piperidine (34).

**ROCKWELL HARDNESS
METER** Instrument used to determine the spherical-pressure hardness
of hard rubber, according to ASTM
D 530-60 T; E 18. The difference is
measured in the depths of indentation
of a sphere of $\frac{1}{4}$ in diameter between
loads of 10 and 60 kg.

RODANIN S 62 2-mercaptoimi-
dazoline (428).

RODFORM Group of accelerators in a dust-free form:
Altax: di-2-benzthiazyl disulphide
Bismate: bismuth dimethyl dithiocarbamate
Butyl Zimate: zinc dibutyl dithiocarbamate
Ledate: lead dimethyl dithiocarbamate
Methyl Selenac: selenium dimethyl
dithiocarbamate
Methyl Tuads: tetramethyl thiuram
disulphide
Methyl Zimate: zinc dimethyl dithiocarbamate
Unads: tetramethyl thiuram monosulphide (51).

RODORUB Trade name for
rubber powder (400).

ROELIG MACHINE Apparatus
used to determine the dynamic
properties of rubber, using forced,
non-resonant, sinusoidal oscillations
between 5–20 c/s.

ROLLED RUBBER Rubber, obtained from various wild rubber
trees, which coagulates on the bark
and is then wound into bales.

ROLLING RESISTANCE
Resistance of a loaded tyre to rolling.
The vehicle horse power absorbed by
the rolling tyres increases markedly
with speed.

$$\frac{\text{Rolling resistance power}}{\text{Wheel load}}$$
$$= \text{Rolling resistance factor (f)}$$

Rolling resistance is a function of
tyre construction, of the compounding (depending particularly on the
tread mix and whether it is hysteretic),
the velocity and road surface.

RONINGER METHOD Process
used to examine the dispersion of
fillers in rubber compounds. A
rubber test piece is dipped for 12–
24 h in molten sulphur at 135°C. The
surface of the hardened test piece is
finely ground and polished. The
surface of the undispersed filler is
examined under a reticulated microscope, which gives a 600% magnification. The values are reproducible and
give a good overall picture of the
degree of dispersion.

ROOT RUBBER Caoutchouc
des herbes. Wild rubber obtained
from the roots of *Landolphia Tholonii*
(Congo), *Mascarenhasia Geayi*, *Mascarenhasia kidroa*. Used to be
obtained by primitive methods, as
was root rubber from the root
nodules of the Kok-Saghyz and Tau
Saghyz in Russia. To obtain rubber
from Composites, the roots were
worked for 4 h in a 2% solution of
caustic soda, then washed with cold
and hot water on a grooved mill.
The mass of tissue and rubber was
suspended and removed from the
cellular material by steam. The
rubber, which floated on the surface,
was purified with steam and processed to crêpe on smooth rolls.

ROSS FLEXER Continuous flexing machine used to determine the

crack growth of a test piece aged at 100°C, gives 5 flexes/min. Flat test pieces of approx. $2 \cdot 5 \times 15 \times 0 \cdot 6$ cm ($1 \times 6 \times 0 \cdot 25$ in) are clamped on one side and are bent over a $9 \cdot 5$ mm ($\frac{3}{8}$ in) bar at an angle of 90°. The free end of each test piece is placed between two rollers to allow an unstretched flexing operation. The test pieces are previously aged for 24 h at 100°C. The number of flexes which increases the size of a slit in the test piece by steps of 100% up to 500% total is determined. ASTM D 1052-55.

ROTATION VISCOMETER A viscometer in which the shearing force between two concentric cylinders is measured. The torsion produced by the liquid on the inner cylinder, rotating at a constant speed is directly proportional to the viscosity at a given shearing force. Instruments based on this principle are the Brookfield and Ferranti viscometers.

ROTOCURE Machine used for the continuous vulcanisation of articles with a large surface area; *e.g.* conveyor belts, floor coverings.

ROTOMILL A continuous internal mixer. Consists of a rotor with rounded ribs in spiral formation which turns in a container. The machine can be filled continuously with rubber and the compounding ingredients through two filling openings. At the lower end the compound is forced out at a side opening. Between the ribs and the wall of the container a shearing effect is produced similar to that in the nip of a mill; the rubber or compound is transported further along the container by the spiral effect of the ribs.

ROUGE DE KASSAI Rouge du Congo. Trade term for wild rubber from *Landolphia Gentilii* (Congo and W. Africa).

ROUGHNESS INDEX Ratio of the surface area given by an electron microscope determination (magnified $\times$ 50 000–75 000) to the filler surface determined, using nitrogen adsorption isotherms.

ROUND-SHOULDERED TYRES Tyres with a wide tread contact area. Have improved cornering power and road holding.

ROYALAC Group of accelerators (246):
133: dithiocarbamate of undisclosed structure. Black powder, s.g. 1·38. For EPDM
134: mixture of 50% zinc dimethyl dithiocarbamate and 50% tetramethyl thiuram disulphide
235: mixture of 2 parts activated dithiocarbamate and 1 part tetramethyl thiuram disulphide.

ROYALENE Ethylene-propylene terpolymer (23).

ROYALENE X-400 Ethylene-propylene terpolymer, with a higher mol. wt. than normal EPT; extended with 100 parts of naphthenic oil (23).

RPA Abbreviation for rubber peptising agent. Group of chemically active plasticising agents or peptisers which act as oxidation catalysts and cause accelerated degradation of the gel structure of rubber with less degree of oxidation than by mechanical plastification. Also effective as reclaiming agents in various reclaiming processes, reduce the processing time and give a smoother sheet with better processing properties.

RPA-2: mixture of 1 part naphthyl-β-mercaptan and 2 parts inert hydrocarbon. Pale yellow flakes, m.p. 55°C, s.g. 0·94. Has less effect on gum stocks with thiazoles and thiurams as accelerators. Has no effect on the ageing properties. Does not cause staining. *Quantity:* natural rubber 0·15–0·6%, SBR 0·25–1%

RPA-3: mixture of 36·5% xylyl mercaptan and 63·5% of inert hydrocarbon. Brown liquid with characteristic mercaptan odour, s.g. 0·92. Soluble in aromatics, insoluble in water, easily dispersed. Has a slight activating effect on gum stocks with thiazoles and thiurams, no effect on ageing properties. *Quantities:* natural rubber 0·1–0·6%, SBR 0·25–1%, reclaiming agent 1–2%. During the production of crêpe rubber on the plantation RPA-3 is added to the latex, before coagulation, as a 20% emulsion in quantities of 0·01–0·1%. It has a strong bleaching effect on the carotenoids of the rubber and is frequently used to produce pale coloured crêpe and sole crêpe, care is needed to prevent pale grey or brown staining. If approx. 0·5% PRA-3 is added to the latex before coagulation, a soft rubber type (pre-softened rubber, peptised rubber, Plastorub) is obtained

RPA-5: mixture of 50% zinc dixylyl disulphide (zinc salt of xylyl mercaptan) and 50% hydrocarbon. Dark brown liquid, s.g. 1·09. Slightly retards vulcanisation, but has no effect on ageing properties. Does not cause staining. *Quantity:* 0·25–1%. *Uses:* plasticising and reclaiming of SBR

RPA-6: pentachlorothiophenol, s.g. 1·72. For natural and synthetic rubber (6).

RPD Diphenyl guanidine (5).

RR 10 N Viscous, brown resorcinol resin. Antioxidant which causes slight staining (43).

RRC Abbreviation for Rubber Reserve Company; a subsidiary of Reconstruction Finance Corporation. Founded to build up the synthetic rubber industry in the USA. Was discontinued in 1945 when the ORR took control of the government's synthetic rubber programme.

RRIC Abbreviation for Rubber Research Institute of Ceylon.

RRIM Abbreviation for Rubber Research Institute of Malaya.

RRI TUNNEL Tunnel smoke house developed by the Rubber Research Institute of Malaya. The sheets were transported on carriers running on tracks into the smoke in the tunnel.

RSS Abbreviation for ribbed smoked sheet.

RTA Abbreviation for Rubber Trade Association.

RTAJ Abbreviation for Rubber Trade Association of Japan.

RTANY Abbreviation for Rubber Trade Association of New York.

R TENSILAC 40, 41 Accelerators of undisclosed composition (obsolete) (6).

RTV Abbreviation for room temperature vulcanisation.

RTV Group of silicone elastomers which may be vulcanised at room temperature (198).

RUBARIT Blend of 40% unvulcanised synthetic rubber with 60% mineral powder. Used as an additive for asphalt in road construction. Disperses easily in asphalt and improves the toughness, ductility, elasticity and bonding (115).

RUBATOR Group of accelerators (397):
DOTG: di-o-tolyl guanidine
DPG: diphenyl guanidine
DTMT: tetramethyl thiuram disulphide
FQ: mixture of basic accelerators
H-7: hexamethylene tetramine
MAT: mixture of mercaptobenzthiazole and zinc diethyldithiocarbamate
MBT: 2-mercaptobenzthiazole
MBTS: di-2-benzthiazyl disulphide
ZMBT: zinc-2-mercaptobenzthiazole.

RUBBER ABSTRACTS The most extensive reference journal in the rubber field. Founded 1923 and appeared until 1951 under the title *Summary of Current Literature*. Published by the Rubber and Plastics Research Association of Great Britain (formerly Research Association of British Rubber Manufacturers), monthly.

RUBBER ACCELERATOR Canadian produced accelerator.
ZE: zinc diethyl dithiocarbamate
ZM: zinc dimethyl dithiocarbamate.

RUBBER AND PLASTICS RESEARCH ASSOCIATION OF GREAT BRITAIN RAPRA. Up to 1961 existed as the Research Association of British Rubber Manufacturers (RABRM). Founded 1919 for advancing the scientific knowledge and progress in the British rubber industry. The association is financed by donations from the industry and a subsidy from the Department of Scientific and Industrial Research. Now at Shawbury, Shrewsbury, Shropshire, the institute was based at Croydon (London) from 1922–45. The field of research includes chemical and technical problems in the rubber field and, since 1961, in the plastics field. The association has over 100 members.

RUBBER AS A thermally treated rubber resin compound obtained in the Soviet Union from the silk plant (*Asclepia syriaca*). Softener for dienes.

RUBBER BEARING PLANTS A great number of plants yield rubber, gutta-percha and rubber-like substances. The ability to form rubber occurs in plants of various families within the higher levels of the plant world, in dicotyledons, spermatophytes of the angiosperm type. Rubber occurs primarily in the sap containers in the bark, and also in leaves, twigs, branches, roots and fruits. In some families without sap tubes, *e.g. Loranthaceae* and *Hyppocrateaceae*, rubber-like substance occurs in the bark or leaves or fruits in long sap cells (idioblasts). In the sap tube system a differentiation is made between systems which have segments (sap containers) and systems without segments (sap tubes). The segmented systems which occur in root rubber bearers are formed by joining of separate cells at each end and the merging of outward growth through anastomoses. The unsegmented sap tube system is formed by outgrowing and dichotomous branching of sap cells formed in the seeds. The formation of rubber depends on the characteristics of the dicotyledons, thus in plants of the *Parthenium* species, the phylogenetically younger forms (*latifolium* and *angustifolium*) produce the most rubber, while the older forms (*marioloides* and *brevifolium*) contain only a small amount in cells which have lignified. Rubber does not occur in gymnosperms and monocotyledons.

Lit.: K. Heyne, *De Nuttige Planten van Nederlansch Indie*, Buitenzorg, 1927.

G. Hubner, *Kautschuk*, Berlin, 1934. *Encylopedie Technologique du Caoutchouc*, Paris, since 1937.

K. Memmler, *Handbuch der Kautschuk-Wissenschaft*, Leipzig, 1930.

J. v. Wiesner, *Die Rohstoffe d. Pflanzenreiches*, 4 vols, Leipzig, 1928.

G. Klein, *Handbuch der Pflanzenanalyse*, Bd. 3. Teil 2, Vienna, 1932.

M. Ulmann, *Wertvolle Kautschukpflanzen des Gemässigten Klimas*, Berlin, 1951.

E. De Wildemann, *Rubb. Rec.* Int. Rubber Congress met Tentoonstelling, Batavia, 1914, p. 1.

K. Heyne, *Rubb. Rec.*, Batavia, 1914, p. 41.

H. Jumelle, *Rubb. Rec.*, Batavia, 1914, p. 33.

RUBBER BONDING An intermediate layer is necessary when bonding rubber, *e.g.* to metals, ceramics, glass, wood or leather; this layer must be capable of adhering to both the rubber and the material being bonded.

RUBBER CRUMB Crumbed rubber. Self-adhesion in powder or crumb form must be prevented by chemical or mechanical means; since 1930 numerous processes have been developed to achieve this, though most have found no commercial application. Among varied substances suggested as a coating for the individual particles are:

1. Colloids, *e.g.* proteins and starches.

2. Solids such as magnesium oxide, calcium carbonate, talc, tricalcium phosphate, sulphur and carbon black.

3. Substances which are soluble in latex, *e.g.* zinc ammonium phosphate, magnesium bicarbonate, sodium silicate, sodium nitrite, the sulphides, carbonates, sulphates and silicates of ammonia or the alkali metals and water soluble salts, *e.g.* of magnesium, aluminium, calcium, barium and zinc.

4. Dusting agents added during a spray drying process, in particular kieselguhr which is suitable because of its low density and large surface area.

5. Substances which coat the latex particles and are transformed into a synthetic resin or form an insoluble precipitate during the drying process. Alternatively the surface nature of

the particles can be changed by vulcanisation or similar reactions.

Hopkinson was the first to obtain a fine, tacky powder by spraying latex in a hot stream of air. In this process it was the spray-drying method which was critical. The most important processes are: De Schepper process. Latex is sprayed onto a conveyor belt, dried in a hot chamber and mixed with a small quantity of zinc stearate. BP 392 592. Pulvatex (Stam process). Originally produced by adding fine powders during spray drying. The high content (up to 30%) of non-rubber constituents was a disadvantage, however. According to a new process zinc ammonium phosphate is added —as a protective agent—to concentrated latex before it is powdered; the rubber powder is dusted with magnesium carbonate. By this process powders passing a 200 mesh (ASTM) sieve can be produced. BP 388 341, 437 230, 438 249, 439 777, 445 591, 446 560.

Mealorub process (meal of rubber). Process developed by van Dalfsen. Latex, preserved with ammonia, is mixed with a dispersion of a curing system, heated for $2-2\frac{1}{2}$ h at 80°C, and coagulated with formic acid. While wet from being washed, the crumbly coagulum is shredded to a fine powder, then dried in tumbler driers or on drying belts. The surface vulcanised powder is not tacky. Curing system dispersion: sulphur 2%, zinc oxide 1·5%, diethyl dithio-carbamate 0·2%, mercaptobenz-thiazole 0·4%, 10% solution of borax caseinate 2%, water 6%. Holl P: 47 833 (1940).

SBR black powder process. Latex is mixed with a dispersed black, *e.g.* 50% EPC and diluted to 10–12%. The two components are coagulated together with zinc chloride solution, and the precipitated powder washed and dried.

Uses: primarily in asphalt compounds for road building (5% on the asphalt), moulded articles.

Lit.: R. J. Noble, *Latex in Industry*, 2nd ed., New York, 1953, 377 (bibliog.).

TN: Harcrumb (169)
Pulvatex
Rodorub (400).

RUBBER DIBROMIDE

$(C_5H_8Br_2)_x$. White to yellowish powder with 70·1% bromine. Soluble in chloroform and carbon tetrachloride. Obtained by the complete saturation of rubber hydrocarbon with added bromine. The reaction is used to determine the rubber hydrocarbon content. It was formerly thought to be the tetrabromide $(C_{10}H_{16}Br_4)_x$.

RUBBER FORMOLITE

Yellow powder. Insoluble in water, swells in carbon disulphide and pyridine. Obtained by reacting formaldehyde with a rubber solution in the presence of sulphuric acid.

RUBBER GROWERS' ASSOCIATION

19, Fenchurch Street, London, EC3. Founded 1907 to pool interests of rubber planters. Today it has over 400 companies as members, a capital of around £100 000 000 and controls, through its members, over $\frac{3}{4}$ million hectares planted with rubber trees. The shareholding areas lie in Malaya, Ceylon and Indonesia.

RUBBER HYDROCHLORIDE

$(C_5H_8.HCl)_n$. White solid, softening p. 50–130°C, 28–30% chlorine (theoretically 33·9%). Rubber can add one molecule of hydrogen chloride per C_5H_8 group. Production is achieved by passing dry hydrogen chloride gas into latex rubber, dissolved in toluene, benzene, chloroform, dichloroethane or tetrachloroethane, or into solid rubber under pressure. The use of damp hydrogen chloride can result in isomerisation or cyclisation, giving addition compounds which decompose easily. Rubber hydrochloride is stable, elastic, tear resistant, resistant to moisture, oils, fats, chemicals, weak acids and alkalis; foils and threads can be stretched by heating. When heated with pyridine, hydrogen chloride is evolved and the rubber is reclaimed as isorubber. Partially hydrochlorinated rubber with 25–26% chlorine may be vulcanised and can be used for oil resistant products.

Uses: include packing foils for food, water resistant coverings, clothing, umbrellas, textiles such as filter cloths and stockings, yarn, brushes, wallpaper and metal/rubber adhesives.

TN: Pliofilm
Pliolastic
Tensolite.

Lit.: as for chlorinated rubber.

RUBBER HYDROFLUORIDE Mixture of $(C_5H_8 . HF)_n$ and $(C_5H_8 . H_2F_2)_n$. Tough, thermoplastic product with approx. 25% fluorine. Obtained by treating rubber with 50–85% hydrofluoric acid. When heated most of the fluorine is removed. If boron hydrofluoride is used, products similar to balata are obtained.

Uses: adhesives for rubber/metal bondings.

RUBBER INDIA The official publication of the Indian Rubber Industries Association, Bombay. Founded 1949; monthly.

RUBBERINE Factice made from blown linseed oil, vulcanised with sulphur and extended with Vaseline and asphalt.

RUBBERINE GEL Mixture of 60% lecithin, 30% glycerides and mineral salts. Antioxidant (167).

RUBBERISED HAIR An elastic layer, *e.g.* of coconut fibres, horse hair or pigs' bristles, bound together with natural or synthetic latex.

RUBBER MANUFACTURERS' ASSOCIATION 444, Madison Avenue, New York 22. Founded 1929 by amalgamation of the Rubber Institute and the Rubber Association of America. An organisation of American rubber trade associations with the function of an independent institute. Originally it produced its own RMA codes for rubber which served as the basis for all consignments to the USA. The types specified in these codes, the packing regulation and standard samples are listed in the so-called 'Green Book' and are internationally recognised; in July 1962 this publication was renamed 'International Standard of Quality and Packing for Natural Rubber Grades'.

RUBBER NITROSATE Reaction product of natural rubber or a terpene, with nitrogen tetroxide (N_2O_4); similar to the nitrosite.

RUBBER NITROSITE Addition product of natural rubber or a terpene and nitrogen trioxide (N_2O_3); varies in composition according to the conditions under which it is formed. Yellowish powder. Soluble in acetone, ethyl acetate and dilute alkalis, insoluble in benzene and ether. Can be used for the quantitative determination of rubber.

RUBBEROID A factice made from cotton-seed oil vulcanised with sulphur.

RUBBEROL Group of odorant and deodorant products for solid rubber (43).

RUBBER PHOSPHINE COMPOUNDS (Saturated) reaction products of rubber dibromide with triphenyl phosphine and triethyl phosphine. Insoluble in benzene and chloroform. The triphenyl phosphine compound is elastic, soluble in water and strongly dissociable, but cannot be hydrolysed.

RUBBER RESEARCH INSTITUTE OF CEYLON RRIC; Dartonfield Estate, Agalawatta. Founded by the plantation industry 1910. This was one of the first rubber institutes, to further the production of natural rubber. In 1930 was officially recognised by the government as the Rubber Research Scheme and since 1951 has been known under its present name. Publications: *Quarterly Journal* (formerly *Quarterly Circular*), *Annual Review*, advisory circulars.

RUBBER RESEARCH INSTITUTE OF MALAYA RRIM; PO Box 150, Kuala Lumpur. Founded 1925, research institute for natural rubber. The major work of this institute is the production of clones which give a high yield and good disease resistance, improvement of rubber production, development of new rubber types and basic research into the botanical, phytopathological, chemical and technological fields. New clones cultivated by the RRIM yield an annual harvest of up to 3000 kg/ha. Publications: *Journal of the Rubber Research Institute of Malaya, Planters' Bulletin, Annual Report*.

RUBBERSCOPE A simple apparatus developed by Firestone, used for a quick, comprehensive determination of the ageing properties of raw rubber and to detect the presence of too great a quantity of copper and manganese in the inferior qualities. The method also indicates when an excess of sulphuric acid has been used as a coagulant. The rubber test piece is exposed to the rays of an infra-red lamp under specified conditions and the surface condition (tacky places) judged visually, or the increase in plasticity determined; the results can be correlated with oxidation, copper, manganese and sulphuric acid contents.

RUBBERSOL Group of odorant and deodorant products for rubber articles made from latex (43).

RUBBER STICHTING Dutch rubber research institute, Delft. Founded 1936, under the co-ordination of the International Rubber Research Board. Until 1956 was under the direction of Dr R. Houwink. It was a branch of the Indonesian Research Institute in Bogor and was closed in 1956 by the Indonesian Government which had financed it since the independence of Indonesia.

RUBBER/TEXTILE BONDING The poor affinity of synthetic fibres to rubber necessitates the improvement of adhesion of rubber impregnated textiles and fabric carcase plies. A two component system of latex and protein or thermosetting resins may be alternately used with polyisocyanates. With the two-component system, the latex effects the adhesion to the rubber, while the other components assist the textile adhesion. Either natural latex, butadiene-styrene latex or butadiene-styrene-vinyl pyridine latex is used. When proteins are used, 5–50% casein or haemoglobin is added to the latex, according to the type of textile. Resorcinol-formaldehyde resin is used for the resin component, 5–40% on the dry rubber content of the latex (100 parts resorcinol, 40–80 parts formaldehyde). While textiles processed with latex and protein need only to be dried, thermal hardening is necessary at 120–130°C if resin is used.

RUBBER THIOGLYCOLATE
Reaction product of rubber with thioglycolic acid. A glass-like solid at room temperature, at 75–80°C it is a viscous mass.
Lit.: Ber, 1932, 65, 1349.

RUBBER THREAD Produced from either solid rubber, or latex. The old method of production, developed by Hancock and Brockedon [BP 11 455 (1846)] has hardly been altered, and consists of adhering a vulcanised sheet to a cylinder, using shellac, and cutting it into strips with rectangular cross-sections. During the spinning process, the latex compound is injected into a coagulation bath through a glass tube; the threads are then dried and vulcanised. Similarly, heat sensitive latex may be injected into a hot bath. A latex process was also developed by Hancock [BP 549 (1838)] in which a screw was dipped in latex, and the latex remaining in the grooves dried and peeled off in the form of threads.

RUBBER TYPES The natural rubber types recognised in international trade are listed by the Rubber Manufacturers Association, New York, in the 'International Standards of Quality and Packing for Natural Rubber Grades' (valid from July, 1962, known as the 'Greenbook'). Former editions, *e.g.* 1957, appeared under the title 'Type Descriptions and Packing Specifications for Natural Rubber Grades used in International Trade'. The latest edition describes seven types with thirty-one International Standard Qualities.
Ribbed smoked sheets. Exclusively coagulated, dried and smoked sheets. Cuttings or sheets which have been stuck together or reveal bubble formation are not admissible, nor are smooth or air dried sheets. *Qualities:* RSS Nos. 1X, 1, 2, 3, 4, 5, 6.
Pale crêpes. Produced from the fresh coagulum of natural rubber latex under controlled uniform production conditions by sheeting on crêping rolls. The types are divided according to the thickness; thick pale crêpe and thin pale crêpe. The former is produced by rolling together sheets of dried thin crêpe. *Qualities:* thick pale crêpe Nos. 1X, 1, 2, 3, thin pale crêpe Nos. 1X, 1, 2, 3.
Estate brown crêpes. Produced from rubber scrap of good quality (lumps, panel scraps, shell scraps) by rolling and washing on wash-roller machines. The use of scraps from the earth, smoked scraps and slabs is disallowed. *Qualities:* thick brown

crêpe and thin brown crêpe Nos. 1X, 2X, 3X.

Thin brown crêpes (remills). Produced in so-called remilling factories from slabs (thick coagulum slabs), obtained from the smallholders and from wet or unsmoked sheets, lumps, and scraps. Scraps which have been contaminated by contact with earth are disallowed. *Qualities:* Nos. 1, 2, 3, 4.

Thick blanket crêpes (ambers). Production as for thin brown crêpes in remilling factories, but thicker (6–10 mm). *Qualities:* Nos. 2, 3, 4.

Flat bark crêpes. Produced from scrap of a much lower quality including pieces which have become contaminated with earth. *Qualities:* standard flat bark crêpe and hard flat bark crêpe.

Pure smoked blanket crêpe. Produced from smoked sheets which have been stuck together in blocks and from cuttings by rolling and washing in the crêpe rollers. Besides the international types others are classified as: technically classified rubber, skim rubber, air dried sheets, SP rubber (superior processing rubber), master batches with red clay, (red clay masterbatches), partially purified crêpe (PP crêpe), peptised rubber, cyclised rubber masterbatches, rubber powder, and crumbed rubber.

RUBBONE Oxidised, depolymerised natural rubber produced by catalytic oxidation at approx. 80°C by oxygen or air. Cobalt, manganese, copper and lead linoleates or naphthenates can be used as catalysts. The oxidising process can be carried out on premasticated rubber in an internal mixer. After clarification and evaporation of the solvents a viscous yellow to orange red mass is obtained. When heated in the absence of oxygen, rubbone polymerises to a tough, non-tacky oil, similar to polymerised wood oil. If heated in air it gives a hard, brittle, resinous product. The product can be classified according to solubility as follows:

Rubbone A, soluble in aromatics, insoluble in acetone and alcohol
— B, soluble in aromatics and acetone, insoluble in alcohol
— C, soluble in aromatics, acetone and alcohol.

When sulphur is added, rubbone can be vulcanised to thermoplastic resins and converted to chlorinated and hydrochlorinated products which form a basis for alkali-resistant colours. A moulding powder can be produced from vulcanised Rubbone C.

Uses: include adhesives, bonding agents for grinding wheels, hot lacquers for food tins, printing colours, impregnation for electrical insulations, colours, paints; the lacquers are resistant to high temperatures and chemicals.

RUBENAMID
C: N-cyclohexyl-2-benzthiazyl sulphenamide
M: oxydiethylene-2-benzthiazyl sulphenamide (397).

RUGISOMETER An instrument developed by M. Mooney to determine the surface of calendered uncured rubber sheets. BP 608 293, USP 2 417 988.

RUKE Former trade term for natural rubber from the Congo.

RULACEL Chlorinated rubber produced from concentrated latex. Types 5, 10, 20, 50, 125, 500, 1000.

The index indicates the mean viscosity (in cp) of a 20% solution in toluene at 25°C (312).

RUSSELL EFFECT A property, discovered by S. Russell, of many organic and inorganic substances, which, after exposure to light rays, form an image on a photographic plate when exposed in the dark. Raw rubber reveals this effect which is caused by peroxides; vulcanised rubber shows the effect to a slight degree.

RUST A brown desposit on the surface of sheet rubber; caused by dried serum constituents and the development of yeasts, bacteria and fungi on the serum constituents. The deposit is usually visible only after the sheet has been stretched or scraped with a blunt object (piece of wood, or the end of a pencil), and is then revealed as a fine, rust-coloured powder.

RYPHAN Japanese produced rubber hydrochloride in film form.

RZ-50-A 50% solution of di-methylcyclohexylamine dibutyl di-thiocarbamate in Cellosolve. Brown liquid, s.g. 0·96. Ultra-fast accelerator for natural and synthetic rubbers. Used for vulcanisation at room temperature.
Uses: adhesives, coating compounds (5).

RZ-100 Dimethyl cyclohexyl-ammonium dibutyl dithiocarbamate (5).

S

S-70 Liquid polysulphide rubber cured at room temperature to a flexible solid mass by adding hardening agents.

S-145 Mould release agent based on a fluorocarbon. Is not transferred onto the mouldings, does not migrate.

SA Abbreviation for Sharples accelerator; group of accelerators:
SA 52: tetramethyl thiuram disulphide
SA 57: zinc dimethyl dithio-carbamate
SA 62: tetraethyl thiuram disulphide
SA 66: selenium diethyl dithio-carbamate
SA 67: zinc diethyl dithiocarbamate
SA 77: zinc dibutyl dithiocarbamate (4).

SAF Abbreviation for super abrasion furnace.

SAG Abbreviation for special auto grade; carcases of old tyres which have been cut into small pieces, the tread and beads having been removed. They may contain rubber from the sidewalls.

SAGHYZ Sagis, Sakkis. Rubber-like plant excreta which were used as a chewing gum in parts of Russia.

SAHAGUN, BERNARDINO DE 1500–1590, b. Sahagun, d. Mexico. Went to Mexico in 1529 as a priest. In his writings on the History of New Spain he mentions the Mexican rubber tree 'ulequahuitl' and its

product 'ulli'. His works were concluded in 1569 in the Aztec language.
The work was confiscated from him
by the church as it was considered
dangerous and it was distributed
among several monasteries. It was
returned to him several years later
only so that he could translate it into
Spanish.

SALICYL ANILIDE
$C_6H_5.NH.CO.C_6H_4.OH$. Odourless, white crystals, m.p. 136–138°C.
Soluble in alcohol, chloroform, ether,
benzene, dilute alkalis, sparingly
soluble in water.
Uses: fungicide, effective in removing
fungi from sheet rubber.
TN: Shirlan.

SALICYLIC ACID
$C_6H_4(OH)COOH$. Fine, odourless,
white crystals, m.p. 156°C, sublimates
on careful heating, decomposes on
being heated rapidly into carbon
dioxide and phenol, s.g. 1·443.
Soluble in hot water, alcohol, ether,
fatty oils, sparingly soluble in cold
water. Scorch retarder at processing
temperatures in compounds containing acidic accelerators. Has a slight
activating effect on thiazoles, thiazolines and thiurams at curing temperatures. Has no effect on ageing or
modulus. Non-staining.
Uses: in natural rubber, SBR, neoprene W types and reclaim.
Quantity: 0·25–0·75% ($\frac{1}{4}$–$\frac{1}{2}$ the
accelerator quantity).
TN: Retarder W (contains dispersing
    agent) (6)
Retarder TSA (with stearic acid)
Vulcosal.

SANTICISER Group of plasti-
cisers for plastics and synthetic
rubbers (SBR and NBR):
1:     p-toluene sulphonamide

1H: N-cyclohexyl-p-toluene
       sulphonamide
2:     p-toluene sulphonamide
3:     N-ethyl-p-toluene
       sulphonamide
7:     m-toluene sulphonethyl amide
8:     blend of o- and p-N-ethyl
       toluene sulphonamides
9:     blend of o- and p-toluene
       sulphonamides
10:    o-cresyl-p-toluene sulphonate
B16:   butyl phthalyl butyl glycolate
E15:   ethyl phthalyl ethyl glycolate
H: cyclohexyl-p-toluene sulphon-
       amide
M10:   blend of o-, m- and p-cresyl
       benzene sulphonates
M17:   methyl phthalyl ethyl glycol-
       ate
127:   n-butyl benzene sulphonamide
140:   cresyl diphenyl phosphate
141:   octyl diphenyl phosphate
160:   butyl benzyl phthalate
409:   adipic acid polyester
601:   50:50 blend of dioctyl phthal-
       ate and dioctyl decyl phthalate
602:   50:50 blend of dioctyl phthal-
       ate and didecyl phthalate
603:   N-ethyl-o- and p-toluene sul-
       phonamides (5).

SANTOBRITE LATEX Concen-
trated latex preserved with 0·3%
sodium pentachlorophenate and 0·1–
0·2% ammonia.

SANTOCURE Group of
accelerators:
Santocure: N-cyclohexyl-2-benz-
           thiazole sulphenamide
26:        2-(2:6-dimethyl-4-
           morpholinethio)benz-
           thiazole
DT: blend of N-cyclohexyl-2-
           benzthiazole sulphen-
           amide with diphenyl
           guanidine and di-2-benz-
           thiazyl disulphide

419

MOR: 2:4-morpholine mercaptobenzthiazole

MOR-90: 90% N-oxydiethylene-2-benzthiazyl sulphenamide and 10% benzthiazyl disulphide

NS: N-tert-butyl-2-benzthiazyl sulphenamide

RF1: blend of N-cyclohexyl-2-benzthiazole sulphenamide and 2-(2:4-dinitrophenyl thio)benzthiazole. Pale yellow powder, s.g. 1·50. Soluble in warm chloroform. Similar in its properties to Santocure, but safer and cures more slowly. *Uses:* in compounds with reinforcing blacks (5).

SANTOFLEX Group of antidegradants (5):

1 P: N-phenyl-N'-isopropyl-p-phenylene diamine. Antiozonant

13: N-1:3-dimethyl butyl-N'-phenyl-p-phenylene diamine

17: N,N'-bis(1-ethyl-3-methyl pentyl)-p-phenylene diamine

35: blend of 65% dihydrotrimethyl phenyl quinoline and 35% diphenyl-p-phenylene diamine

75: blend of 75% N,N'-diphenyl-p-phenylene diamine and 25% 6-dodecyl-1:2-dihydroxy-2:2:4-trimethyl quinoline. Dark flakes, s.g. 1·13. Antioxidant for NR and SR. Blooms at quantities above 0·4%

77: N-isopropyl-N'-phenyl-p-phenylene diamine, s.g. 1·01–1·07

217: N,N'-bis(1:4-dimethyl pentyl)-p-phenylene diamine

9010: blend of phenyl-β-naphthylamine and diphenyl-p-phenylene diamine

AW: ethoxy dihydrotrimethyl quinoline

B: dihydrotrimethyl phenyl quinoline

BX: blend of 85% dihydrotrimethyl phenyl quinoline and 15% diphenyl-p-phenylene diamine

CP: N-cyclohexyl-N'-phenyl-p-phenylene diamine

DD: 6-dodecyl-1:2-dihydro-2:2:4-trimethyl quinoline

DPA: diphenylamine-acetone condensation product

GP: N-cyclohexyl-N'-phenyl-p-phenyl-p-phenylene diamine

HP: blend of phenyl-β-naphthylamine and diphenyl-p-phenylene diamine in the ratio 1:2. Dark grey powder, softening p. above 88°C, s.g. 1·23. Soluble in chloroform, acetone and ethyl acetate. Effective against flex cracking, oxidation and heat, more effective than phenyl-β-naphthylamine alone. Also protects against copper and manganese

IP: N-isopropyl-N'-phenyl-p-phenylene diamine, s.g. 1·14–1·17

R: poly-1:2-dihydro-2:2:4-trimethyl quinoline.

SANTOMERSE Group of alkyl naphthalene sodium sulphonates. Wetting agents, emulsifiers, stabilisers for latex (5).

SANTOVAR Group of antioxidants for NR, SR and latices:

A: 2:5-ditert-amyl hydroquinone

O: 2:5-ditert-butyl hydroquinone (5).

SANTOWHITE Group of anti-oxidants for NR, SR and latices:
Santowhite: 4:4-butylidene bis-3-methyl-6-tert-butyl phenol
54: alkylated phenol
CI: di-β-naphthyl-p-phenylene diamine
CM: di-o-cresol monosulphide
Crystals: 4:4′-thiobis(3-methyl-6-tert-butyl phenol)
L: thiobisdisec-amyl phenol
MK: reaction product of 6-tert-butyl-m-cresol and SCl$_2$. Dark, viscous liquid, m.p. 20–30°C, s.g. 1·06. Gives good protection against normal ageing. Non-staining, used in light-coloured articles. *Quantity:* 1–2%
powder: 4:4-butylidene bis-3-methyl-6-tert-butyl phenol (5).

SAP CONTAINER The plant sap which contains the rubber can occur, according to the species, in sap containers, sap cells or sap arteries.

Sap containers: formed by fusion of a large number of cells, through resorption of separating walls and partial resorption of side walls, to produce a coherent tube system onto which new cells are added continuously.

There is no uniform sap-container system extending through the whole system. As the bark thickens, systems develop which are not connected to the previous systems; several single systems formed by the cambium of the twigs, are joined via the trunk to the roots. Occurrence: *Hevea* and *Manihot* types.

Sap arteries: a highly branched arterial system found throughout the plant body. Develops with the branching of two cells in the nucleus form, which subsequently reveal an unlimited capacity for growth.

Fission takes place only of the nuclei contained in the tube, but there is no formation of separating walls. The cells, therefore, grow into the surrounding tissues and form systems throughout the entire plant enabling the movement of the sap throughout the arterial system. Occurrence: many *Euphorbiaceae, Apocyaneae, Moraceae,* and *Asclepiadaceae, Ficus elastica, Castilla* types, rubber lianes.

Sap cells: isolated sap producing cells scattered in the cell tissue, and hardly different from other cells. When an incision is made into the plant the sap can only come from the pierced cells. Plants with sap cells cannot be tapped; rubber must be obtained by destruction of the tissue. Sap cells occur primarily in *Compositeae* and *Loranthaceae.* In *Compositeae* they occur in the leaves, twigs and roots; in the *Loranthaceae* they are restricted to the fruits, *e.g.* in *Parthenium argentatum* (Guayule), *Taraxacum Kok-Saghyz* and mistletoe.

SAPIUM Member of the *Euphorbiacae* family, group of rubber producing trees giving various excellent quality products similar to that of *Hevea* rubber. The trees are difficult to tap because of the hard bark, and need long pauses for recuperation of the sap, *i.e.* 1–4 months between tappings. The latex normally co-agulates spontaneously.

SAPIUM RUBBER Orinoco scrap. Trade term for the product obtained from *Sapium Jenmani* (British Guiana). The sap flowing out of the cuts in the bark is allowed to dry onto the bark and is then rolled into bales or pressed into blocks or flat slabs. Contains approx. 87–95%

pure rubber, 2–4% resins and 2·5–6% proteins.

SARAN

$$\left[-CH_2-\overset{\displaystyle Cl}{\underset{\displaystyle Cl}{\vert}}\overset{\vert}{C}-CH_2-\overset{\displaystyle }{\underset{\displaystyle Cl}{\vert}}CH- \right]_n$$

Copolymer of 85% vinylidene chloride and 15% vinyl chloride. Non-flammable, thermoplastic material, m.p. approx. 170°C, softening p. 120–140°C, s.g. 1·61, mol. wt. approx. 20 000. Resistant to chemicals such as solvents and acids, sensitive to ammonia, ether and dioxane.
Uses: include fibres, foils, coverings. Saran may also be obtained in latex form, s.g. 1·30, total solids 56%, pH 6·5–7·5, viscosity 20–25 cp, particle size 0·08–0·15 μ (61).

SARELON Synthetic fibre based on groundnut protein (60).

SATRA Abbreviation for the Shoe and Allied Trades Research Association, Rockingham Road, Kettering, Northants, England.

SAVELON Synthetic fibre produced from soya protein.

SBR Styrene-butadiene rubber, in the nomenclature of ASTM D 1418.

SBUD 45% aqueous solution of sodium dibutyl dithiocarbamate.

SCCO Blend of butadiene-acrylonitrile latex with tar, and with the addition of softeners and catalysts. Resistant to oils, fats, benzine and other fuels.
Uses: filler for expansion joints in concrete runways, particularly for jet aircraft (115).

SCF Abbreviation for super conductive furnace.

SCHADT PROCESS A process, originating from the early days of the rubber industry, which was used for producing rubber from latex. The latex was poured onto flat plates and dry steamed; the rubber was then smoked while still containing all the serum constituents. The process had no economic importance; at the time it was thought that good rubber could be obtained only by the original Para method and most of the processes evolved were modifications of this.

SCHIDROWITZ, PHILIP 1872–1960. Studied in London, Zurich and Berne; originally worked in the field of wines and spirits but from 1911 was active solely in the rubber industry. Undertook basic research into the processing of latex, inventing a process for making foam rubber from latex [BP 1111 (1914)], and the vulcanisation of latex. He was responsible for over 500 publications and patents on practically all aspects of rubber and vulcanised rubber. For about 40 years he regularly wrote the 'Views and Reviews' column of the *Rubber Journal*. In 1940, together with H. P. Stevens, he was awarded the Colwyn Medal of the Institute of the Rubber Industry; he was himself a founding member of the Institute in 1921.

SCHOPPER-DALEN MACHINE Machine used to determine the stress/strain properties of vulcanised rubber as rings or dumb-bell test pieces.

SCHRAMM APPARATUS One of the early instruments used to determine the resistance of insulating materials to glowing heat.

SCHRAMM-ZEBROWSKI APPARATUS Instrument used to determine resistance to glowing heat.

SCORCH Premature start of vulcanisation of a compound during mixing or shaping; time or temperature effect.

SCORCHGUARD O Magnesium oxide used for polychloroprene; an aliphatic compound retaining the activity of the magnesium oxide. Rod or pellet form. Reduces incorporation time by up to 15%.

SCORZONERA Branch of a Composite family found in the USSR and used there as a source of rubber (from the roots).
S. tau-saghyz Lipschitz and Bosse (Tau-Saghyz); discovered by Saretzki in the Kara-tau mountains, has a slow growth rate and only reaches its max. rubber content after 5 years, the content then being over 30%. It is obtained from plantations.
S. acanthoclada French (Teke-Saghyz); occurs naturally in the central Asian mountains at 2200–3900 m. The crop is obtained only from rich wild clumps of trees, the plants being difficult to cultivate.

SCOTT FLEXING MACHINE Apparatus used to determine the ply separation in test pieces; this is caused by dynamic fatigue and occurs in textile layers bonded by rubber, *e.g.* tyres, V-belts. The test pieces are flexed at 160 c/min over a bar at an arc of 135°C, oscillating under a bending load of 45·4 kg. ASTM D 430-59.

SCOTT PLASTIMETER Parallel plate plastimeter with a constant but variable load. Has a flat upper platen and the test piece is laid on a piston of approx. 29 mm diameter. After 20 min preheating at 70°C the test piece is compressed from 40 mm in diameter with a load, normally of 10 kg. The surface pressure is 1·5 kg/cm^2. Deformation is read off in 1/100 mm after 60 sec compression.

SCOTT TESTER A machine used to determine the stress/strain properties of vulcanised rubber in ring or dumb-bell shaped test pieces.

SCR Blends of various waxes and antioxidants (100).

SCR Styrene-chloroprene rubber, in the nomenclature of ASTM D 1418.

SCRAP Panel scrap, Sernamby. Rubber strips which coagulate on the tapping surface and dry after the flow of rubber from an incision in the bark has ceased. Before each new tapping they are removed and processed, alone or blended with cup lumps to give 'brown' types.

SCURAX Retarder of undisclosed composition; acts as accelerator in polychloroprene (20).

SDD Sodium dimethyl dithiocarbamate.

SDK Abbreviation for Sozialpolitische Vereinigung der Deutschen

Kautschukindustrie (Hanover). (Socio-political association of the German rubber industry.)

SE Group of silicone rubber types (198).

SEA CABLE The first sea cable was laid in 1851 under the Straits of Dover between the southern tip at Dover and Sangatte, near Calais, and came into use in 1851. Gutta-percha was used as an insulating material. A cable laid in 1850 could not be used because of distortion of signals during transmission.

SEACRUMB Trade name for crumb rubber made by the Heveacrumb process.

SEAT NO. 116 Japanese produced black which has not been classified.

SEDC Abbreviation for selenium diethyl dithiocarbamate.

SEEDINE Mixture of higher fatty acids. Activator and softener (314).

SEEKAY Synthetic waxes based on chloronaphthalene, m.p. 68–135°C. Soluble in benzene, trichloroethylene, ether and carbon tetrachloride.
Uses: to decrease the inflammability of rubber articles (60).

SEIBERLING, FRANK A. 1860–1955. Was known as the 'Little Napoleon of rubber'. Founded the Goodyear Tire and Rubber Company (Akron, Ohio) in 1898, with a loan of \$3500.

SEKUNDA TYPES Trade term for rubbers of an inferior quality.

SELECTION Selective growing of rubber trees to develop properties desirable for culture, *e.g.* high production rate, good resistance to disease, resistance to wind breakage, good bark renewal, rapid growth.

SELENIUM Grey powder, m.p. 217°C, b.p. 688°C, s.g. 4·80, at wt. 78·96. Soluble in carbon disulphide and benzene, insoluble in water and alcohol. Secondary vulcanising agent, increases cure speed, improves ageing and mechanical properties in sulphurless compounds and compounds with a low sulphur content. (See table, p. 425.) Slow accelerator for ebonite. Also used as a catalyst for the digestion of rubber and vulcanised rubber by sulphuric acid in the quantitative estimation of nitrogen: Kjeldahl (BS 903: 1950, 1673: Part 2: 1954).
TN: Ancasal (22)
Vandex (51)
Doptax (blend with DPG, MBT and PBN)
Lit.: G. J. v.d. Bie, *Rubb. Chem. & Techn.*, 1949, **22,** 1134.

SELENIUM DIBUTYL DITHIO-CARBAMATE
$[(C_4H_9)_2.N.CS.S]_4Se$. Dark red liquid, s.g. 1·11, mol. wt. 896. Accelerator.
Uses: in isocyanate bonding agents for bonding natural rubber and synthetic rubbers to polyurethanes and metals.
TN: Novac (89)
Novac A-13 (1:3 mixture with benzthiazyl disulphide) (85).

Compounding examples	Rubber		Ebonite	
	I	II	I	II
Selenium	1–0·5	0·25	2	1
Tetramethyl thiuram disulphide	2–3	—	—	—
Zinc dimethyl dithiocarbamate	—	0·25	—	—
Dibenzthiazyl disulphide	—	1	—	1–2
Sulphur	—	2	40–45	40–45
Zinc oxide	5	5	—	2

SELENIUM DIETHYL DITHIO-CARBAMATE

$[(C_2H_5)_2.N.CS.S]_4Se$. Orange-yellow powder, m.p. 66–68°C, s.g. 1·32, mol. wt. 672. Soluble in carbon disulphide, benzene and chloroform, insoluble in water, dilute alkalis, benzine. Ultra-fast accelerator with 14·3% sulphur available for vulcanisation. Curing agent, primary or secondary accelerator in natural rubber, SBR, nitrile rubber and butyl rubber; zinc oxide or litharge is necessary. Non-staining. Has a low critical temperature and can cause scorching.

Quantity: approx. 10% higher than for selenium dimethyl dithiocarbamate.

Uses: wire and cable insulations, heat resistant articles.

TN: Accel SL (16)
Accelerator 66 (6)
Ethyl Selenac (51)
Ethyl Seleram (120)
P(Se)D-70 (in polyisobutylene) (282)
SA 66-1 (4)
Selazate (23).

SELENIUM DIMETHYL DITHIOCARBAMATE

$[(CH_3)_2N.CS.S]_4Se$. Orange-yellow powder, m.p. 144–167°C, s.g. 1·57, mol. wt. 559·8. Insoluble in water and dilute alkalis, sparingly soluble in carbon disulphide, benzene and chloroform. Ultra-fast accelerator, contains 17·2% sulphur available for vulcanisation.

Quantity: Curing agent 2–4% in NR and SR, 3–5% in butyl. Primary 0·3–1% in NR and SBR with 0·75–3% sulphur and 0–1% thiazole; 1–2% in butyl with 2–1% sulphur and 0·5–2% thiazole. Secondary 0·1–0·3% in NR and SBR, with 1–3% sulphur and 1–1·5% thiazole.

Uses: in natural rubber, SBR and butyl rubber for sulphurless cures and accelerating sulphur cures, also for heat resistant articles.

TN: Methyl Selenac (51).

SEMD Abbreviation for selenium dimethyl dithiocarbamate.

SEMPERPREN Polyurethane foam material based on a polyester (302).

SENTARUB Trade name for a high modulus rubber obtained from a skim latex with a nitrogen content of 2–3·5%. Produced by creaming and coagulating together fresh latex and centrifuged skim of approx. 5–6% DRC. (The name is derived from the Sentang plantation, E. Sumatra.) (122).

SEQUESTERING AGENTS
Compounds which form an anionic complex with polyvalent metal ions. *Uses:* as, for example, stabilisers in latex; copper and manganese inhibitors.

SERINGAL (Plural: Seringuaes). Areas used for rubber production in Brazil. The wooded areas varied in size between a few and 100 km^2 and were bought from the government. The owners and workers on these plantations were called 'Seringueiros'.

SERINGA RANA Term used in Peru for *Sapium Marmieri*, which gives an inferior quality rubber.

SERINGUEIRA BARRIGUDA Local term for *Hevea spruceana*. Grows south of the Amazon. An attempt was made to grow it in SE Asia by grafting onto *Hevea Brasiliensis*.

SERINGUEIRA BRANCA (Branca: white). Local term for *Hevea Brasiliensis* var. *latifolia*. So called in Brazil because of its white bark. Has broad leaves and normally grows near a river.

SERINGUEIRA PRETA (Preta: black). Local term for *Hevea Brasiliensis* var. *angustifolia*. So called in Brazil because of its dark-coloured bark.

SERNAMBY Former trade term for rubber obtained in Brazil by spontaneous coagulation of latex, after which the coagulated rubber was partially matured by protein decomposition during the long drying period. In French speaking areas panel scraps, which dry on the tapping surface when the flow of latex from the tree has ceased, are still known as sernamby.

SETSIT Group of accelerators for latex:
5: activated dithiocarbamate. Reddish-brown liquid, s.g. 1·01. Soluble in acetone and alcohol, soluble in water in the ratio 1:2, partially soluble in benzine, benzene and chloroform. Ultra-fast accelerator. *Quantity:* 1–3% on the dry rubber content in natural latex, 2–4% in SBR latex
9: activated dithiocarbamate. Brown liquid which is miscible with water. Ultra-fast accelerator for films made from natural and synthetic latices, adhesives, foam rubber
51: activated dithiocarbamate. Pale brown liquid, s.g. 1·00–1·04. Miscible with water. For natural and synthetic latices, activator for thiazoles (51).

SEXTOLPHTHALATE Dimethyl cyclohexyl phthalate. Softener for chlorinated rubber and plastics (315).

SGL Soviet leaves-gutta-percha. Gutta-percha produced in the Soviet Union from the leaves of *Eucommia ulmoides* by extraction with dichloroethylene and precipitation, m.p. 57°C.

SHAPE FACTOR (According to Kimmich.) The ratio of the loaded to the unloaded surface of a rubber body. For a cylinder:

$$F = \frac{D}{4H}$$

SHAPE RETENTION The ability of materials to retain their form under certain static or dynamic strains or surrounding conditions (temperature, dampness, solvents).

SHAPE RETENTION DURING HEATING (According to Marens.) The shape retention of hard polymers and plastics is characterised by the temperature up to which a test piece retains its form under a given static strain. DIN 53 458.

SHARSTOP Short-stop for emulsion polymerisation. 204, 40% solution of sodium dimethyl dithiocarbamate, 268, 50% solution of potassium dimethyl dithiocarbamate (120).

SHAWINIGAN SRF gas black (316).

SHAWINIGAN ACETYLENE BLACK CF gas black (316).

SHEET BALATA Balata sheets approx. 2·5 cm thick and weighing approx. 30 kg.

SHEET RUBBER Smoked sheet, airdried sheet, RSS. Rubber sheets approx. 2·5–3·5 mm thick and approx. 0·75 to 1·5 kg in weight (35 × 50 cm to 35 × 120 cm); usually have a diagonal diamond pattern on the lower surface, running approx. at an angle of 50° to the sheet. To produce these sheets fresh latex is diluted to 15–16% rubber content, a small quantity of sodium bisulphite is added to prevent enzymic discoloration and the latex is coagulated in large tanks with formic acid (400–500 ml 1% acid/kg dry rubber) or acetic acid (400–500 ml 2% acid/kg). By inserting 90–120 aluminium plates prior to coagulation, sheets of coagulum approx.

2·5 cm thick are obtained. In small holdings each sheet is coagulated separately in a small, flat aluminium pan. After 3–4 h a sufficiently solid coagulum is formed to be freed from the larger part of the serum by rolling between several pairs of rollers. In small holdings rolling is done by hand but in large factories there are so-called sheeting machines which each have 4–6 pairs of rollers. The rollers of each pair run at the same speed. The last roller pair, which has spiral-shaped grooves, imprints the pattern which facilitates the removal of the water, forced out by syneresis and speeds up the drying process because of the increase in surface area. Sheeting machines in use include the Guthrie-Cadet (4 pairs), Aristo (5 pairs), Braat, Huttenbach (6 pairs). A considerable portion of the serum constituents remains in the rubber and causes the development of mould and bacteria. The sheets are therefore preserved by smoking; this practice can also be attributed to some extent to the old belief that the quality of the rubber is improved by smoking. Sufficient preservation may be achieved by the use of p-nitrophenol and other fungicides. For smoking and drying, smoking chambers (*e.g.* Subur, RRIC, Peweja types) or smoking tunnels (*e.g.* the RRIM tunnel) are used, in which the sheets are hung on fixed or movable frames. Drying takes 3–6 days, on the first day temperature is maintained at 45°C, being increased on the following days to 60°C. By suitable control of air admission a temperature gradient may be achieved in the tunnel. The smoke is usually created by burning damp *Hevea* wood in an oven outside the tunnel. In recent years airdried sheet has also been produced by the introduction of warm air; these sheets have a pale, yellowish-brown colour and are frequently called LC (light-coloured) sheets. The dried sheets are sorted, compressed into bales and packed in sheets of equal quality and coated with a talc dispersion. The finished bales then have a weight of min. 224 lb and max. 250 lb (102–113 kg).

SHEETY CREPE On the west cost of Sumatra (Tapanuli) small-holding rubber is produced. The fresh, relatively soft coagulum is put several times through a hand opera-ted sheeting mill (roller pairs of 8–15 cm diameter running at even speed) with an ever-decreasing nip between the rollers. It is stretched rather highly. A thin, lacy, crêpe-like sheet is produced which is then dried in air.

SHELF LIFE The time for which an unvulcanised compound may be kept without change of properties.

SHELLAC White or yellow, odourless resin, s.g. 1·15. Disperses easily in rubber; improves the dimen-sional stability of extruded and calendered articles. During vulcanisa-tion combines with approx. 3% of its weight of sulphur so that the formulation must be adjusted. Acidic accelerators are necessary, since alkaline accelerators react with shel-lac. It increases the hardness and modulus, but causes some decrease in tensile strength and elongation at break. Improves the electrical proper-ties, solvent resistance and heat ageing.
Quantity: 10%.
Uses: flooring, footwear, cable sheaths.

SHELLFLEX Group of naph-thenic hydrocarbons. Non-staining softeners, extenders and processing aids for NR, SBR and CR (2).

SHELL PROCESS Process for producing butadiene from dichloro-butane.

SHELL S Butadiene-styrene latices (2).

SHINKO Japanese produced rayon staple fibre (337).

SHODDY Jargon formerly used for reclaim; now rarely used.

SHORE HARDNESS Testing according to Shore A, C and D. The resistance to indentation by a trun-cated cone indentor (A or C) or by a spherical indentor (D), used to determine the hardness of soft rubber; in general the Shore A hardness meter is used.
Specifications: ASTM D 676
BS 903: 20: 1950
DIN 53 505.

SHORE SCLEROSCOPE Ap-paratus formerly used to determine rebound resilience, using a vertically falling truncated steel cone. Height of falling 25·12 cm, energy 52 316 erg, impact 1–2 ms.

SHORT-STOP Chemical which, added to a polymerisation reaction acts as a free radical scavenger and stops the reaction at a predetermined conversion of monomer to polymer. For example, in the free radical emulsion polymerisation of buta-diene or its copolymerisation with styrene to produce SB rubber, the polymerisation rate falls to an uneconomically low value at con-versions much in excess of 70%. Also, at higher conversions, the probability of obtaining branched and gel polymer is increased. Examples of suitable short-stops for this reaction are sodium (and ammonium) diethyl dithiocarbamates; hydroquinone, though an efficient radical trap, is no longer used because it stains the finished polymer.

SI Abbreviation for swelling index.

SICCOACTIVEX Thiocarbanilide; alternatively, blends in equimolecular ratio.
11: with zinc oxide
12: with DPG and zinc oxide
13: with DPG
23: with DOTG (430).

SIFLOX Highly active, pure silica with a high bulk factor. Particle size 10–30 μ. Reinforcing white filler for articles subjected to arduous conditions (317).

SILASTIC Series of silicone rubber types; partially vinyl-substituted polydimethyl siloxanes and fluorinated silicone rubber. Good physical and dielectric properties and good low temperature resistance, some types are serviceable between -70 and 300°C. Low compression set, low water absorption, resistant to oxygen, ozone, hot oils, chemicals, and the weather. The various types are available compounded (7).

SILASTOMER Silicone rubber (318).

SILCAFILL Vulkasil. Light reinforcing silica filler for NR and SR.

Grades:
C: precipitated silica with a low calcium silicate content, approx. 82% precipitated SiO_2, s.g. 1·95
S: precipitated silica, approx. 85% SiO_2, s.g. 2·0. More active than Grade C. May be used for transparent compounds. *Quantity:* 10–100% with the addition of filler activators; needs 5–10 parts stearic acid
Sol 30%: aqueous colloidal silica dispersion, 30% SiO_2, s.g. 1·2, pH 8–9. Used in latices (43).

SILENE EF Finely precipitated calcium silicate. White reinforcing filler (134).

SILFIT Pale-coloured, semi-active silicate filler for NR and SR, s.g. 2·6, 90% SiO_2, 7% Al_2O_3, quartz: kaolinite ratio, 4·5:1. Also suitable for translucent vulcanisates (387).

SILICEOUS EARTH Pale, semi-reinforcing silicate filler for NR and SR, s.g. 2·6. 85% SiO_2, 10% Al_2O_3, quartz: kaolinite ratio 3:1. Suitable for translucent products (387).

SILICODERM Silicone ointment for protecting the skin against irritant chemicals. Resistant to water, dilute acids and alkalis, alcohol, acetone and other solvents. The ointment can be used both prophylactically and therapeutically for dermatitis (43).

SILICONE Organic polysiloxanes of the general formula $(-R_2SiO-)_n$, in which the silicon atoms are bound by oxygen bridges while the remaining valencies carry alkyl groups, aryl groups or hydrogen. They are produced by hydrolysis of dialkyl silicon dichlorides and condensation of hydroxyl compounds (silanols) to dialkyl polysiloxanes (silicones). Depending on the original substance used and the polymerisation conditions chain-like, ring-structure or branched products with different properties may be obtained. They may be oily or fatty (*e.g.* silicone oils, lubricants, sealants, impregnants), elastic and rubber-like (*see* silicone rubber), or resinous.

SILICONE ELASTOMERS
Silicone rubbers produced by ICI (66).

SILICONE MOULD RELEASE AGENTS
Aqueous emulsions of silicone oils which, after spraying onto hot moulds, form a thin silicone film capable of repelling rubbers and plastics, and thus facilitating removal of articles from the moulds. The films withstand constant temperatures of 200°C.

SILICONE OIL AK
Group of liquid dimethyl polysiloxanes with viscosities up to 1 million cSt; the classification number indicates the viscosity. Good electrical properties, high flash p., stable viscosity/temperature relationship.
Uses: mould release agents, antifoam agents, lubricants (359).

SILICONE OIL PEROXIDE PASTE
Crosslinking agent for silopren compounds.
Types:
BP 50: mixture of 50% dibenzoyl peroxide, 35% silicone oil and 15% water, s.g. 1·11. Soluble in acetone, ethyl

acetate, methylene chloride and benzene, sparingly soluble in alcohol. Cures are by heating under pressure, in hot air or open steam. Unsuitable for curing without pressure. *Quantity:* 3–6·5% for Silopren B
CL 40: mixture of 40% dichlorodibenzoyl peroxide and 60% silicone oil. White paste, s.g. 1·18. Soluble in benzene, ethyl acetate, methylene chloride and acetone, sparingly soluble in alcohol, insoluble in water. For crosslinking Silopren compounds without pressure. *Quantity:* 4–6% for Silopren B (43).

SILICONE RESINS
Silicone polymers with a crosslinked structure. Obtained by polycondensation of a mixture of hydroxymonoalkyl and hydroxydialkyl siloxanes

$$R.Si(OH)_3 + R_2Si(OH)_2 \rightarrow$$

$$\begin{array}{cc} R & R \\ | & | \\ -Si-O-Si-O- \\ | & | \\ O & R \\ | \\ -Si- \end{array}$$

Uses: include heat resistant lacquers, impregnations, insulating materials, jointings.

SILICONE RUBBER

$$\left[\begin{array}{c} CH_3 \\ | \\ -Si-O- \\ | \\ CH_3 \end{array} \right]_n$$

Long-chain polymers produced by polycondensation of hydrolysed dimethyldichlorosilane and methylphenyldichlorosilane. The starting materials are produced by the reaction of methylene chloride with metallic silicon (silicon-copper alloy) at 250–350°C and separation of the different chlorosilanes by fractional distillation. Hydrolysis of dimethylchlorosilane gives a product which is condensed to a polysiloxane at approx. 130°C, in the presence of a catalyst, *e.g.* phosphoroxychloride, sulphuric acid.

$$2CH_3Cl + Si(Cu) \longrightarrow (CH_3)_2SiCl_2$$
$$R_2SiCl_2 + 2H_2O \longrightarrow R_2Si(OH)_2 + 2HCl$$
$$nR_2Si(OH)_2 \xrightarrow[130°C]{-H_2O} (R_2SiO)_n$$

A high degree of purity is necessary in the dimethyldichlorosilane as impurities such as methyltrichlorosilane or silicon tetrachloride, cause crosslinking and therefore decrease the plasticity and elastic properties, while trimethylchlorosilane is a chain-stopper and decreases the mol. wt.

Polysiloxanes are colourless, plastic masses without the properties of rubber. They are reinforced by silicate fillers, oxides or carbonates, and cross-linked by a free radical reaction, using 1–3% of a peroxide. Benzoyl peroxide, dicumyl peroxide, ditert-butyl peroxide and dichlorobenzoyl peroxide are suitable curing agents. The last-named may be used to give vulcanisates free from porosity when cured without pressure. The speed of the peroxide reaction can be accelerated by alkali titanates. Vulcanisation takes place technically in two stages: heating under pressure to achieve thermal dimensional stability, and post-curing in hot air at 150–250°C. Increased reactivity in the product may be obtained by introducing into the structure vinyl siloxane groups which may be vulcanised with far smaller quantities of peroxide or by sulphur and accelerators. Silicone compounds may be processed cold.

Phenyl modified silicone

Vinyl modified silicone

Silicone rubber has moderate physical properties and higher shrinkage during vulcanisation than other elastomers. It has high resilience, excellent low temperature resistance (dimethyl polysiloxane to $-60°C$, methyl phenyl polysiloxane to $-100°C$), excellent resistance to high temperatures (up to 200°C, or for a short time up to 325°C) and shows only slight changes in properties over its working temperature range. It has good electrical properties, excellent resistance to ageing, ozone and discoloration on oxidation, and is resistant to mineral oils, alcohols and ketones. It swells in benzene hydrocarbons and chlorinated hydrocarbons, and is not resistant to high pressure steam. Nitrile groups introduced into the silicone rubber improve the oil resistance.

The products are marketed as silicone gum bases and loaded masterbatches.

Uses: include mechanical goods for use at low and high temperatures,

seals, electrical insulation, cable sheathing, insulating bands with glass-fibre fabrics, conveyor belts in the food industry. Tyres produced from silicone rubber and glass fibre by the US Rubber and the Dow Corning Corporations can resist temperatures from −45°C to 260°C. The glass fibres are bonded together with a silicone adhesive. Such tyres are claimed to be particularly useful for supersonic aircraft.
TN: GE Silicone Rubber
Rhodorsil
Silopren
Silastic
Silastomer
Silicone elastomer.

SILIN Pale-coloured reinforcing fillers:
Silin Al: precipitated aluminium silicate hydrate. 56% SiO_2, 10% Al_2O_3, s.g. 1·95, bulk density 170 g/l. *Uses:* in synthetic rubbers
Silin Ca: precipitated calcium silicate hydrate. 64% SiO_2, 16% CaO, s.g. 2·12, bulk density 180 g/l, particle size 0·04 micron. *Uses:* in natural and synthetic rubbers.

SILLITINE Group of white, semi-reinforcing fillers with a high silica content (70–75% quartz, 25–30% kaolin), for NR and SR. Free from rubber poisons, easily dispersed, improve extrusion and calendering properties, effective diluents for reinforcing fillers.
Types: N (particle size up to 15 μ), Z (up to 6 μ), sillicolloid (up to 3 μ) (412).

SILOPREN Silicone rubber with good electrical and rubbery properties over a wide temperature range. Used for insulating materials, seals, heat resistant rollers, expanded and sponge rubbers, vibration damping units.
B: colourless, plastic material, s.g. 0·98. Soluble in aliphatic and aromatic hydrocarbons, insoluble in alcohol and water. Has excellent resistance to high and low temperatures, exceptionally low permanent set, excellent ageing properties, good resistance to light and oil, good electrical properties. Benzoyl peroxide is used as the curing agent and filler must be used.
RS: degassed dimethyl polysiloxane with few reactive groups; compounds cure quickly and have very low shrinkage.
SVA: silicone rubber in the form of a masterbatch containing silica (43).

SILOPREN COMPOUND AL-50 Silopren base compound ready for vulcanisation and consisting of 100 parts Silopren B, 42 parts Silopren filler A, 3·4 parts silicone oil/peroxide paste (BP 50). Translucent, easily processable sheets, s.g. 1·16. The product may be modified by addition of fillers and additives (43).

SILOSET RUBBER Group of liquid or paste-form silicone elastomers which may be cast or painted on, and which may be cold cured (60).

SILTEG Aluminium silicate; white, reinforcing filler. Type AS 5 for stable compounds with a high hardness, Type AS 7 for hard compounds (144).

SILTHIAN Reaction products of vinyl chlorosilane and hydrogen sulphide in the presence of pyridine. Added to natural rubber or synthetic

types containing butadiene (at least 25% butadiene) to improve the re-inforcing effect of fillers. When heated during the mixing process (120–200°C), hydrogen sulphide is liberated, and is partially absorbed by the elastomer while a simultaneous reaction occurs between the vinyl silane and the filler. Increases tensile strength and modulus reduces com-pression set and hysteresis. Effective with all fillers.
Quantity: 0·5–10% of the filler weight.

SINSEN Japanese produced, partially hydrolysed polyacryloni-trile fibre.

SIPALIN Group of softeners:
AOC: dicyclohexyl adipate
AOM: dimethyl cyclohexyl adipate
MOI: methylisopropyl adipate
MOM: blend of dimethyl cyclohexyl and methyl adipates.

SIR Abbreviation for styrene-isoprene rubber in the nomenclature of ASTM D 1418.

SIX Abbreviation for sodium isopropyl xanthate.

SKA- Russian produced alde-hyde rubber.

SKA RUBBER Russian pro-duced Buna; butadiene from alcohol.

SKB RUBBER Russian produced Buna rubber; butadiene from the cracking of petroleum:
SKB: lithium catalyst. For use at low temperatures
SKB 35: sodium catalyst
SKB 50: sodium catalyst.

SKBM Collective term for poly-butadiene rubbers produced (over a lithium catalyst) in the USSR. Resist-ant to low temperatures.

SKD Russian produced cis-1:4-polybutadiene: SKD-10, liquid poly-mer, mol. wt. 2000–10 000. Contains 10% carboxyl groups.

SKELLYSOLVE Group of sol-vents for substances such as rubber and plastics:
B: n-hexane
C: n-heptane
D: n-octane
H: light benzine
R: solvent benzine
PA: diethyl glycol phthalate
PM: dimethyl glycol phthalate
REA: acetyl ethyl ricinoleate (319).

SKEP Russian produced ethylene-propylene copolymer.

SKEPT Russian produced ethylene-propylene terpolymer with undisclosed third component. Glass transition temperature -57 to $-67°C$ depending on ethylene : propylene-ratio, unsaturation 0·3–2%.

SKF- Russian produced fluoro-carbon rubber:
26: 26% fluorine
32: 32% fluorine.

SKI Russian produced synthetic cis-1:4-polyisoprene.

SKI-3 Russian produced syn-thetic cis-polyisoprene rubber. De-veloped 1957–58, using a modified complex catalyst (Ziegler type), s.g. 0·93, Mooney 50–75, cis-1:4 90–98%, trans-1:4 1–7%. Is practically identi-cal to Ameripol SN. The mol. wt. is lower than that of the product obtained using a lithium catalyst.

SKID Russian produced isoprene-butadiene rubber.

SKIM LATEX The serum left after centrifuging latex. Dry rubber content approx. 3–8%.

SKIMMING Application of a coating to frictioned textiles.

SKIMMINGS Porous sheets of poor commercial quality, produced from scrap latex foam.

SKIM RUBBER Rubber produced from the skim, *i.e.* the serum left after centrifuging natural latex, which contains 3–8% rubber. Has a relatively high s.g. and a high protein content so that the compounds produced from it have a strong tendency to scorch. Vulcanisates are harder and less flexible than usual. The considerable variation in quality limits its use. Its strong tendency to scorch is aggravated by the use of acidic accelerators; if basic accelerators are used in the compounds the scorching tendency can be decreased. Treatment of the skim latex with proteolytic enzymes (pancreatin, trypsin) decreases the nitrogen content to that of sheet rubber and a product with normal curing characteristics is obtained. Decomposition may be activated by adding 20–30% calcium ions; this also reduces the fairly high copper content. Treatment is at 38–45°C. Vulcanisates are harder with a lower modulus and higher elongations than those based on RSS. The great variability can be reduced by blending with an equal amount of SBR, when the slow cure rate of SBR is also improved.

Prepared skim rubber with low protein content is marketed under the following classifications: ACT 3, Firestone (Liberia); BRUPA, RCMA; Sentarub, RCMA.

SKLIRO Mixture of wool fatty acids, m.p. 50–55°C, iodine No. 27, acid No. 135, saponification No. 165. Softener and dispersing agents. Can be used instead of stearic acid and pine oil tar. Facilitates the dispersion of blacks, activates thiazole accelerators; improves flex cracking resistance.

Uses: tyres, footwear, belting, hoselines, mechanical goods (22).

SKMS Russian produced synthetic rubber; butadiene copolymers: 2: butadiene-methylstyrene rubber

	Skim	Trypsin skim	RSS	Skim latex
Acetone extract, %	8·3	7·3	2·1	10·3
Nitrogen, %	2·4	0·4	0·4	2·3
Ash, %	0·9	0·3	0·3	2·2
Rubber KW, %	77	90	94	
Mooney Scorch, min (ACS-I compound)	4–7	9–10	14–15	
Rubber content, %				7
Total solids, %				11

10: low temperature resistant butadiene-methylstyrene rubber, 90/10

30: butadiene 2-methylstyrene rubber

30A: 30% methyl styrene

30AM: 30% methyl styrene. Cold polymer extended with oil

30 ARKM: modified, oil extended cold 70:30 butadiene-styrene

50: butadiene-methylstyrene rubber with a high methylstyrene content. Reinforcing resin.

SKMVP Type classification for Russian produced butadiene-vinyl-pyridine copolymers; the index number indicates the content of vinyl pyridine as a percentage.

Types:
10, 15, 25, 40, hot polymers (50°C). Give vulcanisates with low hysteresis and excellent abrasion resistance
10A, 15A, 25A, 40A, cold polymers (5°C). Give vulcanisates with low hysteresis and excellent abrasion resistance.

SKN Russian produced nitrile rubber. Acrylonitrile content: SKN-18 17–20%, SKN-26 27–30%, SKN-40 36–40%. All types contain 1·8–3% Neozone D as stabiliser.

SKP Type classification for Russian produced polypiperylene rubbers.

SKS Russian produced butadiene-styrene copolymers:
SKS-10: 10% styrene. Used at low temperatures
SKS-30: 30% styrene. Hot polymer (equivalent to SBR 1000)
SKS-30A: 30% styrene. Cold polymer (equivalent to SBR 1500)

SKS-30AM: 30% styrene. Cold polymer extended with oil (equivalent to SBR 1700)
SKS-30M-15: 30% styrene. Polymer extended with oil
SKS-50: 50% styrene.
SKS-90: 90% styrene. Reinforcing resin
Types −30P, −30U, −30K, −30S, −30SR, −50, −50I, −50GP are latices with 30% and 50% styrene.

SKSM Russian produced butadiene-α-methyl styrene-copolymer.

SKT Russian produced heat resistant polydimethyl siloxane rubber.

SKTV Russian produced vinyl ethyl dimethyl siloxane rubber.

SKV Type classification for a polybutadiene rubber produced (over a potassium catalyst) in the USSR. Has better low temperature resistance than SKBM.

SKVF-30A Russian produced butadiene-vinyl furan copolymer.

SKVP Russian produced butadiene-vinyl pyridine copolymer. Polymerised at 50°C. Vulcanisates have high abrasion resistance and low hysteresis. The index number indicates the percentage of vinyl pyridine.

SKYPRENE Group of sulphur unmodified polychloroprene rubbers (422).
B-31: medium crystallisation, Mooney (50 ML 1 + 4, 100°C): 34–42. *Use:* cables, insulations, extrusion. Comparable with: Neoprene WM-1, Baypren C 211, Butaclor MC-31, Denka Chloroprene M-30

G-40 S: fast crystallisation, Mooney 75–89. *Use:* adhesives. Comparable with: Neoprene AD-20, Baypren C-320, Butaclor MA-40S, Denka Chloroprene A-90

G-40 T: fast crystallisation, Mooney 90–105. *Use:* adhesives. Comparable with: Neoprene AD-30, Baypren C-330, Butaclor MA-40T, Denka Chloroprene A-100

G-41 H: fast crystallisation, Mooney 75–89. *Use:* adhesives. Comparable with: Neoprene AC-soft, Baypren C-321

G-41 K: fast crystallisation, Mooney, 90–105. *Use:* adhesives. Comparable with: Neoprene AC-medium, Baypren C-331, Butaclor MA-41K, Denka Chloroprene TA-95.

Y-31: medium crystallisation, Mooney 90–110. *Use:* hoses, mechanicals, adhesives. Comparable with: Neoprene WHV-100, Baypren C-230, Butaclor MH-31.

SLAB Sheets of coagulum, 2–15 cm thick, in various sizes and often up to 50 kg in weight. Produced in several parts of Indonesia by coagulation of undiluted latex in wooden cases, canisters, large tins, etc. or in holes and grooves in the ground which have been lined with clay. Materials such as alum, formic acid, sulphuric acid, acidic plant saps, fermented coconut juice and urine, are used as coagulating agents. Initially the slabs have a slimy surface which dries to a dirty crust after a few days. The weight of the slabs is often deliberately falsified by addition of stones, sand, iron, bark; the slabs also contain scraps and cup lumps. Slabs are usually sold to the remilling factories by Chinese traders and processed by washing on heavy crêpe rollers to blankets (thick brown crêpe) or thin browns.

SLUDGE The residue in centrifuges and storage containers; mainly magnesium ammonium phosphate, also lutoids, Frey–Wyssling particles and small quantities of rubber. The term also includes all dirty precipitates from the latex on sieves, in vats and tanks.

SM- 61 and 62. Silicone mould release agents (198).

SM LATEX (SM: styrene modified.) A natural latex modified with styrene and containing 23% bound styrene.
Uses: give a higher resistance to compression in foam rubber.

SMR Abbreviation for Standard Malaysian Rubber. Technically classified standard rubber in the form of compact bales of max. weight 70 lb. Grading is according to the amounts of impurities rather than by colour:
Types:
SMR 5: rubber of a high quality, max. impurities 0·05%
SMR 5 L: as SMR 5 but a lighter colour
SMR 10: max. impurities 0·10%
SMR 20: max. impurities 0·20%
SMR 50: max. impurities 0·50%.
SMR EQ: extra quality, light colour, max. impurities 0·02%
SMR 5CV: constant viscosity rubber, Mooney 60 ± 5
SMR 5LV: low viscosity rubber, Mooney 50 ± 5

SNO-BUGGY Vehicle with the largest tyres yet developed (300 cm high, 120 cm wide); for use in loose snow, sand and other soft surfaces. The tyre pressure is 4–25 lb/in^2, depending on the load. The vehicle is equipped with 8 tyres. Each wheel has its own electric power and a separate steering system. Developed in the USA by R. G. Tourneau, Inc.

SOCABUTYL French produced butyl rubber (320).

SOCAL U1 Ultra-fine precipitated calcium carbonate (321).

SOCTEX CD Concentrated latex with a low ammonia content (0·15%). Rubber dry content min. 60%, total solids min. 62%, copper 0·001% max. on the rubber, manganese 0·0005% max. on the rubber.
Uses: for foam rubber and adhesives.

SODIUM ACETATE
$CH_3COONa.3H_2O$. Colourless crystals to white powder, m.p. 58°C, s.g. 1·45. Soluble in water and alcohol.
Uses: retarder in NR and SR, with the exception of butyl rubber.

SODIUM ALGINATE Yellowish powder. Soluble in water and diluted alcohol (max. 30%).
Uses: thickening and creaming agent for latex.

SODIUM AZIDE NaN_3; sodium salt of nitrogen hydracid N_3H. Produced by passing nitrous oxide into molten sodium amide:

$$NaNH_2 + N_2O \rightarrow NaN_3 + H_2O$$

White solid, may be melted without decomposing and explodes when strongly heated. Soluble in water. Was recommended as a preserving agent for latex in a concentration of 0·01% combined with sequestrol and ammonia. Sodium azide is effective as a bactericide and prevents the formation of volatile fatty acids.

SODIUM BENZOATE
C_6H_5COONa. White crystalline powder, m.p. 250°C, s.g. 2·8. 1 g soluble in 1·8 ml water, 75 ml alcohol.
Uses: added to latex, sometimes with sodium nitrite, for protection against rust, *e.g.* in emulsion paints. BP 652 892.

SODIUM BICARBONATE
$NaHCO_3$, s.g. 2·2. Blowing agent for sponge rubber with an open cell structure. To improve its dispersion in rubber and increase its rate of decomposition, sodium bicarbonate is used primarily as a 50% (approx.) dispersion in a light mineral oil; s.g. 1·3. Sodium bicarbonate decomposes above 70°C, liberates carbon dioxide and water; the reaction is complete above 300°C. The recommended vulcanisation temperature is 150°C.
Quantity: 1–15% of the dispersion, 1·7 parts stearic acid to 1 part of dispersion.
Uses: in natural rubber, neoprene, butyl rubber, SBR and NBR. Has no effect on ageing properties, does not stain or bloom, activates acidic accelerators slightly.
TN: Unicel S (50% dispersion) (6).

SODIUM CASEINATE
(Nutrose). White powder, water soluble.
Uses: as for casein.

SODIUM CYCLO HEXYL ETHYL DITHIOCARBAMATE
$(CH_2)_5CH.N(C_2H_5).CS.S.Na$. Yellowish, hygroscopic crystalline powder, s.g. 1·25. Soluble in water, alcohol, ethyl acetate, acetone, sparingly soluble in methylene chloride

and carbon tetrachloride, insoluble in benzene. Ultra-fast accelerator. Suitable for thin walled rubber articles which are vulcanised in aqueous accelerator solution, and for thin walled latex articles. Acts as an activator in latex compounds with insoluble ultra-fast accelerators. Vulcanisation temperature 80–95°C. Vulcanisates have a moderate hardness. Does not discolour, is odourless and non-toxic. Suitable for white and coloured articles.

Quantity: aqueous accelerator solutions 1–2%, solid rubber and latex 0·5–1%, activator in latex compounds 0·2–0·4%.

TN: Accelerator WL (408).
Vulkacit WL (43).

SODIUM DECYL BENZENE SULPHONATE Wetting agent, stable in neutral, alkaline and acidic solutions. Used in latex compounds for the impregnation of tyre cord and textiles. Stability of the compounds is increased by approx. 25%, the wetting time decreased by approx. 30%.

Quantity: 0·1–0·5% on the latex.
TN: Santomerse D (powder form)
S (30% aqueous solution) (5).

SODIUM DIBUTYL DITHIO-CARBAMATE

$(C_4H_9)N.CS.S.Na$. Yellow liquid stabilised with hydrazine (47% aqueous solution); s.g. 1·09. Primary, ultra-fast accelerator for latices of NR, NBR, CR and dispersed reclaims. In natural latex active above room temperature, but tends to cause prevulcanisation. However, it has only a slight effect on stability. Has less effect in SBR than in NR, and less in NBR and CR than in SBR; very fast curing in dispersed reclaim above 80°C. Activates thi-

azoles. Vulcanisates have good ageing properties and a high modulus.

Quantity: NR 0·5–2% with 1–2% sulphur, coating compounds up to 4%. SBR 0·5–2% with 1–3% sulphur, cements 3–4%,
NBR 0·5–2% with 1·5–2% sulphur,
CR 1–2% with 1·5% sulphur (combination with thiurams gives films with a high tensile strength),
dispersed reclaim 1–2%.

Uses: include latex articles, coating compounds, adhesives.

TN: Butyl Namate (solution) (51)
Pennac SDB (47% solution) (120)
Robac SBUD
SBUD (47% solution)
Tepidone (6)
Vulcacure NB (47% solution) (57).

SODIUM DIETHYL DITHIO-CARBAMATE

$(C_2H_5)_2N.CS.S.Na$. White, crystalline powder, m.p. approx. 95°C, s.g. 1·36. Soluble in water, sparingly soluble in benzene and chloroform, solubility in rubber approx. 0·25%. Ultra-fast accelerator. A low vulcanisation temperature is possible when used in combination with diethylammonium diethyl dithiocarbamate. Exposure to light rays can cause discoloration. May be used in solid rubber, natural latex, and SBR latex (in latex as a 10% aqueous solution).

TN: Eveite L (59)
Namate (obsolete) (51)
Nocceler SDC (274)
Super Accélérateur 1500 (20)
Vulcafor SDC (60).

SODIUM DIMETHYL DITHIO-CARBAMATE

$$H_3C{\diagdown}\atop H_3C{\diagup} \hspace{-0.2em} N{-}CS{-}S.Na$$

Soluble in water. Ultra-fast accelerator, short-stop for emulsion polymerisation.
TN: Accélérateur soluble Lat. 4 Fl (19)
Sharstop 204 (40% aqueous solution)
Thiostop N (40% aqueous solution)
Vulnopol NM (57).

SODIUM FLUOROTITANATE
Can be used as a gelling agent for natural latex and SBR latex in the production of foam rubber and other articles. USP 2 472 054 (1949).

SODIUM ISOBUTYL NAPHTHALENE SULPHONATE Yellowish powder, s.g. 1·43. Soluble in water and most organic solvents. Wetting agent for latex, particularly for impregnation purposes. Prevents the formation of condensation droplets during open steam heating, release agent.
Quantity: 5–10 parts of an aqueous 10–20% solution.
TN: Erkanto. BX (43).

SODIUM ISOPROPYL XANTHATE $(CH_3)_2CH.OCS.S.Na$. White powder, s.g. 2·11. Soluble in water. Ultra-fast accelerator, zinc oxide is necessary.
Quantity: 2%.
Uses: latex articles, self-vulcanising latex compounds.
TN: Accélérateur soluble Lat. 5 (19)
Accélérateur rapide 5 R Special (19)
Super accélérateur 6000 (20)
Vulcafor SPX (60).

SODIUM LAURYL SULPHONATE
$CH_3(CH_2)_{10}OSOONa$. White to yellowish crystals. Soluble in water. Gives neutral reaction.

Uses: wetting and dispersing agent for fillers in latex compounds.
TN: Irium.

SODIUM-2-MERCAPTO-BENZTHIAZOLE

Accelerator (obsolete).
TN: R-23 (5).

SODIUM PENTACHLORO-PHENATE C_6Cl_5ONa. Brownish-yellow powder, s.g. 2·0, pH (1% solution) 9·0. Preservative against bacteria and mildew. In natural rubber, in a concentration of 0·3–0·4%, in concentrated latex with a low ammonia content (low ammonia latex) 0·3% with 0·1% ammonia.
TN: Santobrite (5)
Dowicide G (61).

SODIUM PENTACHLOR-THIOPHENATE Reclaim agent for SR and NR, especially for SBR/NR mixtures.
TN: Peptazin NA (428).

SODIUM PENTAMETHYLENE DITHIOCARBAMATE
$(CH_2)_5N.CS.S.Na$. Yellowish white powder with characteristic odour, m.p. 280°C, s.g. 1·4. Soluble in water, insoluble in organic solvents. Ultra-fast accelerator. Piperidine content 38%. A certain amount of zinc oxide is necessary for activation. The addition of stearic acid is not absolutely necessary but advisable. Magnesium carbonate has a slightly retarding effect. Vulcanisation temperature 100°C. Good ageing properties.
Uses: production of rubber thread from latex.

439

Quantity: in gum compounds 0·25–0·5% with 3–5% sulphur.
TN: Robac SPD (311)
Vulcafor SPD (60).

SODIUM-o-PHENYL PHEN-ATE White powder with faint odour, s.g. 1·307. Preservative for latices, NR and SR.

SODIUM POLYACRYLATE Aqueous solutions of sodium polyacrylate are highly viscous liquids, pH 10–11, stable in alkaline environments, unstable below pH 6. Thickening agent and protective colloid for latices with high filler loadings. Used in dipped articles, impregnations, coating compounds.
Quantity: 2–4% (15% solution) on the latex.
TN: Texigel SPA 4 (163)
Thickener 110, 215 (344).

SODIUM SILICOFLUORIDE Na_2SiF_6. White, granular powder, s.g. 2·68. Soluble in 150 parts cold and 40 parts warm water, insoluble in alcohol.
Uses: gelling agent for latex compounds in the production of foam rubber.

SOFTACK Viscous liquid, s.g. 0·97. Softener, tackifier, processing aid. Retards the blooming of sulphur in unvulcanised compounds, gives improved ageing properties, has no effect on vulcanisation.
Quantity: 1–6% (114).

SOFTENER, ANTISTATIC Softener which prevents the occurrence of static electricity.

SOFTENING Process used to increase the softness of polymers. External softening: addition of various softeners by mixing with the polymer. Internal softening: inclusion of mobile groups in the molecular chain by copolymerisation.

SOFT RESIN H 60% solution of a water soluble urea-formaldehyde condensation product. Softener (32).

SOLARISATION EFFECT In 1839 N. Hayward discovered that sunlight causes a surface hardening of rubber/sulphur mixtures. The patent on this process [USP 1090 (1839)] was taken out by C. Goodyear but he did not succeed in achieving a permanent proofing by the method.

SOLE CREPE Sole crêpe is the only form of raw rubber delivered by the plantation for direct use. Production is the same as for standard crêpe of good quality and colour. As sole crêpe must be as white as possible it is normally coagulated twice or bleached. During the so-called precoagulation the unstable yellow fraction (lutoids, Frey–Wyssling particles) is precipitated by addition of 1/6–1/3 of the normal quantity of acid and separated from the remainder of the rubber. Then the remaining quantity of acid is added and coagulation completed in the normal manner. With white latices, *e.g.* LCB 1320, which normally give a pale-coloured crêpe, the yellow fraction may be bleached instead of being precoagulated, using 0·05–0·1% xylyl mercaptan (RPA-3) as bleaching agent. Sometimes both bleaching and precoagulation are carried out. When xylyl mercaptan is used an exact quantity must be added: too high a concentration gives a rubber which is too soft, and if formalin is used as a preserving agent and large amounts of iron are present in the diluting water, a grey to brown discoloration may occur which only becomes visible weeks after production. The crêpe is dried under diffused light in the second stage of drying, as the bleaching effect is only small when drying is in the dark.

The sheets are milled more intensively than is usual with standard crêpe, as each must be 1/32 in thick and free from holes. By carefully plying together 2, 4, 5, 6 individual sheets the sole crêpe is built up to 1/16, 1/8, 5/32 in, and so on. The sheets are smoothed and plied, initially on a table using a hand roller. After this they are pressed together in a calender, sometimes after prewarming, when the self-tack of the sheets is sufficient to ensure a strong adhesion. The sheets are cut to plates of 13 by 36 in and packed in wooden cases containing 49–98 sheets, depending on the thickness. Nine shoe soles may be produced from one plate. (1/16 in = 0·16 cm, 1/8 in = 0·32 cm, 3/16 in 0·48 cm, 1/4 in = 0·64 cm, 5/32 in = 0·40 cm; 13 by 36 in = 33 by 91·4 cm.)

SOLIDAGO Goldenrod, solotarnik. Perennial of world-wide distribution; occurs as approx. 125 different types. Contains rubber in the leaf parenchyma. Rubber content: 1·9–4·6% of the total solids, in hybrids, up to 12%. Suggested by T. A. Edison for rubber production in the USA; is used in Russia.

SOLID RUBBER TYRES Huge tyres, trapezium-shaped in cross-section, for lorries which carry heavy loads at low speed, also for stationary vehicles. The metal/rubber bond is achieved using isocyanates, cyclised rubber, hard rubber or a galvanised brass interlayer.

SOLITHANE 113 Liquid polyurethane rubber which may be vulcanised at room temperature (208).

SOLKA-FLOC Pure cellulose powder, s.g. 1·58. Processing aid and filler; improves dimensional stability, decreases shrinkage, increases the hardness and reduces s.g. of compounds. By reducing the 'nerve' of the compound it assists shaping (322).

SOLPRENE Trade name for synthetic rubbers polymerised using the solution process (130).

SOLPRENE X Stereospecific SBR with 75% butadiene and 25% styrene. Produced by solution polymerisation with lithium as a catalyst. Has good mixing, extruding, calendering and moulding characteristics. May be blended with all normal rubber types. Vulcanisates have good resistance to abrasion and low temperatures, good flex cracking resistance with an otherwise moderate tensile strength and tear resistance (130).

SOL RUBBER The benzene soluble part of rubber.

SOLUX NS Phenolic antioxidant for natural and synthetic rubber. Non-staining, non-discolouring (6).

SOLVAR Group of partially hydrolysed polyvinyl acetates (323).

SOLVATION Swelling, gelling and dissolving of a polymer or resin by a solvent or softener.

SONNTAG MACHINE
Machine used to investigate the dynamic fatigue characteristics of rubber bonded parts.

SORBITP Butyl naphthalene sulphonate. Wetting agent for latex (69).

SORBOPRENE Polyurethane foam material (325).

SOREDUR Group of polyester resins (324).

SOREFLON Polytetrafluoroethylene (326).

SOVALOID Group of aromatic hydrocarbon condensation products. Softeners and plasticisers for natural and synthetic rubbers (327).

SOVATEX A Anionic wetting agent. Accelerates the coagulation of latex (178).

SOVIDEN Russian produced vinyl chloride-vinylidene chloride copolymers.

SOWPREN Russian produced polychloroprene; equivalent to neoprene.

SOXINOL Group of Japanese produced accelerators (328):
808: butyraldehyde-aniline condensation product
A: thiocarbanilide
D: diphenyl guanidine
DM: dibenzthiazyl disulphide
DT: di-o-tolyl guanidine
M: mercaptobenzthiazole
PM: dinitrophenyl benzthiazyl disulphide
PX: zinc ethyl phenyl dithiocarbamate
SD: o-tolyl diguanidine

TS: tetramethyl thiuram monosulphide
TT: tetramethyl thiuram disulphide
Z: zinc mercaptobenzthiazole.

SPAN Group of non-ionic surface active substances; sorbitol fatty acid esters:
20: sorbitan monolaurate
40: sorbitan monopalmitate
60: sorbitan monostearate
62: sorbitan monostearate
65: sorbitan tristearate
80: sorbitan mono-oleate
85: sorbitan trioleate (97).

SPANDEX Elastic fibre based on polyurethane.
TN: Lycra (6)
Vyrene (246).

SPARBUNA A highly polymerised oil extended Buna rubber produced in Germany in 1943.

SPD Abbreviation for sodium pentamethylene dithiocarbamate.

SPDX Group of accelerators (100):
Standard: lead(phenyl aminoethyl)-phenyl dimethyl dithiocarbamate
A: lead-(o-tolyl aminoethyl)-o-tolyl dimethyl dithiocarbamate
G: lead dithiocarbamate with 25% mineral oil
GH: mixture of lead salts of a dithiocarbamate blend and a diphenyl ethylene diamine radical
GL: blend of a mixture of lead salts of complex dithiocarbamates with a diphenyl ethylene diamine radical
L: lead dithiocarbamate type.

SPECIFIC GRAVITY The ratio of weight to volume. Determined using Mohr's scales, from the loss of weight of a test piece in water. With sponge or foam rubber, the s.g. is measured by a suspension method. Specification: DIN 53 550.

SPECIFIC HEAT The specific heat of vulcanisates is a cumulative property of the specific heats of the raw rubber and the compounding ingredients.

SPECIFIC WEIGHT Natural rubber 0·92, SBR, hot type (50°C) 0·94, SBR, cold rubber (5°C) 0·94, neoprene 1·23, nitrile rubber 75:25 0·96, nitrile rubber 60:40 1·00, butyl rubber 0·91, thiokol 1·35.

SPECKS Coarse or poorly dispersed compounding ingredients on the surfaces or the cut surfaces of vulcanisates and compounds.

SPENKEL Polyurethane prepolymers for paints (329).

SPF Abbreviation for super processing furnace black.

SPHERON Trade name for a group of channel blacks (139). 3 and 4: HPC, 6: MPC, 9: EPC, C, I and N: CC.

SPILLER'S RESIN Traditional term for a resinous oxidised rubber. Described in 1865 by J. Spiller; his analysis gave 27·5% oxygen.

S PLY 10 An adhesive for rubber/ p.v.c. based on a graft polymer of 45% NR and 55% methyl methacrylate, as a 4·5% solution in 94 parts toluene and 6 parts ethanol. Produced by solution polymerisation.

S POLYMER 60 Styrene-isobutylene copolymer. Thermoplastic product. Soluble in most solvents. *Uses:* in erasers, coatings; substitute for factice (178).

SPRAYED RUBBER LS rubber, latex sprayed rubber, whole latex rubber. A rubber containing all the serum constituents. Produced by Hopkinson's process, in which ammonia stabilised latex is sprayed from a fast rotating disc fixed to the ceiling of a truncated, pyramid-shaped chamber. On introducing hot air into the chamber the finely divided latex immediately dries and falls to the ground as white flakes. A sponge-like mass is formed which may be compressed into solid blocks. The rubber has a moisture content of 0·5–1·5%. Because of the serum constituents the nitrogen content and the acetone and water extracts are higher than those of coagulated rubber. The rubber is more hygroscopic, harder and cures more quickly than coagulated rubber. The power required for plasticising is greater. In 1930 production was approx. 150 tons per month, but it is now produced only in small quantities by the US Sumatra Rubber Estates, Kisaran (Sumatra).

SPREADING Process for coating textiles with a highly viscous rubber solution or latex compound, by use of a spreading machine.

SP RUBBER Superior processing rubber. Special rubber with excellent extruding and processing properties. Produced by blending vulcanised and fresh latex prior to coagulation. Extruded articles have excellent dimensional stability. Five types are produced: SP smoked sheet, SP crêpe and SP air dried sheet are made from a mixture of 20% vulcanised and 80% fresh latex coagulated and processed by the normal method; PA 80, a concentrated form, is made from 80% vulcanised and 20% fresh latex, and the crumbs produced are dried and compressed into blocks; SP brown crêpe is made from 1 part wet PA 80 crumbs and 3 parts wet scrap rubber, processed on a mill into thin brown crêpe (the process has been developed by the Rubber Research Institute of Malaya).

In production the latex is mixed with a dispersion of a curing system, heated by steam to 82°C and the temperature maintained at 82–85°C for 2 h. After cooling, the latex is decanted to remove excess vulcanising ingredients and sediment. The vulcanised latex is blended with 20% fresh latex in the desired ratio and coagulated with formic acid. It is dispatched in paper sacks.

SPRUCE, RICHARD Botanist, *Hevea* specialist, 1817–1893. From 1849–61 conducted botanical research in tropical S. America; described *Hevea spruceana*.

SRA-1 Blend of mercaptobenzthiazole and diphenyl guanidine (6).

SRA-2 Blend of zinc-2-mercaptobenzthiazole and zinc diphenyl guanidine benzthiazyl mercaptide. Accelerator (obsolete) (6).

SRD Abbreviation for Synthetic Rubber Division; successor to the Office of Rubber Reserve (ORR). Controls state synthetic rubber factories in the USA.

SRF Abbreviation for semireinforcing furnace black.

SRF 3 SRF gas black with a high modulus (139).

SRF-HM Abbreviation for semireinforcing furnace black with a high modulus.

S RUBBER Trade name for styrene-butadiene copolymers. Hot and cold rubbers, black masterbatches, oil extended rubbers and latices produced by the Shell Chemical Corp., USA (2).

SS TEST Simulated service test. Testing method which simulates practical service conditions.

STABELAN Stabilisers based on organic zinc and barium salts, for vinyl polymers and copolymers (127).

STABELITE ESTERS Group of tackifiers; esters of hydrogenated resins:
3: triethylene glycol ester. Brown liquid, s.g. 1·08.
10: glycerine ester. Hard, brittle, brown resin, m.p. 80–85°C, s.g. 1·07.
Uses: tackifiers for natural rubber, synthetic rubbers, latices, adhesives and cements.

STABILISATOR Group of antioxidants. AN, reaction product of α-naphthylamine and glucose, AR, N-phenyl-β-naphthylamine (364).

STABILISATOR H 116 Almost colourless, soft resin, m.p. 38–48°C, s.g. 0·925. Stabiliser, *e.g.* for polyisobutylene, rubber, polyethylene, polyvinyl ether (217).

STABILITE Group of antioxidants for rubber and latex:

Stabilite: diphenyl ethylene diamine
Alba: di-o-tolyl ethylene diamine
FLX: mixture of diphenyl ethylene diamine and diphenyl-p-phenylene diamine. Black granular solid, s.g. 1·12. Used in NR and SR to protect against flex cracking. Discolours
L: diphenyl propylene diamine (100)
White: polyalkyl phenol. Brown liquid, s.g. 0·91. For NR and SR. Non-staining and non-discolouring. Protects against heat and oxygen.

STABLEX Group of stabilisers and wetting agents for latex:

A: mixture of higher alcohols. White powder, s.g. 0·73. Soluble in water. For synthetic and natural latices. Has no effect on vulcanisation, improves the penetration into textiles
B: mixture of higher alcohols. Brown powder, s.g. 1·00–1·05. Soluble in water; added to latex as a 40% solution. Used in natural latex; has no effect on vulcanisation
G: polyalkylated benzene sulphonate. Brown powder. Soluble in warm water; added as a 40% aqueous solution. For natural and chloroprene latices. Has no effect on vulcanisation, improves the penetration into textiles (309).

STAFLEX Group of softeners and stabilisers:

BDP: butyl decyl phthalate
BR: butyl ricinoleate
BRA: butyl acetyl ricinoleate
CP: amyl decyl phthalate
DBEA: dibutoxyethyl adipate
DBES: dibutoxyethyl sebacate
DBP: dibutyl phthalate
DIODA: octyl decyl adipate
DIODP: octyl decyl phthalate
DIOP: diiso-octyl phthalate
DOA: di-2-ethyl hexyl adipate
DOP: di-2-ethyl hexyl phthalate
DOS: dioctyl sebacate
DOZ: di-2-ethyl hexyl azelate
IX: methoxyethyl ricinoleate
IXA: methoxyethyl acetyl ricinoleate
KA, KB, KD: high mol. wt. phthalates
LA: mixture of sebacic and adipic acid esters (330).

STAFOAM Group of polyurethane foam materials (66).

STAMICOL Dutch produced plastic based on polyethylene polysulphide.

STANDARD TEMPERATURES ASTM D 1349 suggests the following possible temperatures (°C) for the testing of elastomers: -75, -55, -40, -25, -10, 23, 50, 70, 100, 125, 150, 175, 200, 225, 250 (with variations of $\pm 2°C$).

STANVACID Types PC, W, T. Activators; substitutes for stearic acid (178).

STANVAPINE **B** Bituminous tar. Tackifier (178).

STA-SOL Lecithin concentrate made from soya. Antioxidant (197).

STATEX Trade name for a group of furnace blacks.
Types:
93: HMF, 125: ISAF, 160: SAF, 930: HMF with 20% oil, B: FF, K: FF of fine particle size, M: FEF

445

(MAF), R: HAF, RH: HAF black for use in polybutadiene compounds; has better dispersion properties than normal HAF (159).

STAUDINGER, HERMANN

Dr, 1881–1965. Professor of organic chemistry at Karlsruhe, Zurich and Freiburg. Established the science of macromolecular chemistry. From 1951, director of the State Institute for Macromolecular Chemistry, Freiburg. He was the first to recognise that various natural substances, which had been termed micelle colloids, reacted as radical compounds, and he described them as macromolecules, establishing the theoretical foundation and the high molecular structure of rubber and cellulose. His main research was into Kekulé's theory on principal valency bonds. His work includes research into ketene (from 1905), autoxidation, pyrethrin synthesis, the explosion reaction, oxalyl chloride, aliphatic diazo compounds, synthetic pepper and coffee aromas. From 1921 he undertook research into macromolecular substances and colloids, rubber, cellulose, starch, plastics, isoprene (over 300 publications). Received many honours and was awarded the Nobel prize in 1953. Publications: *Ketene*, 1902, *Anleitung zur organischen qualitativen Analyse*, 1948, *Tabellen für allg. und anorg. Chemie*, 1947, *Organische Kolloidchemie*, 1947, *Makromolekulare Chemie und Biologie*, 1947. Editor of *Makromolekulare Chemie*.
Lit.: *Rubber J.*, 149, 1967, p. 49.

STAUDINGER INDEX

Viscosity constant used in the determination of the mol. wt. of high polymer materials.

STD-X Mixture of phenyl-β-naphthylamine and dinaphthyl-p-phenylene diamine. Grey powder, s.g. 1·20. Antioxidant (286).

STEARIC ACID

$CH_3(CH_2)_{16}.COOH$. White, granular powder, m.p. 69–70°C, b.p. 383°C, s.g. 0·838. Soluble in benzene, chloroform, carbon tetrachloride, amyl acetate and hot alcohol. Technical stearic acid consists of a blend of stearic acid with palmitic acid; m.p. 55–56°C.
Uses: the fatty acid most widely used as a vulcanisation activator, softener.

N-STEAROYL-p-AMINO-PHENOL

$$HO-\!\!\!\bigcirc\!\!\!-NH-\overset{\overset{\textstyle O}{\|}}{C}-C_{15}H_{31}$$

General purpose antioxidant. White powder, m.p. 131–134°C.
TN: Suconox 18 (333).

STEEL CORD TYRES Tyres in which steel cords are used in place of the usual textile (cotton, rayon, nylon). Depending on the size of the tyre, 1·5–2·8 km of fine steel wire are used per tyre. The heat absorption is 5–10°C lower than with normal tyres. Such tyres have excellent resistance to external mechanical damage.

STENOL Emulsifier based on cetyl and stearyl alcohols.

STERENE Group of activators and softeners; mixtures of higher fatty acids (331).

STEREOSPECIFIC POLY-MERISATION Polymerisation of dienes to give products with a specific stereostructure by using stereospecific catalysts such as lithium, butyl lithium or Ziegler catalyst types, *e.g.* a mixture of triethylene aluminium and titanium tetrachloride. Various catalysts can be used but each is chosen particularly for the type of polymer required. Air, moisture and polar materials (alcohols, ketones, esters, ethers) may deactivate or destroy the catalysts. The choice of catalysts must be made empirically. Examples of stereospecific polymers are cis-1:4-polyisoprene and cis-1:4-polybutadiene.

STEREOSPECIFIC POLYMERS Diolefins and α-olefins may be polymerised stereospecifically, *i.e.* to a uniform spatial structure, using a Ziegler catalyst. The physical properties of polymers are determined largely by the molecular structure.

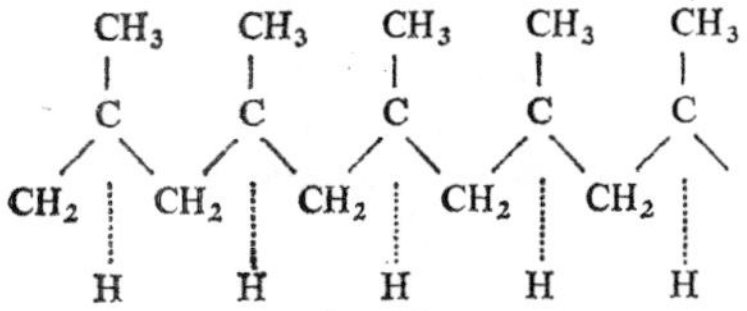

Isotactic

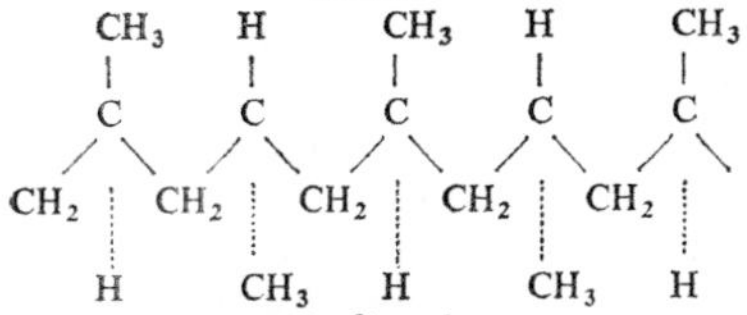

Syndiotactic

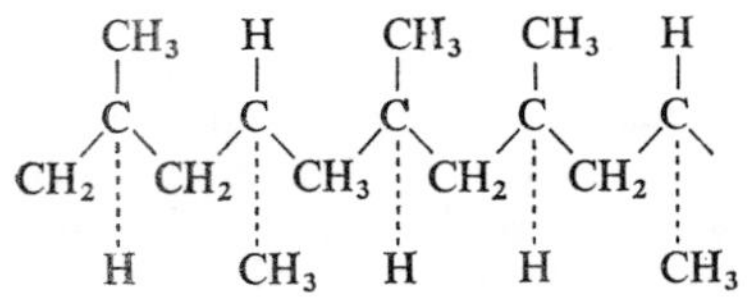

Atactic

In the addition polymerisation of diolefins the molecules may join on in a 1:4, 1:2, or 3:4 position in either a cis or trans configuration (cis-trans isomerism) of the linear

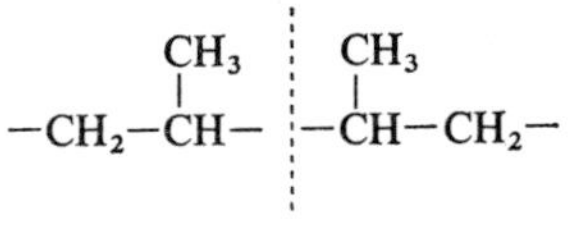

Head-to-head combination

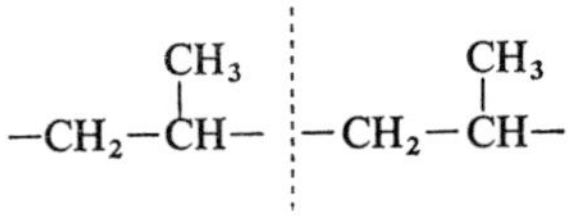

Head-to-tail combination

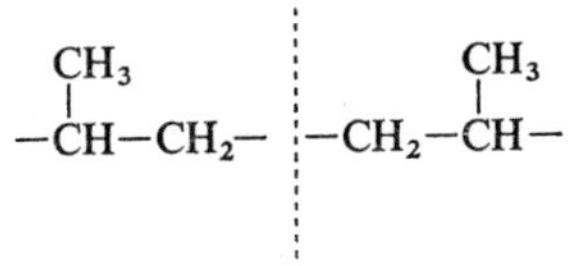

Tail-to-tail combination

chain. Natural rubber is a cis-1:4 polymer; this configuration is generally the aim in synthetic rubbers. Branching reactions are usually undesirable in long-chain polymers. Although the 1:2 and 3:4 additions also result in linear polymers, crosslinking may occur at vinyl side groups whereas this is only possible with a double bond in a 1:4 structure, or an α-methylene group. Under suitable reaction conditions (*e.g.* structure of the mixed catalysts; solvents), the Ziegler catalysts and, in the case of 1:4-polyisoprene, lithium also, give a high percentage of almost pure cis or trans polymer. With the poly-α-olefins, which are classed as plastics, optical isomerism also occurs, based on the spatial arrangement of the subsidiary groups. These isomers are classified as: isotactic polymers with substituents spatially in the same plane, syndiotactic polymers in which substituents alternate in two planes, atactic polymers with substituents distributed irregularly, stereoblocks which are fairly long linear chains of isotactic and atactic units.

Units of the α-olefins can be joined either head-to-head, head-to-tail or tail-to-tail, *e.g.* crystalline polypropylene has a head-to-tail structure.

Stereospecific plastics show high crystallisation. The best-known product is the highly crystalline polypropylene which is an isotactic polymer. Such polymers have m.p. considerably higher than other polymers of the same origin (sometimes exceed 250°C).

STERLING Trade name for a group of furnace and thermal blacks. *Types:*
99: FF, 105: conductive FF, FT: FT, L: HMF, LL: HMF, MT: MT, NS: non-staining SRF, R: SRF, S: SRF, SO: FEF (MAF), V: GPF (139).

STEROX Group of polyoxyethylene esters and thioethers. Surface active materials.

STEVENSON PLAN Scheme for limiting the production and export of natural rubber from British colonies so as to maintain the supply and demand equilibrium and the stability of the price at 1s. 9d./lb. The areas planted before 1914 came into production during World War I. Because of the lack of shipping space the productive areas had to store a tremendous amount of stock. Consequently a huge excess arose, despite increased demand in the post-war period. In October 1921 the British Colonial Office founded a subsidiary

committee with Sir (later Lord) James Stevenson as chairman. The recommendations of this committee came into force in November 1922 and specified that each plantation was allowed to export only a specified percentage of its normal production. Any excess in the amount exported was liable to a high export tax. The production quota was fixed quarterly on the basis of prices and world reserves. The Dutch areas did not cooperate in this plan, and increased production. Between 1922 and 1925 the price in New York rose from 17·5 cents/lb to 72·5 cents/lb. The restrictions were finally lifted by the British Government in November 1928.

STIBIOX-GI Antimony trioxide (417).

STIMULATION As early as 1912 it was discovered that scraping off the outer bark of rubber trees increased latex production. An attempt was subsequently made to increase production by using chemicals and compounds, such as mineral oils and plant oils, wood ash, sodium nitrate, cow dung; alone or mixed with sulphur and copper sulphate (*e.g.* Newbark, Solar Vim). It was then discovered in Indochina that injecting the tree with copper increased production. The best results were achieved with synthetic growth hormones (2:4-D and 2:4:5-T) which could give a rise in production of up to 60%. Nowadays a 1% mixture of n-butyl esters of 2:4:5-T is used, blended with 5 parts palm oil and 3 parts petroleum. This compound is spread above the tapped surface over the breadth of the used bark each month, or in a strip 7·5 cm long beneath the cut, and applied once in 6 months.

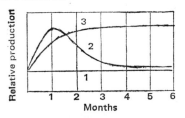

1, normal production; 2, once every 6 months beneath the tapping incision; 3, once a month above the tapping incision.

ST. JOE FLEXOMETER
Machine used to determine the flexing fatigue and heat build-up of rubber under dynamic strain during rotation under pressure and eccentric deviation. A cylindrical test piece of 37.5×37.5 mm forms a link between two parallel horizontal plates. The movement of the upper plate, driven at 875 rpm and with a load of 180–260 kg, is transmitted onto the lower plate by the interposed rubber test piece. A horizontal force pushes the lower plate to one side in a deflection of approx. 4.4 mm eccentrically and imposes a bending strain on the test piece. A sudden reduction in the horizontal force indicates the start of total fatigue of the test piece. ASTM D 623-58.

STOCK BLENDER   Device used on mills to interchange and turn the sheets.

STOKES' LAW   The rate of descent of a spherical particle, of diameter d and density D_1, in a medium of density D_2, viscosity η is:

$$V = \frac{d^2(D_2 - D_1)g}{18\eta}$$

where g is the gravitational constant. This formula is the mathematical basis for the concentration of latex by creaming or centrifuging.

STORMER VISCOMETER
Rotation viscometer used to determine the viscosity of latex and latex compounds.

STRAINER   Extruder with a large screw diameter and changeable sieve plates in its die-head to remove impurities, such as grit, from rubber compounds.

STRAIN TEST   Determination of the elongation of a rubber test piece under a constant load. With the NRPRA strain tester a load of 5 kg is used, ensuring an elongation of less than 100% so that crystallisation is avoided. Used to determine the rate of cure of TC rubber and the vulcanisation characteristics of compounds.

STRESS/STRAIN CURVE
Graphical representation of the tensile stress (kg/cm^2) as a function of the strain ($\%$).
Specification: DIN 53 504.

STRUCTURAL UNIT   The smallest chemical group which repeats itself periodically in macromolecules.

STRUKTOL   Group of processing aids for rubbers and plastics (379):
30:        combination of synthetic resins with stabilised process oils. Special softener for the homogenising of rubber compounds
3033:   combination of an emulsion softener based on

hydrophilic fatty acid esters with polar tackifying resins. Has restricted migration

A 50: combination of assorted zinc soaps of high mol. wt. organic acids. Plasticiser.

IB 531: combination of amine derivatives on an inorganic carrier. Vulcanisation auxiliary for injection/transfer moulding and LC vulcanisation

Neopast: paste of two-thirds zinc oxide and one-third dispersing aid

W 33: emulsion softener based on hydrophilic fatty acid esters and process oils, bound to an active filler

WB 212: emulsion softener based on hydrophilic fatty acid esters bound to an active filler

WB 300: polyester-softener; mixed high mol. wt. aliphatic and aromatic esters. Insoluble in aliphatic hydrocarbons and mineral oils. Used in nitrile rubber and polychloroprene

WB 700: dispersible zinc oxide with 9% dispersing agent

WB 900: mixture of two-thirds magnesium oxide and one-third dispersing aid

WK 6: emulsion softener based on hydrophilic fatty acid esters and process oils

SU: preparations from soluble sulphur, insoluble in carbon disulphide with dispersing aids, under the influence of the electrostatic charge of the sulphur powder.

SU 95: powder. 95% soluble sulphur, 5% dispersing agent

SU 105: paste, 50% soluble sulphur

SU 106: paste, 50% insoluble sulphur

SU 108: powder, 75% insoluble sulphur

SU 120: powder, 83·3% soluble sulphur.

STYPHEN I Styrene-phenol condensation product with tri(α-methylbenzyl)phenol. Pale yellow, viscous liquid. Soluble in the usual rubber solvents, insoluble in water. Antioxidant for white and pale-coloured compounds.
Quantity: 0·5–1% (61).

STYRALOY Butadiene-styrene copolymer with rubber-like properties. Resistant to acids and bases (61).

STYRENE $C_6H_5.CH=CH_2$, vinyl benzene, phenyl ethylene. Colourless to yellowish, oily liquid, m.p. $-33°C$, b.p. 145–146°C, s.g. 0·909 (20°C). Soluble in alcohol, ether, methanol, acetone and carbon disulphide. Isolated by Bonastre in 1831. Production is from ethyl benzene by catalytic dehydrogenation. Polymerises at 200°C to polystyrene, a hard product. When polymerised into stretched polymer chains the phenyl group produces stiffening. The most important use is copolymerisation with butadiene. Up to 50% styrene, the products are rubbery, between 70–85% the copolymer has a resinous nature.
Uses: plastics, copolymers.

STYRENE PHENOL RE-ACTION PRODUCT Pale yellow liquid, s.g. 1·08. Soluble in petroleum hydrocarbons, alcohols and esters. General purpose, practically non-discolouring antioxidant

for natural and synthetic rubbers and latices. Stabiliser for SBR.
Quantity: 1%.
TN: Age-Rite Spar (51)
Styphen I (61)
Wing-Stay S (115).

STYRENOL Aqueous emulsion of a plasticised styrene-butadiene copolymer used in impermeable paints for porous materials such as plaster, cement, bricks (332).

STYREX Group of acrylonitrile-styrene copolymers. Resistant to chemicals, solvents and the effects of temperature (61).

STYRITE Plastics based on polystyrene (61).

STYRON Group of thermoplastic polystyrenes (61).

SUBUR SMOKE HOUSE Chamber smoke house for sheet rubber. Consists of 4 or 8 chambers in which the sheets are hung on frames; each day's production is treated in a separate chamber. Smoking time: 3–5 days.

SUCONOX Group of antioxidants. 4, N-butyroyl-p-aminophenol, 9, N-pelargonyl-p-aminophenol, 12, N-lauryl-p-aminophenol, 18, N-stearyl-p-aminophenol (333).

SULFENAX Group of Czech produced accelerators (428):
CB/30: N-cyclohexyl-2-benzthiazyl sulphenamide
MOB: 2:4-morpholinyl mercaptobenzthiazole.

SULFOLE B-8 Tertiary dodecyl mercaptan. Modifier in emulsion polymerisation (130).

SULPHENAMIDE Group of organic compounds with the general structure

For different accelerators of this group R_1 is usually the benzthiazyl radical while R_2 and R_3 are hydrogen or an alkyl group. Both may be replaced by a cyclic group.

SULPHOXYLATE PROCESS Emulsion polymerisation of SBR cold rubber with sodium formaldehyde sulphoxylate, iron sulphate and ethylene diamine tetra-acetic acid as the activating system.

SULPHUR The most important vulcanising agent for rubber since its use for the purpose was discovered by Goodyear. The quantity used in a vulcanisate varies considerably, depending on the compounding ingredients, between approx. 0·5–4% and in hard rubber between 30–50%. The solubility of sulphur in rubber is approx. 1% at room temperature and increases with temperature rise. At 80°C solubility is approx. 4·5% and at 120°C, approx. 10%. In compounds with a high sulphur content supersaturated solutions are formed on cooling, when excess sulphur crystallises out. If the sulphur blooms from uncured compounds the tack of the surface is greatly reduced, and adhesion becomes much more difficult; insoluble sulphur (μ form) is therefore used for special processes. Compounds and undercured vulcanisates with a high quantity of free sulphur, and also hard rubber containing free sulphur, are liable to have a sulphur bloom.

Quality control tests: the minimum degree of fineness is 100% through a US standard sieve of 100 mesh, 93% through a US standard sieve of 325 mesh, or a chancel grade of 60–80. Minimum purity is 99·5%, the ash content must be below 0·5%. Sulphur must be free from acid because of its retarding effect.

Analysis of vulcanisates: sulphur is classified as:

Total sulphur, the total sulphur content, independent of the chemical form or origin

Added sulphur, sulphur which is elemental and added to a compound to facilitate vulcanisation

Free sulphur, chemically unbound elemental sulphur

Extractable sulphur, the total amount of sulphur in a compound which may be extracted by acetone or acetone/chloroform and includes the free elemental sulphur and the sulphur contained in soluble organic compounds

Bound sulphur, the sulphur present in a chemically bound form in rubber hydrocarbons

Organically bound sulphur, the sulphur which is bound in vulcanisation accelerators and organic fillers, *e.g.* factice

Inorganically bound sulphur, the sulphur which is bound in inorganic fillers and other substances, *e.g.* sulphates, sulphides.

Because of the large number of authors who have written on the subject, there are various interpretations of the individual forms of sulphur.

Specifications: ASTM D 297
BS 903: B6-B10: 1958, BS 1673: 2: 1954 DIN 53 561.

SULPHUR CHLORIDE S_2Cl_2, sulphur monochloride, disulphur dichloride. Reddish yellow, oily fuming liquid with a suffocating smell and irritating effect on the mucous membranes, b.p. 133–141°C, s.g. 1·68–1·71. Soluble in alcohol, benzene and carbon disulphide, dissolves at room temperature approx. 65% sulphur. Decomposes in water with the formation of sulphur dioxide, sulphur and hydrochloric acid.

Uses: cold curing of thin walled articles by dipping in solutions of sulphur chloride in carbon disulphide or benzene (A. Parkes, 1878) or in steam (Abbot, 1878). Production of white factice and organic sulphide. Traces of iron, which are harmless in hot vulcanisation, accelerate the ageing of cold cured vulcanisates.

SULPHUR THIOCYANATES Sulphur dithiocyanate, $S(CNS)_2$, and sulphur monothiocyanate, $S_2(CNS)_2$. According to LeBlanc and Kröger (1925), the thiocyanates can be used in solution in carbon disulphide, for cold curing rubber.

Lit.: *Kolloid Zeit*, 1923, **33**, 267, *ibid.* 1926, **31**, 205.
DRP 408 306 (1925).

SUNDEX Group of softeners and plasticisers for natural rubber, SBR, IIR, NBR, reclaim and blends. Extender for neoprene (50–100%):

41:	blend of a high petroleum fraction with an asphalt type
53:	aromatic oil. Dark, viscous liquid, s.g. 0·9834
85:	highly aromatic oil. Dark, viscous liquid, s.g. 1·0167
170:	aromatic oil. Dark, highly viscous liquid, s.g. 0·9874
1585:	aromatic oil. Dark, viscous liquid, s.g. 0·9937 (155).

SUNOLITE Group of wax compounds. Antioxidants giving protection against ozone and weathering to rubbers under static strain (125).

SUN PROCESS AID Group of processing and extending oils for natural and synthetic rubbers (155).

SUNPROOF Blend of waxes, m.p. 65–70°C, s.g. 0·92. Soluble in benzene, insoluble in water and acetone. Protective agent against light and inhibitor of atmospheric cracking under static conditions, for natural rubber, SBR and acrylonitrile-butadiene copolymers. Also used as a protective agent against freezing. Blooms slowly in uncured compounds and vulcanisates when the quantity exceeds 0·5%.
Quantity: 1–2%, as protective agent against freezing 0·5% (23).

SUNTHENE OILS Group of naphthenic, non-staining processing oils of different viscosity (155).

SUPARAC Mixtures containing piperidine pentamethylene dithiocarbamate. Standard: 25 parts with 75 parts clay, Z: 25 parts with 75 parts zinc oxide.

SUPERACCELERATEUR
Trade name for a series of accelerators (19, 20):
3 RN: zinc ethyl phenyl dithiocarbamate
481: tetramethyl thiuram disulphide
500: tetramethyl thiuram monosulphide
501: tetramethyl thiuram disulphide
1105: zinc ethyl phenyl dithiocarbamate

1500: sodium diethyl dithiocarbamate
1505: zinc diethyl dithiocarbamate
1555: zinc pentamethylene dithiocarbamate
1605: zinc dimethyl dithiocarbamate
3010: diethylammonium diethyl dithiocarbamate
4000: solution of sodium dibutyl dithiocarbamate
4005: zinc dibutyl dithiocarbamate
5010: piperidine pentamethylene dithiocarbamate
6000: sodium isopropyl xanthate
6005: zinc isopropyl xanthate.

SUPERLOID Compound of algin and ammonium alginate. Stabiliser and thickening agent for latex (236).

SUPER MULTIFLEX Precipitated calcium carbonate coated with a surface active material. Particle size 0·03 microns (153).

SUPER NATSYN Further development of the Goodyear polyisoprene 'Natsyn' (115).

SUPERUB Crumb rubber (Nigeria).

SUPRALEN RCH 100 High mol. wt. low pressure polyethylene. Excellent impact resistance, good resistance to high and very low temperatures exceptionally low abrasion resistance, good electrical properties. Stable to acids, alkalis and chemicals (334).

SUPRASEC Group of isocyanates:
D: polyisocyanate for crosslinking alkyd resins
F: isocyanate adduct

M: isocyanate with a high b.p. and low vapour pressure at normal processing temperatures; s.g. 1·18. Cross-linking agent for solid polyurethane rubber (60).

SUPREX American kaolin type.

SURCOPRENE Cyclised rubber; hard resin produced by cyclisation of NR in acidic conditions. Soluble in aliphatic and aromatic hydrocarbons.
Uses: in the lacquer industry, printing inks (399).

SUREX Wax-like antioxidant (336).

SURFACE ACTIVE MATE-RIALS Substances which reduce surface tension, increase wetting ability and ability for emulsifying, dispersing and penetration. According to their effect they are termed wetting, emulsifying, dispersing, penetrating or foam agents, stabilisers, detergents, and so on. On the basis of their ionisation properties, surface active materials are classed as anionic, cationic, non-ionic and amphoteric. Anionic surface active materials ionise in solution to give a small positive and a large negative ion. The negative, usually a lipophilic ion is the effective constituent. Aliphatic and aromatic sulphonates and sulphates of *e.g.* fatty oils, acid esters, acidic amides, alcohols, esters, phenols, alkyl aryl sulphonates belong to this group. With the cationic surface active materials, the large cation is the active component; this group includes quarternary ammonium, sulphonium or phosphonium salts, and aliphatic amines. The non-ionic surface active materials include polyethers, hydroxyalkyl ester, esterised polyalcohols, and can be used with both cationic and anionic surface active materials. The amphoteric surface active materials, *e.g.* C-cetyl betaine, alkyl betaine, may be ionised according to their surroundings.

Surface active materials find frequent use in rubber processing and particularly, in latex processing as: (a) stabilisers to prevent premature coagulation due to salts, pigments, dehydration or mechanical effects, (b) wetting and penetrating agents to decrease the surface tension of textiles, paper, and similar materials, and increase penetration, (c) dispersing and emulsifying agents in the production of latex compounds, (d) foaming agents in the production of latex foam, (e) release agents to prevent sticking in moulds, (f) to prevent the adhesion of uncured compounds. Anionic surface active materials are the most widely used in latex processing because latex is negatively charged. If good stability is required against electrolyte concentrations or high pressure, then non-ionic or amphoteric substances are preferred. Cationic substances have a limited use in Positex, or synthetic latices with a positive charge because of the unstabilising effect on normal latex, *e.g.* cetyl pyridinium bromide can be used to change the latex charge. Most surface active materials are produced under registered trade names.
Lit.: J. v. Alphen, *Gummichemikalien*, Berliner Union, Stuttgart, 1956.
J. Wilson, *British Compounding Ingredients for Rubber*, W. Heffer, Cambridge, 1963.

SURFEX MM Precipitated calcium carbonate, coated with a resin.

Particle size 1–5 microns. Has a low oil absorption. Filler, *e.g.* for vinyl resins, polyesters (153).

SURGICAL GLOVES Rubber gloves were introduced for surgical purposes by the American doctor W. S. Halstead, in 1889–90 in Baltimore. The idea was inspired because the chlorine used in sterilising instruments caused rashes on the hands of the theatre nurses. Because the sample gloves produced under patent by Goodyear could be easily sterilised, Halstead tried them out during operations. In 1898 the seamless surgical glove was introduced at the Surgical Congress by Professor Friedrich of Leipzig. In 1890 the Leipzig Rubber Factory (previously Jul. Marx, Heine and Co.) made the first attempt to produce rubber gloves using a dipping process.

SURLYN A Ethylene copolymer with covalent and ionic bonds (6).

SUS Abbreviation for Saybolt Universal Seconds, the unit of Saybolt viscosity determined using the Saybolt viscometer.

SUSPENSION POLYMERISA-TION Pearl polymerisation. The monomers are held in suspension in a non-solvent, usually water, by purely mechanical effect without an emulsifier and polymerise catalytically as large drops. To prevent the polymerised particles from sticking together suspension stabilisers are added, *e.g.* calcium salts, oxides, silicates, organic colloids as gelatine, polyvinyl alcohol, which are only slightly soluble; they can be easily washed away afterwards. Pearl polymerised polymers generally contain small amounts of impurities

compared with products made by other processes.

SUSPENSO Precipitated calcium carbonate. Particle size 1–5 microns (153).

SUSTANE Group of antioxidants (352a):
BHA: 3-tert-butyl-4-hydroxyanisole
BHT: 2:6-ditert-butyl-4-methyl phenol.

S-VALUE Black value. Measure of the light absorption of carbon blacks; characterises the depth of colour. Is determined by rubbing the black in oil and comparing it on an arbitrary scale with standard blacks. The scale runs in the opposite direction to a nigrometer index and may be compared with it. High S values indicate a high depth of colour (light absorption) and correspondingly low reflection.

SVEDOPRENE Swedish produced polychlorobutadiene during World War II.

SVERIGES GUMMITEK-NISKA FORENINGS Swedish Association of Rubber Technology, Hamngatan 12, Stockholm 7.

SVITPRENE Czechoslovakian produced chloroprene rubber consisting primarily of 2-chlorobutadiene–1:3. Obtained by addition of hydrogen chloride to vinyl acetylene. Polymerisation occurs in two stages: first the α-polymer is formed, this is then converted to the insoluble μ-polymer by crosslinking. Production is by emulsion polymerisation with sulphur as a modifier, persulphate as catalyst and oleic soaps (Type 203) or a mixture of oleic soaps and resins

(Type K) as emulsifier. Thiuram is used as a stabiliser and phenyl-β-naphthylamine as antioxidant.

S/V PRODUCT 2243 Wax-like antioxidant (327).

SWELLING Various solvents, vapours and gases cause rubber to swell, the amount of swelling depending on the type of swelling agent, the type of rubber, ingredients added and the temperature and time. The amount of swelling is determined by the increase in weight or volume under specified conditions. ASTM D 471, D 1460, DIN 53 521.

SYL–KEM 21 Trade name for pentamethyl disiloxane methyl methacrylate (7).

SYNCHROFLEX Precision cogged belts made from Vulkollan with a steel inner ply; transmit the force from a motor to a coupling (337).

SYNERESIS Spontaneous deflation of colloids due to, for example, particle enlargement, agglomeration of particles, change of charge; the enclosed liquid is forced out from between colloid particles. The effect can be used commercially to dehydrate emulsion polymers. In rubber coagulum, synersis is caused by a large agglomeration of the particles.

SYNERGEX Alkyl thiourea. Accelerator (39).

SYNERGISM The co-reaction of two or more substances which boost each other in their effects and whose total effect is greater than the sum of the individual effects.

SYNPEP N Complex ether. Pale red liquid with aromatic odour, s.g. 1·1, pH 10·4, viscosity 20–30 poise (20°C). Antioxidant for NR, SR, reclaim, latex and dispersions, also a processing aid, secondary use as an activator. Improves ageing properties, decreases heat build-up, shortens the mixing cycle, activates most accelerators, prevents gel formation in SBR.
Quantity: NR and SBR 1–2%, NBR 2% (39).

SYNPOL Trade name for butadiene-styrene copolymers of the Texas-US Chemical Co. Index number conforms with the ASTM code but the following are also available:

8000:	non-staining hot polymer with a high styrene content
8150:	50% HAF black. Oil free
8151A:	masterbatch with 52% HAF black. Finely dispersed and with 10% highly aromatic oil
8200:	pale-coloured, non-staining type extended with oil
8201:	pale-coloured, non-staining type extended with oil
8201A:	pale-coloured cold polymer extended with 50% naphthenic oil
8250:	50% HAF black. Oil free
8251:	50% HAF black with 25% highly aromatic oil
8253:	non-staining masterbatch with a finely dispersed FEF black (60%) and naphthenic oil (37·5%)
8253A:	oil extended masterbatch. Contains 60% FEF black and 37·5% naphthenic oil
8254:	masterbatch with HAF black
8254A	oil extended masterbatch with 75% HAF black and 37·5% highly aromatic oil

8255A: oil extended masterbatch with 75% HAF black and 50% highly aromatic oil

8266: masterbatch with ISAF black

EBR: cis-1:4-polybutadiene. Emulsion polymer containing non-discolouring stabiliser. 37·5 phr, highly aromatic oil

X-274: hot rubber for use in articles which come into contact with food (338).

SYNTEX Aqueous dispersions of natural rubber, synthetic rubbers, reclaim and resins; also compounded latices (339).

SYNTHETIC RESIN AP Condensation product of acetone, phenols and formaldehyde, m.p. 80–85°C. Tackifier. Gives a slight increase in the hardness of vulcanisates. Raw material used for the production of lacquers and adhesives based on rubber and nitrocellulose (189).

SYNTHETIC RUBBER Common term, in fact incorrect, for chemically different polymers with rubber-like properties. Since the term rubber applies to the original product, *i.e.* polyisoprene, the term synthetic rubber should strictly be used only for synthetic cis-1:4-polyisoprene. For polymers, with rubbery properties, which do not have an identical structure to natural rubber, the terms artificial rubber, chemical rubber or synthesised rubber have been suggested. In modern usage the term rubber no longer refers to a chemically defined product but relates to physical properties, all synthetic polymers with properties similar to those of natural rubber may therefore be classified as synthetic rubbers.

SYNTHOMER Special types of butadiene-styrene and butadiene-acrylonitrile latices (405).

SYNVARITE Group of phenol-formaldehyde resins.
Types:
RC-6, 12, 16, 18, 20. Reinforcing fillers (340).

SYNVAROL Group of urea-formaldehyde resins.
Types:
TN, TUN, TUD. Reinforcing fillers (340).

T

2:4:5-T Abbreviation for 2:4:5-trichlorophenoxyacetic acid.

T-50 TEST Method for determining the degree of vulcanisation of rubber. The crosslink density and the amount of bound sulphur are closely related to the behaviour of rubber at low temperatures. A stretched rubber test piece is frozen at approx. −70°C, allowed to relax, then heated slowly. T-50 is that temperature at which a 50% retraction of the original stretched length occurs. The T-50 value of raw rubber lies at about 18°C. It decreases with increasing vulcanisation. The following strains are necessary since useless results are obtained when the strain is too low: gum stocks, latex films 700%, low loaded compounds 500%, highly loaded compounds 350%.

$$L\text{-}50 = \frac{L_0 + L_1}{2}$$

where L_0 is the length of unstretched test piece, L_1 is the length of strained test piece and L-50 is the length at T-50.
Specification: ASTM D 599-55.

TA Russian produced thermal black made from acetylene.

TACKIFIERS Materials used to increase the tackiness of compounds, synthetic resins, colophony, abietic acid, dehydroabietic acid and Koresin.

TACRYL Swedish produced polyacrylonitrile fibre.

TACTENE cis-1:4-polybutadiene with more than 85% cis-1:4 content.
Grades:
1200: Mooney viscosity 27. For compounds with a high black loading. Non-staining stabiliser. May be processed alone
1202: Mooney 37. Contains some copolymerised styrene, otherwise as 1220. Additive for polystyrene to improve impact resistance
1220: almost colourless. Mooney viscosity 45. For blends with NR and SBR
1250: high mol. wt., 98% cis-1:4, contains 50 parts of highly aromatic oil. Mooney viscosity 30. Mainly for tyre treads
1251: like 1250 but with 50 parts naphthenic oil. Used in footwear (303).

TALALAY, DR JOSEPH ANTON Rubber technologist, 1882–1961. Co-founder of the German Rubber Society. Undertook research into latex technology, developed the Talalay process for foam rubber.

TALALAY PROCESS A process for the production of foam rubber without mechanical foaming. The latex is foamed by the catalytic decomposition of hydrogen peroxide then rapidly frozen. The frozen structure is coagulated by permeation with carbon dioxide. The foam rubber has an even structure. USP 2 432 353.

TALANDOHA Local term for *Landolphia Richardiana* (Madagascar) and the wild rubber produced from it.

TALC $Mg_6(OH)_4(Si_8O_{20})$, Talcum, steatite, soap stone. Soft, white powder, fatty to the touch, s.g. 2·6–2·9, refractive index 1·57, hardness 1. Non-reinforcing filler widely used because of its excellent electrical insulating properties, dusting agent.

TALCULIN 1060 Liquid detackifying agent for mixing and extruding processes, may be used instead of talc. Does not form by-products during vulcanisation. Used as a dust-free process, gives the products a shiny surface. Applied by dipping, spraying or painting (42).

TALL OIL Natural blends of approx. 30% resinous acids (mostly abietic acid) and approx. 55% fatty acids (oleic, linoleic and linolenic acids) with approx. 12% non-saponifiable constituents (phytosterol, hydrocarbons), s.g. 0·96, iodine No. 180, saponification No.

458

169. Obtained as an alkali soap by evaporation of the sulphite liquor. A dark, obnoxious smelling liquid is obtained by acidifying with sulphuric acid and is further purified by distillation; sulphate pitch is left as a residue.
Uses: occasionally as a softener in rubber compounds.

TAMATAVE RUBBER Former trade term for a rubber of good quality (Madagascar) shipped from the port of Tamatave.

TAMBAQUI SERINGA Local term for *Hevea microphylla Ule* (the Rio Negro region, Brazil). Yields a small amount of latex, which is suitable for rubber production. It is possible that two other types, known as barriguda and sarapo, are a variety of the *Hevea microphylla Ule.*

TAMOL Dispersing agents for latex compounding, emulsion polymerisation (33).

TAPARU Sernamby of Cameta, Seringeirana. Trade term used in the Amazon area for the product of *Sapium taburu Ule.* The latex was frequently mixed with that of *Hevea Brasiliensis.* In its pure form contains approx. 90·5% pure rubber.

TAPPING The latex is collected from the trees by making an incision at an angle of approx. 35° running from the top left to the bottom right of the trunk; the sap vessels are thus cut at a right angle. The latex vessels are in the bark and increase in the direction of the cambium. The incision must be made as deeply as possible, but must not wound the cambium as this retards renewal, causing nodules to form over the wound. The trees are ready for tapping at approx. 5–7 years. The criterion for this is a specific circumference of the trunk, *e.g.* 50 cm at 1 m height. The first cut is made, according to the tapping method used, at a height of between 75 and 150 cm, and leads into a straight channel running down the bark. At the bottom end of the main channel a small aluminium spout is inserted into the bark to direct the sap into the collecting cups. Special tapping knives of various shapes are used for the purpose. The best time of day for tapping is around 6 am as the turgor is highest then. Only in the case of morning rainfall is the tapping left until later in the day. The flow of latex stops after a few hours. The latex is usually collected between 10 and 11 am. The trees often go on dripping for some time afterwards. Finally a small quantity of latex dries onto the bark (scrap, curly scrap, panel scrap) and is collected the next morning before tapping. At each new incision approx. 1 mm of the bark is cut which means 18–20 cm is used per year (180 tapping days). As soon as the bark has been exhausted on one side of the tree, the other side is used. Within a few years the wounded surface becomes covered with new bark and may be tapped again. When cultivating new clones, attention is then paid to the development of a rapidly healing bark. Each clone reacts specifically to the intensity and frequency of the tapping cycle used in the different systems. New plantations are therefore tapped during the first three to five years using a mild system (67%) before the intensity is increased to 100%.

TARAXACUM Member of the *Compositeae* family of rubber yielding dandelion plants; these are grown in

Russia for the production of rubber and are also occasionally used in the USA.

T. kok-saghyz Rodin (Kok-Saghyz). Discovered by E. L. Rodin on the Tien Schan plateau east of Alma Alta. Grows at temperatures of 6–30°C with a max. at a daytime temperature of 24°C. The rubber occurs in the roots in quantities of 15–25%. *Production:* 500–1100 kg/ha. Because the roots have sap arteries, the latex may be obtained in water as a dilute, stabilised suspension which is concentrated by centrifuging. The rubber is obtained by maceration with dilute alkali. *T. megalorhizon Hard. Mzt.* (*T. hybernum Stec., T. Gymnanthum DC*) *Krim-Saghyz.*

TAU-SAGHYZ *Scorzonera Tausaghyz, L et B. Compositeae;* a semi-bush-like plant with an age span of several years. Discovered in 1930 in Kasachtan (Kara Tau Highland). The roots contain up to 30% rubber. There are five known forms.

TBC Abbreviation for p-tert-butyl catechol.

TBS Tetrabutyl phenyl salicylate. Antioxidant (7).

TC Abbreviation for thiocarbanilide (diphenyl thiourea).

TCP Abbreviation for tricresyl phosphate.

TC RUBBER Technically classified rubber may be natural rubber of any international commercial quality, the vulcanisation properties of which also are determined by the producer

and marked on the bales. The classification is indicated by rings: red for slow, yellow for medium, and blue for fast curing. Testing is by use of the ACS-1 standard compound. A test piece is cured for 40 min at 140°C, then stretched by applying a load of 5 kg. The elongation produced is measured after one minute (strain test). The following values are used for the classification:

TC class	For producers	For consumers
Blue	55–73	55–70
Yellow	73–85	67–91
Red	85–103	79–103

The Mooney viscosity also used to be given, but was discarded because of the variation in values on storage. Annual production of TC rubber is about 55 000 tons. TC rubber is suitable for use in both small and medium plants. The original idea of technical classification is beginning to develop today towards the production of a uniform product in any one class. Yellow rubber seems to be in greatest demand by consumers.

Lit.: *Gummi, Asbest, Kunststoffe*, 1961, **14**, 396.

TDI Abbreviation for 2:4-toluene diisocyanate.

TDT TEST Abbreviation for tensile deformation/temperature test. The determination of the changes on applying a tensile stress, as a function of the temperature.

TEAR RESISTANCE Strength. The force in kg/cm required for a standard test piece to resist further tearing.

Strip test piece: 6 mm wide strip with a knife incision at one end. **DIN 53 507.**

Angle test, according to Graves: testing of a 2 mm thick, angled test piece with a 1 mm incision at the inner angle.

ASTM D 624 (Type C), DIN 53 515; 53 507, BS 903: 25.

Bow-shaped test piece (crescent tear

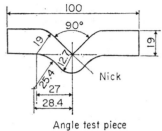

Angle test piece
DIN 53.515
ASTM D 624-54 (Type C)

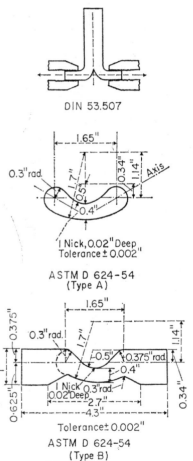

DIN 53.507

I Nick, 0.02" Deep
Tolerance ± 0.002"

ASTM D 624-54
(Type A)

Tolerance ± 0.002"

ASTM D 624-54
(Type B)
BS 903:25:1950

test): ASTM D 624 (Types A and B), BS 903: 25: 1950.

Foam substances: DIN 73 575.

TEAR STRENGTH The force necessary to cause a test piece to break when stretched under specific given conditions, in relation to the cross-section of the unstretched test piece. Expressed in kg/cm^2 or lb/in^2. Tensile strength is a criteria of the quality of a vulcanisate, characterising it, but not necessarily indicating its suitability for a particular use. The value depends on the composition of the compound and is influenced by sulphur and the accelerator. Reinforcing fillers result in an increase in tensile strength whereas substances such as inert fillers, softeners, reclaim generally cause a decrease. ASTM D 412; E-6, BS 903: 14: 1950, DIN 53 504.

TECHNOFLON Fluoro-elastomer having good resistance to chemicals and high temperature; continuous use up to 200°C, short use up to 300°C. Modulus 100% 20–80 kg/cm^2, tensile strength 180–250 kg/cm^2, elongation at break 200–400%, Shore A 55–85° (59).

TECQUINOL Technical hydro-quinone (193).

461

TEDC Abbreviation for tellurium diethyl dithiocarbamate.

TEEPOL Types 410, ED 31, CH 31. Group of wetting, cleaning and emulsifying agents based on higher alkyl sulphates and alkyl aryl sulphonates. Type CH 31, with the highest activity, is an aqueous solution of C_8–C_{13} alkyl benzene sulphonates and an octyl phenol condensate.
Uses: wetting agent in latex compounds, detergent for scrap rubber (2).

TEG Aluminium hydroxide gel, alumina gel. White reinforcing, bulky filler for natural and synthetic rubbers.
Uses: hose lines, sealing rings, soles.

TEFLON Polytetrafluoroethylene. Teflon 100 X: perfluorocarbon resin. Properties similar to those of Teflon, but has a lower melt viscosity (6).

TEKE-SAGHYZ *Scorzonera acanthoclada Franch.* Discovered as a rubber plant and produced in 1931 in Russia. A semi-bush, the roots contain 5–6% each of rubber and resin. Large numbers grow wild in the Turkestan mountain area.

TELCONAX Substitute for gutta-percha and balata. According to BP 322 208, a mixture of rubber and a particularly pure form of bitumen in specified proportions with possible addition of gutta-percha or balata.
Uses: for the insulation of land cables from damp, and to protect lead cables against electrolytic and chemical corrosion.

TELLURIUM Element with non-metallic properties. Grey powder, m.p. 450°C, s.g. 6·26, at. wt. 127·61. Soluble in concentrated sulphuric acid. Secondary vulcanising agent; improves physical properties, ageing properties, heat resistance and, particularly, steam resistance. Effective in compounds which are either sulphurless or have an extremely low sulphur content. Decreases the porosity in thick articles and improves the dimensional stability of extruded articles cured in open steam. Suitable for latex compounds.
Grades:
A: very good heat resistance
B: excellent ageing properties
C: latex compound.
TN: Ancatel (22)
Telloy (51)
PYD-80 (80% masterbatch with polyisobutylene) (282).

	A	B	C
Tellurium	0·5	0·5	0·5
Tetramethyl thiuram disulphide	3–3·5	0·5–0·25	—
Zinc dimethyl dithiocarbamate	—	—	1
Dibenzthiazyl disulphide	—	1–1·5	—
Sulphur	—	1·25–0·5	0·6
Zinc oxide	5	5	2

TELLURIUM DIETHYL DITHIOCARBAMATE

[(C$_2$H$_5$)$_2$N.CS.S]$_4$Te. Orange yellow powder, m.p. 110–120°C, s.g. 1·44, mol. wt. 720·6. Soluble in benzene, carbon disulphide, chloroform, partially soluble in alcohol and benzine, insoluble in water. Ultra-fast accelerator. Contains 13·3% sulphur available for vulcanisation.

Quantity: Primary, NR, SBR and nitrile rubber, 0·3–1% with 0–1·5% thiazole and 3–1% sulphur, butyl rubber 1–2% with 0–2% thiazole and 2–1% sulphur. Secondary, NR, SBR and nitrile rubber 0·1–0·3% with 1–1·5% thiazole and 3–1% sulphur, butyl rubber 0·5–1% with 0·5–1% thiazole and 2–1% sulphur.

Uses: natural rubber, SBR, nitrile rubber, butyl rubber; for air bags, proofed textiles, extruded articles, inner tubes, moulded articles.

TN: Butalene TEL (19)
Tellurac
Ethyl Tellurac (51)
PTD-75 (75% with polyisobutylene) (282)
Van Hasselt ETEL (355).

TELOMER A polymer, or product, consisting of one or more polymerisable monomers, neutralised at the end with certain groups or atoms, *e.g.* H(C$_2$F$_4$)$_n$CH$_2$OH, H(CH$_2$—CH$_2$)$_n$CCl$_3$. The monomer to be polymerised is called the taxogen or Compound A and the compound forming the neutral parts of the end groups is the Telogen or Compound YZ. The schematic reaction is:

$$nA + YZ \rightarrow Y(A)_nZ$$

The degree of polymerisation is normally low, and the reaction is known as telomerisation.

TEMEX Group of organic barium zinc compounds. Heat and light stabilisers for vinyl polymers and copolymers (191).

TEMPERATURE COEFFICIENT OF VULCANISATION

For gum stock 1·86, loaded soft rubber 2·17, hard rubber 2·50.

TENAMENE Antiozonant used particularly for butadiene-styrene copolymers (183).
Grades:

1: butyl-p-aminophenol
2: N,N'-disec-butyl-p-phenylene diamine
3: 2:6-ditert-butyl-4-methyl phenol
4: N,N'-bis(1:4-dimethyl pentyl)-p-phenylene diamine
5: N,N'-diisopropyl-p-phenylene diamine
20: butyl hydroxyanisole
30: N,N'-di-2-octyl-p-phenylene diamine
31: N,N'-di-3-(5-methyl heptyl)-p-phenylene diamine (Eastozone 31)
32: N,N'-dimethyl-N,N'-di(methyl propyl)-p-phenylene diamine (Eastozone 32).

TENNIS BALL A ball, made from two figure eight-shaped pieces of calendered sheet and covered with felt; close control of the properties is maintained. Weight: min 56 g, max. 58·5 g, diameter 63·5 mm, resilience or rebound height: min. 134·6 cm, max. 147·3 cm falling from a height of 254 cm onto concrete at 20°C, deformation under pressure: min. 7·4 mm, max. 8·0 mm, at a pressure of 8·17 kg and at 20°C.

TENOX Group of antioxidants for natural and synthetic rubbers (193).

Grades:
- BHA: butyl hydroxyanisole. White, wax-like mass, m.p. min. 48°C. For articles which come into contact with food
- BHT: butyl hydroxytoluene. White crystalline granules, m.p. 69–70°C, b.p. 265°C, s.g. 1·048. For articles which come into contact with food
- HQ: hydroquinone
- PG: propyl gallate
- V: blend of BHA and BHT.

TENSIDE A group of surface active materials.

TENSILAC Group of accelerators:
- 39: ethylidene aniline
- 40: acetaldehyde-aniline condensation products
- 41: condensation product of molar quantities of acetaldehyde, aniline and formaldehyde (obsolete)
- 50: acetaldehyde-aniline-p-toluidine condensation product (202).

TENSILE STRENGTH The stress in kg/cm^2 at max. tensile strain (*i.e.* break) related to the original cross-section of the test piece. The test is performed on rings according to the Schopper–Dalen process, or on dumb-bells. During the test, the elongation at break, the stress/strain curve and the hysteresis loop may be determined. ASTM D 1456, BS 903: A2: 1956, DIN 53 504.

TENSOLITE Rubber hydrochloride stretched to three to six times its original length at 85–100°C, cooled while under strain.
Uses: foils, fibres (115).

TENSORUB A special type of natural rubber which cures rapidly and has a high modulus, high tensile strength, good ageing properties and improved abrasion resistance (400).

TEPA Abbreviation for tetraethylene pentamine.

TERGAL Polyethyleneterephthalate fibre (165).

TERGITOL Group of sodium salts from the higher alcohol sulphonates. Wetting agents for latex, to improve absorption during the impregnation of rayon cord and other textiles. Also used as release agents (146).

TERITAL Polyethyleneterephthalate fibre (59).

TERLURAN ABS polymer produced by BASF (32).

TERPOLYMER A polymer of three different monomers linked irregularly in the molecule.

TERVAN 2800 Mixture of a paraffin wax with a polyisobutylene, m.p. 60°C. Softener and protective agent against light ageing.

TESLAN Polyvinyl fluoride (6).

TET Abbreviation for tetraethyl thiuram disulphide.

TETA Abbreviation for triethylene tetramine.

TETD Abbreviation for tetra-ethyl thiuram disulphide.

TETOL Group of thickening agents for latex (22).

TETRA Abbreviation for carbon tetrachloride.

TETRABUTYL THIURAM DISULPHIDE

$$C_4H_9{\scriptstyle\diagdown}\underset{C_4H_9{\scriptstyle\diagup}}{}N-\overset{\overset{S}{\|}}{C}-S-S-\overset{\overset{S}{\|}}{C}-N{\scriptstyle\diagdown}\underset{{\scriptstyle\diagup}C_4H_9}{C_4H_9}$$

Brown lumps, m.p. 20°C, s.g. 1·05. Powerful accelerator.
TN: Tobac TBUT (311).

TETRABUTYL THIURAM MONOSULPHIDE

$$C_4H_9{\scriptstyle\diagdown}\underset{C_4H_9{\scriptstyle\diagup}}{}N-CS-S-CS-N{\scriptstyle\diagdown}\underset{{\scriptstyle\diagup}C_4H_9}{C_4H_9}$$

Brown liquid, s.g. 0·9777. Soluble in acetone, benzene, ethylene dichloride, insoluble in water. Accelerator with a delayed action. Gives fast-cures in natural rubber and medium-fast cures in SBR with a low modulus. In NR a marked delayed action occurs at processing temperatures. Prevents scorching in SBR. Used as a primary accelerator or secondary with an aldehyde-amine, guanidine or thiazoles; zinc oxide and sulphur are necessary in normal quantities, fatty acids are not essential but may be added. Because of its strong delayed action it is unsuitable for compounds which have a low sulphur content. Does not stain or discolour, disperses easily.
Quantity: natural rubber, primary 0·75% with 2% sulphur, secondary 0·375% with 1–25% benzthiazyl disulphide and 2·5% sulphur.
Uses: moulded articles, inner tube.
TN: ESEG (with clay) (23)
Pentex
Pentex flour (with 87·5% clay) (23).

TETRACHLORO-p-BENZO-QUINONE

Chloroanil, tetrachloro-quinone, 2:3:5:6-tetrachloro-1:4-benzo-quinone. Yellowish flakes or monoclinic crystals, m.p. 290°C, s.g. 1·97. Sparingly soluble in alcohol and ether, insoluble in water. Very fast vulcanising agent in sulphurless cures, for natural rubber, NBR and SBR, in butyl rubber it acts as an activator for p-quinone dioxime with diaminodiphenyl methane as a scorch inhibitor. Gives very good heat resistance in NBR when ageing articles in contact with lubricants and oils which contain sulphur. Causes practically no stain or discoloration. Used in raw rubber as a stiffening agent.
TN: QCE (307), Vulklor (23)
Actor CL (16).

TETRACHLOROETHANE
$CHCl_2 . CHCl_2$, acetylene tetrachloride, tetrachloroether. Colourless liquid, b.p. 146·3°C, m.p. −43·8°C, s.g. 1·60. Miscible with alcohol and ether, immiscible with water.
Uses: non-flammable, solvent for rubber, cellulose acetate, resins, fats, oils, waxes, sulphur.

TETRACHLOROETHYLENE

C_2Cl_4, perchloroethylene, ethylene tetrachloride. Colourless liquid, m.p. $-22°C$, b.p. $1.21°C$, s.g. 1.631 ($15°C$). Miscible with benzene, alcohol, ether, chloroform. Solvent for NR, SR and cements.

TETRACHLOROSALICYL ANILIDE

Odourless white powder, m.p. $162°C$, s.g. 1.08. Bactericide and fungicide for NR, SR and latices.
TN: Irgasan BS 200 (227).

TETRAETHYLAMMONIUM HYDROXIDE

$(C_2H_5)_4NOH$. Activator for rubber and latex.

TETRAETHYLENE GLYCOL MONOSTEARATE

$C_{17}H_{35}COO(CH_2CH_2O)_4H$, m.p. $35–40°C$, s.g. 0.971 ($25°C$). Softener.

TETRAETHYLENE GLYCOL DICAPRYLATE

$(C_7H_{15}.COO.CH_2CH_2O$
$.CH_2CH_2)_2O$,
s.g. 0.985 ($25°C$). Softener.
TN: DP-200 (187).

TETRAETHYLENE GLYCOL DISTEARATE

$(C_{17}H_{35}COO.CH_2CH_2$
$.CH_2CH_2)_2O$,
m.p. $33°C$. Softener.

TETRAETHYLENE PENT-AMINE

$NH_2(CH_2CH_2NH)_3CH_2CH_2NH_2$.
Pale yellow, viscous, hygroscopic liquid, b.p. $332°C$, s.g. 0.998.
Uses: activator for catalysts in the emulsion polymerisation of synthetic rubbers and in the vulcanisation of foam rubber.

TETRAETHYL THIURAM DISULPHIDE

$$C_2H_5 \diagdown N-\overset{\overset{S}{\|}}{C}-S-S-\overset{\overset{S}{\|}}{C}-N \diagup C_2H_5$$
$$C_2H_5 \diagup \qquad\qquad\qquad \diagdown C_2H_5$$

White to yellowish, odourless powder, m.p. $65–73°C$, s.g. (pure) 1.17, (commercial product) $1.2–1.3$, mol. wt. 296. Soluble in benzene hydrocarbons, chlorinated solvents, acetone and benzine, insoluble in water and dilute alkalis. Ultra-fast accelerator with 10.8% sulphur available for vulcanisation. Active above $121°C$, the critical temperature is lower in NR than in SBR. Activity increases with increasing sulphur content in the compound. May be used as a primary or secondary accelerator with aldehyde-amines, guanidines, or dibenzthiazyl disulphide. In NR the vulcanisation range is small with normal quantities of sulphur but increases with decreasing sulphur content. The range in increased by addition of dibutyl-ammonium oleate. Has a tendency to cause scorching in NR. In SBR the vulcanisation range is wider and compounds which have either low sulphur or are sulphurless have excellent ageing properties. Non-staining and non-discolouring. Acts as a booster for thiazoles and thiazolines; $3–5\%$ zinc oxide is necessary. Disperses easily because of its low m.p. Use as a masterbatch is recommended. Vulcanisation temperature: normal levels $120–145°C$, sulphurless compounds $140–160°C$. The sulphur content may be varied widely to give an even state of cure. In latex it has only a slight effect on stability and prevulcanisation. Compounds containing sulphur may set-up in storage.
Uses: mainly in brightly coloured articles, surgical and household

goods, impregnations, mechanical goods, insulations.

Quantity: rubber, primary 0·1–1% with 3–1% sulphur, or 2–3% with 0·25–0·5% sulphur; secondary 0·05–0·25% with 1–1·5% dibenzthiazyl disulphide and 3–2% sulphur; sulphurless 2–4% with 0–0·5% mercaptobenzthiazole, for good ageing properties 0·5–0·3% with 1–1·5% dibenzthiazyl disulphide, 0·5% tellurium and 1·25–0·5% sulphur. SBR, primary 0·2–0·5% with 1–2% sulphur, secondary 0·2% with 1% dibenzthiazyl disulphide and 2% sulphur, sulphurless 2–4%. Butyl rubber 1·5–2% with 0·5% mercaptobenzthiazole and 2% sulphur. Nitrile rubber 0·1–1% with 1–2% sulphur, or as a sulphurless compound with 2–5%. Neoprene 0·5–2% with 0·5–2% sodium dibutyl dithiocarbamate or 0·1–0·5% piperidine pentamethylene dithiocarbamate. Natural latex, fairly fast cures, 0·5–1% with 2–0·5% sulphur; vulcanisates have good ageing properties and a high modulus with a low sulphur content.

TN: Aceto TETD (25)
Accélérateur TE
Accélérateur rapide DTET (419)
Anacide ET (22)
Antacol
Ethyl Thiurad (5)
Ethyl Thiram (120)
Ethyl Tuads (51)
Ethyl Tuex (23)
Eveite T (59)
PeTD-70 (70% blend with Polyisobutylene) (282)
Robac TET (311)
SA 52-6 (with tetramethyl thiuram disulphide) (4)
SA-62, 62-O, 62-1, 62-9 (4)
Superaccélérateur 481 (20)
T.E.T.D.
TETD
+ Thiuram DSE (6)

Thiuram E (6)
Van Hasselt TET (355)
Vondac TET (433)
Vulcafor VII (60)
Vulcafor DAUF (with MBTS) (60)
Vulcafor TET (60)
Vulcaid TET (3)
Vulcarite 125 (dispersion) (57).

TETRAETHYL THIURAM MONOSULPHIDE

$$\begin{array}{c} C_2H_5 \\ \diagdown \\ C_2H_5 \diagup \end{array} N - \underset{\underset{S}{\|}}{C} - S - \underset{\underset{S}{\|}}{C} - N \begin{array}{c} \diagup C_2H_5 \\ \diagdown C_2H_5 \end{array}$$

Yellow powder, m.p. 103–114°C, s.g. 1·4. Soluble in benzene, acetone, chloroform, chlorinated solvents, carbon disulphide, insoluble in water, dilute alkalis and benzine. Ultrafast accelerator, similar to tetraethyl thiuram disulphide in its effect. Normally used with a thiazole. Suitable for natural rubber, nitrile rubber, SBR and neoprene.

Quantity: NR, NBR and SBR, primary 0·3–1% with 0·1% thiazole and 1–3% sulphur, secondary 0·1–0·3% with 1–1·5% thiazole and 3–2% sulphur, neoprene W 0·5–1% with 0·1% sulphur.

Uses: include floor coverings, foam rubber, soles.

TN: Ethyl Unads (51).

TETRAFLEX Group of softeners for synthetic rubbers and thermoplasts.

Grades:
DBP: dibutyl phthalate
DIOP: diiso-octyl phthalate
DOP: di-n-octyl phthalate
R-122: mixture of alkyl phthalates (237).

TETRAFLUOROETHYLENE

CF_2CF_2. Colourless gas, b.p. −76°C, shows little reaction. Polymerises to polytetrafluoroethylene (Teflon).

TETRAHYDROFURAN

C_4H_8O, 1:4-epoxy butane, b.p. 64–66°C, s.g. 0·889. Miscible with water, ethanol, ketones, esters.

Uses: solvent for natural and synthetic rubbers, polyvinyl chloride.

TETRAHYDROFURFURYL OLEATE

$C_{17}H_{33}COOCH_2O.C_4H_7$, m.p. −25°C, b.p. 240°C (5 mm), s.g. 0·921 (25°C). Softener.
TN: Plastolein 9250 (196).

TETRAHYDROFURFURYL PALMITATE

$C_{15}H_{31}COOCH_2O.C_4H_7$, b.p. 195°C (1·5 mm). Softener.

TETRAHYDRONAPHTHA-LENE

Tetralin. Colourless liquid, b.p. 207°C, s.g. 0·970 (20°C). Miscible with benzene, alcohol, chloroform, acetone, ether, petroleum ether.
Uses: solvent for NR and SR, resins, oils, waxes.

TETRAMETHYL THIURAM DISULPHIDE

White, odourless powder, m.p. 142–155°C, s.g. 1·29–1·31, mol. wt. 240. Soluble in benzene hydrocarbons, carbon tetrachloride, carbon disulphide, chloroform, acetone, ether, insoluble in water, dilute alkalis, alcohol and benzine. Ultra-fast accelerator with 13·3% sulphur avail-able for vulcanisation. May be used in natural rubber and synthetic polymers. Active above 121°C, with a fairly low critical temperature, but relatively good processing safety which may be influenced by the composition of the compound. The critical temperature is increased by even the smallest amount of dibenzthiazyl disulphide and the start of vulcanisation may be considerably retarded. Boosted by basic accelerators such as aldehyde-amines and guanidines. Used as a primary accelerator in the presence of sulphur and as a secondary accelerator with thiazoles, also as a curing agent when sulphur is absent. When sulphur is used the quantity may be varied over a wide range according to the properties required. Sulphurless compounds, or compounds with low sulphur content have excellent ageing properties and high resistance to dry or moist heat. Compounds may be vulcanised in a press, in steam or in hot air. Tetramethyl thiuram disulphide is easily dispersed. Non-staining and non-discolouring. Reduces cure time to a minimum; because it is extremely adaptable gives a good compromise between processing safety and cure rate. To obtain the best physical properties, 3–5% zinc oxide is necessary, also 0·5–1% stearic acid. If the zinc oxide content is low (0·5–1%) transparent products may be produced. Low concentrations of tetramethyl thiuram disulphide used in compounds, makes master batching necessary (1:10). Production and processing should be carried out on cooled mills, calenders and extruders. Vulcanisation temperatures: compounds containing sulphur 120–145°C, sulphurless compounds 140–160°C.

Quantity: natural rubber, primary 0·1–1% with 3–1% sulphur, secondary 0·05–0·25% with 1–1·5% dibenzthiazyl disulphide and 2·75–1·75% sulphur. Sulphurless compounds 2·5–4% with 0·5% mercaptobenzthiazole. For good ageing properties 0·5–0·25% with 1–1·5% tellurium and 1·25–0·5% sulphur. SBR, primary 0·3% with 2% sulphur, secondary 0·15% with 1% dibenzthiazyl disulphide and 2% sulphur, sulphurless compounds 3–4%. Butyl rubber 1–1·25% with 0·5% mercaptobenzthiazole and 2% sulphur. Nitrile rubber 0·5–2·5% with 0·1–3% sulphur and a thiazole as booster, sulphurless compounds 2·5–5% with dipentamethylene thiuram tetrasulphide as booster. Natural latex 0·5–1·5% with 0·5–2% sulphur.

Uses: include white, brightly coloured and transparent articles, articles in which a fast cure and good heat stability is required, extrudates, articles which come into contact with food (drinking water tubing, bottleseals, rubber rings for preserved fruits), gum stocks, impregnations, textile coatings, sponge rubber, household goods, mechanical goods, inner tubes, footwear, bathing caps, hot water bottles, floor coverings, cables, hard-rubber.

TN: Accel TMT (16)
Accélérateur TB (19)
Accélérateur rapide DTMT (419)
Accélérateur rapide TN (19)
+ Accelerator 52
Accelerator Thiuram (408)
Aceto TMTD (25)
Amizen TMTD (395)
Ancazide ME (22)
Cyuram DS (21)
DIDIDI
DTMT
Eveite 4 MT (59)

Kure Blend MT (50% master batch with SBR) (31)
Leda TMT (247)
Methyl Thiram (120)
Methyl Tuads (51)
Nocceler TT (274)
PmTD-70 (70% master batch with polyisobutylene (282)
Rapidex GR (390)
Robac TMT (311)
Roberts Thiram
Rodform Methyl Tuads (51)
SA-52, SA 52-O, SA 52-1 (4)
SA 52-2, SA 52-3, SA 52-6, SA 52-9
Soxinol TT
Superaccélérateur 501 (20)
Thiurad (5)
Thiuram
Thiuram 16
+ Thiuram DS
Thiuram DSM (5)
Thiuram M (6)
TMT
TMT-Henley (123)
TMTD
Tuads (51)
Tuex, Tuex Naugets, Tuex Powder (23)
Van Hasselt TMT (355)
Vondac TMT (433)
Vulcacure TMD (57)
Vulcadote TM
Vulcafor TMT (60)
Vulcaid 888 (3)
+ Vulcator TH (364)
Vulkacit Thiuram (43)
Wobezit Thiuram (367)
+ XKA (23)
Dispersions:
Naugatex 235 (23)
Vulcarite 122 (57)
Vulkacit Thiuram Dispersion 50 (43)
With 2-mercaptobenzthiazole:
Accelerator 108 (23)
Accelerator 108 PDR (23)
Butylaccelerator 21 (6)
Tuads Blends (51)
Vulkacit MT/C (2:1) (43).

TETRAMETHYL THIURAM MONOSULPHIDE

$$\begin{array}{c} H_3C \\ H_3C \end{array}\!\!\!>\!N\!-\!\underset{\underset{S}{\|}}{C}\!-\!S\!-\!\underset{\underset{S}{\|}}{C}\!-\!N\!<\!\!\!\begin{array}{c} CH_3 \\ CH_3 \end{array}$$

Pale yellow, non-hygroscopic powder, m.p. 104–107°, s.g. 1·40 mol. wt. 208. Soluble in acetone, benzene, chloroform, methylene chloride, ethylene dichloride and ethyl acetate, sparingly soluble in alcohol and carbon tetrachloride, insoluble in benzine and water. Relatively safe ultra-fast accelerator with slight delayed action, particularly in natural rubber compounds. Active above 121°C. Gives fast cures in SBR without scorching problems. In NR has a short, peaky vulcanisation range with a normal sulphur content. If the sulphur content is reduced, or 0·25–1% dibutyl ammonium oleate added, the cure range is considerably lengthened. Compounds with a low sulphur content (under 0·5%) have excellent ageing properties. Zinc oxide is necessary in normal quantities (5%), fatty acids are not essential, but may be used. The sulphur content may be varied widely, depending on the properties required in the final product. Lead oxide acts as a retarder. Can be used as a primary accelerator, or secondary with guanidines or aldehyde-amines, is an excellent secondary accelerator when used with thiazoles. Non-staining and non-discolouring on contact. May cause a slight yellowish stain in uncured white articles, but this disappears after vulcanisation. Its use as a master batch is recommended because of its high activity and the low amount used. Vulcanisation temperature: 130–145°C.

Quantity: NR, primary 0·15–0·30% with 2–3% sulphur, for good ageing properties 2–3% with 0·5–0·25% sulphur, secondary 0·1–0·25% with 1% mercaptobenzthiazole or dibenzthiazyl disulphide and 2–5% sulphur. SBR, primary 0·25–1·5% with 2–3·5% sulphur, secondary 0·1–0·75% with 1–2% dibenzthiazyl disulphide 1–3% sulphur. Butyl rubber, 1–2% with 1–2% sulphur, boosted by thiazoles (compounds are free from scorch, vulcanisates have good ageing properties). NBR 0·1–3% with 1–2% dibenzthiazyl disulphide and 0·5–2% sulphur (good ageing properties, scorch free). Neoprene, 0·5–1% with 1–3% guanidine and 0·5–1% sulphur, boosted by 0·1–0·5% NA-22 (gives fast cures). Reclaim, 0·15–0·5% with 2–4% sulphur (compounds become very scorchy when a thiazole is also added and it is necessary to use zinc oxide as a retarder).

Uses: notably natural rubber, SBR, neoprene W, butyl rubber, NBR, reclaim, transparent, light-coloured and white articles, articles which come into contact with food, dipped goods, bathing caps, household goods, insulation, mechanical goods, textile coatings, impregnations, sponge rubber, footwear.

TN: Accel TS (16)
Accélérateur TM (19)
Accélérateur rapide 500 (20)
Accélérateur rapide TM (19)
Accélérateur rapide MTMT (419)
Accelerator Thiuram MS (408)
Aceto TMTM (25)
Ancazide 1 S (22)
Ancazide ME (22)
Cyuram MS (21)
Eveite MST (59)
Eveite MTS (59)
Kure Blend MS (50% masterbatch with SBR) (31)
Monex, Monex Powder, Monex Naugets (23)

Mono Thiurad (5)
Morfex 55 (23)
Nocceler TS (274)
Pennac MS (120)
Robac TMS (311)
Rodform Unads (51)
Soxinol TS
Super Accélérateur 500 (20)
Thionex (6)
Thiuram MSM
TMTM
TMTM-Henley (123)
Unads (51)
Van Hasselt TMTM (355)
Vulcadote TB
Vulcafor MS (60)
Vulcaid 222 (3)
Vulkacit Thiuram MS (43)
Blends with zinc-2-mercaptobenz-thiazole:
Bantex M (5)
with 2-mercaptobenzthiazole:
Morflex (23)
with zinc-2-mercaptobenzthiazole:
Morflex 33 (23)
with a reaction product of butyraldehyde-aniline carbon disulphide:
Vulcapont (6).

TETRAMETHYL THIURAM TETRASULPHIDE

$$H_3C{\searrow}N-CS-(S_4)-CS.-N{\swarrow}CH_3$$
$$H_3C{\nearrow} \qquad\qquad\qquad {\nwarrow}CH_3$$

Pale grey powder, m.p. above 90°C, s.g. 1·52. Very active accelerator and curing agent with 32 % sulphur available for vulcanisation. Active above 110°C. Good processing safety in the absence of sulphur. Compounds with sulphur have strong tendency to scorch and are unsuitable for normal use. Sulphurless vulcanisates have excellent ageing properties and no stain.
Quantity: primary 0·6–2% without sulphur, secondary, with tetramethyl

thiuram monosulphide and without sulphur.
TN: Tetrone (6)
TMTT.

TEXAS Trade name for carbon blacks.
109: channel black giving high abrasion resistance
E: EPC
M: MPC (341).

TEXIGEL SPA 4 Aqueous solution of sodium polyacrylate. Has 13·5–16% total solids, pH 10–11, viscosity 25 000–40 000 cp (Brookfield); stable above pH 6. Thickening agent and protective colloid for latex. *Quantity:* 2–4% (163).

TEXIN Polyurethane elastomers, s.g. 1·25. Suitable for moulding and extrusion.
Grades:
912: Shore A Hardness 87–93
355: Shore D Hardness 49–55
480: Shore A Hardness 77–83 (263).

TEXOFOR Fatty alcohol ethylene oxide reaction products. Al and Bl. Yellowish, wax-like mass. Soluble in water. Non-ionic stabilisers for latex.
Quantity: 0·1% (342).

TEXOWAX Group of polyoxyethylenes (342).

TG-10 Russian produced thermal black with a specific surface area of 10 m²/g. Similar to HT.

THBS Abbreviation for trihydroxybutyrophenone.

THERMAL DEGRADATION Used only for the processing of Buna S. The rubber is cut into strips and plasticised by oxidation at 150°C at normal pressure, high pressure or with circulating air. The degree of degradation depends on the temperature, rate of change of air, time and pressure. Heavy metal salts and iron salts act as catalysts.

THERMAX MT. Medium fine thermal black (343).

THERMILO F Organic polysulphide (33).

THERMOFLEX Trade name for group of antioxidants (6):
Thermoflex: di-p-methoxydiphenylamine (obsolete)
A: mixture of 50% phenyl-β-naphthylamine, 25% di-p-methoxy-diphenylamine and 25% diphenyl-p-phenylenediamine.Grey, granular substance, s.g. 1·20. Has no effect on plasticity or vulcanisation. Gives excellent protection against heat, normal oxidation and flex-cracking. *Quantity:* 1–2%
B: secondary aromatic amines (obsolete)
C: mixture of aromatic amines.

THERMOPLASTIC RUBBER Developed by Shell Chemical Company, USA. Elastomers, block copolymers which behave like vulcanisates at room temperature but which are formable, like thermoplastics, at higher temperatures. The transformation is not due to chemical reaction, as is shown by its reversibility, but rather the network is of physical origin due to van der Waal's bonds which break down at elevated temperatures. The stability of the network probably contributes to the high strength and elongation at break. Usable rubbery properties are apparent in the temperature range −40 to approx. 60°C.
Uses: general application, according to the specific properties.
TN: Cariflex TR
TR-B (2).

THERMOPRENE Named by H. L. Fisher. A tough, thermoplastic hydrocarbon substance, produced by cyclisation of rubber using various processes. Thermoprenes are high mol. wt. products with the same composition as rubber $(C_5H_8)_x$, but with fewer unsaturated bonds. To produce a thermoprene, 100 parts or rubber with 10 parts sulphonyl chloride or a sulphonic acid, are passed through a mill at 125–135°C. A code is used to differentiate between the different types: GP, similar to gutta-percha, HB similar to hard balata, SL similar to shellac. Thermoprenes are used, according to the Vulcalock process, for bonding rubber to metal.

THERMOVYL P.V.C. fibre produced by the dry-spinning process from a solution of p.v.c. in acetone and carbon disulphide (165).

THIATE Group of accelerators (5):
A: thiohydropyrimidine
B: trialkyl thiourea, reddish brown fluid, s.g. 1·04
E: trimethyl thiourea.

THIAZOLE

Colourless to yellowish liquid, b.p. 115–118°C, s.g. 1·20. Soluble in alcohol, ether and many organic materials, sparingly soluble in water. Heterocyclic compound. Derivatives of thiazole, *e.g.* mercaptobenzthiazole, benzthiazyl disulphide are used to a large extent as accelerators.

THIAZOLINES

$$HC \overset{S}{\underset{H_2C-N}{\diagup\diagdown}} CSH$$

Dihydrothiazoles, group of accelerators, *e.g.* 2-mercaptothiazoline.

THIAZOLYL MERCAPTO-SUCCINATES Group of chemical peptisers, according to USP 2 725 382/1955.

THICKENER Solutions of sodium polyacrylate. Types 110 and 215. Thickening agents and protective colloids for latex (344).

THICKENING AGENTS Materials used to increase the viscosity of latex compounds; include bentonite, colloidal kaolin, alginate, methyl cellulose, ammonium and sodium polyacrylates, casein, gelatin.

THIO Abbreviation for thiocarbanilide.

2:2'-THIOBIS(4-ALKYL-5-THIAZOLE CARBOXYLIC ACID) Scorch resistant accelerator for NR and butyl rubber. Gives vulcanisates with high tensile strengths and high elongation at break. Compounding examples: 5% zinc oxide, 3% sulphur, 1% stearic acid, 0·3% diphenyl guanidine, 0·7% accelerator; cure, 60 min at 135°C. Lit.: USP 2 746 970 (1956).

4:4'-THIOBIS(6-TERT-BUTYL-3-METHYL PHENOL)

4:4'-tert-bis-6-tert-butyl-m-cresol. Pale grey, crystalline powder, m.p. 140°C, s.g. 1·079. Non-staining antioxidant for NR, SR and latices. Gives good, general protection for light-coloured articles and protects against heat ageing in chloroprene compounds.
Quantity: 1–2%.
Uses: light-coloured goods and articles which come into contact with lacquers and protective coatings, and which may be discoloured by other antioxidants.
TN: Santowhite crystals (5).

4:4'-THIOBIS(6-TERT-BUTYL-2-METHYL PHENOL) 4,4'-thiobis(6-tert-butyl-o-cresol). White powder, m.p. 124°C, s.g. 1·084. Non-staining, non-discolouring antioxidant for NR, SR and latex.
TN: Ethyl Antioxidant 736 (88).

THIOBISDISEC-AMYL PHENOL Dark, viscous liquid, m.p. max. 0°C, s.g. 0·98 (50°C). Non-staining, non-discolouring antioxidant with medium power for latex. May be easily emulsified and has no effect on the stability of the latex.
TN: Santowhite L (5).

2:2'-THIOBISMETHYL BUTYL PHENOL

473

2:2'-thiobis(4-methyl-6-tert-butyl phenol). White, crystalline powder, m.p. 82–88°C. Soluble in alcohols, hexane, acetone, toluene, ethylene dichloride and mineral oils. Non-staining antioxidant effective against normal oxidation and heat degradation in natural rubber, polyethylene and polystyrene. Suitable for white and light-coloured articles.
Quantity: 0·25–1·5%.
TN: AC-6 (15)
TBP 6 (380)
CAO 4 (15)
CAO 6 (15).

1,1'-THIOBIS(2-NAPHTHOL)

White powder, m.p. 215°C. Antioxidant.
TN: CAO-30 (15).

THIOCARBAMIDE

$H_2N.CS.NH_2$, thiourea, m.p. 182°C, s.g. 1·405. Soluble in ether. Forms addition compounds with many organic materials and inorganic salts.
Uses: accelerator, in the production of aminoplasts with formaldehyde, activator for dithiophosphates.

THIOCARBANILIDE

$$2C_6H_5.NH_2 + CS_2 \rightarrow$$
$$S=C \begin{cases} NH.C_6H_5 \\ NH.C_6H_5 \end{cases}$$

(sym. diphenyl thiourea, 1:3-diphenyl-2-thiourea, THIO, DPTH, DPTU. m.p. 148–158°C, s.g. 1·26–1·32. Soluble in acetone, ethyl acetate, methylene chloride, sparingly soluble in benzene, alcohol and carbon tetrachloride. One of the first organic accelerators; discovered by G. Oenslager, 1906. A relatively slow accelerator with a fairly low critical temperature and a slight tendency to cause scorching. It is not sensitive to changes in the sulphur content of the compound. Zinc oxide and fatty acids are necessary. Boosts thiazole accelerators but gives the best results when used as the sole accelerator. Unvulcanised compounds may be stored.
Quantity: natural rubber 2–5% with 3·5–5% sulphur, neoprene latex 0·5%.
Uses: include thin walled articles, hot-air hose, adhesives, latex articles, neoprene latices, booster for dithiophosphates.
TN: A-1 (6)
Accelemal
Accélérateur L (19)
Accélérateur LL (19)
Accelerator 14 (6)
Accelerator BB (14)
Activit
Anchoracel (22)
Anvico
Eveite TC (59)
Excellerex
NCC (obsolete) (23)
Noceller C (274)
Nurac (366)
Siccoactivex (430)
Socinol A
Thip
Velocite (356)
Vulcafor IV (60)
Vulcafor TC (60)
Vulcanol NCA (6)
Vulcarite 123 (50% dispersion) (57)
Vulcogène (420)
Vulkacit CA (43)
Vulkator I (364).

THIOGUTT Vulcanisable polysulphide rubber, similar to Thiokol,

with excellent resistance to swelling.
Types: A (powder), aqueous suspensions AH, AT, AS.
Uses: seals, moulded articles, hoses which must be oil and benzene resistant. (345).

THIOHYDROPYRIMIDINE

Structurally a tautomer of 4:4:6-trimethyl-2-mercaptodihydropyrimidine and 4:4:6-trimethyl-2-thiotetrahydropyrimidine. Crystalline powder, m.p. 250°C, s.g. 1·12. Slightly soluble in chloroform and alcohol, insoluble in water, dilute alkalis, carbon disulphide, benzine and benzene. Accelerator for neoprene, alone or in combination with other accelerators; gives good processing safety. 1% gives a fast cycle for press or open air cures. With low concentrations, guanidines and sulphur act as boosters.
Quantity: 0·5–1% with 0·5–0% sulphur.
Uses: include continuous vulcanisation, cable sheathings, wire insulations.
TN: Thiate A (51).

THIOKOL Group of thioplasts in solid and liquid forms, also trade name for various other products (208).
Grades:
A: solid polymers obtained from 1:2-dichloroethane and Na_2S_4 with 85% sulphur. A peptiser such as a guanidine, tetramethyl thiuram disulphide or benzthiazyl disulphide is necessary for processing
AP: powder form of Grade A.
F (obsolete): identical to Grade FA. Probably made from a mixture of ethylene dichloride and dichloroethyl formaldehyde acetal

FA: obtained from 1:2-dichloroethane, dichloroethyl formaldehyde-acetal and a polysulphide of the formula $Na_2S_{1.8}$ with 47% sulphur. A peptiser is necessary, as for Type A
LP-2: (LP = liquid polymer). Brown liquid, s.g. 1·27, mol. wt. approx. 4000, pH 6–8, viscosity 40 000 cp
LP-3: brown liquid, s.g. 1·27, mol. wt. approx. 1000, pH 6–8, viscosity 700–1200 cp
LP-32: comprises 99·5 mol. % bis-(chloroethyl) formal and 0·5 mol. % trichloropropane. Brown liquid, s.g. 1·27, mol. wt. 4000, pH 6–8, viscosity 35 000–40 000. Has a lower degree of crosslinking than Grade LP-2
LP-38: liquid polythiodithiol polymer, bis(2-chloroethyl)formal and 0·5 mol. % trichloropropane. Low mol. wt.
LP-205: polysulphide polymer for use at low temperatures.
The liquid types LP-2 and LP-3 are a blend of 98:2 of dichloroethyl formaldehyde-acetal and 1:2:3-trichloropropane, they are crosslinked to a certain degree. *Uses:* protective paints, impregnation, jointing. LP-3 is also used as a softener and plasticiser for Type ST and in combination with epoxy resins and formaldehyde resins for soft to hard castings
N (obsolete): similar to GR-P. Obtained from ethylene and propylene dichlorides
PR-1: obtained from ethylene dichloride and dichloroethyl formaldehyde-acetal with the probable addition of trichloropropane and Na_2S_2 as polysulphide. Chain scission as for Type ST
R: plastic containing sulphur
ST: obtained from dichloroethyl formaldehyde-acetal with the addition of a trihalide, probably

1:2:3-trichloropropane as a crosslinking agent, and a polysulphide of the formula $Na_2S_{2.25}$ By scission of some of the polysulphide groups the chain length is reduced, processability improved. Contains 40% sulphur

VA-7: liquid polysulphide, s.g. 1·45, viscosity 5000–10 000 cp. *Uses:* curing agent, *e.g.* for natural rubber, SBR, NBR, urethanes

ZL-190: polysulphide

ZR-454: type with a high resistance to swelling by fuels, toluene, benzene, ethyl acetate, methyl ethyl ketone and other solvents. Pale brown mass, s.g. 1·41. Preplasticised, chemical softeners are not necessary. Has excellent resistance to the effects of weather, but other properties are only moderate. *Uses:* primarily for high resistance to solvents, fuel hose, printing rollers, printers' blankets, seals.

LP-33: 99·5 mol. % bis(2-chloroethyl)formal and 0·5 mol. % trichloropropane. Brown liquid, s.g. 1·25, mol. wt. 1000, pH 6–8, viscosity 1300–1500 cp. Has a lower degree of branching than Grade LP-3.

The following products are not polysulphides:

RD: butadiene-acrylonitrile rubber

TP-90B: high mol. wt. polyether. Pale brown liquid, s.g. 0·967. Softener and plasticiser for natural and synthetic rubber and vinyl resins

TP-95: high mol. wt. polyether ester. Pale brown liquid, s.g. 1·013, b.p. approx. 350°C (4 mm). Softener and plasticiser for natural rubber, SBR, NBR, neoprene, Thiokol ST and vinyl resins

Butyl rubber	Degree of unsaturation	Mooney
035	0·6–1	38–47
150	1·0–1·4	41–49
165	1·5–2·0	41–49
215	1·5–2·0	41–49
217	1·5–2·0	61–70
218	1·5–2·0	71–80
265	1·5–2·0	41–49
267	1·5–2·0	61–70
268	1·5–2·0	71–80
325	2·1–2·5	41–49
365	2·1–2·5	41–49

ZL-239: polyurethane prepolymer for foam products.

THIOKOL LATICES Aqueous dispersion of thioplasts, s.g. 1·3–1·45, particle size 2–15 μ. Particles have no charge.
Grades:
MF: obtained from ethylene dichloride and dichloroethyl formaldehyde-acetal
MX: obtained from ethylene and propylene dichlorides
WD-2: obtained from dichloroethyl formaldehyde-acetal and trichloropropane
WD-6: 67:33 blend of polyethylene disulphide and polypropylene disulphide. Film former.
For film formation the addition of 0·5–1% tetramethylene disulphide, or heat treatment is necessary. Vulcanised with 5–10% zinc oxide.
Uses: corrosion resistant paints, bonding agents (208).

THIOKOL METHOD Thiokol bent loop test. A test for determining the brittleness of a test piece, bent to a bow shape under strain, at very low

476

temperatures (-40 to $-50°C$), over a period of 5 h. ASTM D 736.

THIONITE Japanese produced thioplasts.

THIOPLASTS Polysulphide rubber, organic polysulphides. Linear rubber-like elastic or thermoplastic polycondensation products of alkali polysulphides with aliphatic dihalides (methylene, ethylene, propylene dichlorides, dichloroethyl ether, glycerol dichloride, dichloroethylene formaldehyde-acetal) according to the equation:

$$R.Cl_2 + Na_2S_x \rightarrow (-R-S_x-)_n$$

with $x = 2, 3$ or 4, thus:

$$n(Cl-CH_2-CH_2-Cl)$$
$$+ n(Na_2S_4)$$
$$\rightarrow (-CH_2-CH_2-S_4-)_n$$
$$+ 2nNaCl$$

The end groups are —SH groups, or because the dihalides may be hydrolysed on reaction with polysulphides, they may be OH groups. Thioplasts are produced by the reaction of the components in an aqueous phase with magnesium or barium hydroxide as a dispersing agent. Fairly large, uncharged spherical particles are produced ($2-15$ μ) which separate out by sedimentation. Vulcanisation must be regarded as further condensation and may be achieved with zinc oxide, zinc peroxide or lead peroxide, with a possible addition of stearic acid as modifier for the rate of vulcanisation. For the liquid types ethoxy resins, phenol formaldehyde resins, aldehydes are among the chemicals used, together with amines. During mixing the thioplasts are not degraded but various guanidines, tetramethyl thiuram disulphide and benzthiazyl disulphide have a strong peptising effect due to the scission of the disulphide bonds. The addition of trihalides during processing causes some crosslinking and improves the dimensional stability. By introducing a reducing reaction during production, thioplasts may also be obtained in liquid form. These are miscible with *e.g.* aromatic and chlorinated hydrocarbons, esters, and ketones, and may be hardened to solid products by further condensation.

Thioplasts vary in their properties according to the type of halogen used and the number of sulphur atoms in the polysulphides. They have excellent resistance to aliphatic and aromatic solvents, oxidation, ozone, the effects of weather, water, and temperatures as low as $-50°C$. However, the end products have only moderate physical properties. A disadvantage is the unpleasant odour which occurs on heating because of the various cyclic thio compounds. Processing of the thioplasts must take place below $70°C$. Both processing and physical properties may be improved by addition of SRF black and cross blending with natural rubber.

Uses: include corrosion resistant coatings, cable sheathings, hose lines, seals, impregnations, covers for printing rollers, adhesives, vulcanising agents for natural and synthetic rubbers (liquid types). Thioplasts appear on the market in solid, powder and liquid forms.

TN: Ethanite (Belgium)
GR-P (US, former TN)
Hikatol (Japan)
Hydrite (Japan)
Novoplas (Britain)
Perduren (Germany)
Resinite (USSR)

Thiogutt A (Germany)
Thiokol (USA)
Thiolatex (France)
Thionite (Japan)
Vulcaplas (Britain).

THIOSAN Thiazole/dithiocarbamate blend. Yellowish powder, s.g. 1·46. Soluble in benzene, chloroform, carbon disulphide and alcohols, partially soluble in water.
Uses: primarily as fast accelerator for textile coatings, footwear, cables, sponge rubber, hot vulcanisable adhesives (5).

THIOSTOP Short-stop for the emulsion polymerisation process.
Grades:
K : 50% aqueous solution of potassium dimethyl dithiocarbamate
N : 40% aqueous solution of sodium dimethyl dithiocarbamate (23).

THIURAM Trade name for a group of thiuram accelerators produced by various manufacturers.
Grades:
16:     tetramethyl thiuram disulphide (TMTD)
DS: TMTD
DSE: TMTD
DSM: TMTD
E: TMTD
M: TMTD
MSM: tetramethyl thiuram monosulphide
P.25:   tetramethyl thiuram tetrasulphide.

THIURAM SULPHIDES
Group of sulphur containing compounds related to the dithiocarbamates, and with the general formula R_2NCS. Ultra-fast accelerators. Thiuram disulphides are formed by the direct oxidation of soluble dithio-

carbamates with hydrogen peroxide, ammonium persulphate or sodium nitrite. The thiuram monosulphide is obtained by desulphurisation of the disulphide using sodium cyanide, and the polysulphide by the reaction of certain dithiocarbamates with sulphur monochloride. The di- and polysulphides act as curing agents in the absence of sulphur.

THIXCIN Hydrogenated castor oil. Thickening agent for Hypalon paints (9).

THIXON Group of bonding agents for metal/rubber bonds (346, 127).

THIXOTROPY (From the Greek thixis: to touch; and tropa: change). Term usually applied to the change in the properties of a substance on contact or mechanical action. It was put forward by T. Peterfi (*Arch. Entw. Mech. Organe*, 1927, **112**, 389), and treated as a rheological property by H. Freundlich (*Thixotropy*, Paris, 1935). The term is used to indicate the labile condition between the sol and gel phases in colloidal solutions which form a gel or become viscous on standing and which are liquefied by mechanical action. The thixotropic effect occurs because the anisometric particles of the solid phase form an irregular honeycombed structure in which they are only linked at certain single positions by electrolytic or Van der Waal's forces. A relatively weak mechanical force is sufficient to destroy the structure and thus reduce the viscosity. A certain period of inactivity is necessary to bring about restitution of the original structure.

THOMSON, ROBERT WILLIAM 1822–1873. In 1845, invented the first inflated tyre (BP 10 990) consisting of a sailcloth tube impregnated with rubber and with a leather cover. He suggested that the inner tube could consist of several tubes, that a steel reinforced tread could be used, and that inflatable tyres would be used for trains. His patent was almost forgotten, presumably because of the shortage of rubber at the time and the small volume of traffic.

TI-CAL Mixture of titanium dioxide and calcium sulphate, s.g. 3·13. Pigment for rubber compounds.

TILDEN, SIR WILLIAM A. 1842–1926. Professor of chemistry at Mason College, Birmingham, and at the Royal College of Science, onetime pupil of A. W. von Hofmann. Pioneer in the fields of terpene chemistry and synthetic polyisoprene. Produced isoprene by pyrolysis of turpentine (1882) and terpenes, d-limonene and terpilene (1884). Also produced synthetic polyisoprene by spontaneous polymerisation (1892). In 1882 assigned the formula C_5H_8 to isoprene and confirmed the structural formula.

TIMONOX Finely dispersed antimony oxide (416).

TINIUS OLSON PLASTI-VERSAL MACHINE Machine for determining the stress/strain properties of rubber.

TINTING STRENGTH Measurement which characterises the colouring power of carbon blacks; partly dependent on particle size.

Measurement is by comparison with a standard compound.

TINUVIN P Substituted benztriazole. White powder, m.p. 132°C, s.g. 1·38. Protective agent against ultra-violet rays, for NR and SR (227).

TIOLAN Synthetic fibre based on casein.

TIP Small, round eraser, approx. 7 mm long on the end of pencils.

TITANIUM DIOXIDE Titanic acid anhydride. White powder, m.p. 1775°C, s.g. 3·84. Occurs as alpha and beta titanic acid. Since 1908 has been obtained by the decomposition of Ilmenite with sulphuric acid followed by hydrolysis.
Uses: white pigment in rubber and plastics, for protection against dulling of artificial silk, as a colour pigment, and to render enamels opaque; particles of 1/5000 mm diameter have the best whitening power.

TKhM Russian produced scorch retarder; mixture of trichloromelamine and barium sulphate. White powder, m.p. 140°C.

TLACHTLI Game played with rubber balls in ancient Mexico.

TLARGI Abbreviation for The Los Angeles Rubber Group Inc., a branch of the Division of Rubber Chemistry of the American Chemical Society, founded 1927.

TM- Russian produced thermal black. The index number denotes the specific surface area in m^2/g. Types TM-14 and TM-30.

TMDB Polystyrenes, modified with other polymers (137).

TMN Russian produced thermal black obtained from methane as a by-product of acetylene production.

TMS Abbreviation for tetramethyl thiuram monosulphide.

TMTD Abbreviation for tetramethyl thiuram disulphide.

TMTM Abbreviation for tetramethyl thiuram monosulphide.

TMT Abbreviation for tetramethyl thiuram disulphide.

TODI Abbreviation for 3:3'-dimethyl-4:4'-diphenyl diisocyanate (148).

p-TOLUENE SULPHONYL HYDRAZIDE

$$CH_3-\!\!\left\langle\bigcirc\right\rangle\!\!-SO_2-NH-NH_2$$

m.p. 100–110°C, s.g. 1·42. Blowing agent.
TN: Celogen TSH (23).

α-o-TOLYL DIGUANIDINE

White, odourless powder, m.p. 144°C, s.g. 1·17. Soluble in ethyl alcohol and methylene chloride, sparingly soluble in acetone, ethyl acetate and water, insoluble in carbon tetrachloride and benzene. A medium-fast basic accelerator giving good processing safety. Vulcanisates are odourless and tasteless. Suitable for products which come into contact with food. Curing is possible without zinc oxide, but its full effect is only realised when zinc oxide is present. Suitable as a booster for mercaptobenzthiazole, dibenzthiazyl disulphide, cyclohexylbenzthiazyl sulphenamide, tetramethyl thiuram disulphide and zinc ethyl phenyl dithiocarbamate. Curing should be at temperatures above 140°C. A certain amount of stearic acid should be added in the form of a master batch to improve dispersion. Can also be blended with a filler or used with an oil paste.
Quantity: primary, 1–1·5% with 2·75–4% sulphur, for very fast cures up to 4%. Secondary, 0·1–0·5% with 0·7–1·2% mercaptobenzthiazole and 1·5–3% sulphur, 0·1–0·2% with 0·25% tetramethyl thiuram disulphide and 2–2·4% sulphur.
Uses: jar sealing rings, bottle caps, drinking water tubing, vulcanisable cements, erasers.
TN: Accélérateur 80 (20)
Eveite 1000 (59)
Nissim **SD**
Noceller **BG** (274)
Sopanox (5)
Soxinol **SD**
Vulkacit 1000 (43)
Wobezit 1000 (367).

2:4-TOLYLENE DIISO-CYANATE

Toluene-2:4-diisocyanate. Colourless to yellowish liquid, m.p. 21·7°C, b.p. 250°C, s.g. 1·22. Soluble in aromatic and chlorinated aromatic hydrocarbons, esters, ethers, acetone, nitrobenzene. Reacts with moisture.

Causes irritation to the skin and mucous membranes.

Uses: production of polyurethane elastomers, crosslinking agent for polyurethane, adhesives, lacquers, bonding agent for metal/rubber bonds. The dimer is particularly suitable as a crosslinking agent as the two isocyanate groups are blocked and only activated on vulcanisation, thus improving the stability of the compounds.

TN: Hylene T (6)

Hylene TM (80% with 20% 2:6-toluene diisocyanate (6)

Vulcafor VCN (60)

Mondur TDS and TD 80 (80% 2:4-T) (263)

Desmodur T (blend of T-2:4 and T-2:6) (43)

Desmodur TT (T dimers) (43)

Nacconate 100

Nacconate 80 (80:20% blend of 2:4 T and 2:6 T)

Nacconate 65 (65:35% blend of 2:4 T and 2:6 T).

p-(p-TOLYLSULPHONYL-AMIDO)DIPHENYLAMINE

$CH_3 . C_6H_4 . SO_2NH_2 . C_6H_4 . NH_2 . C_6H_5$

Grey powder, m.p. min. 135°C, s.g. 1·28. Soluble in acetone, benzene and ethylene dichloride, slightly soluble in hot water and hot alkaline solution, insoluble in benzine and cold water. Antioxidant for rubber and latex, with a slightly discolouring effect; slightly activates the curing speed. A white bloom occurs on vulcanisates if more than 0·4% is used. Gives moderate protection against heat, good protection against oxygen, copper, manganese, and free chlorine in textiles which have been proofed with neoprene and which would otherwise attack the rubber. Reduces the plasticity of unvulcanised neoprene compounds.

Uses: in natural rubber, neoprene, and SBR for textile proofing impregnation and mechanical goods.

Quantity: for optimum effect in articles where blooming is unimportant 1–1·5%, otherwise below 0·4%.

TN: Aranox

Naugatex 510 (50%)

Naugatex 510B (aqueous dispersion) (23).

TOMARKIN TYRES All-weather tyres. The tread contains dispersed, prevulcanised elastomer particles of a greater hardness than that of the tread; this gives a firm grip on smooth surfaces, *e.g.* ice.

TOPANOL Group of antioxidants.

Grades:

A: 2:4-dimethyl-6-tert-butyl phenol

M: N,N'-disec-butyl-p-phenylene diamine

O: 2:6-ditert-butyl-4-methyl phenol

OC: 4-methyl-2:6-ditert-butyl phenol.

TORNESIT Chlorinated rubber (10).

TORQUEMADA, JUAN DE Compiler of 'Libros Rituales y Monarchia Indiana con el Origen y Guerras de los Indios Occidentales, de sus Poblacones Descubriemiento, Conquista, Conversion y Otras Cosas Maravillosas de la Mesma Tierra', Sevilla, 1615. This work referred to shoes, clothes, head coverings and other rubber waterproof articles worn by the Mexican Indians, the rubber being obtained from the sap of a plant, probably *Castilla elastica* and known as Olli or Ulli, a corruption of the native name.

TORR-GERICKE PROCESS
Electrical coagulation of rubber; developed by Torr-Gericke Rubber, in the Belgian Congo. The process has no technical or economic advantages.

TPG Abbreviation for triphenyl guanidine.

TPPP Abbreviation for tetrapotassium pyrophosphate.

TPX POLYMER Poly-4-methylpentene, high temperature polyolefin, softening p. 240°C, s.g. 0·83 (60).

TRADESCANT, JOHN
British explorer who brought the first piece of gutta-percha to Europe. He described it as 'mazer wood (grained wood); which can be shaped after being heated in hot water'. The term 'wood' is not surprising as even Sir William Hooker, director of Kew Gardens in the nineteenth century, thought that a gutta-percha slab was wood because of the characteristic graining.

TRAGACANTH Endosperm of the seeds of *Ceretonia Siliqua*. Yellow, odourless and tasteless powder, pH approx. 6·3, contains approx. 29% galactan, 58% mannan, 2·75% pentosanon, 5·3% albuminoids. Soluble in warm water. Creaming agent for latex (349).

TRANSFER MOULDING Process by which the rubber compound is transferred under the action of heat and ram pressure from the transfer cavity through an orifice (transfer port) into the shaped mould cavity, where vulcanisation takes place. The process differs from injection mould-

ing in that the rubber compound is placed in a separate mould cavity, not in the heated barrel of a machine. *Uses:* particularly for rubber/metal bonded units.

TRANSITION TEMPERATURE
The temperature boundary shown by thermoplastic, non-crosslinked plastics, at which a permanent, plastic deformation takes place.

TRANSITION TEMPERATURE
The temperature at which changes take place in the physical properties of a polymer.

TRANSPARENT RUBBER
Developed by F. Jones, 1923. To achieve high transparency it is necessary to use accelerator and antioxidants which dissolve completely in the rubber and a grade of zinc oxide which mixes and disperses easily. By using a very fine grade only a small quantity of zinc oxide must be used to activate the accelerators and obtain satisfactory properties. The cure time also affects the transparency. If the period is too short then the vulcanisates are cloudy because the particles have had insufficient time to dissolve properly; too long a cure time causes dark discoloration.
Examples of compounding ingredients are:

	A	B
Pale crêpe	100	100
Stearic acid	0·5	0·5
Zinc oxide	0·5	0·5
Aldehyde-amine antioxidant	0·75	0·75
TMTM	0·02	0·25
Zinc MBT	0·65	—
Sulphur	1·50	1·50

Cure: 5 min at 153°C. 'A' has better ageing properties, whereas 'B' has less taste.

TRAUBE PROCESS Process for the concentration of latex by creaming. BP 226 440 (1924) and DRP 414 210 (1923).

TRAUMATICIN A solution of one part gutta-percha in 9 parts chloroform.
Uses: in surgery and dentistry to close small wounds.

TREAD RIPPLE Formation of standing waves in a tyre tread; generated by impact of the tread pattern with the road. Occurs at high speeds and results in an unsmooth action. Can be overcome in the tyre construction by use of stiff treads and in use by inflation to high pressures. A serious problem in the design of racing tyres.

TRED Styrene-butadiene copolymer used to reinforce natural and synthetic rubber, in compounds requiring a low s.g., high hardness, good flex cracking and abrasion resistance, good ageing properties and high volume stability; it improves the dispersion of fillers and the surface finish, assists extrusion and reduces die-swell in extruded articles: soluble (Tred 50) or swollen (Tred 85) in benzene, toluene, solvent naphtha, carbon tetrachloride and chloroform, insoluble in water, alcohol ether, acids and alkalis:
Tred 50: 50% bound styrene, s.g. 1·00. Easily processed, can be mixed either on a mill or in an internal mixer at normal processing temperatures. Partially unsaturated,

therefore additional sulphur and accelerators are necessary
-65: 65% styrene, s.g. 1·02
-85: 85% bound styrene, m.p. 54°C, s.g. 1·04. Mixed in an internal mixer or on a mill, both of which are steam heated, processing temperature 100–120°C.
Uses: shoe soles, microcellular soles, cables, mechanical goods, floor coverings, extrusion aid (5).

TREFOIL Phenolic resin (350).

TRENYLENE Triphenyl guanidine (21).

TREVIRA (High strength). Polyester fibre, *e.g.* for conveyor belts, fire extinguishing hoses, V-belts, transformer belts (217).

TRIANGULAR ACCELERATION A three-component accelerator system, recommended by the Vanderbilt Laboratory, with mercaptobenzthiazole, benzthiazyl disulphide and, as the third component, zinc dimethyl dithiocarbamate, zinc diethyl dithiocarbamate or tetramethyl thiuram disulphide. Changes in the quantities of the individual components result in a greater range of cure speed, and scorch. The first two components give a wide vulcanisation range and good ageing properties, the third component ensures a cure throughout. Benzthiazyl disulphide controls the tendency to scorch.
Lit.: *The Vanderbilt Rubber Book*, 1958.

TRIBOLUMINESCENCE Light emission by solid crystalline bodies, due to mechanical action. The effect may be observed by stretching rubber compounds extended with barytes or chalk, or by removing rubber which is sticking to a surface, as, for example, by removing an insulating tape.

TRIBUTOXYETHYL PHOSPHATE ($C_4H_9OC_2H_4)_3PO_4$, m.p. $-70°C$, b.p. 200–232°C (4 mm), flash p. 224°C, s.g. 1·02. Softener for plastics, plasticiser for NBR. Gives excellent flexibility at low temperatures and improves heat stability. Vulcanisates have a high tensile strength and elongation at low hardness. Improves the flame resistance.
TN: KP-140 (184).

TRIBUTYL ACONITATE Yellowish liquid, b.p. 190°C (3 mm), s.g. 1·02–1·04. Softener for NBR and other elastomers. Gives excellent low temperature flexibility.

TRI-n-BUTYL CITRATE ($C_4H_9OCOCH_2)_2C(OH)$
.$COOC_4H_9$, m.p. $-85°C$, b.p. 170°C (1 mm), s.g. 1·05. Insoluble in water. Softener for cellulose and polyvinyl plastics, NR and SR, improves low temperature flexibility.
TN: Citroflex 4 (155).

TRI-p-TERT-BUTYL PHENOL PHOSPHATE
$[(CH_3)_3C.C_6H_4]_3PO_4$, m.p. 95°C, b.p. (5 mm) 320°C, flash p. 275°C. Softener.
TN: Dow 77 (7).

TRIBUTYL PHOSPHATE
$(C_4H_9)_3PO_4$. Clear liquid, m.p. $-80°C$, b.p. 177°C (27 mm), flash p. 146°C, s.g. 0·98. Solubility in water

0·6%, miscible with most organic solvents. Softener for plastics, antifoam agent.

TRICHLOROETHYLENE
$CHCl=CCl_2$, trichloroethene, acetylene trichloride. Colourless liquid, m.p. $-73°C$, b.p. 86·7°C, s.g. 1·465, mol. wt. 131·4. Miscible with chloroform, alcohol, benzene, ether and carbon disulphide. Solvent for rubber, resins, oils, fats, waxes, cellulose esters.
TN: Tri
Trilene
Tri-Clene
Trichloran
Trichloren
Triline
Trimar
Chlorylen
Trethylene
Westrosol, etc.

TRICHLOROMELAMINE
$C_3H_3N_6Cl_3$. Vulcanisation retarder, particularly for compounds containing a mercapto accelerator and blends with antimony sulphide. Is normally extended with barium sulphate.
TN: Retarder TCM and TCM-25 (containing 75% blanc fixe) (3).

TRICHLOROMETHANE SULPHONYL CHLORIDE
White powder. Soluble in most organic solvents.
Uses: curing agent for natural rubber and SBR in the presence of sulphur (166).

N-TRICHLOROMETHYL MERCAPTO-4-CYCLOHEXANE-1:2-DICARBOXIMIDE N-trichloromethyl thio tetrahydrophthalimide. Odourless crystals, m.p. 173°C, s.g. 1·74. Soluble in chloroform and

benzene, insoluble in water. Fungicide and bactericide for vulcanised rubber and vinyls.
TN: Captan
Orthocide 406
SR-406
Orthocide 50
Vancide 89 (51).

2:4:5-TRICHLOROPHEN-OXYACETIC ACID 2:4:5-T; synthetic hormone for plant growth. Colourless crystals, m.p. 153°C. Soluble in alcohol, insoluble in water. Has a stronger effect than 2:4-dichlorophenoxyacetic acid.
Uses: stimulation of latex production in rubber trees, the butyl ester being preferred; a 1% mixture with an inert oil, or a mixture with 5 parts raw palm oil and 3 parts inert oil, is spread on the bark beneath the tapping surface. Gives a 25–50% increase in latex production, but the latex usually has a lower rubber content. In higher concentrations it is used to kill old trees which no longer give an economic yield; these die about two months after application.

TRICRESYL PHOSPHATE $(CH_3C_6H_4)_3PO_4$; p-form, m.p. 78°C, b.p. 340°C, s.g. 1·18. Soluble in alcohol, chloroform, benzene, ether and acetic acid, sparingly soluble in water. Softener for plastics. The o-isomer is toxic (colourless, oily liquid, b.p. 410°C) and is no longer used.
TN: Celluflex 179-C
Lindol (152)
Kronitex (184)
PX-917 (305).

TRICROTONYLIDENE TETRAMINE Condensation product of crotonaldehyde and ammonia. Dark brown, viscous oil,

s.g. 1·02. Soluble in benzene, alcohol, ethyl acetate, acetone, carbon tetrachloride, methylene chloride, insoluble in water and benzine. Medium-fast aldehyde amine accelerator for mechanical goods containing large amounts of reclaim, also used in ebonite (without zinc oxide) and as an emulsion for latex. Zinc oxide is necessary in soft rubber articles but fatty acids are not essential. Give low modulus and good ageing properties. Unsuitable for hot air cures, except when used in ebonite. Causes dark staining and is suitable only for dark-coloured articles. Vulcanisates have a characteristic odour and taste and are unsuitable for articles which come into contact with food.
Quantity: mechanical goods, 0·75–1·5% with 3–4% sulphur, reclaim compounds 0·5–1% with 1·5–2% sulphur, highly loaded black compounds 2% with 3·5% sulphur, ebonite 1–1·5% with 35–50% sulphur.
TN: Accelerator CT-N (408)
Vulkacit CT-N (43)
Crotonaldehyde ammonia.

TRI-2:3-DIBROMOPROPYL PHOSPHATE Flame retardant for plastics.
Uses: polystyrene and urethane foams, polyesters, acrylic resins, epoxy resins, p.v.c.

TRIDIMETHYL PHENYL PHOSPHATE $[(CH_3)_2C_6H_3]_3PO_4$, m.p. −35°C, b.p. 255–295°C (10 mm) flash p. 233°C, s.g. 1·55. Softener for plastics.
TN: Trixylenyl phosphate
Celluflex 179-A (152).

TRIETAL Russian produced neutral salt of triethanolamine and o-phthalic acid.

TRIETHANOLAMINE

$N(CH_2.CH_2.OH)_3$. Viscous, pale yellow liquid, m.p. 277°C (150 mm), s.g. 1·1. Vulcanisation activator in compounds which contain reinforcing silica, aluminium silicate or calcium silicate, also in compounds containing white factice mercaptobenzthiazole and diphenyl guanidine, *e.g.* in erasers.
TN: Ankoltet (22)
Polymel Actisil.

TRIETHYL CITRATE

$(C_2H_5OCOCH_2)_2C(OH)$
$.COOC_2H_5$.
Colourless liquid with faint odour, m.p. −55°C, b.p. 150°C (3 mm), s.g. 1·136. Solubility in water 6%. Softener.
TN: Citroflex 2 (155).

TRIETHYLENE GLYCOL

$HO.CH_2CH_2O.CH_2CH_2O$
$.CH_2CH_2OH$.
Colourless, odourless hygroscopic liquid, b.p. 285°C, s.g. 1·127 (15°C). Miscible with water, benzene, toluene and alcohol. Activator for NR and SR in compounds containing white reinforcing filler. Softener in plastics.

TRIETHYLENE GLYCOL DIACETATE

$(CH_2OCH_2CH_2OCOCH_3)_2$, m.p. below −60°C, b.p. 170–197°C (620 mm), s.g. 1·115 (25°C). Soluble in water. Softener.

TRIETHYLENE GLYCOL DIBENZOATE

$C_6H_5CO(OCH_2CH_2)_3OCOC_6H_5$, m.p. 57°C, b.p. 223–237°C (61 mm), s.g. 1·168 (25°C). Softener.
TN: Benzoflex T-150.

TRIETHYLENE GLYCOL DICAPRYLATE

$(CH_2OCH_2.CH_2OCOC_7H_{15})_2$. Clear liquid, b.p. 243°C (5 mm), s.g. 0·973. Softener for natural and synthetic rubbers and vinyl resins, particularly for use at low temperatures.
TN: RC Plasticiser TG-8 (110).

TRIETHYLENE GLYCOL DI(2-ETHYL BUTYRATE)

$C_5H_{11}CO(OC_2H_4)_3OCOC_5H_{11}$, b.p. 256°C (5 mm), s.g. 0·995 (20°C). Softener.
TN: Flexol 3GH (146).

TRIETHYLENE GLYCOL DIPELARGONATE

$(CH_2OCH_2.CH_2OCOC_8H_{17})_2$. Clear, light-coloured liquid with mild odour, b.p. 251°C (5mm), s.g. 0·964. Processing aid and low temperature softener for NR, SR and vinyl resins.
TN: RC Plastikator TG-9 (110) Plastolein 9404 (194).

TRIETHYLENE GLYCOL DIPROPIONATE

$(CH_2OCH_2CH_2OCOC_2H_5)_2$. Clear liquid, m.p. below −60°C, b.p. 140°C, s.g. 1·066. Slightly soluble in water (6·5%). Softener for NR, SR, vinyl and cellulose resins.

TRIETHYLENE TETRAMINE

TETA. $H_2N(CH_2CH_3NH)_3H$. Viscous liquid which is miscible with water; b.p. 277·5°C, s.g. 0·982.
Uses: with a peroxide in the emulsion polymerisation of SBR cold rubber (Redox system), curing agent for polyacrylic esters and copolymers (1–3% with 0·5–2% sulphur; improves oil and heat resistance), curing agent for polysiloxane at room

temperature, for the production of foam rubber from natural and synthetic latices (0·1–0·5% on the rubber considerably improves the cell structure).

TRI(2-ETHYL HEXYL)PHOSPHATE

$[C_4H_9CH(C_2H_5)CH_2]_3PO_4$. Clear liquid, m.p. −80°C, b.p. 230°C (10 mm), s.g. 0·926 (20°C). Softener. *TN:* Flexol TOF.

TRIETHYL PHOSPHATE

$(C_2H_5)_3PO_4$, ethyl phosphate. Colourless liquid, m.p. −56°C, b.p. 215–216°C, s.g. 1·07 (20°C). Soluble in alcohol, ether and water. Softener.

TRIETHYL TRIMETHYLENE TRIAMINE

$(C_2H_5N{=}CH_2)_3$; reaction product of ethylene chloride, formaldehyde and ammonia. Dark brown viscous liquid, s.g. 1·10. Soluble in water. Accelerator for NR, SBR and their latices, and for chloroprene adhesives, has a medium to long curing range. Stiffens unvulcanised NR compounds. Has a fairly low critical temperature, gives fast curing compounds when used with thiazoles, thiurams and guanidines, is retarded by acidic substances, channel blacks and alumina. Causes a slight staining in white or light-coloured articles, which is bleached on exposure to light. Zinc oxide and sulphur are necessary in the normal quantities, fatty acids are not essential but may be used, causing slight retardation. Has a destabilising effect on latex in the presence of zinc oxide, makes the latex heat sensitive and gel more easily. Stabilises latex foams and reduces shrinkage.

Uses: hot air cures, rubber shoes, textile proofing, impregnation, artificial leather, sponge rubber, latex foam.
TN: Trimene Base (23)
Trimene (mixture with stearic acid) (23)
Vulcaid 777 b (3)
Vulkacit TR (43)
TTT.

TRIFLUOROCHLOROETHYLENE

$CF_2{=}CFCl$, perfluorovinyl chloride. Colourless gas, b.p. −28·4°C.
Uses: monomer for polymerisation to polytrifluorochloroethylene.

TRIGONOX

Group of organic peroxides.
Grades:

A 75:	tert-butyl hydroperoxide (75%)
B:	ditert-butyl peroxide
HM 80:	methyl isobutyl ketone peroxide (80%)
K 70:	cumyl hydroperoxide (70%) (171).

TRIHYDROXYBUTYROPHENONE

THBS. Antioxidant for polyethylene.
Quantity: 0·1%.

TRIHYDROXYTRIETHYLAMINE

Strongly hygroscopic, viscous, yellow liquid with slight ammoniacal odour, m.p. 21·2°C, b.p. 360°C, s.g. 1·124 (20°C). Activator and dispersing agent for natural and synthetic rubbers; improves processability, decreases gas permeability, causes a slight increase in tear resistance.

TRIK

Russian produced triethanolamine derivative.

TRIMATE M Dithiocarbamate accelerator. Yellow liquid, s.g. 1·03–1·07, pH 9·0. Miscible with water. Accelerator for latex compounds cured at low temperatures (51).

TRIMENE Mixture of triethyl trimethylene triamine and stearic acid. Dark brown paste, s.g. 1·03. Accelerator for natural rubber and SBR, has a fairly low critical temperature and a medium to long curing range. Gives fast cures in blends with thiazoles, thiurams and guanidines; is retarded by the acidic ingredients, clays and channel blacks. Causes a staining in light-coloured articles, which fades on exposure to light. Zinc oxide and sulphur are necessary in normal quantities. Has a plasticising effect due to the fatty acid content. May be used in compounds in which crystallisation of the stearic acid is unimportant.
Uses: cellular rubber articles and foam rubber (23).

TRI(α-METHYL BENZYL) PHENOL

$C_6H_2(OH)[C_6H_5CH(CH_3)-]_3$
.—1—2:4:6.
Used in blends with other ingredients as a non-staining antioxidant.
TN: Styphen 1 (7).

TRIMETHYL THIOUREA

White to yellowish crystalline powder, s.g. 1·20–1·26. Soluble in benzene, water, chloroform and toluene. Accelerator for polychloroprene which has not been modified with sulphur (neoprene W types). Gives fast cures with freedom from scorch, low compression set, good physical and ageing properties.
Quantity: 0·7–0·8%.
TN: Montorage TU (5)
Thiate E (51).

TRINITROBENZENE

$C_6H_3(NO_2)_3$, 1:3:5-trinitrobenzene. Colourless flakes, m.p. 123°C, s.g. 1·48. Soluble in alcohol, ether and benzene. Vulcanising agent which does not require addition of sulphur; process according to I. Ostromislensky. Natural rubber and SBR may be vulcanised by heating with approx. 5% trinitrobenzene in the presence of lead monoxide; 1:3-dinitrobenzene and other polynitrobenzenes also have a vulcanising effect.

TRIOCTYL PHOSPHATE

$[C_4H_9CH(C_2H_5)CH_2]_3PO_4$, b.p. 216°C (5 mm), s.g. 0·926. Insoluble in water. Its viscosity is only slightly dependent on the temperature. Softener of synthetic rubbers, polyvinyl compounds and cellulose esters, which must be used at extremely low temperatures, down to −70°C.

TRIOL 10 Mixture of trihydric alcohols. Reddish brown, viscous liquid, m.p. 165°C, s.g. 1·1, pH 8. Dispersing agent for NR and SR.
Quantity: 3–5% on the filler.

TRIOL 124 Abbreviation for 1:2:4-butane triol.

TRIPHENYL GUANIDINE

$$C_6H_5N{=}C{\Large\langle}\begin{matrix}NH.C_6H_5\\NH.C_6H_5\end{matrix}$$

TPG. White powder, m.p. 140–145°C, s.g. 1·10–1·17. Slow vulcanisation accelerator, slower than diphenyl guanidine, minimum vulcanisation temperature 138°C. Hardly used today.
Quantity: 0·8–1·8% with 3–5% sulphur.

Uses: ebonite, extruded and moulded articles.
TN: Accelerator 11 (19)
Accélérateur B (6)
Vulcafor TPG (60).

TRIPHENYL METHANE TRIISOCYANATE $(C_6H_4NCO)_3CH$.

Crystalline, m.p. approx. 90°C, s.g. 1·32. Miscible with benzene, toluene and trichloroethylene. Technical product sold as a 20% solution in dichloromethane. After evaporation of the solvent (b.p. 41°C) a tacky film is produced which is reactive and used for bonding natural rubber, NBR, SBR, neoprene, and to a certain extent butyl rubber to metals and alloys. Bonds have high strengths and are resistant to solvents and heat. Considerably increases the adhesion between rubber and textile made from natural and synthetic fibres.

Uses: conveyor belts, tyres, high pressure hose, impregnations, metal/rubber bonding, self vulcanising cements and adhesives with a high bond strength.
TN: Desmodur R (43)
Mondur TM (263).

TRIPHENYL PHOSPHATE

$(C_6H_5)_3PO_4$. Colourless crystals, m.p. 49–50°C, b.p. 245°C (1 mm), s.g. 1·185. Soluble in benzene, chloroform, acetone and ether, insoluble in water. Softener for synthetic rubbers, particularly for SBR and butadiene-acrylonitrile copolymers.
Uses: in compounds with high elasticity and in compounds which are resistant to oil and benzine.
Quantity: 10–30%, larger quantities cause decreases in hardness and tensile strength.

TRIPROPIONINE Glycerine tripropionate.

TRITACTIC POLYMERS Polymers with three types of steric structure in the main chain; for each polymer unit there are two asymmetric carbon atoms and a transdouble bond. They are formed by the stereospecific anionic polymerisation of esters, which have two conjugated double bonds in the acid part, are crystalline and have a high m.p.

TRITHENE Polytrifluorochloroethylene (373).

TRITOLYL GUANIDINE

$(CH_3.C_6H_4.NH)_2=N.C_6H_4CH_3$. White powder. Accelerator, slightly more active than TPG.
TN: TTG.

TRITON Group of alkyl aryl polyether alcohols. Emulsifiers and stabilisers for natural and synthetic latices. Emulsifiers in emulsion polymerisation.
Grades:
770: aryl polyether sulphate. Stabiliser for nitrile latex
K.60: quaternary ammonium chloride. Sensitiser for latex
R-100: neutral sodium salt of a condensed organic acid. Dispersing agent for carbon blacks, fillers and pigments
W-30: aryl alkyl polyether sulphate. Emulsifier in emulsion polymerisation
X-100: polyether alcohol. Dispersing agent for latex
X-155: polyether alcohol. Emulsifier
X-200: aryl alkyl polyether sulphonate. Emulsifier for acrylates, esters, resins and oils (33).

TRITON OIL Pure paraffin oil (96).

TRL-100 TL Masterbatch of 50:50 natural rubber and red clay stabilised with dinaphthyl-p-phenylene diamine. Produced by coagulating a latex/clay dispersion. A reinforced compound with good dynamic, flexing and ageing properties.

TRS-RUBBER Abbreviation for Terres Rouge sheet. Air dried sheet produced from latex with a higher rubber content than normal. Vulcanises quickly and has a well controlled Mooney viscosity 95 ± 5; modulus 100%, 6·55 ± 0·3 kg/cm² in ACS-1 compound.

TR TEST Abbreviation for temperature retraction test. A test used for the rapid examination of crystallisation effects and viscoelastic properties of elastomers at low temperatures. The temperatures determined, are those at which a previously stretched and frozen test piece retracts 10 and 70% (TR-10 and TR-70). The difference between the 10 and 70% retraction temperatures increases with increasing crystallisation. ASTM D 1329.

TS Abbreviation for total solids.

TSCHUGAEFF-GOLODETZ REACTION The addition of formaldehyde to a solution of rubber in trichloroacetic acid to give a deep violet colour.

TSPP Abbreviation for tetrasodium pyrophosphate.

TTBR Trithiodibutylamine (14).

TTDB Trithiodibutylamine (14).

TTG Abbreviation for tritolyl guanidine.

TTS-5 3,3,3′,3′-tetramethyl-5,6,5′,6′-tetrahydroxyspirobisindane. Accelerator for chloroprene rubber (176).

TTT Abbreviation for triethyl trimethylene triamine (33).

TU Abbreviation for thiourea.

TUBORYL Non-hygroscopic aluminium silicate which is free of water and has a particle size of 2 microns. Contains 65–67% SiO_2, 29–33% Al_2O_3, but no copper or manganese. Light reinforcing filler for natural rubber and other elastomers, particularly effective in butyl rubber; has the same reinforcing effect as SRF. Decreases gas permeability and improves electrical resistivity and water resistance.

TUNU Local term for *Castilla tunu Hemsl* (east coast of Central America).

TURPENTINE Pine oil, pinewood oil, turpentine oil. Colourless to yellow liquid with characteristic odour, b.p. 185–225°C, s.g. 0·925–0·942. Obtained from the scrap wood of American pines (*e.g. Pinus palustris, P. heterophylla*). Consists of a mixture of 65–70% terpineol oil, 10–15% borneol oil and tertiary alcohols. Soluble in most organic solvents, insoluble in water.
Uses: solvent and softener for natural rubber, antifoam agent.

TURPOL Terpene resin. Extender for oil resistant rubber compounds (1).

TURPOL NC 1200 Soft, rubber-like synthetic polymer based on a terpene. Softener and processing aid for neoprene and nitrile rubber, SBR, Thiokol FA and ST, butyl rubber, Hypalon, fluoroelastomers and natural rubber. Has good chemical resistance (1, 147).

TWEEN Group of fatty acid esters of polyoxyethylene sorbitol. Oily liquids. Soluble in water and alcohol. The classification number gives the approx. number of oxyethylene groups. Wetting agents, emulsifiers, mould release agents (43, 97):

20, 21:	monolaurate
40:	monopalmitate
60, 61,	monostearate
65:	tristearate
80, 81:	monooleate
85:	trioleate.

TYGOBOND Resinous adhesive used for bonding vinyl compounds and polyurethane foam to materials such as textiles, metals, wood (351).

TYLAC Types 640, 650, 740, 750, 840, 850, 1640, 1650, 2241, 2430, 2570. Group of butadiene-acrylonitrile latices with unspecified monomer ratio, 41% total solids, pH 8–9, viscosity 30–4500 cp, particle size 600–800 Å, s.g. (latex) 0·98–1·01. *Uses:* dipped articles, textile impregnation, bonding for non-woven fabrics. *Types:* 410, 420, 450A, 3040, 3340, 3500; group of butadiene-styrene latices for general use, styrene content 46% (Type 3340, 56%) (352).

TYLOSE Group of odourless, tasteless, non-toxic methyl celluloses and cellulose glycolates. Soluble in water. Thickening agents for latex; stabilisers, emulsifiers, non-ionic dispersing agents, mould-release agents (217).
Grades:
C: sodium carboxy methyl cellulose
CR: carboxy methyl cellulose
MH: methyl hydroxyethyl cellulose
H: hydroxyethyl cellulose.

TY-PLY Group of bonding agents for metal/rubber bonds; based on rubber hydrohalides.
Grades:
BN: for butadiene-acrylonitrile copolymers
Q: for natural rubber, butyl rubber and SBR
S: for neoprene
UP and BC: for butyl rubber
UP and RD: two part system for natural rubber, SBR, nitrile rubber, and neoprene (172).

TYRE ASSEMBLY The building of a vulcanisable tyre, *e.g.* from rubber compounds, textile, metal.

TYRE NUMBER An indication of the dimensions of the tyres, *e.g.* 6·40–15. The first number is the breadth of the cross-section of the tyre and the second the diameter of the rim in inches.

TYRES The first pneumatic tyre was invented in 1845 by R. W. Thompson (BP 10 990) and consisted of a canvas and a rubber tube with a leather covering. He also designed a model consisting of several tubes surrounded by one covering. In 1867 he patented the solid tyre which was used with success for steam driven vehicles. The pneumatic tyre was not widely used, however, and was forgotten until

J. B. Dunlop rediscovered the hollow tyre in 1888. It consisted of a rubber tube covered with textile and rubber but suffered the disadvantage of having to be bonded to the rim. Tyres which could be removed were described in 1890 by J. K. Welch (BP 14 563) but even then contained a wire framework (bead). The earliest tyres contained woven textile layers, but breaks quickly occurred at places where the yarns crossed and the life of the tyre was thus short. The cord weave was introduced between 1916 and 1919. A tyre consists in principle of a carcase built up from rubberised textile layers together with a wire bead, a soft rubber layer (cushion), a tread and sidewalls. The carcase is made from four or more rubberised cord layers assembled on a building machine so that the directions of the cords cross at a right angle. The bead, consisting of rubber coated wire, is included on both sides between the layers. Between the tread and the carcase a soft rubber strip is interposed which consists of a highly elastic compound strengthened by a breaker. The tread is extruded as a profile from a compound which has high abrasion resistance and is laid on the tyre with the sidewalls. The cylindrical green tyre produced on a tyre building machine is preshaped with the simultaneous insertion of a hot air tube or shaped in a vulcanising press by an in-built Bag-O-Matic. Vulcanisation proceeds in single presses or, for heavy tyres, in vulcanisation autoclaves.

TYSONITE A mixture of gilsonite and vulcanised vegetable oils. Black, thermoplastic material, s.g. 1·04. Improves the ozone resistance of electrically insulating compounds;

litharge should be added to the compounds to improve its ageing properties.
Quantity: up to 200% (51).

U

UCAR ADDITIVE PN-1 Phenyl-α-naphthylamine (112).

UCC L-522 Low viscosity silicone mould release agent used in the tyre industry (112).

UGIKRAL French produced acrylonitrile-butadiene-styrene terpolymers.

UGIPOL French produced butadiene-styrene copolymer with a high styrene content.
Grades:

85:	85% styrene, s.g. 1·04. Hot polymer. Resin soap as emulsifier
R 50:	blend of 50% Ugipol 85 and 50% SBR 1503. Styrene content 53–54%, s.g. 0·85
RH-50:	blend of 50% Ugipol 85 and 50% SBR 1708, styrene content 53–54%, s.g. 0·98.

UITLE Former trade name for a natural rubber from the Congo.

UKARB Group of reinforcing furnace blacks (114).

UKEM PROCESS AID Viscous liquid, s.g. 0·98. Softener and dispersing agent for compounds reinforced with carbon blacks, particularly suitable for furnace blacks (114).

ULE Local term for *Castilla elastica* (Mexico).

ULTEX Diphenylguanidine dibutyl-dithiocarbamate. White powder, m.p. 121–124°C, s.g. 1·14. Ultra-fast accelerator for NR and SR, has a low critical temperature. Boosted by thiazoles, thiurams, and aldehyde-amines. Vulcanisates have a high modulus and good ageing properties. Non-staining and non-discolouring (100).

ULTIPARA Butadiene-styrene copolymer.

ULTO Zinc(phenyl aminoethyl)phenyl dimethyl dithiocarbamate. Accelerator (no longer produced under this trade name) (100).

ULTRACEN Blend of cyanoguanidine with zinc oxide (167).

ULTRAMARINE BLUE Synthetic lapis lazuli, sodium aluminium silicate. Contains sulphur, has a varying composition, approx. formula $Na_7Al_6Si_6O_{24}S_2$. Blue pigment. Infra-red rays are reflected by compounds containing zinc oxide and ultramarine blue. Fast to light.

ULTRASIL VN 3 Highly reinforcing silica. White filler for articles which require high hardness, stability and good abrasion resistance. Also suitable for transparent compounds (144).

ULTRATHENE Trade name for ethylene-vinyl acetate copolymer with rubber-like properties.

ULTRA ZINC DMC Zinc dimethyl dithiocarbamate (5).

UNAFOR Cis-polybutadiene (59).

UNICEL Group of blowing agents:

Unicel:	diazoaminobenzene
100:	N,N′-dinitrosopentamethylene tetramine
ND:	dinitrosopentamethylene tetramine (40%)
NDX:	dinitrosopentamethylene tetramine
S:	dispersion of 50% sodium bicarbonate in oil. Activated by stearic acid, gives a uniform cell structure in sponge rubber with open cells
SX:	dispersion of sodium bicarbonate in oil (6).

UNION RUBBER Narrow erasers, sharpened at each end, which are stuck, half concealed, onto pencil tops.

UNIROYAL New name for United States Rubber Co., which consists of: US Rubber, Dominion Rubber, US Rubber Mexicana, Englebert, North British Rubber.

UNITANE O-220 Titanium dioxide (21).

UNITED 65-SPF Special SPF black for stereospecific polymers. Disperses more easily than normal SPF black (164).

UNIVOL V-314 Lauric acid.

UOP Group of antiozonants
(352 a).

88: N,N′-bis-1-ethyl-3-methyl
pentyl-p-phenylene diamine
288: N,N′bis-1-methyl heptyl-p-
phenylene diamine
588: antiozonant of undisclosed
composition, s.g. 1·01–1·02
688: asymetrically substituted
phenylene diamine
788: antiozonant of undisclosed
composition, s.g. 0·898
Sustane: 2-tert-butyl hydroquinone
monomethyl ether.

UPSIDE DOWN MIXING
With an internal mixer, fillers and
other chemicals are normally added
after the rubber has been partly
masticated. However, if the filler
loading is very high, it is sometimes
an advantage to put in the filler
first and then add the rubber. This
process is known colloquially as the
upside down process.

URANYL ACETATE
$UO_2(CH_3COO)_2.2H_2O$. Yellow,
fluorescent crystals, s.g. 2·89. Polymerisation initiator in the presence
of light.

UREA $CO(NH_2)_2$, carbamide.
White powder or crystals, m.p. 129–
130°C, s.g. 1·32–1·34. Soluble in
water, alcohol, and benzene, slightly
soluble in ether, insoluble in chloroform.
Uses: activator for some blowing
agents, *e.g.* Celogen AZ, activator
for thiazole accelerators, softener for
cellulose.

UREKA Group of accelerators.
Grades:
Ureka: mixture of 2:4-dinitrophenyl
thiobenzthiazole and diphenyl

guanidine. Yellow powder, s.g.
1·26. Soluble in chloroform
and alcohol. Gives fast cures
without showing a tendency to
scorch. Causes discoloration.
Uses: footwear, mechanical
goods
B: mixture of 2:4-dinitrophenyl
thiobenzthiazole and diphenyl
guanidine phthalate
Ureka Base: 2:4-dinitrophenyl thiobenzthiazole
Blend 3816: mixture of 2:4-dinitro-
phenyl benzthiazole, diphenyl
guanidine and diphenyl guanidine phthalate
Blend B: as for Ureka but with
part of the DPG replaced by
Guantal
C: 2-benzthiazyl dithiobenzoate
(obsolete)
DD: mixture of 2:4-dinitrophenyl
thiobenzthiazole and diphenyl
guanidine. Yellow powder, s.g.
1·42. Partly soluble in benzene,
chloroform and dilute acids.
Cures more slowly than Ureka,
but gives greater processing
safety. *Uses:* thick walled articles (solid rubber tyres) and
compounds with furnace blacks
HR: mixture of 2:4-dinitrophenyl
thiobenzthiazole and diphenyl
guanidine. Yellow powder, s.g.
1·41. Similar to Ureka but has a
longer delayed action. Recommended for compounds which
contain furnace blacks. Good
resistance to heat and tearing
White: mixture of 2:4-dinitrophenyl
thiobenzthiazole and diphenyl
guanidine acetate, mercaptobenzthiazole and diphenyl guanidine phthalate. Yellow powder,
s.g. 1·39. Soluble in benzene and
chloroform. Fast accelerator
with a marked delayed action.
Non-staining and suitable for

white and light coloured articles, particularly for butadiene-
styrene copolymers

White F: mixture of di-2-benzthi-
azyl disulphide, hexamethylene
tetramine and diphenyl guanidine. Yellowish powder, s.g. 1·43.
Partly soluble in chloroform
and water. General purpose
accelerator with a delayed action;
recommended for cable sheathing compounds

White FM: mixture of a guanidine
and a thiazole derivative. Yellowish powder, s.g. 1·45. Soluble
in benzene and chloroform. Fast
accelerator with a delayed action.
Suitable for light-coloured articles and cable compounds. Shiny
surfaces are obtained with a
low sulphur content (5).

UREPAN Millable urethane
elastomers. Types 600 and 601;
cross-linkable with isocyanates, for
moulded and open heated products.
Type 640: crosslinked with peroxides
for moulded goods (43).

URETHANE
$NH_2.CO.OC_2H_5$, ethyl carbamate, m.p. 48–50°C, b.p. 182–184°C,
s.g. 1·1. Soluble in water, alcohol,
chloroform, and ether.
Uses: production and modification
of aminoplasts; melted urethane is a
good solvent for various organic
substances.

UREX Modified urea. Activator
for nitrogenous blowing agents, s.g.
1·17 (215).

URYLON A synthetic fibre;
polycondensation product of nona-
methylene diamine and urea, m.p.

240–250°C, s.g. 1·07. The diamine is
produced by the removal of water
from the ammonium salt of azelaic
acid to give the nitrile which is then
reduced by hydrogen. Has high
resistance to water and chemicals
(353).

USTEX Chemically and
mechanically treated cotton fibre
with greater strength than untreated
cotton. Developed by the US Rubber
Co.
Uses: primarily in conveyor belts,
high pressure hose.

UTERMARK PROCESS
Process for concentrating latex by
centrifuging according to BP 219 634
(1923).

UV ADSORBER 9 Physical protective agent against light; 2-hydroxy-
4-methoxy benzophenone (21).

UVF 2932 Antioxidant of undisclosed composition (81).

UVINUL Group of physical protective agents against light; important
because of their high capacity for
ultra-violet ray absorption.
Grades:
400: 2:4-hydroxybenzophenone
D-49: 2:2′-hydroxy-4:4′-dimethoxy-
benzophenone
D-50: 2:2′:4:4′-tetrahydroxybenzo-
phenone
M-40: 2-hydroxy-4-methoxybenzo-
phenone
N-35: acrylonitrile derivative, m.p.
98–99°C
N-38: acrylonitrile derivative, m.p.
140–144°C (85).

V

V- (3, 5, 150, 200); group of Russian produced polyisobutylenes. Index number × 1000 gives the average mol. wt.

VA-7 Liquid alkyl polysulphide with the general formula —R—$(S)_n$—R in which n = approx. 4·5 and R is an aliphatic ether. Curing agent for natural rubber and unsaturated synthetic polymers; has better vulcanisation characteristics than elemental sulphur. Gives good resistance to heat ageing, and high tensile strength. Does not bloom. Tetramethyl thiuram monosulphide or disulphide, benzthiazyl disulphide and N-cyclohexyl-2-benzthiazyl sulphenamide may be used as accelerators. In nitrile rubber litharge and salicyclic acid behave as activators (208).

VALRON ESTERSIL A fine, synthetic, esterified silica. Contains 86–89% silica, s.g. 1·86, particle size 8–10 μ, surface area 310 m²/g. Reinforcing filler for silicone rubber. *Quantity:* 30–70% (6).

VANCIDE Group of fungicides, bactericides and insecticides used in rubber and latex processing.
Grades:
26 EC: lauryl pyridine salt of 5-chloro-2-mercaptobenzthiazole. Used for cotton cloth which is proofed with rubber; 1% is applied directly on the cotton as an aqueous solution
51 Z: synergistic mixture of zinc dimethyl dithiocarbamate

and zinc-2-mercaptobenzthiazole. May be used as a fungicide in neoprene compounds, in concentrations of 1% on the rubber, does not affect the vulcanisation characteristics. The fungicidal effect of the compound is considerably higher than that of the individual components
89: N-trichloromethyl mercapto-4-cyclohexane-1-dicarboximide. Used for vinyl compounds which contain sensitive softeners; a concentration of 2% on the softener is used (51).

VAN HASSELT Group of accelerator and antioxidants (355):
BTN: nickel dibutyl dithiocarbamate
BTZ: zinc dibutyl dithiocarbamate
DBZ: zinc dibenzyl dithiocarbamate
DPTT: dipentamethylene thiuram tetrasulphide
ETU: ethylene thiourea (2'-mercaptoimidazoline)
ETEL: tellurium diethyl dithiocarbamate
ETZ: zinc diethyl dithiocarbamate
MTBi: bismuthdimethyl dithiocarbamate
MTCu: copper dimethyl dithiocarbamate
MTL: lead dimethyl dithiocarbamate
MTZ: zinc dimethyl dithiocarbamate
PBN: phenyl-β-naphthylamine
TET: tetraethyl thiuram disulphide
TMT: tetramethyl thiuram disulphide

TMTM: tetramethyl thiuram
monosulphide
ZPD: zinc pentamethylene
dithiocarbamate.

VANILLIN 3-methoxy-4-
hydroxybenzaldehyde, 4-hydroxy-3-
methoxy-benzaldehyde, m.p. 81–
83°C, b.p. 285°C, s.g. 1·056. Soluble
in alcohol, pyridine, ether, chloroform and carbon disulphide.
Uses: deodorising and perfuming
rubber articles, in particular household goods.
Quantity: 0·01–0·1%.

V-BELTS V shaped, continuous
drive belts made from a rubber-
covered layer; described in 1860 in a
patent by W. Clissold. First used as
a drive belt on a motorcycle in 1902.
In 1915 were used in cars as drive
belts for fans, also for light engines.
Adopted for industrial machinery in
1924.

VELSICOL Group of resinous
hydrocarbons which may be blended
with natural rubber and synthetic
elastomers. Softeners and dispersing
agents for fillers. Suitable for compounds requiring a high electrical
resistance (357).

VELVAPEX Antioxidant of undisclosed composition (159).

VELVETEX General purpose
furnace black (159).

VERCORYL Reinforcing
colloidal kaolin. Filler (926).

VEROXAZIN PROCESS
Redox emulsion polymerisation of
SBR rubber, with diisopropyl ben-
zene hydroperoxide (oxidising agent)

hydrazine (reducing agent), and iron
pyrophosphate and ethylene diamine
tetra-acetic acid (catalysts).

VEROXY SULPHIDE PROCESS
Redox emulsion polymerisation of
SBR cold rubber with diisopropyl
benzene hydroperoxide (oxidising
agent), sodium sulphide (reducing
agent) and the ferric complex of
ethylene diamine tetra-acetic acid
(catalyst).

VERSAMIDES Group of polyamides produced by the condensation of polymerised unsaturated
fatty acids (linoleic, ricinoleic, linolenic and eleostearic acids) with di-
and triamines (ethylene diamine,
diethylene triamine and mixed condensates of ethylene diamine and
dicarboxylic acids, *e.g.* sebacic acid).
Soft to hard resins, m.p. approx.
200°C. Have excellent adhesive
strength.
Uses: include adhesives, coatings,
production of lacquers for flexible,
elastic films, *e.g.* rubber articles,
castable resins, bonding agents, paper
impregnations.

VESTINOL Group of softeners.
AH: dioctyl phthalate
C: dibutyl phthalate
HX: dicyclohexyl phthalate
DB: diglycol benzoate
N: dinonyl phthalate (189).

VFA NUMBER VFA: volatile
fatty acids; denotes the amount of
volatile fatty acids (mainly acetic) in
concentrated latex. Expressed as the
amount of KOH necessary (in g) to
neutralise the volatile fatty acids
present in 100 g dry solids. Extraction
is by water-steam distillation in
Markham flasks. It is used to judge
the quality of latex. ASTM D
1076-59.

VGC Abbreviation for viscosity–gravity constant.

VHS Abbreviation for very high structure (carbon blacks).

VIBRATHANE Group of polyurethanes (23).

VIBRIN Group of polyester resins (23).

VICAT NEEDLE INSTRUMENT Apparatus used to determine the softening p. of service products made from hard rubber and plastics. A steel needle, ground flat underneath, of 1 mm^2 cross-section and loaded with a weight of 5 kg, is placed on a test piece and the depth of penetration measured. The Vicat softening p. is the temperature in °C at which the needle penetrates 1 mm deep into the compressed test piece at a uniform temperature rise of 50°C/hour. The apparatus is contained in a heating chamber.

VINYL ACETATE $CH_3COOCH=CH_2$. Colourless liquid, b.p. 72°C. Polymerised to polyvinyl acetate on exposure to light.

VINYL ACETYLENE $CH\equiv C—CH=CH_2$. Colourless gas, liquefies at 5·5°C. Chloroprene is formed from it by the addition of hydrogen chloride.

VINYL CHLORIDE $CH_2=CHCl$, monochloroethylene. Easily condensed gas, b.p. −14°C, s.g. (liquid) 0·96. Soluble in alcohol. Described by Regnault, 1838. As with other dienes, it may be polymerised easily to linear, high mol. wt. polymers. The polymer is a hornlike mass, and must be plasticised by the addition of softeners, *e.g.* nitrile copolymers; modification with other monomers may decrease the quantity of softener necessary. Has been widely used in industry since 1933.

VINYL COMPOUNDS Compounds containing the group $CH_2CH—$; important because they are easily polymerised to high polymers, and are therefore used in the production of commercial polymers.

VINYL ETHER $CH_2=CH.O.C_2H_5$. Colourless liquid, b.p. 28–31°C, s.g. 0·767–0·771. Miscible with alcohol, ether and chloroform. May be polymerised as a liquid or vapour.

VINYLIDENE CHLORIDE $CH_2=CCl_2$, 1:1-dichloroethylene. Liquid which may be easily polymerised.

VINYLON Japanese produced fibre with properties similar to nylon; made from polyvinyl alcohol.

VINYLSILTHIAN Polymeric reaction product of vinylchlorosilane and hydrogen sulphide. Increases the modulus and improves the physical properties of loaded compounds of natural and synthetic rubbers. When mixed with rubber and filler, heated to 120–200°C in an internal mixer, the curing system is added after the compound has cooled. It is assumed that the improvement in properties of the product is caused by vinylsilthian reacting with the filler. BP 734 457 (55), (246).

VINYON Synthetic polymers based on vinyl copolymers.

Grades:

E: experimental fibre, 1942–45. Used for military purposes

N: 40% acrylonitrile, 60% vinyl chloride

CF: 88% vinyl chloride, 12% vinyl acetate

HH: stable CF fibre

HR: 85% vinyl chloride and 15% ethylene chloride

Supplementary letters indicate the degree of stretching: HST: highly stretched, UST: unstretched. The fibres have been developed since 1936 by the Carbide and Carbon Chemicals Corp. (USP 2 161 766), and have been produced industrially since 1939 by the Viscose Corp., Meadville, Pa, USA.

VIRGIN RUBBER Imperfectly dried rubber with a high moisture content.

VISCOMA Group of waxes which act as protective agents against light, lubricant waxes for cable fillings and yellow paraffins (231).

VISCOMETER An instrument used to measure the viscosity of liquids and plastic substances. The shearing viscometer developed by Mooney is used to determine the plasticity of elastomers.

VISCOSITY The internal friction of a liquid. Dynamic viscosity: for pure viscous or Newtonian liquids

$$\eta = \frac{\tau}{D}$$

where τ is the shearing force and D the shear velocity. The time, t, is measured for a specified volume, v,

to flow through a capillary of radius, r, and length, l, at a pressure, p. Then the Hagen–Poiseuille formula gives

$$\eta = \frac{\pi r^4 p t}{8 l v}$$

The units are the poise (1 poise = 100 cp).

Kinetic viscosity is the dynamic viscosity divided by the density:

$$\eta_k = \frac{\eta}{\rho}$$

The units are the stokes.

Relative viscosity expresses the viscosity relative to a solvent or water. It is also valid for suspensions of solids in liquids

$$\eta_r = \frac{\eta}{\eta_0} = \frac{t}{t_0} = 1 + K . c_v$$

where η is viscosity of the solution, η_0 the viscosity of the solvent, c_v is the volume of the dissolved substance in unit volume, K a constant, t the time taken for the solution to flow, and t_0 the time taken for the solvent to flow.

Structure viscosity. Pure liquids are to be regarded as Newtonian bodies and their $(\tau$–$D)$ graph has a linear form. With polymers this relationship deviates from linear and thus their viscosity is termed apparent viscosity or structure viscosity.

$$D = \frac{t}{\eta} \tau^n$$

$$\eta^* = \frac{\tau^n}{D}$$

Liquids with $n = 1$ exhibit Newtonian flow, those with $n < 1$ and $n > 1$ have a quasi-viscous flow. η^* cannot be expressed in poise.

Specific viscosity. Obtained by a

comparison of the viscosities of solution and solvent.

$$\eta_{sp} = \frac{\eta - \eta_0}{\eta_0} = \eta_r - 1 = \frac{t - t_0}{t_0}$$

For very dilute solutions of high polymers, Staudinger's equation applies

$$\eta_{sp} = K_m c \cdot M$$

where M is the mol. wt., c the concentration in mol. units, for rubber (C_5H_8) and K_m is a constant.

Dilute solution viscosity (DSV). The mol. wt. calculated from the specific viscosity is only an approx. value; to exclude mechanical and structural influences, the dilute solution viscosity is calculated using a series of decreasing concentrations

$$DSV = \frac{\eta_{sp}}{c}$$

The concentration c is given in g polymer per 100 ml solution.

Intrinsic viscosity. Obtained by extrapolating the dilute solution viscosity to $c = 0$; this is equal to the extrapolated inherent viscosity at $C = 0$

$$[\eta] = \left[\frac{\eta_{sp}}{c}\right]_{c=0} = \left[\frac{\ln \eta_r}{c}\right]_{c=0}$$
$$= K \cdot Ma$$

Inherent viscosity:

$$\eta_i = \frac{\ln \eta_r}{c}$$

is far less dependent on the concentration than is the diluted solution viscosity η_{sp}/c and if dilution is sufficient, it is practically independent of the concentration. If the values are extrapolated to $c = 0$, then

$$[\eta] = \left[\frac{\ln \eta_r}{c}\right]_{c=0} = [\eta]$$

is identical to the intrinsic viscosity.

VISCOSITY–GRAVITY CONSTANT VGC. Characteristic used to estimate the properties of processing oils:

$$VGC = \frac{D - 0.24 - 0.022 \log (V_1 - 35.5)}{0.755} \quad \text{(I)}$$

$$VGC = \frac{10D - 1.0752 \cdot \log (V_2 - 38)}{10 - \log (V_2 - 38)} \quad \text{(II)}$$

where D is the density at $15.6°C$, V_1 is the Saybolt viscosity at $98.9°C$, V_2 is the Saybolt viscosity at $37.8°C$. Equation (I) is valid for oils with a viscosity of over 35.5 Saybolt seconds; equation (II) applies to oils of a lower viscosity. As VGC increases, the nature of the oils changes from paraffinic *via* naphthenic to aromatic.

VISCUROMETER Pendulum elastomer. Instrument with an oscillating rotor working on the principle of the Mooney plastimeter. May be used to investigate the whole vulcanisation curve. Developed by B. F. Goodrich.

VISTALON Ethylene propylene rubbers. Esso.

VISTANEX Group of polyisobutylenes.
Code:
LM low mol. wt., MM medium mol. wt., MH medium hard, MS medium soft.
Grades:
LM-MH: white to yellowish, viscous, tacky, semi-solid mass, s.g. 0.92, medium mol. wt. 10 000–11 700 (viscosity method according to Staudinger). Blends easily with oils, waxes, solvents

and polymers in heavy compounds, but cannot be vulcanised. *Uses:* unvulcanised, alone or in blends, *e.g.* for cements, pressure adhesives, jointing compounds. Improves the properties of NR and SR vulcanisates, gives good electrical properties, good resistance to chemicals and ozone, good ageing properties and a low gas permeability, improves tack of compounds

LM-MS: white to yellowish, viscous, tacky, semi-solid mass, s.g. 0·92, medium mol. wt. 8700–10 000 (viscosity method according to Staudinger). Properties and uses as for Grade LM-MH

MM-L80: white to yellowish, tough, rubber-like mass, s.g. 0·92, medium mol. wt. 64 000–81 000 (viscosity method according to Staudinger), inhibitor content less than 0·3%. May be masticated or mixed with other ingredients in an internal mixer. Cannot be calendered or extruded unless mixed with other substances: mixing temperature 120–150°C. Similar to rubber in its properties, *e.g.* tensile strength, elongation, resilience at high temperatures, electrical properties, solubility and mechanical orientation. Because of its saturated nature, its resistance to heat, light and chemicals is better than that of rubber, and the dynamic properties are also improved

MM L-100: white to yellowish, tough, rubber-like mass, s.g. 0·92, medium mol. wt. 81 000–99 000 (viscosity method according to Staudinger), inhibitor content less than 0·3%. Properties and uses as for Grade MM L-80

MM L-120: white to yellowish, tough, rubber-like mass, s.g. 0·92, medium mol. wt. 89 000–117 000 (viscosity method according to Staudinger), inhibitor content less than 0·3. Properties and uses as for Grade MM L-80

MM L-140: white to yellowish, tough, rubber-like mass, s.g. 0·92, medium mol. wt. 117 000–135 000 (viscosity method according to Staudinger), inhibitor content less than 0·3%. Properties and uses as for MM L-80 (131).

VISTEX VISCOSITY Viscosity of a dilute solution of a polymer dissolved directly as latex in a mixture of hydrophobic and hydrophilic solvents, *e.g.* 3 parts xylene, 1 part pyridine. The method was developed so that changes in polymer properties during the production of SBR might be more quickly followed.

VITASAN Tetrasulphide of a complex ether. Brown–yellow liquid. Accelerator (39).

VITON Copolymer of hexyfluoropropylene and vinylidene fluoride with approx. 65% fluorine. White, solid, translucent mass, s.g. 1·85. Soluble in low mol. wt. ketones, fairly low mol. wt. in comparison with other polymers. Is changed to tough, rubber-like products by heating with polyamines (triethylene tetramine), by β or γ rays of high intensity, or by heating with peroxide radicals. Has good physical and elastic properties up to 200°C, is resistant to oils and solvents at high temperatures. Compounds must contain a filler (carbon black or silicate) and an acid acceptor to absorb traces of free acid.

Uses: in the aircraft and car industries and in products which must be heat resistant and chemically and mechanically stable.
Types: Viton A, Mooney viscosity 65; —A—HV, Mooney viscosity 180; B Mooney viscosity 117 (6).

VITRAFIX MC 20% solution of methacrylatochromyl chloride in aqueous acetone and isopropanol. Dark green fluid, s.g. 1·03. Coordination complex.
Use: for bonding materials with vinyl groups to negative surfaces (60).

VLM Abbreviation for very low modulus (carbon blacks).

VLS Abbreviation for very low structure (carbon blacks).

VNIISK Abbreviation for All Union Scientific Research Institute of Synthetic Rubber, Leningrad, Gapsalskaya Street 18, USSR.

VOCON Accelerator of undisclosed composition for EPDM. Yellowish green fluid, s.g. 1·25–1·27 (5).

VOLTEX CC gas black (164).

VONDAC Group of accelerators (433):
TET: tetraethyl thiuram disulphide
TMT: tetramethyl thiuram disulphide
ZDC: zinc diethyl dithiocarbamate
ZBUD: zinc dibutyldithiocarbamate
ZMD: zinc dimethyl dithiocarbamate.

VORANOL CP Series of polypropylene glycol esters; polyurethane prepolymers, s.g. 1·081, mol. wt. 2000–5000 (indicated by the code no.), hydroxyl no. 32–67. Reacts with diisocyanates to give polyurethanes (61).

VUGPT Abbreviation for the Czechoslovakian Research Institute of Rubber and Plastics Technology, Gottwaldov.

VULCABOND Group of bonding agents.
Grades:
T: adhesive for natural and synthetic rubbers to rayon and other textiles; aqueous dispersion from resorcinol-formaldehyde resin with natural latex plus 20% drying substance
TVPN: aqueous dispersion of resorcinol formaldehyde resin and a mixture of rubber latices with 18–19% DRC. Bonding agent for NR and SR to rayon. Like T, but contains some butadiene styrene vinyl pyridine latex
TX: solution of a polyisocyanate in xylene, for bonding rubber to metals, plastics, ceramics (60).

VULCACEL Group of blowing agents (60).
Types:
A: diazoaminobenzene
AN: diazoaminobenzene
B-40: dinitrosopentamethylene tetramine. 40%, s.g. 1·73
BN: stabilised dinitrosopentamethylene tetramine, s.g. 1·40
BN-94: dinitrosopentamethylene tetramine. 75% with stabilisers (60).

VULCAFOR Series of rubber chemicals (accelerators, vulcanising agents and pigments) produced by ICI (60).

Grades:

+I: p-nitrosodimethyl aniline
+II: diphenyl guanidine
+III: triphenyl guanidine
+IV: thiocarbanilide
+V: reaction product of ammonia and an aldehyde
+VI: zinc diethyl dithiocarbamate
+VII: tetraethyl thiuram disulphide
+VIII: diethylamine diethyldithiocarbamate
+IX: zinc isopropyl xanthate
+XII: di-o-tolyl guanidine
+XIII: di-o-xylyl guanidine
+XIV: methylene-p-toluidine
BA: butyraldehyde-aniline condensation product. Oily liquid, s.g. 0·94. Soluble in benzene, benzine and alcohol, insoluble in water, solubility in rubber more than 5%. Medium-fast accelerator with a high critical temperature. Gives a high modulus. Suitable for natural rubber, SBR, compounds containing reclaim, and hard rubber. Boosted by thiazoles. In latex compounds it is slow and causes discoloration
BDN: vulcanising agent. Contains 25% dinitrosobenzene, stabilised on an inorganic base. Pale brown powder, s.g. 1·565. Suitable for unsaturated elastomers, particularly for isobutylene-isoprene copolymers in combination with sulphur and thiuram accelerators, also

for neoprene latex. Gives fast cures, high modulus, good heat stability. Stiffens unvulcanised compounds and thus improves the shape stability of extruded articles, particularly tubing. *Uses:* include hoses for hot substances, extruded articles
BQ: p-quinone dioxime
BQN: 50% p-quinone dioxime, stabilised on an inorganic base. Brown powder, s.g. 1·83. Curing agent for elastomers, particularly for isobutylene-isoprene copolymers in combination with lead oxide or lead peroxide. Gives vulcanisates with good resistance to heat and steam. May also be used in combination with sulphur and other accelerators. *Quantity:* up to 4%. *Uses:* hot-air hose
BSB: tert-butyl benzthiazole sulphenamide
BSM: N-oxydiethylene-2-benzthiazole sulphenamide
BSO: benzthiazole sulphentertoctyl amide
BT: thiuram/thiazole mixture. Pale yellow powder, s.g. 1·44. Accelerator for isobutylene-isoprene copolymers. Gives fast cures at 135–200°C without the risk of scorching. Non-discolouring. *Quantity:* 1·5% with 5% zinc oxide and max. 2·5% sulphur; high sulphur content causes blooming. Fatty acids are not required. 1–2% Vaseline is recommended as a lubricant. Suitable for press moulding, hot air and open steam cures. Unsuitable for

articles which come into contact with food

DA: di-2-dibenzthiazyl disulphide

DAK: mixture of diphenyl guanidine and 2:4-dinitrophenyl benzthiazole sulphenamide

DAT: mixture of 2-mercaptobenzthiazole and diphenyl guanidine tartrate

DAU: mixture of di-2-benzthiazyl disulphide and tetraethyl thiuram disulphide. Yellowish powder, s.g. 1·47. Insoluble in water. Fast curing accelerator with delayed action. Gives good ageing properties. Non-discolouring. Zinc oxide is necessary. Suitable for white and light coloured articles. Unsuitable for latex

DAUF: mixture of dibenzthiazyl disulphide and tetraethyl thiuram disulphide

DAW: mixture of 2-mercaptobenzthiazole and an acid salt of diphenyl guanidine. White powder, s.g. 1·40. Similar to mercaptobenzthiazole in activity but has greater processing safety and a delayed action. Zinc oxide and stearic acid are necessary. Good ageing properties, vulcanises above 125°C

DDC: diethylammonium diethyl dithiocarbamate

DDCN: diethylammonium diethyl dithiocarbamate

DDD: dimethylammonium dimethyl dithiocarbamate

DHC: mixture of 2-mercaptobenzthiazole and a dithiocarbamate. Yellowish powder, m.p. 148–158°C, s.g. 1·51. Soluble in benzene and acetone, sparingly soluble in

alcohol, insoluble in petroleum ether and water. Nonstaining accelerator with a delayed action. Zinc oxide and stearic acid are necessary. *Quantity:* 0·375–0·75% with 1–2·75% sulphur. *Uses:* transparent, white and coloured articles, sponge rubber, textile proofing, footwear, continuous vulcanisation. Unsuitable for latex

DMC: mixture of 2-mercaptobenzthiazole and zinc diethyl dithiocarbamate

DOTG: di-o-tolyl guanidine

DPG: diphenyl guanidine

DPG(LC): DPG in the form of a moist mass for latex

EFA: ethyl chloride-formaldehyde-ammonia condensation product. Dark brown liquid, s.g. 1·1. Accelerator with a medium speed. Increases the modulus and hardness of a product. Little discoloration occurs at optimum cure but a yellow–brown discoloration occurs in overcured articles. *Quantity:* 0·5–1·5%, high concentrations are necessary in compounds containing carbon blacks, as the accelerator is adsorbed. *Uses:* for medicinal articles, articles which come into contact with food; rubber thread, textile proofing, footwear

F: mixture of dibenzthiazyl disulphide and diphenyl guanidine. Pale yellow powder, s.g. 1·46. General purpose ultra-fast accelerator. Non-staining. Zinc oxide

and stearic acid are necessary. *Uses:* particularly suitable for sheathing com-
pounds and cable insula-
tion, and in rubber/cork
compounds, butadiene-
acrylonitrile copolymers,
used in SBR as a medium-
fast accelerator. *Quantity:*
0·5–1% with 2–3% sulphur

FN: mixture of dibenzthiazyl
disulphide, diphenyl guanidine and hexamethylene
tetramine. Fast accelerator
with delayed action, s.g.
1·47. Nonstaining. Specially
developed for cable insulation and sheathing com-
pounds. *Quantity:* in cable
insulation 1·5% with 1%
sulphur, for general use
0·5–1% with 2–2·75% sul-
phur

HBS: N-cyclohexyl-2-benzthi-
azole sulphenamide
I: p-nitrosodimethyl aniline
M: 2-mercaptobenzthiazole
MA: anhydroformaldehyde
aniline
MBT: 2-mercaptobenzthiazole
MBTS: di-2-benzthiazyl disulphide
MS: tetramethyl thiuram mono-
sulphide
MT: anhydroformaldehyde-p-
toluidine
P: piperidine pentamethylene
dithiocarbamate
PT: acetaldehyde-aniline
condensation product
+R: aldehyde-aniline
+Resin: aldehyde-aniline
+RN: acetaldehyde-aniline
condensation product
SDC: sodium diethyl dithio-
carbamate
SPX: sodiumisopropylxanthate
TC: thiocarbanilide

TET: tetraethyl thiuram disul-
phide
TMT: tetramethyl thiuram
disulphide
+TPG: triphenyl guanidine
VCC: solution of 4:4′-diphenyl
methane diisocyanate in
xylene. Brown liquid, s.g.
1·27. Vulcanising agent for
Vulcaprene
VCN: toluene-2:4-diisocyanate
VDC: halogenated naphthol.
Brown powder, s.g. 1·34.
Accelerator for Vulcaprene;
used in combination with
Vulcafor VHM
VHM: aminotriazine derivative.
Yellow, syrup-like liquid,
s.g. 1·20. Vulcanising agent
for Vulcaprene, used in
combination with VDC
VIR: diethylamine diphenyl
dithiocarbamate
ZDC: zinc diethyl dithiocarba-
mate
ZDCL: ZDC paste
ZEP: zinc ethyl phenyl dithio-
carbamate
ZIX: zinc isopropyl xanthate
ZMBT: zinc mercaptobenzthiazole
ZNBC: zinc dibutyl dithiocarba-
mate.

VULCAID Group of accelerators
and antioxidants (3):
27: zinc butyl xanthate
28: dibenzylamine
33: liquid amine reaction product
34: amine-naphthol reaction
product
44: amine-naphthol reaction
product
55: acetaldehyde-aniline reaction product
66: amine reaction product
77: blend of 33 and 66

99:	blend of waxes
111:	butyraldehyde-aniline condensation product
222:	tetramethyl thiuram monosulphide
333:	2-(2:4-dinitrophenyl mercapto)benzthiazole (mixture with diphenyl guanidine)
444:	heptaldehyde-aniline
444B:	heptaldehyde-aniline
555:	2-(2:4-dinitrophenyl mercapto)benzthiazole (mixture with diphenyl guanidine)
666:	dinitrophenyl thiocarbamate
666s:	diphenyl carbamyldimethyl dithiocarbamate
777:	mixture of 55% triethyl trimethylene triamine and 45% stearic acid
777b:	ethylamine-formaldehyde condensation product
888:	tetramethyl thiuram disulphide
999:	thiazole derivatives of aldehyde-amines
BTZ:	zinc dibutyl dithiocarbamate
DPG:	diphenyl guanidine
LP:	lead pentamethylene dithiocarbamate
MBT:	2-mercaptobenzthiazole
MBTS:	di-2-benzthiazyl disulphide
MTZ:	zinc dimethyl dithiocarbamate
P:	piperidine pentamethylene dithiocarbamate
TET:	tetraethyl thiuram disulphide
TMS:	tetramethyl thiuram monosulphide
TMT:	tetramethyl thiuram disulphide
ZDC:	zinc diethyl dithiocarbamate
ZP:	zinc piperidine pentamethylene dithiocarbamate

VULCALOCK PROCESS Process developed by W. C. Geer (B. F. Goodrich) to bond substances such as metals, wood, concrete, to rubber, with or without vulcanisation and using cyclised rubber as an adhesive. The cyclised rubber is produced by heating a mixture of 100 parts rubber and 7·5 parts p-phenol sulphonic acid. Adhesion is unstable above 60°C as the bonding layer softens. USP 1 617 588 (1927).

VULCAMEL TBN Mixture of 33% naphthyl-β-mercaptan and 67% inert wax, m.p. approx. 50°C, s.g. 0·9. Peptiser for natural rubber, butadiene-styrene copolymers and neoprene W. Boosts thiurams and dithiocarbamates. Is ineffective in the presence of blacks. Mixing temperature min. 100°C, has its greatest effect in an internal mixer at 140–150°C. *Quantity:* natural rubber, 0·2–0·6% up to 2% SBR (60).

VULCAN Series of furnace blacks.
Grades:

3:	(HAF) average particle diameter 235 Å
6:	(ISAF) average particle diameter 175 Å
C:	(CF) conductive oil black. Average particle diameter 175 Å
SC:	SCF
XC-72:	antistatic highly conductive furnace black
XXX:	(SCF) highly conductive black. Average particle size 180 Å.

Used for articles which require antistatic properties or which must conduct electricity (aircraft tyres, conveyor belts) (139).

VULCAN COLOURS Group of pigments which are less fast than the Vulcan fast colours.
Quantity: 0·5–2% on the total compound (217).

VULCAN FAST COLOURS Group of rubber pigments important because of their fastness and because they have no effect on vulcanisation or ageing. Do not bloom or stain. Give clear, bright colours with excellent light fastness; particularly suitable for articles which come into contact with sunlight or water. Stable against dilute soda, soap solution, acetic acid, alcohol and benzine.
Quantity: 0·5–2% on the total compound (217).

VULCANINE Inorganic accelerator, used around 1920, developed by T. H. Hewlitt. Consists of equal parts of basic lead carbonate, ammonium carbonate, yellow lead oxide and precipitated sulphur.

VULCANISATION Cure. The conversion of rubber from a predominantly plastic to an elastic condition by three-dimensional crosslinking. The term came from Hancock's associate, William Brockedon, who named it after Vulcanus, the god of fire and sulphur-bearing volcanoes. Goodyear originally used the term 'metallisation'. The reaction, in which bridges or crosslinks form between the single molecules, giving a three-dimensional molecular structure and which imparts characteristic physical properties to the vulcanisate, may be induced by reaction with sulphur or other suitable chemicals, *e.g.* selenium, organic peroxides, nitro compounds, also by high energy rays (ultra-violet, cathode and nuclear rays). The chemistry of vulcanisation depends on the curing system and has not been explained completely. The most widespread method of vulcanisation is with sulphur, the reaction rate being considerably increased by addition of organic accelerators. The reaction is additive and results in the addition of sulphur at double bonds, with the formation of polysulphide bridges and possibly some direct crosslinking without bridges, by a radical mechanism. Vulcanisation with peroxides is also by a radical mechanism; the free radicals formed by decomposition of the peroxide extract hydrogen from side groups and the activated side groups join by crosslinking. High energy rays are also effective by direct activation of side groups.

VULCANISATION RANGE The plateau on which the optimum physical properties of a vulcanisate remain unchanged with increasing cure time.

VULCANISED LATEX Discovered by Ph. Schidrowitz. The curing agents are added to the latex, together with casein, in a paste form. The latex is then vulcanised at 75°C by introducing steam, and the excess ingredients removed by decanting or centrifuging. Concentrated latex may also be vulcanised (then called Revultex), also synthetic latices.
A particular type of rubber (SP rubber) is produced by blending fresh and vulcanised latex, and is particularly important because of its excellent dimensional stability. Films produced from vulcanised latex have poor resistance to oils and solvents

and are more difficult to dry because the surface skin does not allow water to permeate easily. Because of its lower tack, rolling of beads is more difficult and rubber solution is usually necessary.
Uses: dipped articles, cast articles, jointings.

VULCANISING AGENTS
In general, all substances which effect vulcanisation of natural rubber or elastomers. Since its discovery by Goodyear, sulphur has remained the most important vulcanising agent for rubber and butadiene polymers. Many vulcanising agents are of academic interest only. Others are used to give the vulcanisates particular properties, *e.g.* heat resistance; the most important are: sulphur, sulphur chloride, tellurium, selenium, thiuram disulphides, polysulphide polymers, p-quinone dioxime, dibenzoyl-p-quinone dioxime, metallic oxides (zinc, lead, magnesium oxides), dibenzoyl and other peroxides, diisocyanates.

VULCANISATION COEFFICIENT
VC. The quantity of bound sulphur (as a percentage) in relation to the rubber content.

VULCANISING MACHINE
Machine used for the continuous vulcanisation of long lengths, the length being taken over heated curing rollers.
Uses: for floor coverings, conveyor belts, proofed textiles.

VULCANOL
Group of accelerators.
Grades:
NCA: thiocarbanilide
ND: diphenyl guanidine (6, 252).

VULCANOSIN
Group of pigments for dry rubber. Powders and easily dispersed pastes (32).

VULCANOSOL COLOURS
Group of latex pigments which may be dispersed in water and latex (32).

VULCAPLAS
Blend of polysulphide rubber and neoprene (60).

VULCAPLAS
Alkyl polysulphide plastic (60).

VULCAPRENE
Polyurethane. Condensation product of a polyesteramide obtained from ethylene glycol, monoethanolamine and adipic acid, and approx. 5% 1:6-hexamethylene diisocyanate. The production of the polyesteramide is performed without a deficiency of dicarboxylic acid, as for Vulcolan; thus the polyester amide may contain hydroxyl, amino and organic acid end groups which react further with the diisocyanate. The product is a tough, rubber-like, millable mass. It is crosslinked by formaldehyde in the presence of acidic substances, *e.g.* Vulcafor VHM, a complex formaldehyde-amine condensation product and Vulcafor VDC, a halogenated phenol which decomposes to give an acid. Peroxides, quinone dioxime, trinitrotoluene, hexamethylol hexamethyl ether and chromates are among other suitable crosslinking agents. In lacquers, crosslinking is achieved by adding 2:4-toluylene diisocyanate or 4:4'-diphenyl methane diisocyanate. A reaction occurs with the remaining free hydroxyl, amino and organic acid end groups and takes 1–2 h at 125°C, or with higher concentrations, at room temperature.

Grades:

AC: soluble in acetone. Used in lacquers, varnishes, adhesives and bonding agents

PL: 50:50 blend with hydrolysed leather for applications such as textile coating, imitation leather, pressure fabrics, pump diaphragms. Has excellent abrasion resistance (60).

VULCARESAT Group of resin curing agents based on alkyl phenols, for EPT rubber and its compounds (58).

VULCARESEN Group of resin curing agents based on alkyl phenols for NR and SR.

VULCARESOL 315 E Formerly Durophen 219 W. Resin curing agent based on bisphenol. Soluble in butanol and xylene. Particularly effective with 1–2% $SnCl_2.2H_2O$ as activator.

Quantity: 4–6% (58).

VULCARITE Group of accelerators and antioxidants in the form of 50% dispersions (57).

Grades:

100: octylated diphenylamine (Agerite Stalite)

103: 40% piperidine pentamethylene dithiocarbamate

104: hydroquinone monobenzyl ether (Agerite Alba)

106: N-phenyl-β-naphthylamine

108A: 2:2'-methylene bis(4-methyl-6-tert-butyl phenol) (2246)

109: Zenite

111: N,N'-di-β-naphthyl-p-phenylene diamine (Agerite White)

112: Aminox

115: Wingstay S

117: polytrimethyl dihydroquinoline (Agerite Resin D)

120: 2-mercaptobenzthiazole (Captax)

121: dibenzthiazyl disulphide

122: tetramethyl thiuram disulphide

123: thiocarbanilide

125: tetraethyl thiuram disulphide

126: 4:4'-thiobis(disec-amyl phenol) (Santowhite L)

128: dipentamethylene thiuram tetrasulphide (Tetrone A)

129: 2-mercaptoimidazoline

133: styrenated phenol (64)

134: polytrimethyl dihydroquinoline (Flectol H)

136: 2:5-ditert-amyl hydroquinone (Santovar A)

137: 4:4'-thiobis(6-tert-butyl-m-cresol) (Santowhite crystals)

138: 2:2'-methylene bis(4-ethyl-6-tert-butyl phenol) (CAO 425)

141: diphenyl guanidine

144: 40% 2:2'-methylene bis (4-methyl-6-tert-butyl phenol) (246).

VULCASTAB Group of stabilisers, curing, dispersing and mould release agents for latex and rubber compounds.

Grades:

A: sulphates of mixed fatty acid monoglycerides (obsolete)

BX: sodium salt of alkylated naphthalene sulphonic acids

C: 20% aqueous solution of sodium oleyl sulphate

HS: 20% aqueous solution of the sodium salt of a highly sulphonated oil

LS: sodium oleyl-p-anisidine sulphonate. Powder and aqueous paste

LW: condensation product of a fatty alcohol with ethylene oxide

OT: aqueous solution of an alkyl sulphate

PG: 50% aqueous solution of polyethylene glycol. Straw-coloured fluid, s.g. 1·09. Release agent for dry rubber and latex articles. Used at 2–10% concentration

RFA: mixture of surface active substances. Brown fluid, s.g. 1·06. Antitack coating for uncured compounds. *Uses:* sprayed or brushed on as 4% aqueous solution

T: aqueous solution of ammonium polymethacrylate

TM: aqueous paste of cetyl trimethylammonium bromide (60).

VULCAZOL Vulcanisation accelerators.
Grades:
Vulcazol: hydrobenzamide
A: hydrofurfuramide
N: nitroso hydrofurfuramide (363).

VULCOGENE Group of accelerators (240).
Types:
Standard: thiocarbanilide
ND: diphenyl guanidine.

VULJEX Group of prevulcanised natural latices (426).
Grades:
L.5: standard product
L.15: with increased modulus
L.25: for transparent products.

VULKACIT Trade name for vulcanisation accelerators produced by Bayer, Leverkusen.
Grades:
+470: crotonaldehyde-aniline condensation product
576: condensation product of homologous acroleins and aromatic bases. Dark

brown liquid, s.g. 0·99. Soluble in benzene, carbon tetrachloride, methylene chloride, alcohol, ethyl acetate, acetone and benzine, insoluble in water. Medium fast accelerator. Gives a high modulus, high tensile strength and high elasticity. Discolours and is unsuitable for white or coloured articles. Boosted by thiazoles, thiurams and mercapto accelerators. *Uses:* mechanical goods, tyres, inner tubes, conveyor belts, hard rubber and compounds with a high reclaim content

+576 Extra: reaction product of 2-ethyl-3-propyl acrolein and p-heptaldehyde-aniline

774: cyclohexyl ethylammonium cyclohexyl ethyl dithiocarbamate

1000: α-'o-tolyl diguanide

A: acetaldehyde-ammonia

AZ: N-N'-diethyl-2-benzthiazole sulphenamide

+BP: blend of heterocyclic bases with an ointment-like consistency. Medium-fast accelerator

+BZ: derivative of 2-mercaptobenzthiazole

CA: thiocarbanilide

+CT: butyraldehyde-aniline condensation product

CT-N: tricrotonylidene tetramine

CZ: N-cyclohexyl-2-benzthiazole sulphenamide

D: diphenyl guanidine

DB-1: double salt of zinc ethyl phenyl dithiocarbamate and cyclo-hexyl ethylamine. White powder, m.p. 109°C, s.g. 1·30. Soluble in many solvents. Very fast accelerator for NR and SR. Fatty acids are unnecessary. Suitable for hot air and cold cures. Should be added only on prewarming, or processed in two parts. Gives a high modulus and good ageing properties. *Quantity:* latex 0·5–1·5% with 1·5–2·5% sulphur. *Uses:* self-vulcanising compounds and solutions, dipped goods and surgical articles

DM: di-2-benzthiazyl disulphide

DM/C: di-2-benzthiazyl disulphide (surface treated to improve dispersion)

DOTG: di-o-tolyl guanidine

DZ: sulphenamide accelerator

EZ-57: sulphenamide accelerator

F: mixture of di-2-benzthiazyl disulphide and basic accelerators. Yellow powder, m.p. above 150°C, s.g. 1·31. Sparingly soluble in benzene, carbon tetrachloride, methylene chloride, alcohol and acetone, insoluble in water. Accelerator with delayed action and good processing safety. Zinc oxide is necessary, the start of curing may be controlled by fatty acids. Has a good plateau effect and is suitable for all curing methods. Boosted by basic accelerators and compounds (magnesium oxide, magnesium carbonate, alkali reclaim) and by thiurams and dithiocarbamates. Reduction of the sulphur content improves heat resistance.

Quantity: natural rubber 0·6–2% with 0·6–3·5% sulphur, SBR 1·5–1·8% with 0–0·25% tetramethyl thiuram disulphide and 1·8–2·2% sulphur, butadiene-acrylonitrile rubber 1·3–1·5% with 1·5–1·8% sulphur. *Uses:* in natural rubber, nitrile rubber and SBR, for mechanical goods, cables, tyres. Unsuitable for latex compounds and articles which come into contact with food

FR: anhydroformaldehyde-p-toluidine

FZ: mixture of N-cyclohexyl-2-benzthiazole sulphenamide and basic accelerators. Grey powder, m.p. above 83°C, s.g. 1·27. Soluble in benzene and methylene chloride, sparingly soluble in benzine, acetone, alcohol, carbon tetrachloride and ethyl acetate, insoluble in water. Accelerator with a delayed action. Boosted by basic accelerators and basic compounding ingredients, *e.g.* magnesium oxide and carbonate, alkaline reclaim; also by thiurams, dithiocarbamates, and mercapto accelerators. Zinc oxide is necessary while the addition of fatty acids gives a high modulus and enables the sulphur content to be decreased considerably. Effective at normal vulcanisation temperatures and for open steam cures. Gives a high modulus, low heat build-up, good ageing properties in compounds with a low sulphur content. *Quantity:* natural rubber 0·5–1·5% with 1–3·5% sulphur and 0–0·6% secondary accelerator, SBR 0·6–2% with 1·8–4% sulphur and 0–0·3% secondary accelerator, nitrile rubber 1·5% with 1·5–1·8% sulphur. *Uses:* articles under high dynamic strain (tyres, mountings) or with a high loading of reinforcing blacks or reclaim, together with thiurams and dithiocarbamates for the continuous vulcanisation of cables. Suitable for natural rubber, butadiene-styrene and butadiene acrylonitrile copolymers

H: hexamethylene tetramine

Hexine: secondary amine

HX: cyclohexyl ethylamine

J: N,N'-dimethyl-N,N'-diphenyl thiuram disulphide

L: zinc dimethyl dithiocarbamate

LDA: zinc diethyl dithiocarbamate

LDB: zinc dibutyl dithiocarbamate

LS 2: anhydroformaldehyde-aniline

LSH: anhydroformaldehyde-aniline

M: 2-mercaptobenzthiazole

MB: 2-mercaptobenzthiazole

MDA/C: mixture of 2-mercaptobenzthiazole and zinc diethyl dithiocarbamate (surface treated)

Mercapto: 2-mercaptobenzthiazole

Mercapto/C: 2-mercaptobenz-thiazole (surface treated to improve dispersion)

MOZ: benzthiazole-2-sulphenmorpholide

MT: mixture of 2 parts tetramethyl thiuram disulphide and 1 part mercaptobenzthiazole

MT/C: surface treated mixture of 2 parts tetramethyl thiuram disulphide and 1 part mercaptobenzthiazole. Yellow powder, s.g. 1·42. Soluble in acetone and methylene chloride, sparingly soluble in alcohol and benzene, insoluble in water. Good processing safety in compounds of butyl rubber and butyl reclaim. Zinc oxide is necessary, fatty acids are not essential but may be added in small quantities to loaded compounds. Gives a low modulus and good ageing properties. Curing system, *e.g.* for butyl rubber: 2% Vulkacit MT/C, 3% zinc oxide, 2–2·5% sulphur. *Uses:* linings of tubeless tyres, inner tubes, hoselines, extruded and moulded articles

NP: triazine derivative. White powder, m.p. above 168°C, s.g. 1·29. Soluble in water, sparingly soluble in acetone, insoluble in benzene, carbon tetrachloride, ethyl acetate and alcohol. Effective accelerator for chlorobutadiene polymers, particularly for neoprene W types. Addition of sulphur is unnecessary. Compounds with sulphur give stiffer products with lower heat

512

resistance. Gives a high modulus and good ageing properties. Compounding ingredients for curing system: 0·4–0·8% Vulkacit NP, 3–5% zinc oxide, 4–6% magnesia usta, 0–2% antioxidant

P: piperidine pentamethylene dithiocarbamate

P Extra N: zinc ethyl phenyl dithiocarbamate

S: thiuram disulphide

Thiuram: tetramethylthiuram disulphide

Thiuram MS: tetramethyl thiuram monosulphide

TR: ethylene polyamine

U: blend of 2:4-dinitrophenyl mercaptobenzthiazole and diphenyl guanidine. Pale yellow powder, m.p. above 125°C, s.g. 1·25. Soluble in benzene, methylene chloride, ethyl acetate and acetone, sparingly soluble in carbon tetrachloride and water. Medium fast accelerator with good processing safety. Zinc oxide is necessary, fatty acids are not essential, although they are boosters. Has good ageing properties, but causes yellow staining. *Quantity:* for hot air cures 1–1·2% with 2–2·2% sulphur, or 0·6–0·8% with 0·1–0·12% tetramethyl thiuram disulphide and 2–2·2% sulphur. Mechanical goods 1·2–1·3% with 1–1·2% sulphur or 0·8–1·2% with 0–0·2% mercaptobenzthiazole or tetramethyl thiuram disulphide and 2–3% sulphur. Soles 1·5–2% with 2·5–3·5% sulphur.

Uses: in natural rubber, SBR and nitrile rubber; particularly suitable for hot air cures (*e.g.* footwear), mechanical goods and tyres. Not particularly suitable for latex compounds, unsuitable for articles which come into contact with food

UC: mixture of 60% diphenyl guanidine and 40% mercaptoether

WL: sodium cyclohexyl ethyl dithiocarbamate

ZM: zinc-2-mercaptobenzthiazole

ZP: zinc pentamethylene dithiocarbamate (43).

VULKA COMPOUND Fast-curing compound with good flow properties and a delayed onset of cure. Used in direct vulcanisation, *e.g.* for shoe soles.

VULKADUR Group of rubber processing aids based on formaldehyde resins.
Grades:
A: phenol-formaldehyde resin. Yellow powder, m.p. 105–155°C, s.g. 1·3. Soluble in alcohol, acetone and ethyl acetate, sparingly soluble in water, benzine, benzene carbon tetrachloride and methylene chloride. Has a reinforcing and stiffening effect, particularly in butadiene-acrylonitrile copolymers. Improves filler absorption and mould flow of compounds, increases hardness, tensile strength and resistance to swelling at high temperatures, with a corresponding decrease in elongation at break and resilience. Improves metal/rubber bonding with Desmodur R

B: phenol-formaldehyde resin modified with sulphur. Yellowish powder, m.p. 100–120°C, s.g. 1·47. Soluble in alcohol, acetone and ethyl acetate, insoluble in benzine, benzene, methylene chloride, carbon tetrachloride and water. Has a stiffening and reinforcing effect, particularly in butadiene-acrylonitrile copolymers. Increases hardness and tensile strength, resistance to high temperature and swelling of the vulcanisate, while correspondingly decreasing the elongation at break and resilience·

T: 40% aqueous solution of a precondensed resorcinol-formaldehyde resin. *Uses:* in latex impregnating compounds for synthetic fibres (43).

VULKAFORM Group of silicone mould release agents (43)

VULKALENT Vulcanisation retarder.
Grades
A: N-nitrosodiphenylamine
B: phthalic anhydride blended with a dispersing agent (43).

VULKAMETER (Agfa vulkameter). Instrument for drawing the vulcanisation curve automatically. The compound is set up between three plates, the two outer ones subjecting it to shear stress by an excentric drive. The varying force exerted on the middle, static plate, is measured by a dynamometer.

VULKATOR Group of accelerators.
Grades:
I: thiocarbanilide
II: di-o-tolyl thiourea

III: di-o-tolyl guanidine thiosulphate
CPX: 2-mercaptobenzthiazole
DX: di-2-benzthiazyl disulphide
PX: 2-mercaptobenzthiazole
TH: tetramethyl thiuram disulphide
Z: zinc methyl phenyl dithiocarbamate (369).

VULKENE 107 E Reinforced polyethylene with improved resistance to chemicals and heat (198).

VULKOLLAN Polyurethane elastomers. Millable reaction products of a linear adipic acid-ethylene glycol polyester (Desmophen 2000, mol. wt. 2000, hydroxyl No. 50–60) with diisocyanates, *e.g.* 1:5-naphthylene diisocyanate (Desmodur 44), 1:4-butanediol (Type 18): a blend of butanediol and trimethylpropane 60:40 (Type 18:40) or water (Type 18W) is used as crosslinking agents. The percentage of 1:5-naphthylene diisocyanate in the cast types is indicated by the type number. Type 18, Shore Hardness 80°, Type 18:40, 65° Shore, Type 18W, 70° Shore, Type 30, 94° Shore (43).

VULNOC Group of vulcanising agents:
GM: p-quinone dioxime
DGM: p,p′-dibenzoyl quinone dioxime
R: morpholine disulphide (274).

VULT Group of synthetic latices:
N: butadiene-acrylonitrile copolymers
S: butadiene-styrene copolymers (365).

VULTAC Vulcanising agent. Brown, resinous mass, m.p. 45–90°C, s.g. 1·1–1·42. Softener and tackifier

in natural rubber and SBR; improves homogeneity of NR and SBR blends:
1: alkyl phenol monosulphide
2, 3, 4, and 5: alkyl phenol disulphides. Contains 23–28 % sulphur (4)

VULT-ACCEL B, zinc dibutyl dithiocarbamate, **E,** zinc diethyl dithiocarbamate (365).

VULTAMOL Sodium salt of an alkylated naphthalene sulphonic acid. Dispersing agent for latex (45).

VULTEX Vulcanised latex. The rubber particles show a slightly more rapid movement than in fresh latex. Vulpro was chosen as the term for products made from this latex.

VYGENE Trade name for p.v.c. (31).

VYRENE Elastic fibres based on adipic acid esters of ethylene glycol and propylene glycol (246).

W

W 80 Diphenyl guanidine salt of mercaptobenzthiazole (366).

WACKER SILICONES Group of hot and cold curable silicone rubbers, silicone oils and emulsions, pastes, fats and silicone resin solutions (359).

WALLACE HARDNESS METER Instrument used to determine the hardness of rubber. Has a spherical indentor of 1/16 inch. Calibrated in British Standard Hardness Degrees.

WALLACE MICROHARDNESS TESTER Instrument used to determine the microhardness and hardness of rubbers and plastics, according to the recommendations of the ISO. The instrument measures the depth of penetration of an indentor as the difference between a preload and a total load of up to 3500 g. May be used with various indentors and varying loads. The measurement is made in °ISO (British Standard Hardness Degree, International Rubber Hardness Degree). The depth of penetration may be determined to 1/10 micron. Specification: BS 903: A20: 1959.

WALLACE RAPID PLASTIMETER Plastimeter based on the parallel plate principle. A test piece of 1 mm depth is placed between a plate and a pressure foot and is compressed by a load of 10 kg for 15 sec at 100°C. The depth in 0·01 mm is taken as the measure of plasticity. A robust, automatic instrument similar to the Hoekstra steam plastimeter (358).

WASHINGS On washing the various containers used for processing latex, a dilute, impure latex is obtained. By coagulation, small quantities of rubber may be obtained which, together with other scrap, may be processed to low quality rubber.

WATER The water which is used for washing and diluting in rubber production on the plantations must not contain concentrations above the following:

Iron (for crêpe) 1 mg/l
    (for sheet) 3 mg/l
Copper 0·5 mg/l
Manganese 0·5 mg/l
Permanganate No. 20 mg/l
Chalk 50 mg/l
Bicarbonate 300 mg/l
Suspended impurities 10 mg/l, smaller than 325 mesh of the US Standard
pH 5·8–8
Hardness 6°D.

Larger quantities of iron cause patches of discoloration due to flocculation during coagulation; copper and iron act as oxidation catalysts and in small quantities cause tack. Oxidisable organic substances (indicated by the permanganate number) stimulate the growth of bacteria (also fermentation) which may discolour the rubber. Chalk in too large quantities may cause precoagulation in unstable latex, while bicarbonate causes bubbles in sheet. If the water is too acidic precoagulation may occur, water with too high an alkali content results in an increase in the amount of acid necessary for coagulation.

WATER EXTRACT Water extracts more or less completely some non-rubber constituents from raw rubber (quebrachitol, proteins, mineral salts), substances such as emulsifiers, stabilisers and coagulants from synthetic rubber, and substances such as glue, gelatin, starch, certain accelerators, soaps, sulphates, non-volatile constituents of blowing agents (sodium carbonate and sodium chloride) from vulcanisates. The water extract is used to estimate water soluble substances, free acids and alkalis, chlorides, sulphates and

electrical conductivity (presence of electrolytes).

WATER RESISTANCE The ability of rubber to resist water absorption under certain specified conditions of time and temperature; expressed in % volume increase. ASTM D 471.

WAX C Amide wax, m.p. 135–140°C, s.g. 0·99–1·01. Soluble in low boiling organic solvents. Lubricant for synthetic rubber (217).

WAX DPE Polyethylene wax. Mould release agent. May be emulsified in water (217).

WAX PA 520 Polyethylene wax with a relatively low mol. wt. and low melt viscosity. Processing aid for extrusion, reduces heat build-up of vulcanisates under dynamic strain. Stable to light and oxidation. Good electrical properties. Used in NR and SR.
Quantity: 1–6% (217).

WAX VP 7242 Polyethylene wax similar in effect to wax PA 510. Used in compounds requiring high dimensional stability (217).

WEATHERING APPARATUS Apparatus, which by means of comparison, determines the stability of various rubbers to sunlight and rain. A light source is used for irradiation with a spectrum which is similar to that of the sun, but has a greater intensity in the ultra-violet region. Water can be sprayed over the test pieces to imitate rain. The decrease of tensile strength and elongation at break is taken as the criterion; crack formation and blooming are also observed. ASTM D 750-55T.

WEATHER-O-METER Apparatus used to determine the weather resistance of rubber test pieces; similar to the Fade-O-Meter, in which, also, water is sprayed at intervals onto the test pieces.

WEBER, DR CARL OTTO 1860–1904, b. Pforzheim, d. Boston. Studied at the Technical University, Stuttgart, under Fehling. Originally a colour chemist specialising in rubber from 1892, his laboratory in Manchester was soon world famous. Undertook research expeditions to Mexico and Columbia for the Hood Rubber Co. In 1904, emigrated to Boston, where he died in the same year. His research gave great insight into the chemical constituents and reactions of rubber. In 1902 he outlined a theory of vulcanisation depending on pectinisation of the rubber molecule, the vulcanisation process leading to the formation of a continuous line of addition products of sulphur and polyprene and proceeding as a reaction independent of the physical state of the rubber colloid. Published the results of his basic research in *The Chemistry of India Rubber* (London, 1902).

WEBER COLOUR TEST (1902). Test in which brominated rubber reacts with phenols to form violet compounds confirming the presence of natural rubber. The reaction cannot be obtained with reclaimed rubber.

WERNER AND PFLEIDERER INTERNAL MIXER Two different types are available. One as internal mixer without ram and a spiral type three wing rotor, the other type with a ram similar to the Banbury internal mixer.

WESTRON Trade name for tetrachloroethane.

WESTROSOL Trade name for trichloroethylene.

WHEAL RUBBER Factice vulcanised with sulphur, and containing blown maize oil.

WHITBY, GEORGE STAFFORD 1887–1972. Professor of rubber chemistry at Akron University. From 1910 was active in Malaya, from 1923 professor of organic chemistry, in 1929 was appointed director of the Chemistry Section of the Canadian National Research Council, from 1939–42, director of the Chemical Research Laboratories, Teddington. Was responsible for numerous research articles in natural and synthetic rubbers. In 1928 was the first to receive the Colwyn Medal. Editor of *Synthetic Rubber, Plantation Rubber and Testing of Rubbèr*, 1926. Editor of the rubber section of the *International Critical Tables*, 1926, co-editor of the *Journal of the American Chemical Society*, 1934–39.

WICKHAM, SIR HENRY ALEXANDER 1846–1928. Forestry official in British Honduras. Founded his own plantation at the mouth of the Tapajos. Brought 70 000 seeds of *Heavea Brasiliensis* from the high plateau of the Tapajos to England, at the suggestion of Sir Joseph Hooker, director of Kew Gardens, and of the India Office. The India

Office paid £10 per 1000 seeds. After he had twice sent small quantities of seeds to England, he chartered a ship to take him up the Tapajos. The seeds were of the black-barked variety, *angustfolia*, grown on the Tapajos plateau between Tapajos and Moadeira. Approximately 2700 of the well-packed seeds fertilised. In 1876, 38 cases containing 1919 seedlings were sent to Ceylon and were planted in the Botanical Gardens there (Peradeniya and Heneratgoda). In 1876 another 100 of the seedlings brought over by Cross were taken to Ceylon. From Ceylon, the plants were supplied to the plantations of Malaya, Burma, Java, India and Sumatra. Wickham received a knighthood for his work in rubber cultivation. He was later appointed to various other high posts and spent years journeying through Central America, New Guinea and the Pacific Islands; he maintained his interest in the rubber industry and helped to establish several plantations. Was concerned with improving the processing technique for para rubber.

Publications: 'Rough Notes of a Journey through the Wilderness, from Trinidad to Para, Brazil, by Way of the Great Cataracts of the Orinoco, Atabapo and Rio Negro', London, 1872, 'On the Plantation, Curing and Cultivation of Para India Rubber (*Hevea Brasiliensis*). With an account of its Introduction from the West to the Eastern Tropics', London, 1908.

WICKHAM PROCESS Roa process. Modified Brazilian method for coagulating rubber used since the early days of the rubber industry. The latex is placed on an endless belt in a chamber with a counter-current of smoke and is dried and smoked. BP 191 487 (1921) (H. A. Wickham, and Roa Ltd.).

WIEGAND, WILLIAM B. Canadian rubber expert. In 1920 discovered the connection between particle size and the reinforcing effect of carbon blacks. A pendulum which he constructed to demonstrate the Joule effect was named after him.

WILD RUBBER Rubber from non-cultivated plants; obtained from wild clumps of trees in Central and S. America, Africa and Asia. Before World War I, the wild rubber formed a relatively large part of world rubber production. With the exception of small areas planted with Guayule (USA, Mexico, Russia) and Kok-Saghyz (Russia), 98% of current natural rubber comes from *Hevea Brasiliensis*. During World War II a large demand arose in Britain and the USA for all types of wild rubber and in 1942 approx. 100 000 tons were added to the US stockpile. Wild rubber has the disadvantage of being heterogeneous and the mediocre types have a large resin content and are usually impure. Wild rubber had to be washed on rollers before processing. The types were classified according to the botanical or geographical origin.

WILLIAMS, DR IRA American chemist specialising in rubber, 1894–1961. Developed neoprene rubber while working for Du Pont, invented the Du Pont abrasion machine. Awarded the Goodyear Medal in 1946.

WILLIAMS PLASTIMETER Parallel plate instrument used to determine the plasticity of rubber.

The height of a test piece, 15 mm in diameter and 10 mm in height is measured after compression for 60 sec with a load of 5 kg. The pressure plates may be heated and enable measurements to be taken at different temperatures. For plastic recovery the difference is taken between the plasticity No. and the height of the test piece 1 min after the test piece has been removed from the apparatus and allowed to cool.

$$K = yx^n$$

K is the plasticity, y the thickness of the test piece (in mm) after compression, x the time in min, and n a constant. Specification: ASTM D 926-56.

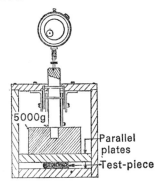

Parallel plates
Test-piece

WILLOUGHBEIA FIRMA
Lianes of the *Apocynaceae*. Provided most of the natural rubber exported from Borneo. Grows on light, rich ground. The lianes were cut into short pieces to obtain the rubber, the latex was then allowed to flow into containers and coagulated by adding salt water. Latex could also be obtained by making an incision in the living stem of the plant. The rubber was marketed in the form of round or pear-shaped pieces, and the quality was frequently debased by addition of mediocre liane rubber or jelutong. At the turn of the century an attempt was made to establish plantations, but was unprofitable. In Sumatra and Borneo the lianes are still planted in small quantities, and the latex is mixed with that of *Hevea* to produce slab rubber.

WING-STAY Group of antioxidants.
Grades:
100: mixture of diaryl-p-phenylene diamines. Brown solid, s.g. 1·20. Antioxidant particularly for SBR. Protects against ozone
200: mixture of diaryl-p-phenylene diamines
250: mixture of alkyl-aryl-p-phenylene diamines, s.g. 1·06. Dark brown fluid
275: mixture of alkyl-aryl-p-phenylene diamines
300: N-1,3-dimethylbutyl-N'-phenyl-p-phenylene diamine
L: phenolic antioxidant, s.g. 1·10
S: styrenated phenol. Brown liquid, s.g. 1·08. Soluble in benzene, toluene, ether, dioxane, methyl alcohol, acetone, ethyl acetate and chloroform, insoluble in water. Non-staining antioxidant; prevents decomposition and discoloration of natural rubber and synthetic rubber by oxygen, heat and light
T: phenolic antioxidant. Brown liquid, s.g. 0·9. For NR, SR and latex (115)
V: alkylated-styrene-phenol resin, s.g. 1·00. For SBR and NBR.

WINNOFIL S Precipitated calcium carbonate, coated with calcium stearate. Reinforcing filler for natural and synthetic rubbers.
Quantity: up to 60% (60).

WIRTSCHAFTSVERBAND DER DEUTSCHEN KAUTS-CHUK-INDUSTRIE Economic Association of the German Rubber Industry. Founded 1950 as successor to the Arbeitsgemeinschaft der Deutschen Kautschuk-Industrie, Frankfurt am Main, Zeppelin Allee 69.

WITAMOL Group of softeners for NR, SR and plastics (410).
Grades:
60: polyglycol ester of fatty acids of intermediate chain length
100 (DOP): dioctyl phthalate
110: phthalic acid ester of C_7–C_{10} alcohols
150 (DINP): diisononyl phthalate
180 (DIDP): diisodecyl phthalate
190 (DITP): diisotridecyl phthalate
320 (DOA): di-2-ethyl hexyl adipate
500 (DOS): di-2-ethyl hexyl sebacate
615: polypropylene glycol adipate.

WITAREX 400 Lubricant (410).

WITAREX TF Activator and dispersing agent based on natural fatty acids. Similar to stearic acid in its chemical and physical effects.

WITCARB Ultra-fine precipitated calcium carbonate. White reinforcing filler. Type P: 0·2 micron, Type P (ultra-fine): 0·055 micron (215).

WITCIZER Group of softeners.
Grades:
Witcizer: dibutyl tartrate
300: dibutyl phthalate
312: di-2-ethyl hexyl phthalate
313: diiso-octyl phthalate
200–201: n-butyl stearate (215).

WITCO Group of channel blacks of the Continental Carbon Co., for which the Witco Chemical Co. is the agent.
1: MPC
6: HPC
12: EPC (215).

WITTEN K 44 Chlorinated paraffin. Extender and softener (410).

WOLCRYLON Staple fibre made from polyacrylonitrile using the wet spinning process (367).

WORLD RUBBER CONSUMPTION, 1972 In 1000 long tons, according to information compiled by the International Rubber Study Group, London.

	NR	SR
USA	650·7	2328·5
Great Britain	174	272·6
France	160·2	289·8
Germany, Federal Republic	193	362·4
Italy	118	220
Netherlands	23·2	56·9
East European countries and China	650	1250
Rest of Europe	201·5	317·5
Australia	46·6	57·7
Brazil	44·2	114·3
Canada	60·3	172·8
Japan	312	588
Rest of the world	1474·4	1656·9
TOTAL	4108·1	7687·4

WORLD RUBBER PRODUCTION, 1972 In 1000 long tons. According to information compiled by the International Rubber Study Group, London

	NR	SR
Malaysia	1324·9	
Indonesia	818·7	
Thailand	336·9	
Sri Lanka	140·3	
Vietnam	20	
Cambodia	15·3	
India	109·1	27·9
Rest of Asia	54·5	
Africa	197	
Brazil	25·8	94·6
Latin America	15	
USA		2455·4
Canada		195·5
Germany, Federal Republic		300
Great Britain		307·1
Italy		200
Netherlands		186·4
Japan		819·3
Australia		41·7
Belgium		60
Czechoslovakia		51·5
South Africa		30
Rest of the world	39.6	2443
TOTAL	3097·1	7212·4

World Rubber Production 1900–1972
in 1000 long tons

	NR	SR1)
1900	45	—
1905	55	—
1910	97·5	—
1915	172·5	—
1920	347·5	—
1925	535	—
1930	837·5	−2)
1935	842·5	−2)
1940	1440	100
1945	255	1200
1950	1890	800
1955	1947·5	1085
1960	2035	1880

	NR	SR
1965	2352·5	3790
1970	3102·5	5855
1971	3077·5	6082·5
1972	3102·5	6525

1. Estimated, as the synthetic rubber production of the Eastern Bloc is unknown, was insignificant before 1960.
2. Production of small quantities of synthetic rubber in Germany (IG Farben), in the USA (Du Pont) and in Russia.

W PROCESS Process for impregnating textiles with natural and synthetic latices. The impregnating material is placed in the form of a foam on one side of the textile and is drawn in by a vacuum. Enables exact control of the depth of impregnation and the thickness of the layer. Developed by the Witco Chemical Co. BP 866 389.

WRAPPER A rubber sheet in which bales are packed.

WURTZEL APPARATUS An instrument which was used to determine the gas permeability of impregnated textiles.

WYEX EPC channel black (40).

WYTOX Group of antioxidants (237):
312: mixture of alkylated aryl phosphites, s.g. 0·98
355: mixture of alkylated aryl phosphites, s.g. 1·01
ADP: octylated diphenylamine, s.g. 0·98
BHT: 2·6-di-tert-butyl-4-methyl phenol.

X

X-28 Technical diphenyl guanidine.

X-51 Synthetic polyacrylonitrile fibre with a small amount of methyl methacrylate, s.g. 1·17. Produced by the wet spinning process. Resistant to boiling, alkalis and acids (21). *TN:* Creslan.

X-1034 R General purpose silicone rubber. Has excellent resistance to high temperatures, low permanent set and water absorption. Usual catalysts are used for crosslinking. Used with di-tert-butyl peroxide, is suitable for thick walled articles (112).

XC Group of pressure sensitive silicone adhesives (7).

XDI Trade name for 4:4′-diphenyl methane diisocyanate (148).

XENON LAMP EXPOSURE TEST Method of assessing resistance of compounds to light using an accelerated test (438).

XEROGEL According to Freundlich, a dried gel, *e.g.* silica gel.

XETAL Positively charged latex containing cationic soaps (also Positex).

XFLX Antioxidant; secondary aromatic amine (100).

XF RESINS Abbreviation for xylene-formaldehyde resins.

X POLYMERS Experimentally produced SBR polymers which were placed at the disposal of industry in the USA for technical application trials at the time of state control of butadiene-styrene rubber production. Over 750 types, including black and filler master batches, were produced and tested, and approx. 65 suitable types adopted for general production as GR-5 types.

X-RAY DIFFRACTION ANALYSIS Roentgen structure. Examination of the lattice structure of a crystal from the diffraction of X-rays. The phenomenon of crystallisation in elastomers can be examined using this method. Katz used the method to confirm the crystalline structure of stretched rubber. Lit.: *Chem. Zeit.*, 1925, **49**, 353.

X TYRES Special tyres produced by Michelin. They have a stiffened tread achieved by a triangular arrangement of the textile fibres at an acute angle to the circumference. Instead of cords, steel wires are put in the cushion of the tyre, and arranged almost in the direction of running of the tyre. These wires stretch far less than cords. The tyre is not noisy and has a low resistance to rolling. Driving comfort is decreased because of the stiffness of the tyre. The Pirelli Cinturato tyre is built on the same lines, but solely with textiles.

XYLENE-FORMALDEHYDE RESIN XF resin. Condensation product of xylene and formaldehyde, obtained with sulphuric acid as a catalyst.

XULITE Technical diphenyl guanidine

XYLYL MERCAPTAN

Brown liquid. Soluble in aromatics. Chemical peptiser for natural rubber and SBR, also reclaiming agent. Has a slight boosting effect on gum stocks containing thiazoles and thiurams, has no effect on ageing properties. Does not stain, is odourless in vulcanisates. Has an oxidising effect on carotenoids in fresh natural latex and is used for bleaching natural rubber (0.01–0.1% of the 36.5% trade product). As a peptiser, 0.15–0.6% is used in natural rubber, 0.25–1% in SBR.
TN: RPA-3 (36.5% with an inert hydrocarbon), RPA-3 conc. (82.5% with an inert hydrocarbon).

Y

YELKIN S Antioxidant of undisclosed composition (368).

YERZLEY OSCILLOGRAPH A mechanical oscillograph used to determine the mechanical properties of vulcanised elastomers in relation to damping of oscillations and insulation from the work of elastic deformation. The determination is carried out under compression or shear strain. ASTM D 945-59.

YOUNG, THOMAS British physicist, 1773–1829. Did research on optics, diffraction phenomena, colour, mathematical physics, statics. The elasticity modulus bears his name.

Z

Z-51 Zinc dithiocarbamate (45).

Z-88-P Blend of a 2-mercaptobenzthiazole derivative and a Schiff's base, m.p. $108°C$. Fast accelerator (obsolete) (5).

ZALBA Condensation product of β-naphthol and aniline. Brown, viscous liquid, s.g. 0.94. Sparingly soluble in aromatic solvents, insoluble in water. Non-staining and non-blooming antioxidant and stabiliser for butadiene-styrene copolymers. *Quantity:* 0.5–2% (6).

ZBD Abbreviation for zinc dibenzyl dithiocarbamate.

ZDBC Abbreviation for zinc dibutyl dithiocarbamate.

ZDC Abbreviation for zinc diethyl dithiocarbamate (369).

ZDMC Abbreviation for zinc dimethyl dithiocarbamate (369).

ZENITE Group of mixed accelerators with zinc-2-mercaptobenzthiazole as the main constituent:
Zenite: mixture of 90% zinc-2-mercaptobenzthiazole and 10% inert hydrocarbon.

Pale yellow, wax-like powder, s.g. 1·53. Medium-fast accelerator, active above 138°C. Boosted by other accelerators. Gives good ageing properties. Has no tendency to scorch. Large quantities act also as a non-staining antioxidant. Suitable for natural rubber, SBR, reclaim, latex, neoprene, NBR. *Quantity:* NR, primary, 0·75–1·5% with 1–4% sulphur and 1–4% stearic acid, boosted by aldehyde-amines, guanidines and thiuram, 2–4% as antioxidants. SBR 1–2% with 1–3% sulphur. In latex has a slight influence on prevulcanisation and stability, 1–1·5% without activation, 2–4% in the absence of other antioxidants, boosted with 0·12–0·25% sodium dibutyl dithiocarbamate, or 0·25–0·5% dipentamethylene thiuram tetrasulphide or piperidine pentamethylene dithiocarbamate

-A: old form of AM, s.g. 1·42

-AM: mixture of 97% Zenite and 3% tetramethyl thiuram monosulphide. Pale yellow powder, s.g. 1·53. Fast accelerator with good processing safety, active above 121°C. Gives good ageing properties. Does not stain or bloom and is suitable for light coloured articles. *Quantity:* natural rubber and reclaim 0·5–1·5% with 1–2·5% sulphur and 1–4% stearic or salicylic acids. For very fast compounds it may be boosted by the addition of tetramethyl thi-

uram monosulphide, an aldehyde amine or a guanidine. Butadiene-styrene polymers, medium-fast acceleration with an excellent processing safety, 1–2% with 1–3% sulphur. Butadiene-acrylonitrile polymers 3–5% with 1–1·5% dipentamethylene thiuram tetrasulphide if a low permanent set is required. Butyl rubber, medium-slow acceleration, 0·75–1·5% with 0·2% sulphur

-B: mixture of 91% zinc-2-mercaptobenzthiazole and 9% di-o-tolyl guanidine. Pale yellow, wax-like powder, s.g. 1·39. Fast accelerator with good processing safety, active above 121°C. Discolours only slightly but unsuitable for pure white articles. *Quantity:* 0·5–1% with 1–2·5% sulphur and 1–4% stearic acid. May be boosted by additional guanidine, aldehyde-amines or thiurams

-Special: zinc-2-mercaptobenzthiazole (6).

ZEOLEX 23 Reinforcing white silicate filler (40).

ZEPD Abbreviation for zinc ethyl phenyl dithiocarbamate.

ZEWA AND ZEWA DIS RESINS Group of processing aids for NR and SR; based on the salts of lignin sulphonic acid, s.g. 1·06–1·28. Plasticisers, dispersing agents for fillers, activators; improve fabrication. Have no effect on the hardness of vulcanisates (370).

ZG 92 Zinc stearate in paste form. Dispersion agent for light and coloured mixes (378).

ZI-102 Inhibitor used to prevent discoloration of white neoprene compounds.

ZIEGLER CHEMISTRY The chemistry of metal–organic compounds. Using these compounds, it is possible to produce stereospecific polymers (cis-1:4-polyisoprene and cis-1:4-polybutadiene), and to polymerise ethylene, propylene, α-butylene and isoprene at normal pressures. The Ziegler catalyst is a compound catalyst of triethylene aluminium and titanium tetrachloride. Apart from titanium, zirconium, vanadium, chromium and other metals, Groups IV to VIII of the Periodic Table are suitable in compound catalysts. Metals of Groups I to III may be used as metal-organic compounds, particularly magnesium, beryllium and boron. The compound catalysts are termed Ziegler catalysts.

ZINCAZOL Yellowish white powder, m.p. 190°C, s.g. 1·25. Insoluble in conventional solvents. Safe accelerator with a uniform power over a large temperature range. Suitable for white and coloured compounds, and for compounds containing up to 25% white factice (336).

ZINC-2-BENZAMIDOTHIO-PHENATE

Grey powder, m.p. 170°C, s.g. 1·32.

Chemically effective peptiser for natural rubber at relatively low temperatures. Active above 65°C, most effective in the absence of other ingredients. Suitable for processing on mills.
Quantity: NR 0·05–0·5%, SBR 0·5–3%.
Uses: compounds which require a high plasticity, *e.g.* sponge rubber.
TN: Noctizer SZ (274)
Noticizer SM (with dibenzamidodiphenyl disulphide) (274)
Pepton 65 (21)
Pepton 65 B (with dispersant) (21).

ZINC CARBONATE Basic carbonate with a variable composition. White, amorphous powder, s.g. 4·45. Soluble in dilute acids and ammonia, insoluble in water and alcohol. Occurs naturally as hydrozincite, calamine, and Smithsonite. Activator, filler and pigment for NR and SBR; has the same effect as zinc oxide.

ZINC DIBENZYL DITHIO-CARBAMATE

White powder, m.p. 165–175°C, s.g. 1·41. Partially soluble in benzene and ethylene dichloride, insoluble in benzine, acetone and water. Ultrafast accelerator with a relatively low critical temperature and a medium vulcanisation range up to 121°C, the range is reduced on increasing the temperature. A very safe processing compound; more active than thiurams. Excellent booster for thiazoles. Suitable for natural rubber

SBR, and their latices and solutions. Zinc oxide and sulphur are necessary in the usual quantities, fatty acids are not essential but may be used. Non-staining and non-discolouring. May be added to latex in paste form or as an aqueous dispersion.

Quantity: zinc dibenzyl dithiocarbamate (I), zinc-2-mercaptobenzthiazole (II), sulphur (III).

| | Natural Rubber | | SBR | |
	Primary	Secondary	Primary	Secondary
I	0·5	0·15	1·0	0·75
II	—	1·0	—	1·5
III	1·5	1·5	2·0	2·0

TN: Ancazite BZ (22)
Arazate (23)
Naugatex 245 (aqueous dispersion) (23)
Naugatex 511 (aqueous dispersion) (23)
NX-Paste 511 (paste form) (23)
Robac ZBED (311)
Van Hasselt DBZ (355).

ZINC DIBUTYL DITHIOCARBAMATE

$$\left[\begin{array}{c} C_4H_9 \\ C_4H_9 \end{array} \!\!\! \diagdown\!\!\!N\!-\!\!\underset{\underset{S}{\|}}{C}\!-\!S- \right]_2 Zn$$

White to yellowish powder, m.p. 105–108°C, s.g. 1·24–1·28. Soluble in aromatic and aliphatic hydrocarbons and chlorinated solvents, insoluble in water and dilute alkalis. Ultrafast accelerator for natural rubber and latex, has a relatively low critical temperature. Safer and cures more slowly than other zinc dithiocarbamates. Medium vulcanisation range below 120°C. Greatly reduced on increasing temperature. Vulcanisation temperature: 95–110°C. Zinc oxide and sulphur are necessary in the usual quantities, is also effective without zinc oxide and may therefore be used for transparent articles made from latex. Fatty acids are not essential but may be used. Non-staining and non-discolouring. May be used, with the addition of dibutylamine, for curing at room temperature. Boosted by thiazoles.

Quantity: primary 0·5–1·5% with 0·5–2% sulphur, secondary 0·15–0·5% with 1–1·5% thiazole and 1·5–2% sulphur. Transparent articles 1·5–2% with 1% sulphur and without zinc oxide.

Uses: latex articles, adhesives, cements, impregnations, proofed textiles; and in compounds containing reclaim.

TN: + Accelerator 77 (4)
Accelerator LDB (408)
Accélérateur 3 RS
Aceto ZDBD (25)
Ancazate BU (22)
Butsan (23)
Butazate (23)
Butyl Ten (mixture with an amine) (51)
Butyl Zimate (51)
Butyl Ziram (120)
Cyzate B (50% aqueous dispersion) (21)
EPTAC 4 (6)
Naugatex 514-A (aqueous dispersion) (23)
Nocceler BZ (274)
NX-Paste 514-A (paste form) (23)
Robac ZBUD (311)
Robac ZBUD Extra (complex with dibutylamine) (311)
Rodform Butyl Zimate (51)
SA 77, 77-0, 77-7 (50% aqueous dispersion), 77-9 (4)
Superaccélérateur 4005 (20)
Van Hasselt BTZ (355)

Vondac ZBUD (433)
Vulcacure ZB (50% dispersion) (57)
+ Vulcafor ZNBC (60)
Vulcaid BTZ (3)
Vulkacit LDB (43)
Vult-Accel B (dispersion) (365)
ZDBC (6).

ZINC DIBUTYL XANTHATE

$$C_4H_9\text{—}O\text{—}CS\text{—}S\text{—}Zn\text{—}S\text{—}$$
$$CS\text{—}O\text{—}C_4H_9.$$

White powder, s.g. 1·40–1·56. Decomposes on heating, tends to decompose slowly on storage and should therefore be stored at low temperatures. Partially soluble in benzene and ethylene dichloride, sparingly soluble in acetone, insoluble in water. Accelerator used in conjunction with booster, preferably dibenzylamine, to give an ultra-fast accelerator combination. Used at low temperature with a wide vulcanisation range. Curing can take place overnight at room temperature or in 20 min at 100°C. Channel blacks, alumina and acidic compounding ingredients act as retarders. The amine combination may be further boosted by thiazoles, thiurams and dithiocarbamates. Zinc oxide and sulphur are necessary in the usual quantities, fatty acids retard and are unnecessary. Processing should take place in two separate compounds, one with and one without sulphur. These and dibenzylamine should be blended when ready for use, although the accelerator should be added only when warming the compounds. The accelerator and dibenzylamine should not be blended before they are put into the compound. In neoprene cements (neoprene W and AC) dibutyl xanthate may be used as the sole activator. Non-staining and non-discolouring.

Quantity: natural rubber and SBR 2% with 2% dibenzylamine, 0–0·25% tetramethyl thiuram disulphide and 2% sulphur, neoprene 4%.
Uses: suitable for natural rubber, SBR and neoprene. Textile proofings, gum stocks, medical articles, bathing shoes and caps, fancy goods, cement which may be cold cured.
TN: Vulcaid 27 (3)
ZBX (23).

ZINC DIETHYL DITHIO-CARBAMATE

$$\left[\begin{array}{c} H_5C_2 \\ H_5C_2 \end{array} \right\rangle N\text{—}C\text{—}S\text{—} \underset{S}{\overset{\|}{}} \right]_2 Zn$$

ZDC. White, odourless, non-hygroscopic powder, m.p. 174–180°C, s.g. 1·47, mol. wt. 363. Soluble in dilute alkalis, carbon disulphide, benzene, chloroform, ethylene and methylene dichlorides, sparingly soluble in acetone and carbon tetrachloride, insoluble in water and benzine. Ultra-fast accelerator. Low critical temperature, medium vulcanisation range at 120°C which decreases with rise in temperature. Vulcanisation temperature: 120–135°C. Boosted by thiazoles. Zinc oxide and sulphur are necessary in normal quantities, fatty acids are not essential but may be used. Has good ageing properties. Does not stain or discolour on contact, damp latex films tend to discolour on handling. Suitable for solid rubber and latex, has only a slight effect on the stability of latex, but tends to cause prevulcanisation.
Quantity: primary 0·5–2% with 3–1% sulphur, secondary 0·1–0·5% with 1–1·5% thiazole or thiuram and 1·75–2·75% sulphur. For excellent

ageing stability in compounds 0·5–0·25% with 1–1·5% dibenzthiazyl disulphide or MBT, 0·5% tellurium and 1·25–0·5% sulphur. Latex (on 100 parts rubber) 0·75–0·5% with 0–0·5% dibenzthiazyl disulphide, 1·25–1% sulphur and 3% zinc oxide.

Uses: latex articles, dipped articles, bathing caps, footwear, impregnations, rubber thread, foam rubber, insulations, adhesives, coloured household articles.

TN: Accel EZ (6)
Accélérateur rapide 3 RS
Accélérateur rapide DEDCZ (419)
+ Accelerator 67-O (6)
Aceto ZDED (25)
Ancazate ET (22)
Cyzate E (50% aqueous dispersion) (21)
Di VII (180)
Ethasan (5)
Ethazate (23)
+ Ethex (6)
Ethyl Zimate (51)
Ethyl Ziram (120)
Eveite Z (59)
Leda ZDC (247)
Naugatex 513 (aqueous dispersion) (23)
Nocceler EZ (274)
NX-Paste 513-B (paste form) (23)
Robac ZDC (311)
Rubber Accelerator ZE
SA 67 (4)
Superaccélérateur 1505 (20)
Superaccélérateur 4005 (20)
Van Hasselt ETZ (355)
Vondac ZDC (433)
Vulcacure ZE (10% ZDC) (57)
Vulcafor ZDC (60)
Vulcafor ZDCL (paste form) (60)
Vulcaid ZDC (3)
Vulkacit LDA (43)
Vult-Accel E (dispersion) (365)
ZDC
ZDEC (369)

ZE
+ Zinc DEC (269, 5)
Mixture with 2-mercaptobenzthiazole:
Vulcafor DHC (60)
Vulkacit MDA/C (43).

ZINC DIMETHYL DITHIO-CARBAMATE

$$\left[\begin{matrix} H_3C \\ \\ H_3C \end{matrix} \hspace{-4pt} \begin{matrix} \\ N-\overset{\displaystyle S}{\overset{\|}{C}}-S- \\ \end{matrix} \right]_2 Zn$$

White to yellowish, odourless, non-hygroscopic powder, m.p. 240–250°C, s.g. 1·65–1·70, mol. wt. 306. Soluble in dilute alkalis, carbon disulphide, benzene hydrocarbons, chlorinated hydrocarbons and acetone, insoluble in alcohol, ethyl acetate, carbon tetrachloride and water. Ultra-fast accelerator. Has a medium vulcanisation range up to 120°C, the range decreases with rise in temperature. Gives fast cures with a fairly low critical temperature. More active than thiurams. The addition of dibenzthiazyl disulphide improves the processing safety. Primary accelerator for cures at 120°C, secondary accelerator with thiazoles, increases modulus and resilience. Vulcanisation temperature: 120–135°C; zinc oxide and sulphur are necessary in normal quantities, fatty acids are not essential but may be used. Gives odourless, non-toxic, non-staining and non-discolouring vulcanisates.

Quantity: natural rubber, primary 0·1–1% with 0–1% benzthiazyl disulphide and 3–1% sulphur, secondary 0·05–0·25% with 1–1·5% mercaptobenzthiazole and 3–1·5% sulphur. Continuous vulcanisation 0·5% with 0·5% mercaptobenzthiazole and 2% sulphur, for excellent ageing properties 0·5–0·25% with 1–1·5% mercaptobenzthiazole, 0·5% tellurium and 1·25–0·5% sulphur. SBR, primary 0·75% with 2% sulphur, secondary 0·5–1·5% with 1·5% zinc mercaptobenzthiazole and 0·5% sulphur. Continuous vulcanisation 1% with 1% benzthiazyl disulphide and 2% sulphur.

TN: Accélérateur rapide DMDCZ (419)
Accel PZ
Accelerator 57-0 (4)
Accelerator L (408)
Aceto ZDMD (25)
Ancazate ME (22)
Di IV (180)
Kure Blend MZ (50% masterbatch) (31)
EPTAC 1 (6)
Leda MTZ (247)
Methasan (5)
Methazate (23)
Methyl Zimate (51)
Methyl Ziram (120)
Naugatex 512 (23)
Nocceler PZ (274)
PMzD-75 (75% masterbatch with polyisobutylene)
Robac ZMD (311)
Rodform Methyl Zimate (51)
Rubber Accelerator ZM
Sharpels Accelerator (SA-)57, 57-0 (4)
57-1 (with 3% mineral oil)
57-7 (50% aqueous dispersion)
57-9
Superaccélérateur 1605 (20)
Super Sulphur No. 1 (51)
Ultra Zinc DMC (5)
Van Hasselt MTZ (355)
Vondac ZMD (433)
Vulcacure ZM (57)
Vulcaid MTZ (3)
Vulkacit L (43)
ZDMC (369)
Zimate (51)
ZM.

ZINC DIXYLYL DISULPHIDE

Zinc salt of xylyl mercaptan. Chemical peptiser, reclaiming agent for SBR. Retards curing slightly, has no effect on ageing properties. Non-staining. Odourless in the final product; trade product is a blend with an inert hydrocarbon.
TN: PRA-5 (6).

ZINC ENANTHATE

$(C_6H_{13}COO)_2Zn$. White, very light, voluminous powder with faint odour, m.p. 136–137°C, s.g. 1·18 (apparent 0·18). Contains 20·1% zinc. Softener and activator. The addition of 3% reduces the mastication time by one half and replaces half the necessary zinc oxide content and all the stearic acid; 5% zinc enanthate is normally equivalent to the zinc oxide/stearic acid and softener contents in most compounds. Has no effect on the ageing properties and tensile strength, increases resistance to flex cracking (380).

ZINC ETHYLENE BIS-DITHIOCARBAMATE

Pale-yellow powder with characteristic odour, decomposition temperature 220°C. Accelerator.
TN: Leda Zineb (247).

ZINC ETHYL PHENYL DITHIOCARBAMATE

$[C_6H_5(C_2H_5)N.CS.S]_2Zn$. White to yellowish powder, m.p. 205–208°C, s.g. 1·46. Soluble in chlorinated and benzene hydrocarbons, sparingly soluble in carbon tetrachloride and acetone, insoluble in alcohol, benzine and water. Safe, ultra-fast accelerator with a fairly wide plateau below 130°C; zinc oxide and fatty acids are necessary for optimum results. Compounds activated by lead oxide have a higher critical temperature than zinc oxide compounds; the addition of mercaptobenzthiazole, dibenzthiazyl disulphide, tetramethyl thiuram disulphide, benzoic acid, N-nitrosodiphenylamine and phthalic anhydride retards the start of vulcanisation and increases processing safety. Boosted by basic accelerators, *e.g.* dibutylamine, piperidine pentamethylene dithiocarbamate and cyclohexylamine. Vulcanisates are odourless, tasteless and non-staining, have good ageing properties and give a high modulus. Suitable for NR and SR.

Quantity: primary 0·3–1 % with 0·3–0·4 % mercaptobenzthiazole and 1·5–2·5 % sulphur, or 0·1 % dibenzthiazyl disulphide and 0·1 % tetramethyl thiuram disulphide, secondary 0·1–0·2 % with 0·8–1·2 % mercaptobenzthiazole or alternative mercapto accelerator, and 2–3 % sulphur, self-vulcanising compounds 3 % with 1·5–2 % basic accelerator and 2·5–3·5 % sulphur.

Uses: latex articles, white or transparent articles, impregnations, textile proofings, dipped articles, cable compounds, self-vulcanising cements, articles used in the food industry.

TN: Accel PX (16)
Accelerator P Extra N (408)
Accélérateur rapide 3 R and 3 RN (19)
Ancazate EPH (22)
Eveite P (59)
Nocceler PX (174)

Riken 200
Soxinol PX (328)
Superaccélérateur 1105 (20)
Vulkacit P Extra N (43)
Vulcafor ZEP (60)
Wobezit P Extra N (367)
double salt with cyclohexyl ethylamine
Vulkacit DB 1 (43).

ZINC ETHYL XANTHATE

$$[C_2H_5-O-CS-S-]_2 \, Zn$$

White powder, s.g. 1·56. Ultra-fast accelerator; very active in the presence of zinc oxide. Vulcanisation temperature 108–125°C, cure: 3 min at 130°C.

TN: + Xanthophone (23).

ZINC ISOPROPYL XANTHATE

$$H_7C_3.O.CS.SZn.S.CS.O.C_3H_7.$$

ZIX. Yellowish powder with a pungent odour, m.p. 145°C, while decomposing, s.g. 1·53. Soluble in benzene, insoluble in water, solubility in rubber approx. 1 %. Ultra-fast accelerator used for curing latex and adhesives at room temperature. May not be used above 110°C because of its tendency to decompose. The use of zinc oxide is recommended; a stabiliser is necessary in latex compounds.

Quantity: 1–2 % with 2·5–4 % sulphur.

Uses: latex and dipped articles, with other accelerators for self-curing adhesives (two part adhesives).

TN: A-C-A-K
Accélérateur rapide 5 R (19)
Accélérateur rapide 5 R Extra (19)
Robac ZIX (311)
Superaccélérateur 6005 (20)
Vulcafor IX (60)
Vulcafor ZIX (60).

ZINC LUPETIDINE DITHIO-CARBAMATE

$[(CH_3)_2C_5H_8N.CS.S]_2Zn$. Pale brown powder, with characteristic odour, m.p. 84–98°C, s.g. 1·55. Soluble in benzene, xylene, toluene, insoluble in water. Accelerator, primarily for use in latex compounds. Very active, will cure latex articles within 4 days at room temperature, zinc oxide is unnecessary. Nonstaining. Articles have a high degree of transparency. Acts more slowly in solid rubber.

Quantity: 0·5–1% on the rubber, with 1·5% sulphur.

TN: Robac Z.L. (311).

ZINC-2-MERCAPTOBENZ-IMIDAZOLE

White powder, m.p. 300°C, s.g. 1·75. Sparingly soluble in conventional solvents, insoluble in water. Antioxidant with a whitening effect. Gives good protection in combination with other antioxidants; optimum effect in compounds with thiurams and dithiocarbamates. Does not stain or bloom.

Quantity: 0·6–1·5%.

Uses: white, coloured and transparent articles.

TN: Antioxidant ZMB
Antigen MBZ
Antioxygène MTBZ (19)
Nocrac MBZ (274)
Permanax Z21 (20)
ZMB.

ZINC-2-MERCAPTOBENZ-THIAZOLE

(zinc benzothiazyl sulphide). White to yellowish, non-hygroscopic powder, decomposes without melting above 200°C, s.g. 1·63–1·65, mol. wt. 398. Soluble in acetone and dilute alkalis, sparingly soluble in alcohol, benzene hydrocarbons and chlorinated solvents, insoluble in water and benzine. Medium fast accelerator with a narrow curing range. Active above 116°C. Accelerating effect is somewhat slower than that of mercaptobenzthiazole, but it has a higher critical temperature. Medium activity at vulcanisation temperatures of latex. Boosted by 0·25–1% thiuram, dithiocarbamate, aldehydeamines or guanidine, also boosted by alkaline reclaim and other basic substances. Zinc oxide and fatty acids are necessary in the usual quantities. Non-staining and does not discolour on contact. Gives excellent ageing properties, especially in sulphurless compounds. Effective as an antioxidant in higher concentrations. Does not bloom. Suitable for NR, SBR, neoprene and latices. Sensitises and has a slight influence on prevulcanisation and stability of latex compounds. Used in press cures or in hot air cures. Unsuitable because of its bitter taste for articles which come into contact with food. Enables the zinc oxide content to be lowered in foam rubber and thus gives more easily controlled gelling.

Quantity: NR and SBR, primary 1·25–2% with 3–2% sulphur and 5% zinc oxide, secondary 0·5–1·5% with 0·25–1% tetramethyl thiuram monosulphide or other boosters. Neoprene 3–5% (has excellent antioxidant in heat resistant compounds; retarding effect on vulcanisation). Latex 1·5% with 1% zinc diethyl dithiocarbamate, 2·5% sulphur and 0·5–1%

531

zinc oxide. Antioxidant 2–4%. Reclaim 0·5–2% with 2–4% sulphur and 0·25–1% DOTG or other basic accelerator.

Uses: for light-coloured and bright articles, footwear, wire insulation, latex articles and foam rubber.

TN: Accélérateur MBTZn (418)
Accélérateur rapide Z 200 (20)
Accelerator ZM (408)
Bantex (5)
Gequisa ZMBT (397)
Kuracap Zinc MNT (241)
Naugatex 188, 255, 503, A, B, C (aqueous dispersion) (23)
Nocceler MZ (274)
Pennac ZT (120)
OXAF (23)
R-23 (5)
Robac MZI (311)
Soxinol Z (328)
Vulcacure ZT (57)
Vulcafor ZMBT (60)
Vulkacit ZM (43)
Zenit (6)
Zenit Special (6)
Zetax (51)
Zinc Ancap (22)
ZMBT
ZML.

Mixtures:
Accelerator 531 (with di-o-tolyl guanidine and anhydroformaldehyde-p-toluidine)
Bantex (with diphenyl guanidine) (5)
Bantex DN (with diphenyl guanidine) (5)
Bantex M (with tetramethyl thiuram monosulphide) (5)
Morfex 33 (ZS-Morfex) (with 25% tetramethyl thiuram monosulphide) (23)
Morfex 55 (with 50% tetramethyl thiuram monosulphide) (23)
Orthex (anhydroformaldehyde-p-toluidine) (23)
Robac MZ 2 (with 9% di-o-tolyl guanidine) (31)

SRA-2 (with zinc diphenyl guanidine benzthiazyl disulphide) (6)
Zenite (10% paraffin (as dispersing aid) (6)
Zenite A = Zenite AM (with 3% tetramethyl thiuram monosulphide) (6)
Zenite B (with 9% di-o-tolyl guanidine) (6)

ZINC METHYL PHENYL DITHIOCARBAMATE

$$\left[\begin{matrix} H_3C \\ C_6H_5 \end{matrix} \!\!>\!\! N\!-\!\overset{\displaystyle S}{\overset{\|}{C}}\!-\!S\!- \right]_2 Zn$$

Colourless powder, m.p. 230°C, s.g. 1·53. Ultra-fast accelerator; zinc oxide is necessary.

Quantity: 0·7–1% with 1·5–2%.

TN: Accélérateur rapide R (19)
Vulkator Z (364).

ZINC OXIDE Zinc white, flowers of zinc, Lana philosophica. Loose, white powder, s.g. 5·4–5·8. Soluble in dilute acetic acid and mineral acids, ammonia and alkali hydroxide solution, insoluble in water. Activator, mainly used with a fatty acid (stearic), zinc stearate is soluble in rubber. Also a filler and white pigment (the latter use was noted by Hoffer in 1882). Until 1930, zinc oxide, zinc sulphide and lithopone, were the major white pigments.

ZINC OXIDE TEST ZOV, zinc oxide viscosity. Quick and sensitive method of determining the chemical stability of *Hevea* latex in relation to its processing. In latex which has been preserved with ammonia, zinc oxide increases the viscosity by dissolving and forming positively charged complex ions, which result in a reduction of colloidal stability.

The solubility depends on the ammonia concentration and is at a maximum at pH 9·4; it also depends on the quantity of ammonium salts in the latex. The fatty acid soaps in the latex increase the thickening and coagulation of the latex by forming insoluble zinc soaps with the dissolved zinc oxide. The zinc oxide stability of concentrated latex depends on the total solids, the age, storage temperature and ammonia concentration. In practice, either the viscosity increase or the time for coagulation may be determined. The ZOV test is suitable for examining the processing properties of latex for foam rubber and other products, *e.g.* for foam rubber with a sodium silicofluoride gelling agent the ZOV should be 600–700 cp. Latex with a ZOV of over 1000 cp is too sensitive to zinc oxide, whereas a ZOV of under 300 indicates too high a stability. The properties of the latex may be improved by adding a stabiliser or a sensitiser.

ZINC PENTACHLORTHIO-PHENATE

$(C_6Cl_5.S-)_2Zn$. Grey white, odourless powder, s.g. 2·33. Practically insoluble in conventional solvents. Chemically effective peptiser, particularly for NR on a mill or an internal mixer. Accelerates mastication above 80°C. Has no effect on the physical properties of the vulcanisates. Non-staining. Excellent reclaiming agent for NR and SBR. Carbon blacks have no effect on the plasticising effect, sulphur and accelerators strongly retard and may nullify the plasticising effect. Also effective in black masterbatches and oil extended types.

Quantity: 0·1–0·4%.
TN: Endor (6)

Peptazin ZN (428)
Renacit IV/GR (43).

ZINC PENTAMETHYLENE DITHIOCARBAMATE

ZPD. White powder, m.p. 220–230°C, s.g. 1·35. Soluble in benzene and chlorinated hydrocarbons, partially soluble in acetone, insoluble in water and benzine. Ultra-fast accelerator with a medium vulcanisation range below 120°C, has a very small range above this temperature. Has a low critical temperature, is more active than thiuram. Excellent secondary accelerator in combination with thiazoles; zinc oxide and sulphur are necessary in normal quantities. Fatty acids are not essential but may be used. Does not discolour or stain on contact. Can be added to latex as a dispersion or paste. In cements it may be added on a cold mill, or stirred into the solution.

Quantity: natural rubber 0·5% with 1·5% sulphur, or 0·15% with 1% mercaptobenzthiazole and 1·5% sulphur. SBR 1% with 2% sulphur or 0·5% with 1·5% thiazole and 2% sulphur.

TN: Accelerator ZP (408)
Carbamate PZ
Kuracap Zinc PD (421)
Naugatex 515 (aqueous dispersion) (23)
Pipazate (23)
Robac ZPD (123)
Robac ZPD Extra (with 53% piperidine) (123)
Superaccélérateur 1555 (20)
Vulkacit ZP (43)
Van Hasselt ZPD (355)

ZPD
ZPD Henley (123)

ZINC-α-PHENYL BIGUANIDE
Ultra-fast accelerator.
TN: Zincazol (336).

ZINC STEARATE
$(C_{17}H_{35}COO)_2Zn$. White, amorphous powder, m.p. 126–130°C, s.g. 1·06. Soluble in benzene, insoluble in water. The commercial product is a blend of zinc salts of stearic and palmitic acids with an excess of zinc oxide, m.p. 120°C, s.g. 1·1. Accelerator activator in place of zinc oxide and stearic acid, particularly suitable for transparent rubber articles as it does not cause clouding. Softener in rubber compounds, widely used as a dusting agent; it dissolves into the rubber during vulcanisation and does not leave traces on the articles. Prevents sticking of individual sheets before vulcanisation. Does not influence tack after vulcanisation. Used as an aqueous dispersion in latex, and as an excellent mould release agent.

ZINC THIOBENZOATE
$(C_6H_5COS)_2Zn.2H_2O$, m.p. 107–117°C. Peptiser for natural rubber. Active at low temperatures.
TN: Robac TBZ (34).

ZINC WHITE Gummidur, red seal FH and red seal granular types of zinc oxide (425).

ZINSPER Blend of zinc salts of fatty acids and resinous acids. Activator and softener for rubber and latex (215).

ZIX Abbreviation for zinc isopropyl xanthate.

ZL Abbreviation for zinc lupetidine dithiocarbamate.

ZLD Abbreviation for zinc lupetidine dithiocarbamate.

ZMD Abbreviation for zinc dimethyl dithiocarbamate.

ZML Mixture of zinc laurate and zinc-2-mercaptobenzthiazole. Accelerator (obsolete) (23).

ZMPD Abbreviation for zinc methyl phenyl dithiocarbamate.

ZOPAQUE Titanium dioxide (371).

ZOV Abbreviation for zinc oxide viscosity, formerly called ZOT test (zinc oxide thickening).

ZPD Abbreviation for zinc pentamethylene dithiocarbamate.

ZS MORFEX (Formerly Morfex 33.) Mixture of zinc-2-mercaptobenzthiazole and 25% tetramethylthiuram monosulphide. Accelerator (23).

ZX 3-5 Phenyl methyl urethane-p-sulphonyl hydrazide. Russian produced blowing agent.

ZXB Zinc butyl xanthate (23).

ZYROX RESIN Condensation product of chlorinated aromatics. Antioxidant (137).

ZYTEL Group of long-chain polyamides, s.g. 1·09–1·14. Resistant to normal solvents, esters, ketones, alkalis and weak acids, non-resistant to phenol, formic and concentrated mineral acids.
Uses: construction parts, *e.g.* bearings, rolls, with a high tensile strength and shock and heat resistance (6).

PRODUCERS AND MARKETING ORGANISATIONS

(1) Minnesota Mining & Mfg Co., Chemical Div., St Paul, Minn., USA
(2) Shell Chemical Co.
(3) Binney & Smith Co., New York 17, N.Y., USA
(4) Sharples Chemicals Inc., Philadelphia 7, Pa., USA
(5) Monsanto Chemical Co., Akron 11, Ohio, USA; Monsanto Chemical Ltd, London, SW1, England
(6) E. I. Du Pont de Nemours & Co. Inc., Wilmington 98, Del., USA
(7) Dow Corning Corp., Midland, Mich., USA
(8) US Gypsum Co., Industrial Div., Chicago 6, Ill., USA
(9) Baker Castor Oil Co., Bayonne, N.J., USA
(10) Hercules Powder Co., Inc., Wilmington 99, Del., USA
(11) American-British Chemical Supplies Inc., New York 16, N.Y., USA
(12) A. Boake, Roberts & Co., Ltd, Abrac Works, Stratford, London, E15, England
(13) Abril Corporation (Gt Britain) Ltd, Golden Mile Works, Bridgend, Glamorgan, Wales
(14) B. F. Goodrich Chemical Co., Cleveland 15, Ohio, USA
(15) Catalin Corporation of America, New York 16, N.Y., USA
(16) Kawaguchi Chem. Ind. Co., Japan
(17) Acralite Co., New York, N.Y., USA
(18) Industrial Nucleonics Corp., Columbus 12, Ohio, USA
(19) Société Anonyme des Matières Colorantes et Produits Chimiques de Saint Denis, Paris, France
(20) Société des Usines Chimiques Rhone-Poulenc, Paris 8, France
(21) American Cyanamid Co., Calco Chem. Div., Bound Brook, N.J., USA
(22) Anchor Chemical Co., Ltd, Clayton, Manchester 11, England
(23) Naugatuck Chemical Div., Uniroyal, Naugatuck, Conn., USA
(24) Warwick Wax Co., Div. of Sun Chemical Corp., Chicago 5, Ill., USA
(25) Aceto Chemical Co., Inc., Flushing 54, N.Y., USA
(26) Arizona Chemical Co., Inc., New York 20, N.Y., USA
(27) Acme Resin Corporation, USA
(28) Newport Industries Inc., Div. of Heyden Newport Chem. Corp., New York 17, N.Y., USA
(29) Allied Chemical Corp., New York 6, N.Y., USA
(30) Glyco Products Co., Inc., Brooklyn 2, N.Y., USA
(31) General Tire & Rubber Co., Chem. Div., Akron 9, Ohio, USA
(32) Badische Anilin- und Soda-Fabrik AG, Ludwigshafen/Rhein, W. Germany
(33) Rohm & Haas Company, Philadelphia 5, Pa., USA; Rohm & Haas Chem. Fabr., Darmstadt, W. Germany
(34) The Borden Co., New York 17, N.Y., USA
(35) Acryvin Corp., Brooklyn, N.Y., USA
(36) The Chemstrand Corporation, New York 1, N.Y., USA

(37) American Monomer Corporation, New York, N.Y., USA
(38) Firestone Plantation Co., Harbel, Liberia
(39) Harsyd Chemicals Inc., Holland, Mich., USA
(40) J. M. Huber Corporation, New York 17, N.Y., USA
(41) Esso Standard Oil Co.
(42) Rhein-Chemie, Mannheim-Rheinau, W. Germany
(43) Farbenfabriken Bayer AG, Leverkusen, W. Germany
(44) Archer–Daniels–Midland Co., Minneapolis, Minn., USA
(45) Advance Solvents Chemical Corp., New York 16, N.Y., USA; Deutsche Advance Produktion GmbH, Marienberg, Post Bensheim, W. Germany
(46) Aeropreen Products Ltd, England
(47) Hammill & Gillespie Inc., USA
(48) Pechiney S. A., Paris, France
(49) Agawam Chemicals Inc., West Springfield, Mass., USA
(50) American Resinous Chemicals Corp., Peabody, Mass., USA
(51) R. T. Vanderbilt Co., Inc., New York 17, N.Y., USA
(52) American Agile Co., USA
(53) Hadogaya Soda KK., Japan
(54) Akron Chemical Co., Akron 4, Ohio, USA
(55) General Mills, Inc. (Chem. Div.), Kankakee, Ill., USA
(56) New Jersey Zinc Co., New York 38, N.Y., USA
(57) Alco Oil & Chemical Corp., Philadelphia 34, Pa., USA
(58) Chemische Werke Albert, Wiesbaden-Biebrich, W. Germany
(59) Montecatini, Soc. Gen. per L'Ind. Mineraria e Chim., Milan, Italy
(60) Imperial Chemical Industries Ltd, London, SW1, England
(61) The Dow Chemical Company, Midland, Mich., USA
(62) Commercial Solvents Co., New York 16, N.Y., USA
(63) The Carborundum Co., Niagara Falls, N.Y., USA
(64) The Glidden Co., Chem. Div., Baltimore 26, Md.; Hammond, Ind., USA
(65) Wallace & Tiernan Inc., Lucidol Div., Buffalo 5, N.Y., USA
(66) American Latex Product Corpn, USA
(67) Società Italiana Resine, Milan, Italy
(68) Société Franco-Belge du Caoutchouc Mousse, Paris, France
(69) Alrose Chemical Co., Pittsburgh 19, Pa., USA
(70) W. H. & F. Jordan Jr. Mfg Co. Inc., Philadelphia, USA
(71) American Alkyd Industries, Carlstadt, N.J., USA
(72) Feedwaters, Inc., USA
(73) Procter & Gamble, Cincinnati 1, Ohio, USA
(74) Stamford Rubber Supply Co., Stamford, Conn., USA
(75) Amber Oils Ltd., London, England
(76) Goodrich–Gulf Chemicals Inc., Cleveland 14, Ohio, USA
(77) Pan American Refining Corp., USA
(78) Amoco Chemicals Corp., Chicago 1, Ill., USA
(79) American Mineral Spirits Co., Chicago 1, Ill., USA
(80) British Titan Products Ltd, Coppergate, York, England
(81) R. W. Greef & Co., Ltd, London, EC2, England; New York 20, N.Y., USA

(82) Colombian International (Great Britain) Ltd, London, EC4, England
(83) American Anode Inc., USA
(84) Ansul Chemical Co., Marinette, Wisc., USA
(85) General Aniline & Film Corp. (Antara Chemicals), New York 14, N.Y., USA
(86) Japan Dyestuffs Mfg. Co., Japan
(87) Clayton Aniline Co., Ltd, Clayton, Manchester, England
(88) Ethyl Corporation, New York 17, N.Y., USA
(89) Herron Bros. & Meyer Inc., New York 1, N.Y., USA
(90) Applied Plastics Co. Inc., USA
(91) Apex Chemical Co., New York 1, N.Y., USA
(92) Acheson Colloids Co., Div. of Acheson Industries Inc., Port Huron Mich., USA
(93) National Dairy Products Co., New York, N.Y., USA
(94) Ciba AG, Basel, Switzerland
(95) Arbonite Corporation, USA
(96) Union Oil Co. of California, Los Angeles 17, Calif., USA
(97) Atlas Powder Co., Wilmington 99, Del., USA
(98) Armour & Co., Chicago 9, Ill., USA
(99) US Industrial Chemicals, Inc., New York 17, N.Y., USA
(100) C. P. Hall Co., Akron 8, Ohio, USA
(101) Toa Synthetic Chemical Co., Japan
(102) Minerals & Chemicals Philipps Corp., Menlo Park, N.J., USA
(103) American Synthetic Rubber Corp., New York, N.Y., USA
(104) Vereinigte Deutsche Metallwerke AG, Frankfurt/M, W. Germany
(105) Charles Eneu Johnson & Co., Philadelphia 47, Pa., USA
(106) American Zinc Sales Co., Subs. of American Zinc, Lead & Smelting Co., Columbus 16, Ohio, USA
(107) Barrett Div., Allied Chemical & Dye Corp., New York 6, N.Y., USA
(108) Baker Chemical Co., USA
(109) United Rubber & Chemical Co., Subs. of United Carbon Co., New York 22, N.Y., USA
(110) Rubber Corporation of America, Hicksville, N.Y., USA
(111) Reichhold Chemicals, Inc., White Plains, N.Y., USA
(112) Union Carbide Corp., New York 17, N.Y., USA
(113) Beacon Chemical Industries Inc., Cambridge, Mass., USA
(114) Hubron Rubber Chemicals Ltd, Failsworth, Manchester, England
(115) Goodyear Tire & Rubber Co., Chem. Div., Akron 16, Ohio, USA
(116) British Industrial Solvents, London, W1, England
(117) US Rubber Reclaiming Co., Buffalo 5, N.Y., USA
(118) Carbon Dispersions, Inc., Newark 12, N.J., USA
(119) Sherwin-Williams Co., Cleveland 1, Ohio, USA
(120) Pennsalt Chemicals Corp. (Rubber Chemicals Department), King of Prussia, Pa., USA
(121) British Enka Ltd, England
(122) Rubber Cultuur Maatschappij Amsterdam, Amsterdam, Holland
(123) Henley & Co., Inc., New York, N.Y., USA
(124) Bunawerke Hüls, Marl, Kr. Recklinghausen, W. Germany

(125) Wilmington Chemical Corp., New York 16, N.Y., USA
(126) Pennsylvania Industrial Chemical Corp., Clairton, Pa., USA
(127) Harwick Standard Chemical Co., Akron 5, Ohio, USA
(128) Bunatak Chemical Co., Malden 48, Mass, USA
(129) Firestone Synthetic Rubber & Latex Co., Akron 1, Ohio, USA
(130) Phillips Petroleum Co., Bartlesville, Okla., USA
(131) Enjay Chemical Co., New York 19, N.Y., USA
(132) Kessler Chemicals Co., Inc., Philadelphia 35, Pa., USA
(133) Shawinigan Resin Corp., Springfield 1, Mass., USA
(134) Columbia–Southern Chemical Corp., Pittsburgh 22, Pa., USA
(135) Byerlyte Co., Cleveland 13, Ohio, USA
(136) Polymel Corp., Baltimore 2, Md., USA
(137) Bakelite Corp., New York 17, N.Y., USA
(138) Société des Produits Azotés, France
(139) Godfrey L. Cabot Inc., Boston 10, Mass., USA
(140) Cadet Chemical Corp., Burt, N.Y., USA
(141) Hagan Chemicals & Controls, Inc., Pittsburgh 30, Pa., USA
(142) Golden Bear Oil Co., Los Angeles 14, Calif., USA
(143) J. & E. Sturge Ltd, Birmingham 15, England
(144) Füllstoff-Gesellschaft Marquart-Wesseling GmbH, Wesseling, Bez
      Cologne, W. Germany
(145) W. Smith Ltd, London, EC3, England
(146) Carbide & Carbon Chemicals Co., New York 17, N.Y., USA
(147) Irvington Varnish & Insulator Co., Irvington 11, N.J., USA
(148) Carwin Corporation, North Haven, Conn., USA
(149) British Celanese Ltd, London, W, England
(150) Johns-Manville Products Co., New York 16, N.Y., USA
(151) British Resin Products Ltd, London, W1, England
(152) Celanese Corp. of America, Chem. Div., New York 16, N.Y., USA
(153) Diamond Alkali Co., Cleveland 14, Ohio, USA
(154) NV Chemische Industrie AKU-Goodrich, Arnhem, Holland
(155) Sun Oil Co., Philadelphia 3, Pa., USA
(155a) Chas. Pfizer & Co., Inc., Brooklyn 6, N.Y., USA
(156) Degussa. Deutsche Gold- und Silber-Scheideanstalt, Frankfurt/M,
      W. Germany
(157) Rhein-Collodium GmbH, Cologne, W. Germany
(158) Continental Carbon Co., New York 16, N.Y., USA
(159) Columbian Carbon Co., New York 17, N.Y., USA
(160) Copolymer Rubber & Chemical Corp., Baton Rouge, La., USA
(161) Courtaulds Ltd, England
(162) Hooker Chemical Corp., Niagara Falls, N.Y., USA
(163) Scott Bader Co. Ltd, London, WC2, England
(164) United Carbon Co. Inc., Charleston 27, West Va., USA
(165) Société Rhodiaceta, Lyon, France
(166) Stauffer Chemical Co., New York 17, N.Y., USA
(167) Lehmann & Voss, Hamburg, W. Germany
(168) Ethelburga Agency Ltd, Kuala Lumpur, Malaysia
(169) Harrisons & Crosfield Latex Ltd, Kuala Lumpur, Malaysia

(170) Durham Raw Materials Ltd, London, EC3, England
(171) Oxydo, Gesellschaft f. Chemische Produkte, Emmerich/Rhein, W. Germany
(172) Marbon Chemical Div., Borg-Warner Corp., Washington, West Va., USA
(173) Osaka Kinzoku Co., Japan
(174) Dewey & Almy Chemical Co., Cambridge 40, Mass., USA
(175) Dartex GmbH, Frankfurt/M, W. Germany
(176) Koppers Co. Inc., Pittsburgh 19, Pa., USA
(177) Detel Products Inc., USA
(178) Standard-Vacuum Oil Co.
(179) Oronite Chemical Co., San Francisco 4, Calif., USA
(180) Metallgesellschaft AG, Frankfurt/M, W. Germany
(181) Dicalite Dept., Great Lakes Carbon Corp., Los Angeles 17, Calif., USA
(182) Bromine Producers Co., Adrian, Mich., USA
(183) Horn-Kem Corp., Long Island City, N.Y., USA
(184) Food Machinery & Chemical Corp., New York 17, N.Y., USA
(185) Phillips Chemical Company, Akron 8, Ohio, USA
(186) Süddeutsche Zellwolle AG, Kelheim, W. Germany
(187) E. F. Drew & Co., Inc., Boonton, N.J., USA
(188) Argus Chemical Corp., Brooklyn 31, N.Y., USA
(189) Chemische Werke Hüls, Marl, Kr. Recklinghausen, W. Germany
(190) Dura Commodities Corp., New York 7, N.Y., USA
(191) National Lead Co., New York 6, N.Y., USA
(192) Dynamit AG vorm. Nobel & Co., Troisdorf/Cologne, W. Germany
(193) Eastman Chemical Products, Inc., Kingsport, Tenn., USA
(194) Rhodes Industrial Corpn., East Hampton, N.Y., USA
(195) Elektrochemisches Kombinat Bitterfeld, VEB, Bitterfeld, E. Germany
(196) Emery Industries, Cincinnati 2, Ohio, USA
(197) A. E. Staley Mfg Company, Illinois, USA
(198) General Electric Co., Schenectady 5, N.Y., USA
(199) American Phenolic Corp., Chicago, Ill., USA
(200) ANIC, Azienda Nazionale Idrocarburi, SpA, Ravenna, Italy
(201) Carter Bell Mfg Co., Springfield, N.J., USA
(202) Roessler & Haslacher Chemical Co., New York, USA
(203) Ferro Chemical Corp., Bedford, Ohio, USA
(204) Resistoflex Corp., Roseland, N.J., USA
(205) Krupp Kohlechemie, W. Germany
(206) The Miner Laboratories Co., Chicago, Ill., USA
(207) Internationale Galalith Gesellschaft, Hamburg, W. Germany
(208) Thiokol Chemical Corp., Trenton 7, N.J., USA
(209) Whiffen & Sons Ltd, London, N.W.1, England
(210) Hardesty Chemical Div., W. C. Hardesty Co., Inc., Belleville 9, N.J., USA
(211) Croda Ltd, Snaith, Goole, Yorkshire, England
(212) Harold Chemical Products Co., USA
(213) Heresite & Chemical Co., Manitowoc, Wis., USA
(214) Hooker Electrochemical Co., New York, N.Y., USA

(215) Witco Chemical Co., New York 16, N.Y., USA
(216) Nederlandsche Optiek- en Instrumentenfabriek, Dr C. E. Bleeker, Zeist, Holland
(217) Farberwerke Höchst, Höchst/M, W. Germany
(218) Harchem Div., Wallace & Tiernan, Inc., Belleville, N.J., USA
(219) Mantz & Co., London, England
(220) Internationale Kunststoffen NV, Holland
(221) S. R. F. Freed Ltd, London, England
(222) Hoffner Rayon Co., Philadelphia, USA
(223) West Virginia Pulp & Paper Co., New York 17, N.Y., USA
(224) Rudolf Fuchs Mineralölwerk KG, Mannheim, W. Germany
(225) International Synthetic Rubber Co., Ltd, London, W1, England
(226) Synthetic Chemicals, Inc., Paterson 4, N.J., USA
(227) J. R. Geigy AG, Basel 16, Switzerland
(228) Skanska Attik Fabriken A.B., Perstorp, Sweden
(229) Union Bay State Chemical Co., Cambridge, Mass., USA
(230) Société Rhovyl, Tronville-en-Barrois, France
(231) E. Schliemann's Export-Ceresin-Fabrik, Hamburg–Billstedt 1, W. Germany
(232) Japan Synthetic Rubber Co., Ltd, Tokyo, Japan
(233) Kleeman Ltd, London, England
(234) Rubber Research Ass. Ltd, Gamin, Israel
(235) Kautex Ltd, Elstree, Herts., England; Kautex-Werke, Bechlinghoven, W. Germany
(236) Kelco Company, Clarke, N.J., USA
(237) National Polychemicals, Inc., Wilmington, Mass., USA
(238) Kenrich Petrochemicals, Inc., Maspeth 78, N.Y., USA
(239) Aberdeen Combworks Co., England
(240) Kirklees Ltd, England
(241) Dickinson & Son, Church, Lancs., England
(242) AKU, Algemene Kunstzijde Unie, Arnhem, Holland
(243) B. X. Plastics Ltd, London, E4, England
(244) SNIA Viscosa, Società Nazionale Industria Applicazioni Viscosa SAP, Milan, Italy
(245) Lansil Ltd, England
(246) United States Rubber Co., New York, N.Y., USA
(247) Leda Chemicals Ltd, London, N18, England
(248) Virginia–Carolina Chemical Corp., Richmond 5, Va., USA
(249) Foote Mineral Co., Philadelphia 44, Pa., USA
(250) J. Wolf & Co., Passaic, N.J., USA
(251) United Ebonite & Lorival Ltd, Little Lever, Bolton, Lancs, England
(252) Société Lorraine–Kuhlmann, Dieuze, France
(253) M. H. Lummerzheim & Co., Ghent, Belgium
(254) Rheinische Olefinwerke, Wesseling/Rh, W. Germany
(255) Merck & Co., Inc., Rahway, N.J., USA
(256) Hardman & Holden Ltd, Manchester 10, England
(257) Marathon Corp., Chem. Div., Rothschild, Wis., USA
(258) Mead Corp., Chillicothe, Ohio, USA

(259) Philadelphia Quartz Co., Philadelphia 6, Pa., USA
(260) Micronizer Processing Co., Inc., Moorestown, N.J., USA
(261) Nopco Chemical Co., Harrison, N.J., USA
(262) Österreichische Stickstoffwerke, Linz/Donau, Austria
(263) Mobay Chemical Co., Pittsburgh 5, Pa., USA
(264) Morton-Withers Chemical Co., Greensboro, N.C., USA
(265) Nationale Aniline Div., Allied Chemical & Dye Corp., New York 6, N.Y., USA
(266) Caldwell Co., Akron 5, Ohio, USA
(267) Spangenberg-Werke GmbH, Hamburg–Eidelstedt, W. Germany
(268) Onyx Oil & Chemical Co., Jersey City, N.J., USA
(269) Neville Chemical Co., Pittsburgh 25, Pa., USA
(270) A. Nova, Legnano, Italy
(271) The Japanese Geon Co., Japan
(272) Nippon Soda KK, Japan
(273) Nixon Nitration Works, Nixon, N.J., USA
(274) Ouchi Shinko Chemical Ind., Japan
(275) Nordiska Gummifabriks A.B., Sweden
(276) Dunlop Plantation Ltd, Malaysia
(277) Nyma NV, Nijmegen, Holland
(278) Lech-Chemie, Gersthofen, Augsburg, W. Germany
(279) Pluess-Staufer, Oftringen, Switzerland
(280) Crown Zellerbach Corp., Chem. Products Div., Camas, Wash., USA
(281) Burke Research Co., USA
(282) Wyrough & Loser, Trenton 8, N.J., USA
(283) Casella Farbwerke, Mainkur, W. Germany
(284) Truscon Laboratories Inc., USA
(285) Luminous Resins Inc., Chicago, Ill., USA
(286) Benson Process Engineering Co., Eden, N.Y., USA
(287) Watson Standard Co., Pittsburgh, Pa., USA
(288) Heyden-Newport Chemical Corp., New York 17, N.Y., USA
(289) Pitt-Consol Chemical Co., Newark, N.J., USA
(290) M. A. Knight, Akron, Ohio, USA
(291) Spies, Hecker & Co., Cologne-Raderthal, W. Germany
(292) Chemische Werke Bärocher, Munich, W. Germany
(293) Sponge Rubber Products Co., USA
(294) Société Nobel Francaise, Paris, France
(295) Plastanol Ltd, Belvedere, Kent, England
(296) Silice & Kaolin, Barbières, France
(297) American Petrochemical Corp., Minneapolis, USA
(298) Expandite Ltd, England
(299) Wyandotte Chemicals Corp., Wyandotte, Mich., USA
(300) Massachusetts Mineral Mfg. Co., USA
(301) Spencer Chemical Co., Kansas City, USA
(302) Semperit, AG, Vienna, Austria
(303) Polymer Corporation Ltd, Sarnia, Ontario, Canada
(304) Hans Heinrich Hütte GmbH, Frankfurt/M, W. Germany
(305) Pittsburgh Coke & Chemical Co., Pittsburgh 19, Pa., USA

(306) Pyleen-Chemie, Düsseldorf, W. Germany
(307) Amecco Chemicals Inc., New York 17, N.Y., USA
(308) Raolin Corporation, USA
(309) Heveatex Corporation, Melrose 76, Mass., USA
(310) Spencer Products Co., Inc., Ridgewood, N.J., USA
(311) Robinson Bros. Ltd, Ryders Green, West Bromwich, England
(312) C. F. Stork & Co., Chemische Industrie NV, Hengelo, Holland
(313) Ketjen-Carbon NV, Amsterdam, Holland
(314) Woburn Chemical Corp., Harrison, N.J., USA
(315) Howards & Sons Ltd, Canada
(316) Shawinigan Products Corp., New York, N.Y., USA
(317) Chemische Fabriken Oker u. Braunschweig AG, Oker/Harz, W. Germany
(318) Midland Silicones Ltd, London, W1, England
(319) Skelly Oil Co., Kansas City 41, Mo., USA
(320) Société du Caoutchouc Butyl, Notre Dame de Gravenchon, France
(321) Solvay & Cie, Paris, France
(322) Brown Corporation, Montreal, Canada
(323) Shawinigan Ltd, London, EC3, England
(324) Svenska Oljeslageri AB, Sweden
(325) Sorbo Ltd, Woking, Surrey, England
(326) Société des Resines Fluorés, France
(327) Socony Vacuum Oil Co., Inc., New York 4, N.Y., USA
(328) Nihon Senryo Seizo, Osaka, Japan
(329) Spencer Kellog & Sons, Inc., USA
(330) Deecy Products Co., Cambridge 42, Mass., USA
(331) Acme-Hardesty Co., New York 17, N.Y., USA
(332) New System Mfg. Co., Ltd, England
(333) Miles Chemical Co., Elkhart, Ind., USA
(334) Mannesmann-Plastic, Düsseldorf, W. Germany
(335) Kalle & Co., Wiesbaden, W. Germany
(336) E. & W. Smith Ltd, London, EC3, England
(337) Continental Gummiwerke AG, Hannover, W. Germany
(338) Texas–US Chemical Co., New York 20, N.Y., USA
(339) Flintkote Co., USA
(340) Synvar Corporation, USA
(341) Sid Richardson Carbon Co., Fort Worth, Texas, USA
(342) Glovers (Chemicals) Ltd, Wortley, Leeds, England
(343) Thermatomic Carbon Co., Sterlington, La., USA
(344) Allied Colloids (Manufacturing) Co. Ltd, London, EC4, England
(345) Anorgana GmbH, Gendorf/Obb, W. Germany
(346) Dayton Chemical Products Laboratories Inc., West Alexandria, Ohio, USA
(347) Titanium Pigment Corp., New York 6, N.Y., USA
(348) New York Hamburger Gummi-Waren Co., Hamburg-Harburg, W. Germany
(349) Tragasol Products Ltd, Hooton, Wirral, Cheshire, England
(350) Bakelite Ltd, London, SW1, England

(351) US Stoneware Co., Akron, Ohio, USA
(352) International Latex Corp., Chem. Div., Dover, Del., USA
(352a) Universal Oil Products Co., Des Plaines, Ill., USA
(353) Toyo Koastu In., Inc., Tokyo, Japan
(354) US Peroxygen Corp., Richmond, Calif., USA
(355) NV Chem. Industrie van Hasselt, Amersfoort, Holland
(356) A. Smith Ltd, Stoneferry, Hull, England
(357) Velsicol Chemical Corp., Chicago 11, Ill., USA
(358) H. W. Wallace & Co., Croydon, England
(359) Wacker Chemie GmbH, München, W. Germany
(360) Colton Chem. Co., Cleveland, Ohio, USA
(361) Société Belge de l'Azote, Belgium
(362) Schering Kahlbaum AG, Berlin, W. Germany
(363) Distilleries de Melle, Melle, France
(364) Silesia, Verein chemischer Fabriken, Saarau, Poland
(365) General Latex & Chemical Corp., Cambridge 39, Mass., USA
(366) Dovan Chemical Corp., New York, N.Y., USA
(367) Film- und Chemiefaserwerk Agfa, Wolfen, VEB, Wolfen, E. Germany
(368) Ross & Rowe Inc., New York 4, N.Y., USA
(369) Chem. Fabr. NV Aagrunol, Groningen, Holland
(370) Zellstoff-Fabrik Waldhof, Mannheim-Waldhof, W. Germany
(371) Chemical & Pigment Co., Cleveland, Ohio, USA
(372) Firestone Plastics Co., Pottstown, Pa., USA
(373) Viking Corpn, Chicago, Ill., USA
(374) Bell Telephone Laboratories Inc., New York 14, N.Y., USA
(375) Kurashiki Rayon Co., Japan
(376) Nova Chemical Corpn, New York 14, N.Y., USA
(377) Shinko Rayon Co. Ltd, Tokyo, Japan
(378) Chemische Fabrik Geller & Tutt, Düren, W. Germany
(379) Schill & Seilacher, Hamburg, W. Germany
(380) Société Organo-Synthese, Neuilly s/Seine, France
(381) Kali-Chemie Stauffer GmbH, Hannover, W. Germany
(382) Henkel u. Cie GmbH, Klebstoffwerk, Düsseldorf, W. Germany
(383) Scholven-Chemie AG, Gelsenkirchen-Buer, W. Germany
(384) Chemomedica, Crantzberg & Co., Vienna, Austria
(385) Deutsche Ölfabrik Dr Grandel, Hamburg, W. Germany
(386) Guthrie Estates Ltd, London, EC3, England
(387) Globus-Werke/Kieselweiss, Neuburg, Donau, W. Germany
(388) Conap Inc., Allegheny, N.Y., USA
(389) Société d'Electrochimie d'Ugine, France
(390) Gomu Yakashui Seizosho, Japan
(391) Zaklady Chemotechnicky Diosit, Lodz, Poland
(392) Nippon Kassai Industries Ltd, Japan
(393) Deutsche Hydrierwerke, Dessau, E. Germany
(394) Yanaginauma Kagaku Keniju-Ji, Tokyo, Japan
(395) Kawanishi Co., Japan
(396) Oderberger Chemische Werke, Oderberg, Czechoslovakia
(397) General Quimica SA, Bilbao, Spain

(398) Thüringer Industriewerk VEB, Rauenstein, E. Germany
(399) Surface Coating Synthetics Ltd, London, England
(400) SOCFIN Co., Ltd, Kuala Lumpur, Malaysia
(401) Ozonair, Paris 20, France
(402) Shamokin Filler Co., Inc., Shamokin, Pa., USA
(403) Boustead Buttery Estates Agency, Kuala Lumpur, Malaysia
(404) Mobil Oil Co.
(405) Synthomer Chemie GmbH, W. Germany
(406) Chemische Fabrik E. Rühl, Friedrichsdorf/Taunus, W. Germany
(407) Shiraishi Kogyo Kaisha Ltd, Osaka, Japan
(408) Naftone, Inc., New York 22, N.Y., USA
(409) Berkshire Chemicals Inc., New York 17, N.Y., USA
(410) Chemische Werke Witten GmbH, Witten/Ruhr, W. Germany
(411) AKZO, formerly Chemische Fabrik Hoesch KG, Düren/Rhld., W. Germany
(412) Chemische Fabrik Franz Hoffmann & Söhne, Neuburg/Donau, W. Germany
(413) Verkaufsvereinigung f. Teererzeugnisse AG, Essen, W. Germany
(414) Sigma Chem. Corp., Lawrence, Mass., USA
(415) Wallace & Tiernan GmbH, Günzburg-Donau, W. Germany
(416) Associated Lead Manufacturers Export Co., Ltd, London, EC2, England
(417) Stibiox-Werk Lang & Co., Braunschweig, W. Germany
(418) Manufacture Landaise de Produits Chimiques, Rion des Landes, France
(419) Fabrications Etudes Conseils pour l'Industrie Chimique, Paris, France
(420) Etablissements Kuhlman, Paris, France
(421) Sandoz AG, Basel, Switzerland
(422) OMYA GmbH, Cologne, W. Germany
(423) Gebr. Giulini GmbH, Ludwigshafen/Rhein, W. Germany
(424) NV Chemische Verfstoffenfabriek v/h L. Th. Ten Horn, Maastricht, Holland
(425) Zinkweiss-Forschungsgesellschaft GmbH, Oberhausen, Rhld., W. Germany
(426) Maatschappij Dr van Roon en Co. Chemische Fabriek en Laboratorium NV, Waalwijk, Holland
(427) Knapsack AG, Knapsack bei Cologne, W. Germany
(428) Chemapol, Aussenhandelsunternehmen für Ein- und Ausfuhr Chemischer Produkte und Rohstoffe, Prague, Czechoslovakia
(429) Chemische Fabrik Oscar Neynhaber & Co., A. G. Loxstedt/Bremerhaven, W. Germany
(430) Société Normande pour l'Industrie Chimique, Rouen, France
(431) Albright & Wilson (Mfg) Ltd, London, SW1, England
(432) The Rubber Regenerating Co., Ltd, London, England
(433) Fabriek van Chemische Produkten Vondelingenplaat NV, Rotterdam, Holland
(434) Distugil, Paris, France
(435) Daicel Ltd, Japan

(436) Th. Goldschmidt AG, Essen, W. Germany
(437) Staatsmijnen, Holland
(438) Petrofina (Gt Britain) Ltd, London, England
(439) Heraeus Quarzlampen Ges., Hanau, W. Germany
(440) Aziende Colori Nazionali Affini, Milan, Italy
(441) Distillers Co., Ltd, Chemical Division, London, England
(442) Toyo Soda Manufacturing Co., Ltd, Tokyo, Japan
(443) Norac Co., Inc., 405 S. Motor Avenue, Azusa, Calif. 91702, USA